Carpentry
Level Two

Trainee Guide

 Pearson

Boston   Columbus   Indianapolis   New York   San Francisco   Upper Saddle River
Amsterdam   Cape Town   Dubai   London   Madrid   Milan   Munich   Paris   Montreal   Toronto
Delhi   Mexico City   São Paulo   Sydney   Hong Kong   Seoul   Singapore   Taipei   Tokyo

NCCER
Chief Executive Officer: Don Whyte
President: Boyd Worsham
Chief Operations Officer: Katrina Kersch
Director of Product Development: Tim Davis
Carpentry Curriculum Project Manager: Jamie Carroll, Rob Richardson
Senior Production Manager: Erin O'Nora
Senior Manager of Projects: Chris Wilson

Managing Editor: Natalie Richoux
Desktop Publishing Manager: James McKay
Digital Content Coordinator: Rachael Downs
Art Manager: Kelly Sadler
Production Specialists: Gene Page, Eric Caraballoso
Editors: Graham Hack, Jordan Hutchinson

Writing and development services provided by S4Carlisle Publishing Services, Dubuque, IA
Project Manager: Barb Tucker
Writers: Michael B. Kopf
Art Development: S4Carlisle Publishing Services

Permissions Specialists: Kim Schmidt, Karyn Morrison
Media Specialist: Genevieve Brand
Copy Editor: Michael H. Toporek

Pearson
Director of Alliance/Partnership Management: Kelly Trakalo
Content Producer: Alexandrina B. Wolfe
Assistant Content Producer: Alma Dabral
Digital Content Producer: Jose Carchi
Senior Marketing Manager: Brian Hoehl

Composition: NCCER
Printer/Binder: LSC Communications
Cover Printer: LSC Communications
Text Fonts: Palatino and Univers

Credits and acknowledgments for content borrowed from other sources and reproduced, with permission, in this textbook appear at the end of each module.

Copyright © 2013, 2005, 2001, 1995 by NCCER, Alachua, FL 32615, and published by Pearson, New York, NY 10013. All rights reserved. Printed in the United States of America. This publication is protected by Copyright and permission should be obtained from NCCER prior to any prohibited reproduction, storage in a retrieval system, or transmission in any form or by any means, electronic, mechanical, photocopying, recording, or likewise. For information regarding permission(s), write to: NCCER Product Development, 13614 Progress Blvd., Alachua, FL 32615.

27 2022

Perfect bound ISBN-13: 978-0-13-418622-1
ISBN-10: 0-13-418622-2

Preface

To the Trainee

If you're ready to nail down a career in construction, consider carpentry. Carpenters make up the largest building trades occupation in the industry and those with all-around skills are in high demand. Carpenters are involved in many different kinds of construction activities, from building highways and bridges to installing kitchen cabinets.

Carpenters construct, erect, install, and repair structures and fixtures made from wood and other materials. Depending on the type of construction, size of company, and other factors, carpenters may specialize in one or two activities or may perform many different tasks. Each carpentry task is somewhat different, but most involve the same basic steps: working from blueprints, laying out the structure, assembling the structure, and checking the work afterward. Having good hand-eye coordination, an attention to detail, and the ability to perform math calculations will help you as you progress through your carpentry training.

We wish you success as you continue your training in the carpentry craft. There are more than one million people employed in carpentry work in the United States, and as most of them can tell you, there are many opportunities awaiting those with the skills and desire to move forward in the construction industry.

New with *Carpentry Level Two*

Carpentry Level Two showcases a new instructional design and features a streamlined teaching approach which now includes breaking information into objectives and sub-objectives. The new format enhances the learning experience by presenting concepts in a clear, concise, and easily digestible manner. This version features a new training order and incorporates the latest tools, techniques, and technology of the trade. In addition, *Carpentry Level Two* offers two different career paths: residential and commercial carpentry. This provides learners with the opportunity to explore both specialized areas. This updated edition does not include the Cabinet Fabrication module. However, Roofing Applications and Exterior Finishing are now offered as electives for students on the commercial construction path, and Commercial Drawings and Suspended Ceilings are now offered as electives for students on the residential construction path.

We invite you to visit the NCCER website at **www.nccer.org** for information on the latest product releases and training, as well as the *Breaking Ground* digital newsroom and Pearson's NCCER product catalog.

Your feedback is welcome. You may email your comments to **curriculum@nccer.org** or send general comments and inquiries to **info@nccer.org**.

NCCER Standardized Curricula

NCCER is a not-for-profit 501(c)(3) education foundation established in 1996 by the world's largest and most progressive construction companies and national construction associations. It was founded to address the severe workforce shortage facing the industry and to develop a standardized training process and curricula. Today, NCCER is supported by hundreds of leading construction and maintenance companies, manufacturers, and national associations. The NCCER Standardized Curricula was developed by NCCER in partnership with Pearson, the world's largest educational publisher.

Some features of the NCCER Standardized Curricula are as follows:

- An industry-proven record of success
- Curricula developed by the industry for the industry
- National standardization providing portability of learned job skills and educational credits
- Compliance with the Office of Apprenticeship requirements for related classroom training (CFR 29:29)
- Well-illustrated, up-to-date, and practical information

NCCER also maintains a National Registry that provides transcripts, certificates, and wallet cards to individuals who have successfully completed a level of training within a craft in NCCER's Curricula. *Training programs must be delivered by an NCCER Accredited Training Sponsor in order to receive these credentials.*

Special Features

In an effort to provide a comprehensive, user-friendly training resource, we have incorporated many different features for your use. Whether you are a visual or hands-on learner, this book will provide you with the proper tools to get started in the Carpentry trade.

Introduction

This page is found at the beginning of each module and lists the Objectives, Performance Tasks, Trade Terms, and Required Trainee Materials for that module. The Objectives list the skills and knowledge you will need in order to complete the module successfully. The Performance Tasks give you an opportunity to apply your knowledge to the real-world duties that carpenters perform. The list of Trade Terms identifies important terms you will need to know by the end of the module. Required Trainee Materials list the materials and supplies needed for the module.

Special Features

Features provide a head start for those entering the Carpentry field by presenting technical tips and professional practices from craftworkers in various disciplines. These features often include real-life scenarios similar to those you might encounter on the job site.

Color Illustrations and Photographs

Full-color illustrations and photographs are used throughout each module to provide vivid detail. These figures highlight important concepts from the text and provide clarity for complex instructions. Each figure reference is denoted in the text in *italics* for easy reference.

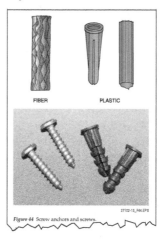

Figure 44 Screw anchors and screws.

Notes, Cautions, and Warnings

Safety features are set off from the main text in highlighted boxes and are organized into three categories based on the potential danger of the issue being addressed. Notes simply provide additional information on the topic area. Cautions alert you of a danger that does not present potential injury but may cause damage to equipment. Warnings stress a potentially dangerous situation that may cause injury to you or a co-worker.

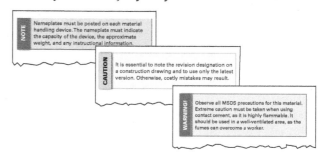

Going Green

Going Green looks at ways to preserve the environment, save energy, and make good choices regarding the health of the planet. Through the introduction of new construction practices and products, you will see how the "greening of America" has already taken root.

Wind Farms

The US Department of Energy states that wind farms may be a major source of power as we approach the year 2030. Their analysis shows that up to 20 percent of US power needs could be handled by wind power, which would reduce pollution to the same extent as taking 140 million cars off the road.

Did You Know?

The Did You Know? features offer hints, tips, and other helpful bits of information from the trade.

Did You Know?
Balloon Framing

Balloon framing is frequently used in hurricane-prone areas for gable ends. In fact, this type of framing may be required by local building codes.

Step-by-Step Instructions

Step-by-step instructions are used throughout to guide you through technical procedures and tasks from start to finish. These steps show you not only how to perform a task but also how to do it safely and efficiently.

Step 1 Select the proper-size toggle bolt and drill bit for the job.

Step 2 Check the toggle bolt for damaged or dirty threads or a malfunctioning wing mechanism.

Step 3 Drill a hole completely through the surface to which the part is to be fastened.

Step 4 Insert the toggle bolt through the opening in the item to be fastened.

Step 5 Screw the wings onto the end of the toggle bolt, ensuring that the flat side of the wing is facing the bolt head.

Trade Terms

Each module presents a list of Trade Terms that are discussed within the text and defined in the Glossary at the end of the module. These terms are denoted in the text with **bold, blue type** upon their first occurrence. To make searches for key information easier, a comprehensive Glossary of Trade Terms from all modules is located at the back of this book.

on at least 1½" of wood. In platform construction, the ends of all the joists are fastened to a header joist, also called a band joist.

Joists must be doubled where extra loads require additional support. When a partition runs parallel to the joists, a double joist is placed underneath. Joists must also be doubled around openings in the floor frame for stairways, chimneys, etc., to reinforce the rough opening in the floor. These additional joists used at such openings are called trimmer joists. They support the headers that carry short joists called tail joists. Double joists should be spread where necessary to accommodate plumbing.

In residential construction, floors traditionally

Review Questions

Review Questions reinforce the knowledge you have gained and are a useful tool for measuring what you have learned.

Review Questions

1. The construction worker most likely to become knowledgeable about many trades is the _____.
   a. electrician
   b. carpenter
   c. plumber
   d. mason

2. The group that is responsible for the enforcement of apprenticeship standards and also ensures that proper apprentice training is conducted is the _____.
   a. contractor
   b. Occupational Safety and Health Administration
   c. Department of Commerce
   d. Apprenticeship Committee

3. The term used to describe the overall behavior and attitude expected in the workplace is _____.
   a. absenteeism
   b. responsibility
   c. tardiness
   d. professionalism

4. The *Code of Federal Regulations (CFR)* 29:30 specifies requirements for _____.
   a. on-the-job safety classes
   b. supervisory training
   c. fall prevention procedures
   d. approved apprenticeship programs

5. The purpose of the Youth Apprenticeship Program is to _____.
   a. make sure all young people know how to use basic carpentry tools
   b. provide job opportunities for people who quit high school
   c. allow students to start in an apprenticeship program while still in high school
   d. make sure that people under 18 have proper supervision on the job

6. The work of a crew of craftworkers and laborers is usually directed by a(n) _____.
   a. lead carpenter
   b. apprentice supervisor
   c. general contractor

7. The foundation of an effective safety program is _____.
   a. holding frequent safety meetings
   b. using a process of hazard recognition, evaluation, and control
   c. rewarding workers who don't suffer injuries
   d. developing a master accident prevention campaign

8. Which of the follow statements about SkillsUSA is *not* true?
   a. It promotes understanding of the free enterprise system.
   b. It has more than 500,000 active members.
   c. It emphasizes high ethical standards and pride in the dignity of work.
   d. It consists of 54 state and territorial associations.

9. A combined total of 8,000 hours on-the-job and classroom training is needed for a carpentry apprentice to _____.
   a. become a master carpenter
   b. complete a degree in construction technology
   c. advance to journeyman
   d. receive a competency certificate

10. An important purpose of OSHA is to _____.
    a. catch people breaking safety regulations
    b. make rules and regulations governing all aspects of construction projects
    c. ensure that the employer provides and maintains a safe workplace
    d. assign a safety inspector to every project

11. Which of the following is *not* an advantage provided to students by SkillsUSA membership?
    a. Teamwork and leadership development
    b. Income-earning opportunities
    c. Community service opportunities
    d. Networking with potential employers

NCCER Standardized Curricula

NCCER's training programs comprise more than 70 construction, maintenance, pipeline, and utility areas and include skills assessments, safety training, and management education.

Boilermaking
Cabinetmaking
Carpentry
Concrete Finishing
Construction Craft Laborer
Construction Technology
Core Curriculum:
    Introductory Craft Skills
Drywall
Electrical
Electronic Systems Technician
Heating, Ventilating, and
    Air Conditioning
Heavy Equipment Operations
Highway/Heavy Construction
Hydroblasting
Industrial Coating and Lining
    Application Specialist
Industrial Maintenance
    Electrical and Instrumentation
    Technician
Industrial Maintenance
    Mechanic
Instrumentation
Insulating
Ironworking
Masonry
Millwright
Mobile Crane Operations
Painting
Painting, Industrial
Pipefitting
Pipelayer
Plumbing
Reinforcing Ironwork
Rigging
Scaffolding
Sheet Metal
Signal Person
Site Layout
Sprinkler Fitting
Tower Crane Operator
Welding

Maritime

Maritime Industry Fundamentals
Maritime Pipefitting
Structural Fitter

Green/Sustainable Construction

Building Auditor
Fundamentals of Weatherization
Introduction to Weatherization
Sustainable Construction
    Supervisor
Weatherization Crew Chief
Weatherization Technician
Your Role in the Green
    Environment

Energy

Alternative Energy
Introduction to the Power
    Industry
Introduction to Solar
    Photovoltaics
Introduction to Wind Energy
Power Industry Fundamentals
Power Generation Maintenance
    Electrician
Power Generation I&C
    Maintenance Technician
Power Generation Maintenance
    Mechanic
Power Line Worker
Power Line Worker: Distribution
Power Line Worker: Substation
Power Line Worker:
    Transmission
Solar Photovoltaic Systems
    Installer
Wind Turbine Maintenance
    Technician

Pipeline

Control Center Operations,
    Liquid
Corrosion Control
Electrical and Instrumentation
Field Operations, Liquid
Field Operations, Gas
Maintenance
Mechanical

Safety

Field Safety
Safety Orientation
Safety Technology

Management

Fundamentals of Crew
    Leadership
Project Management
Project Supervision

Supplemental Titles

Applied Construction Math
Careers in Construction
Tools for Success

Spanish Translations

Basic Rigging
    (Principios Básicos de
    Maniobras)
Carpentry Fundamentals
    (Introducción a la
    Carpintería, Nivel Uno)
Carpentry Forms
    (Formas para Carpintería,
    Nivel Trés)
Concrete Finishing, Level One
    (Acabado de Concreto,
    Nivel Uno)
Core Curriculum:
    Introductory Craft Skills
    (Currículo Básico:
    Habilidades Introductorias del
    Oficio)
Drywall, Level One
    (Paneles de Yeso, Nivel Uno)
Electrical, Level One
    (Electricidad, Nivel Uno)
Field Safety
    (Seguridad de Campo)
Insulating, Level One
    (Aislamiento, Nivel Uno)
Ironworking, Level One
    (Herrería, Nivel Uno)
Masonry, Level One
    (Albañilería, Nivel Uno)
Pipefitting, Level One
    (Instalación de Tubería
    Industrial, Nivel Uno)
Reinforcing Ironwork, Level One
    (Herreria de Refuerzo,
    Nivel Uno)
Safety Orientation
    (Orientación de Seguridad)
Scaffolding
    (Andamios)
Sprinkler Fitting, Level One
    (Instalación de Rociadores,
    Nivel Uno)

Acknowledgments

This curriculum was revised as a result of the farsightedness and leadership of the following sponsors:

Abbott Construction Inc.
ABC Eastern Pennsylvania Chapter
ABC Northern California Chapter
ABC South Texas Chapter
Bowden Contracting Company, Inc.
Brasfield & Gorrie
The Haskell Company
HB Training and Consulting
Mountain Home High School Career Academies
PCL Construction
Suwannee-Hamilton Technical Center

This curriculum would not exist were it not for the dedication and unselfish energy of those volunteers who served on the Authoring Team. A sincere thanks is extended to the following:

John Ambrosia
Howard Davis
Owen Carpenter
Vincent Console
Curtis Haskins
Hal Heintz
Jeff Henry
Mark Knudson
Bob Makela
Tim Mosley
Kendall Purvis
Rob Underwood

NCCER Partners

American Fire Sprinkler Association
Associated Builders and Contractors, Inc.
Associated General Contractors of America
Association for Career and Technical Education
Association for Skilled and Technical Sciences
Carolinas AGC, Inc.
Carolinas Electrical Contractors Association
Center for the Improvement of Construction Management and Processes
Construction Industry Institute
Construction Users Roundtable
Construction Workforce Development Center
Design Build Institute of America
GSSC – Gulf States Shipbuilders Consortium
Manufacturing Institute
Mason Contractors Association of America
Merit Contractors Association of Canada
NACE International
National Association of Minority Contractors
National Association of Women in Construction
National Insulation Association
National Ready Mixed Concrete Association
National Technical Honor Society
National Utility Contractors Association
NAWIC Education Foundation
North American Technician Excellence
Painting & Decorating Contractors of America
Portland Cement Association
SkillsUSA®
Steel Erectors Association of America
U.S. Army Corps of Engineers
University of Florida, M. E. Rinker School of Building Construction
Women Construction Owners & Executives, USA

Contents

Module One
Commercial Drawings

Describes how to read and interpret a set of commercial drawings and specifications. (Module ID 27201; 25 Hours) *Elective for Residential Path*

Module Two
Cold-Formed Steel Framing

Describes the types and grades of steel framing materials, and includes instructions for selecting and installing metal framing for interior and exterior walls, loadbearing and nonbearing walls, partitions, and other applications. (Module ID 27205; 15 Hours)

Module Three
Exterior Finishing

Covers the various types of exterior finish materials and their installation procedures, including wood, metal, vinyl, and fiber-cement siding. (Module ID 27204; 35 Hours) *Elective for Commercial Path*

Module Four
Thermal and Moisture Protection

Covers the selection and installation of various types of insulating materials in walls, floors, and attics. Also covers the uses and installation practices for vapor barriers and waterproofing materials. (Module ID 27203; 7.5 Hours)

Module Five
Roofing Applications

Describes how to properly prepare the roof deck and install roofing for residential and commercial buildings. (Module ID 27202; 25 Hours) *Elective for Commercial Path*

Module Six
Doors and Door Hardware

Describes the installation of metal doors and related hardware in steel-framed, wood-framed, and masonry walls, along with their related hardware, such as locksets and door closers. A discussion on the installation of wood doors, folding doors, and pocket doors is also presented. (Module ID 27208; 20 Hours)

Module Seven
Drywall Installation

Describes the various types of gypsum drywall, their uses, and the fastening devices and methods used to install them. Also contains detailed instructions for installing drywall on walls and ceilings using nails, drywall screws, and adhesives. A discussion of fire- and sound-rated walls is also presented. (Module ID 27206; 15 Hours)

Module Eight
Drywall Finishing

Describes the materials, tools, and methods used to finish and patch gypsum drywall. A discussion of both automatic and manual taping and finishing tools is presented. (Module ID 27207; 17.5 Hours)

Module Nine
Suspended Ceilings

Describes the materials, layout, and installation procedures for many types of suspended ceilings used in commercial construction, as well as ceiling tiles, drywall suspension systems, and pan-type ceilings. (Module ID 27209; 15 Hours) *Elective for Residential Path*

Module Ten
Window, Door, Floor, and Ceiling Trim

Describes the different types of trim used in finish work and focuses on the proper methods for selecting, cutting, and fastening trim to provide a professional finished appearance. (Module ID 27210; 25 Hours)

Module Eleven
Cabinet Installation

Provides detailed instructions for the selection and installation of base and wall cabinets and countertops. (Module ID 27211; 10 Hours)

Glossary

Index

Note: *Carpentry Level Two* has two possible career paths: Commercial Framing and Finishing, and Residential Framing and Finishing. For trainees following the Commercial path, Module Three (27204) Exterior Finishing, and Module Five (27202) Roofing Applications, are electives. The modules shown on the Commercial course map below are required for this path. For trainees following the Residential path, Module One (27201) Commercial Drawings, and Module Nine (27209) Suspended Ceilings, are electives. The modules shown on the Residential course map below are required for this path.

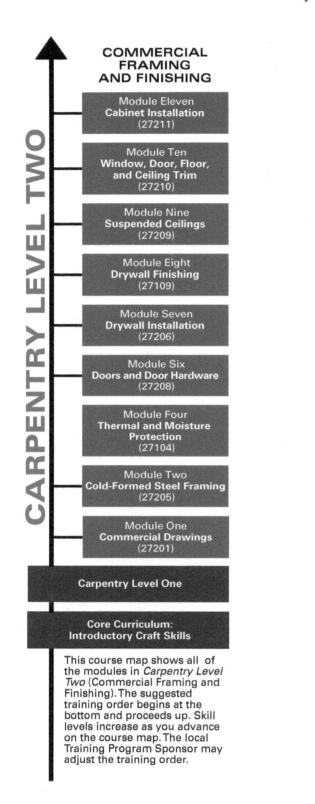

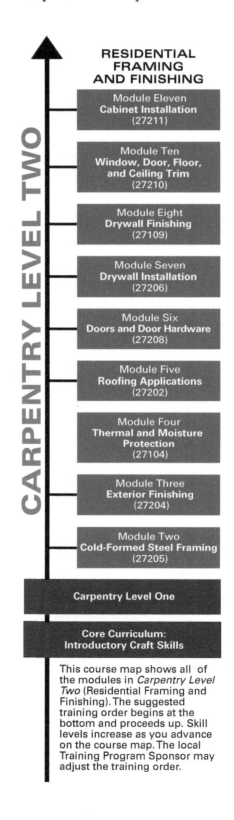

Commercial Drawings

OVERVIEW

Reading and interpreting drawings and specifications are essential to construction work. The construction drawings will tell you everything about a structure, from where to place the building on the site to how to construct an exterior wall. The specifications provide details on materials and construction methods. An effective and productive carpenter can read and interpret construction drawings.

Module 27201

Trainees with successful module completions may be eligible for credentialing through NCCER's National Registry. To learn more, go to **www.nccer.org** or contact us at **1.888.622.3720**. Our website has information on the latest product releases and training, as well as online versions of our *Cornerstone* magazine and Pearson's product catalog.

Your feedback is welcome. You may email your comments to **curriculum@nccer.org**, send general comments and inquiries to **info@nccer.org**, or fill in the User Update form at the back of this module.

This information is general in nature and intended for training purposes only. Actual performance of activities described in this manual requires compliance with all applicable operating, service, maintenance, and safety procedures under the direction of qualified personnel. References in this manual to patented or proprietary devices do not constitute a recommendation of their use.

Copyright © 2013 by NCCER, Alachua, FL 32615, and published by Pearson, New York, NY 10013. All rights reserved. Printed in the United States of America. This publication is protected by Copyright, and permission should be obtained from NCCER prior to any prohibited reproduction, storage in a retrieval system, or transmission in any form or by any means, electronic, mechanical, photocopying, recording, or likewise. To obtain permission(s) to use material from this work, please submit a written request to NCCER Product Development, 13614 Progress Blvd., Alachua, FL 32615.

27201 V5

From *Carpentry, Trainee Guide*. NCCER.
Copyright © 2013 by NCCER. Published by Pearson. All rights reserved.

27201
Commercial Drawings

Objectives

When you have completed this module, you will be able to do the following:

1. Identify the types and uses of commercial construction drawings and schedules.
    a. Compare and contrast residential and commercial construction drawings.
    b. Describe the purpose of a civil drawing.
    c. Describe the use of architectural drawings and schedules.
    d. Describe the use of structural drawings.
    e. Describe the purpose of mechanical, electrical, and plumbing drawings.
    f. Compare drawings from two different disciplines.
2. Define the use of specifications and how they are referenced.
    a. Describe the format of specifications.
    b. Explain how specifications are written.

Performance Tasks

Under the supervision of your instructor, you should be able to do the following:

1. Locate 10 items contained in a set of commercial drawings. (The instructor will select the 10 items.)
2. Examine a drawing to cross-reference the accuracy of dimensions from architectural to structural drawings.
3. Using an instructor-provided schedule, identify various criteria necessary for interpretation.
4. Using an instructor-provided shop drawing, interpret key aspects of that drawing.

Trade Terms

Beams
Benchmark
Callouts
Civil drawings
Columns
Contour lines
Detail drawings
Elevation drawings
Field notes

Girders
Invert
Isometric drawings
Joists
Landscape drawings
Liability
Plan view
Riser diagram
Schedule

Code Note

Codes vary among jurisdictions. Because of the variations in code, consult the applicable code whenever regulations are in question. Referring to an incorrect set of codes can cause as much trouble as failing to reference codes altogether. Obtain, review, and familiarize yourself with your local adopted code.

Industry-Recognized Credentials

If you're training through an NCCER-accredited sponsor, you may be eligible for credentials from NCCER's Registry. The ID number for this module is 27201. Note that this module may have been used in other NCCER curricula and may apply to other level completions. Contact NCCER's Registry at 888.622.3720 or go to **www.nccer.org** for more information.

Contents

Topics to be presented in this module include:

1.0.0 Commercial Construction Drawings and Schedules 1
   1.1.0 Commercial vs. Residential Drawings 2
      1.1.1 Requirements for Commercial Plans 2
      1.1.2 Commercial Drawing Contents 3
      1.1.3 Reading Commercial Drawings 5
   1.2.0 Civil Drawings and Schedules 6
   1.3.0 Architectural Drawings and Schedules 9
      1.3.1 Floor Plans 9
      1.3.2 Schedules and Details 10
      1.3.3 Elevation Drawings and Sections 12
   1.4.0 Structural Drawings 13
      1.4.1 Foundation Plans 18
      1.4.2 Framing Plans 19
      1.4.3 Shop Drawings 20
   1.5.0 Mechanical, Electrical, and Plumbing Drawings 20
      1.5.1 Mechanical Drawings 20
      1.5.2 Electrical Drawings 22
      1.5.3 Plumbing Drawings 24
   1.6.0 Comparing Drawings from Different Disciplines 27
2.0.0 Specifications 30
   2.1.0 Format of Specifications 30
   2.2.0 How Specifications Are Written 30
      2.2.1 Qualifications 32
Appendix A Material Symbols 38
Appendix B Topographic Symbols 40
Appendix C HVAC Symbols 41
Appendix D Electrical Symbols 42
Appendix E Plumbing Symbols 43

Figures

Figure 1	Modern buildings	2
Figure 2	Drawing set	4
Figure 3	Drawing lines	5
Figure 4	Site plan	7
Figure 5	Contour map of a hill	8
Figure 6	Contour lines used for grading	8
Figure 7	Roof plan	11
Figure 8	Door schedule	12
Figure 9	Finish schedule	13
Figure 10	Detail drawing	14
Figure 11	Elevation drawing	14
Figure 12	Structural drawing	16
Figure 13	Roof framing plan	17
Figure 14	Grid lines for structural plan	18
Figure 15	Footing/pier drawing and schedule	18
Figure 16	Structural steel notations	19
Figure 17	Structural steel framing diagram	19
Figure 18	Rebar sizes and placement	21
Figure 19	HVAC mechanical plan	23
Figure 20	Electrical plan (1 of 2)	25
Figure 20	Electrical plan (2 of 2)	26
Figure 21	Sanitary plumbing plan (1 of 2)	27
Figure 21	Sanitary plumbing plan (2 of 2)	28
Figure 22	Building information modeling	28
Figure 23	Example of the *MasterFormat*™ 2012 organization	31

This page is intentionally left blank.

Section One

1.0.0 Commercial Construction Drawings and Schedules

Objective

Identify the types and uses of commercial construction drawings and schedules.
a. Compare and contrast residential and commercial construction drawings.
b. Describe the purpose of a civil drawing.
c. Describe the use of architectural drawings and schedules.
d. Describe the use of structural drawings.
e. Describe the purpose of mechanical, electrical, and plumbing drawings.
f. Compare drawings from two different disciplines.

Performance Tasks 1 through 4

Locate 10 items contained in a set of commercial drawings. (The instructor will select the 10 items.)

Examine a drawing to cross-reference the accuracy of dimensions from architectural to structural drawings.

Using an instructor-provided schedule, identify various criteria necessary for interpretation.

Using an instructor-provided shop drawing, interpret key aspects of that drawing.

Trade Terms

Beams: Loadbearing horizontal framing members supported by walls or columns and girders.

Benchmark: A known elevation on the site used as a reference point during construction.

Callouts: Markings or identifying tags describing parts of a drawing; callouts may refer to detail drawings, schedules, or other drawings.

Civil drawings: A drawing that shows the overall shape of the building within the confines of the site. Also referred to as site plans.

Columns: Vertical structural members that support the load of other members.

Contour lines: Imaginary lines on a civil drawing that connect points of the same elevation. Contour lines normally never cross one another.

Detail drawings: Drawings shown at a larger scale in order to show specific features or connections.

Elevation drawings: Drawings showing a view from the front, rear, or side of a structure.

Field notes: A permanent record of field measurement data and related information.

Girders: Large steel or wood beams supporting a building, usually around a perimeter.

Invert: The lowest point of a pipe through which liquid flows.

Isometric drawings: Three-dimensional drawings in which the object is tilted so that all three faces are equally inclined to the picture plane.

Joists: Horizontal wood or steel members supported by beams; joists support floor sheathing.

Landscape drawings: A drawing that shows proposed plantings and other landscape features.

Liability: An obligation or responsibility, typically financial.

Plan view: A drawing that represents a view looking down on an object.

Riser diagram: Type of drawing that depicts the layout, components, and connections of a piping system.

Schedule: Tables that describe and specify the various types and sizes of construction materials used in a building. Door schedules, window schedules, and finish schedules are the most common types.

A large part of the construction market is commercial and industrial construction. A wide variety of commercial buildings can be constructed (*Figure 1*). In order to build commercial structures, a craftsperson must be able to understand and interpret the architect's plans and drawings. Commercial plans and drawings are usually more complex than residential plans. This module introduces commercial drawings. The basic principles, such as scaling and dimensioning, are the same as those that apply to residential drawings. However, commercial structures will have more dimensions to interpret.

Figure 1 Modern buildings.

1.1.0 Commercial vs. Residential Drawings

Commercial or industrial construction work is often called heavy construction since heavy equipment and materials are used for these jobs. Heavy equipment includes cranes, hoists, and graders. Heavy materials include steel and concrete. In most cases, commercial projects are larger and more complicated than single-family residential projects. They require a greater variety of construction techniques, equipment, and materials.

Consequently, plans and drawings for a commercial project are also more complicated than residential plans and drawings. Commercial structures have many different uses. Safety and environmental requirements must also be considered. The complexity of the commercial project is reflected in the complexity of the plans.

1.1.1 Requirements for Commercial Plans

There are several reasons why most commercial construction plans are more detailed than residential plans. First, the structures are usually larger and more expensive to build. Another major consideration is legal liability. The contractor's legal liability is far greater in commercial construction because there are more applicable codes, ordinances, and regulations. Other reasons that commercial plans are more complex include the following:

- The architectural plans, drawings, schedules, and specifications are legal documents. Many state, local, and federal agencies demand greater detail in commercial construction drawings to substantiate any legal disputes.

- Code restrictions and safety requirements for commercial and industrial buildings are far more complicated than for residential construction. More detailed drawings are used to make certain that all codes and local ordinances are met.
- The size of a commercial building requires a greater number of drawings, sections, details, and schedules, with more detail required to correlate the various parts of the structure.
- The materials used in commercial construction call for more detailed information on construction techniques, especially for structural steel.

The set of construction drawings for a commercial project may have 50 or 60 drawings, plus associated schedules. Specialty subcontractors will generally receive a partial set of the plans relating to their work. At least one complete set of plans will be kept in the field office for reference.

1.1.2 Commercial Drawing Contents

Construction drawings consist of several different types of drawings assembled into a set (*Figure 2*). Each type of drawing is assigned a letter. For each type, there may be several drawings, which are then numbered. For example, the first few electrical drawings would be numbered E1, E2, and E3. A complete set of commercial construction plans typically includes the following drawing types:

- Civil – C
- Life Safety – LS
- Architectural – A
- Structural – S
- Mechanical – M
- Electrical – E
- Plumbing – P
- Fire Protection – FP

The exact content of the set will vary, depending on the type and size of the job and local code requirements. For example, a set of drawings for

Brick upon Brick

For nearly 40 years, the Empire State Building was the tallest building in the world. Approximately 10 million brick were used in its construction. It was completed in 1931. Today, it is likely that curtain walls containing brick facades would be used in place of individual bricks.

Despite the enormity of the project, construction of the Empire State Building was completed in about 15 months. One of the methods used to reduce construction time was to have trucks dump the brick down a chute, instead of dumping them in the street. The chute led to a large hopper from which the brick were then dumped into carts and hoisted to the location where they were needed. The innovative technique eliminated the backbreaking work of moving brick from the pile to the bricklayer using a wheelbarrow.

27201-13_SA01.EPS

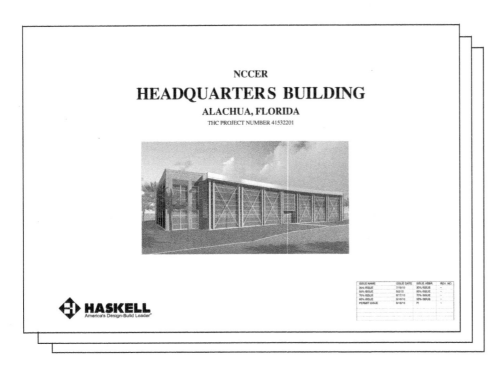

Figure 2 Drawing set.

a commercial office building may include landscape drawings; these would be denoted with the letter L. In some sets of commercial drawings, the site plans are called civil drawings and are marked C1, C2, C3, and so forth. Specifications and schedules may be referenced or included with each type of drawing. *Figure 3* shows common lines used on construction drawings.

As was introduced in Level One, object lines show the main outline of the structure. Dimension lines and extension lines are used to indicate the size of an object. Leaders are used to connect the notes to features on the drawing. Arrows terminate extension lines and leaders. The dimension is noted above the dimension line or in a break.

Cutting plane lines show the location of a section view of that particular part of the structure. Designers use various formats to indicate sections. Letters or letter/number combinations are usually used to identify the section. The section drawing is marked with the corresponding letter/number. Section views are placed either on the same page as the main drawing or on a separate page where other section views are located. Section views and symbols, such as those shown in *Appendix A*, are used to represent materials in section.

Symbols provide another graphic indication for different types of materials. There are many commonly used symbols. However, there are no standardized symbols for specific materials. The best-known reference for standard symbols, lines, and abbreviations is *Architectural Graphic Standards*, prepared by the American Institute of Architects. Large commercial projects usually have a legend commonly located on the index sheet or on separate sheets. The legend lists the symbols and abbreviations used throughout the set of drawings. *Appendix A* includes common material symbols.

Safeguarding the Drawing Set

Always treat a drawing set with care. It is best to keep two sets—one for the office and one for field use. Never remove original drawings or sheets from the original permit set of drawings in the office. Removal of original permit drawings from the office may result in their loss or damage. It can also cause delays and lost time for others who need to use or reproduce the drawings. The field set, also called as-built drawings, is marked up throughout the construction. The as-built drawings, requests for information (RFIs), addenda, and other information are archived for record retention purposes.

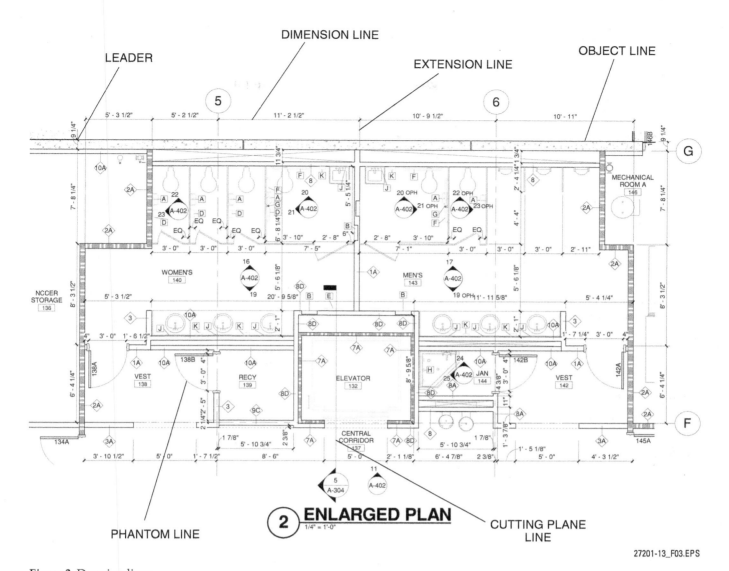

Figure 3 Drawing lines.

1.1.3 Reading Commercial Drawings

When reading a commercial drawing, it is best to follow a step-by-step process to avoid confusion and to identify the important details. Use the following list as a guide when reading a set of commercial drawings:

Step 1 Carefully review all drawings to gain a general impression of the shape, size, and appearance of the structure, including the site plan, floor plans, and exterior elevation drawings.

Step 2 Read the project specifications to gain a sense of the overall scope of the project. Remember that in cases of conflicting information, specifications take precedence over the drawings, and structural drawings take precedence over architectural drawings.

Step 3 Begin correlating the floor plans with the exterior elevation drawings, making certain that the parts appear to fit together logically. Pay particular attention to the General Notes on the sheets; they provide a good overall understanding of the structure.

Step 4 Next, review the wall sections and determine the partition types (materials; load-bearing or nonbearing) and construction details or procedures.

Step 5 Review the structural drawings. Determine the foundation requirements and the type of structural system. Turn back to the site plan, floor plans, wall sections, and elevation drawings as often as needed to relate the structural and architectural drawings.

Step 6 Review all details on the architectural and structural drawings. Carefully consider all items that may require special construction procedures.

Step 7 Review all interior elevation drawings and try to get a clear picture of what the interior of the building will look like.

Step 8 Review the finish schedule.

Step 9 Review the mechanical, electrical, and plumbing (MEP) drawings.

If some part of a drawing is not clear, check with your supervisor. There may be a logical explanation for the item in question, or the plans may be incorrect. It is best to find an answer before construction begins. The architect should clarify and resolve all conflicts in writing so there will not be any confusion later in the project.

> **NOTE**
> It is important to cross-reference drawings from all disciplines (mechanical, electrical, plumbing, and so on) to gather all relevant information. All relevant information will not be found on one specific drawing. A foreman should be aware of where the information is located.

1.2.0 Civil Drawings and Schedules

The most common type of civil drawing is the site plan. A site plan locates the structure within the confines of the building lot. The site plan clearly shows the building's orientation (*Figure 4*). The building is usually shown by the size of the foundation and the distances to the respective property lines.

A commercial building needs a site that is appealing, convenient for customer traffic, and in harmony with other commercial structures in the area. Commercial site plans show many of the same features as residential drawings, but are far more detailed. They usually include details on site improvement features, existing and finish contour lines, paving areas, and site access. Since a commercial structure serves the public, the architect, owner, and zoning board should note and specify all site changes. Commercial site plans will also show the location of public utilities.

When looking at a commercial site plan for the first time, notice the features that also appear on residential location plot plans, including the following:

- Survey data such as directional arrow, property lines, and the structure's relationship to other features
- Geographic data such as lot corner elevations, existing and proposed contours, and landscaping features
- Building features such as floor elevations, exterior wall positioning, roof overhang, and drainage

Features on the commercial site plans that may not be found on residential plans include the following:

- Details of paved areas, including walkways, driveways, and parking areas
- Details on grading and other site work
- Notations that refer to details on the site plan or elsewhere in the drawings
- Position of existing and new utility lines
- Referenced details, sections, and elevation drawings that pertain to site preparation features
- Symbols or notations detailing materials types, sizes, and positions

Another primary purpose of the site plan is to show the unique surface conditions, or topography, of the lot. The topography of a particular lot may be shown right on the site plan. For projects in which the topography must be shown separately for clarity, a grading plan is used.

The topographical information includes changes in the elevation of the lot such as slopes, hills, valleys, and other variations in the surface. *Figure 5* shows the contour map of a hill. These

Drawing Revisions

When a set of drawings has been revised, always ensure that the most up-to-date set is used for all future work. Either destroy the old, obsolete drawing or else clearly mark on the affected sheets "Obsolete Drawing—Do Not Use." A good practice is to remove the obsolete drawings from the set and file them as archive copies for possible future reference.

Also, when working with a set of construction drawings and written specifications for the first time, thoroughly check each page to see if any revisions or modifications have been made to the original. Doing so can save time and expense for all concerned.

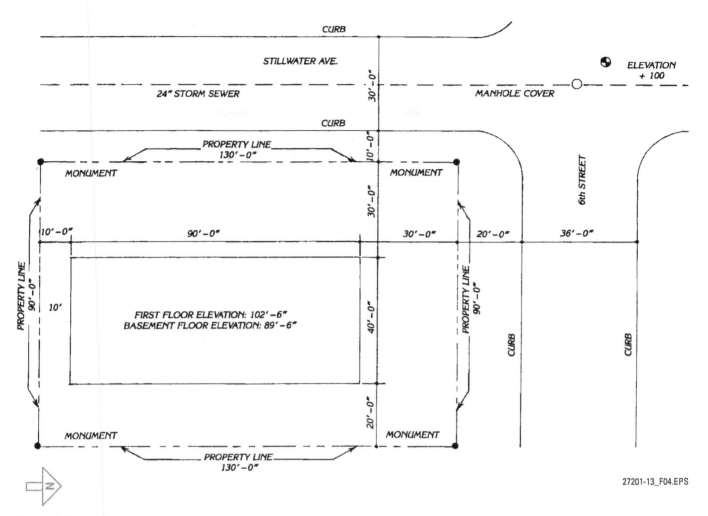

Figure 4 Site plan.

changes in the surface conditions are shown on a site plan by means of a contour, which is a line connecting points of equal elevation. An elevation is a distance above or below a known point of reference, called a datum. This datum could be mean sea level, or may be an arbitrary plane of reference established for the particular building.

Two important characteristics of the contour need to be observed when reading a site plan:

- Contours are continuous and frequently enclose large areas in comparison to the size of the building lot. For this reason, contours are often drawn from one edge to the other edge of the site plan.
- Contours normally do not intersect or merge together. The only exception to this rule is in the case of a vertical wall or plane.

The existing contour is shown as a dashed line; the new or proposed contour is shown as a solid line (*Figure 6*). Both are labeled with the elevation of the contour in the form of a whole number. The spacing between the contour lines is at a constant vertical increment, or interval. The typical interval is 5 feet, but intervals of 1 foot are not uncommon for site plans requiring greater detail, or where the change in elevation is gradual.

Site layout involves establishing a network of control points on a site that serve as a common reference for all construction. The exact locations of these control points are marked at the site and recorded in the field notes as they are made. Annotating control point location reference data in the field notes is important for two reasons. First, it makes it possible to locate a point should it become covered up or otherwise hidden. Second, it makes it possible to reestablish a point accurately if the marker is damaged or removed. There are three basic categories of control points:

- *Primary control points*—These points are used as the basis for locating secondary control points and other points on the site. Primary control points are located where they are accessible and protected from damage for the duration of the job. Primary vertical control points can

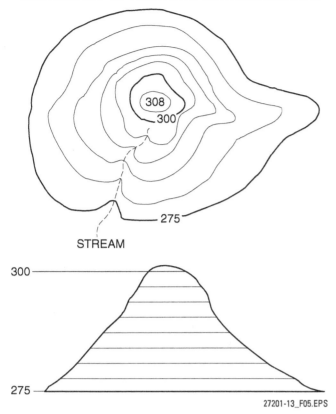

Figure 5 Contour map of a hill.

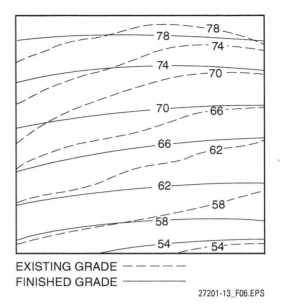

Figure 6 Contour lines used for grading.

be located and marked on many kinds of permanent and immovable objects such as fire hydrants and power poles. Primary control point markers are typically established by a registered licensed surveyor. They are commonly referred to as monuments or **benchmarks**.

- *Secondary control points*—Additional control points located within the job site aid in the construction of the individual structures on the site. Secondary control points typically are marked by a hub stake surrounded by protective laths or fencing. A surveyor's tack, with a depression in the center of the head, is driven into the top of the hub stake to locate the exact point.
- *Building layout or working control points*—These points are usually located with reference to the secondary control points. These are the points from which actual measurements for construction are taken. Building layout points are used to locate the corners of buildings and building lines. They usually are marked with a hub and a related marker stake. In addition to serving as hub markers, these stakes are also used to mark line or grade and other information for center lines, offset lines, slope stakes, etc. Additional information regarding the use of control points is included in the Carpentry Level Four module titled *Site Layout I*.

A benchmark is commonly noted on the site plan with a physical description and its elevation relative to the datum. The benchmark will be established and recorded by a registered licensed surveyor. (Surveyors must be bonded and licensed, and should be familiar with the area where the survey is being performed.) For example: "Northeast corner of catch basin rim—Elev. 102.34'" might be a typical benchmark found on a site plan. When individual elevations, or grades, are required for other site features, they are noted with a + and the grade. Grades vary from contours in that a grade has accuracy to two decimal places, whereas a contour is expressed as a whole number.

The site plan shows topographic features including trees, bodies of water, and ground cover. It will also show man-made features such as roads, railroad tracks, and utility lines. Typical topographic symbols are shown in *Appendix B*.

Sophisticated site plans showing utilities and drainage services often require a legend. The legend is similar to that on the architectural drawings, listing the different symbols and abbreviations found in the particular group of site plans.

As always, review material symbols, dimensions and scales, and fundamental construction techniques before attempting to understand a full set of plans. Carefully review the site plan and get an overall concept of the work required. Look at the contours to determine where excavation will be required. For example, if a point is at an existing elevation of about 68' and a finish elevation of 70', 2' of fill will be required. It is often helpful to divide the site plan into sections or by grid lines to fully understand the amount of work required on the site.

Site plans are drawn using any convenient scale. This may be ⅛" = 1'–0", or it may be an engineering scale, such as 1" = 20'–0".

Larger projects have several civil drawings showing different scopes of related or similar work. A couple such plans are the drainage and utility plans. Drainage plans detail how surface water will be collected, channeled, and dispersed on- or off-site. Utility plans show locations of the water, gas, sanitary sewer, and electric utilities that will service the building. Drainage and utility plans illustrate, in plan view, the size and type of pipes, their length, and the special connections or terminations of the various piping.

The elevation of a particular pipe below the surface is given with respect to its invert. This is typically noted with the abbreviation for invert and an elevation, for example: "I.E. 12.34'." The inverts are shown at the intersections of pipes or other changes in the continuous run of piping, such as a manhole, catch basin, sewer manhole, and so on. Inverts are usually given for piping that has a gravity flow or pitch. Using benchmarks, contours, or spot elevations, you can quickly calculate the distance of the piping below the surface and the direction of the flow.

With projects of a more sophisticated nature, separate drawings showing various site improvements may be needed for clarification. Site improvements may include such items as curbs, walks, retaining walls, paving, fences, steps, benches, and flagpoles.

Paving and curbing plans show the various types of bituminous, concrete, and brick paving and curbing, as well as the limits of each. Use this information to calculate the area and measurements of paving and curbing. Review the legend symbols to understand where one material ends and another begins. Do not make assumptions. Details show subsurface sections. They show the thickness of the paving and the substrate.

Location, Location, Location

Some site plans include a small map, called a locus, showing the general location of the property in respect to local highways, routes, and roads.

1.3.0 Architectural Drawings and Schedules

The architectural drawings are usually labeled with page numbers and begin with the letter prefix of A. These drawings contain general design features of the building, room layouts, construction details, and materials requirements. The architectural drawings include the following:

- Floor plans
- Wall sections
- Door and window details and schedules
- Elevation drawings
- Special application details, including finish details and schedules

Architectural drawings are the core of any set of drawings.

1.3.1 Floor Plans

Floor plans of both commercial and residential structures show various floor levels as if a horizontal cut had been made through the structure, resulting in an overhead view of each floor. One common use of the floor plan is to calculate the area of each room and other areas represented on the floor plan. This information is needed to determine the amount of floor sheathing, floor covering, and other material required. For most rooms, it is simply a matter of multiplying the length and width dimensions to determine the area of the room in square feet. If the room is not a perfect rectangle, however, the calculation must be done by dividing the room into pieces, and then adding the square footage of the pieces.

The most noticeable difference between commercial and residential floor plans is the amount of detail. Commercial plans contain more details of room use, finishes, wall types, sound transmission, and fire retardation. In fact, most commercial floor plans will incorporate a legend or finish schedule to specify the various interior wall types shown on the plan. The drawing scale is normally ¼" = 1'–0" for floor plans. The detailing instructions usually include the following types of information:

- Numerous callouts specifying sectional views and details
- Room assignment designations by function or number
- Detailed dimensioning of all visible parts of the structure
- Finish designations referenced to schedules

On commercial plans, little is left to chance for several reasons. First, the construction itself is varied and complex. Second, construction must meet all commercial code specifications. Finally, different contractors will be working on the same job. Since there are many ways to accomplish the same task, the requirements are specified in detail to achieve consistency throughout the structure.

Each door or window on a floor plan for a commercial building is typically accompanied by a number, letter, or both. This number/letter is an identifier that refers to a door or window schedule that describes the corresponding door by size, type of materials, or model number for the specific door. Schedules are discussed in more detail in the next section Schedules and Details.

When supplied, roof plans (*Figure 7*) provide information about the roof slope, roof drain placement, and other pertinent information related to the roof. Where applicable, the roof plan may also show information on the location of air conditioning units, exhaust fans, and other ventilation equipment.

To help clarify unclear parts of the drawing, notes may be included on commercial floor plans. This is particularly true when any feature differs from one area to another. In most instances, these notes are important not only because they show variations, but also because they detail the responsibilities of those involved in the construction. A plan note may read "Furnished by owner," "Refer to structural drawings," or "Not in contract."

For large buildings, the architect often will divide the floor plan into sections by grid lines. The grid is the same for all floors of the building. Using the grid allows the architect and engineer to locate features very specifically anywhere in the building. The grid lines are useful for locating features that are repeated on one floor or from one floor to another. The grid markings on the floor plan also reappear on the structural drawings where they locate footings and columns.

1.3.2 Schedules and Details

Since a commercial building has so many parts and functions, it is impossible to draw all of the features on one drawing. Many features are described in notes on the actual drawings. The drawings may become cluttered if too many notes appear. To prevent such a problem, schedules, detail drawings, and written specifications are used to describe the construction materials and procedures required.

In most cases, the architectural drawings include schedules for doors (*Figure 8*), windows, and interior wall finishes. A schedule is actually

A Museum of American Architecture

The Athenaeum of Philadelphia is a museum of American architecture and interior design. The work of over 1,000 American architects from 1800 to 1945 is collected and available for research. The museum holds 150,000 drawings, 50,000 photographs, and many other documents. Most of the drawings are from Philadelphia, but the collection also includes drawings for buildings in most of the states and several countries. The Athenaeum was designed in 1845 by John Notman. It is one of the first Philadelphia buildings built of brownstone.

27201-13_SA02.EPS

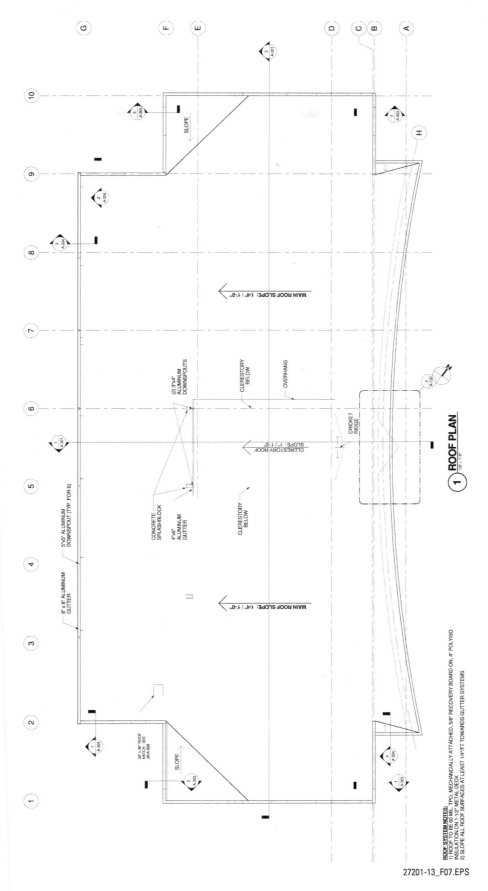

Figure 7 Roof plan.

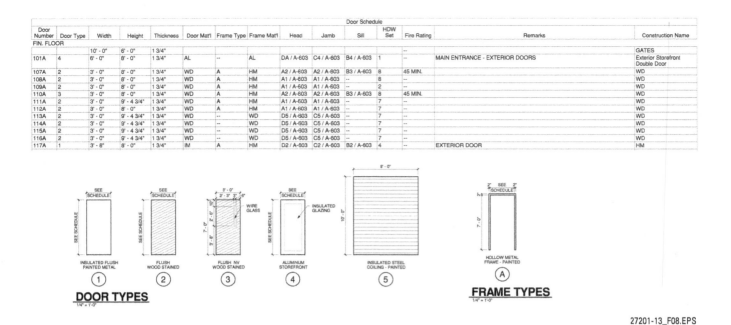

Door Number	Door Type	Width	Height	Thickness	Door Mat'l	Frame Type	Frame Mat'l	Head	Jamb	Sill	HDW Set	Fire Rating	Remarks	Construction Name
FIN. FLOOR		10' - 0"	6' - 0"	1 3/4"								--		GATES
101A	4	6' - 0"	8' - 0"	1 3/4"	AL	--	AL	DA / A-603	C4 / A-603	B4 / A-603	1	--	MAIN ENTRANCE - EXTERIOR DOORS	Exterior Storefront Double Door
107A	2	3' - 0"	8' - 0"	1 3/4"	WD	A	HM	A2 / A-603	A2 / A-603	B3 / A-603	8	45 MIN.		WD
108A	2	3' - 0"	8' - 0"	1 3/4"	WD	A	HM	A1 / A-603	A1 / A-603	--	8			WD
109A	2	3' - 0"	8' - 0"	1 3/4"	WD	A	HM	A1 / A-603	A1 / A-603	--	2			WD
110A	3	3' - 0"	8' - 0"	1 3/4"	WD	A	HM	A2 / A-603	A2 / A-603	B3 / A-603	8	45 MIN.		WD
111A	2	3' - 0"	9' - 4 3/4"	1 3/4"	WD	A	HM	A1 / A-603	A1 / A-603	--	7			WD
112A	2	3' - 0"	8' - 0"	1 3/4"	WD	A	HM	A1 / A-603	A1 / A-603	--	7			WD
113A	2	3' - 0"	9' - 4 3/4"	1 3/4"	WD	--	WD	D5 / A-603	C5 / A-603	--	7			WD
114A	2	3' - 0"	9' - 4 3/4"	1 3/4"	WD	--	WD	D5 / A-603	C5 / A-603	--	7			WD
115A	2	3' - 0"	9' - 4 3/4"	1 3/4"	WD	--	WD	D5 / A-603	C5 / A-603	--	7			WD
116A	2	3' - 0"	9' - 4 3/4"	1 3/4"	WD	--	WD	D5 / A-603	C5 / A-603	--	7			WD
117A	1	3' - 8"	8' - 0"	1 3/4"	IM	A	HM	D2 / A-603	C2 / A-603	B2 / A-603	4	--	EXTERIOR DOOR	HM

Figure 8 Door schedule.

a chart or table that provides detailed information corresponding to various parts of the drawings. Common features of schedules include:

- A reference mark, letter, or number, which corresponds to markings on the drawings
- The desired manufacturer for a specific item
- Information on item part numbers, sizes, special finishes, and hardware requirements

Door and window schedules are designated by a number or letter on the drawings. The same letter or number is duplicated in the schedule, with a brief description of the item. Typical door and window schedules must show their relationship to the floor plan.

A finish schedule (*Figure 9*) references a location by a room number noted on the schedule; the drawing usually references the schedule with a symbol or note. Finish schedules for commercial buildings will provide detailed lists of finishing materials and their application, plus information about floors, walls, base trim, ceilings, and molding.

Certain areas of a floor plan, elevation drawings, or other drawings may be enlarged for greater clarity. These enlargements, called detail drawings, are drawn to a larger scale (*Figure 10*). Detail drawings can be found either on the sheet where they are first referenced, or grouped together on a separate detail sheet included in the set of drawings. These drawings are important sources of information for the contractor and craftsperson.

1.3.3 Elevation Drawings and Sections

Elevation drawings on commercial construction drawings are similar to residential elevation drawings, but provide more information. Exterior elevation drawings (*Figure 11*) provide views of the building from each major orientation, as well as references for section views. Elevation drawings are normally drawn to the same scale as the floor plans.

Some interior elevation drawings provide vertical dimensions for interior work, materials lists, and construction details. They are important sources of information for built-in cabinets, shelving, finish carpentry, or millwork items. The scale of the drawing depends on the detail required. This may be as small as ¼" = 1'–0", or as large as ¾" = 1'–0".

In a set of commercial construction drawings, there are many wall sections to show the different types of exterior and interior walls. The interior sections detail the construction of each wall type, such as curtain walls, partitions, loadbearing walls, fire-resistant walls, and noise-reduction walls. The floor plans incorporate a legend that specifies the interior wall type. The wall section drawing provides the necessary detail. Wall sections usually provide the following details:

- Construction techniques and material types
- Stud types and placement criteria
- Fire ratings of various materials, which is measured in terms of hours of resistance
- Sound-barrier placement or materials
- Insulation applications and materials
- In-wall features such as recesses or chases

Room No.	Room Name	Floor	Base	N	S	E	W	CLG	Remarks
FIN. FLOOR									
101	MAIN LOBBY	PAV	PAV	GYP/PT	GYP/PT	GL	--	EXP2	ALT: STAINED CONCRETE FLOOR
102	RECEPTION	PAV	PAV	--	--	--	GYP/PT	EXP2	ALT: STAINED CONCRETE FLOOR
104	SOUTH GALLERY	PAV	PAV	--	GYP/PT	GYP/PT	GYP/PT	EXP2	ALT: STAINED CONCRETE FLOOR
105	NORTH GALLERY	PAV	PAV	GYP/PT	--	GYP/PT	GYP/PT	EXP2	ALT: STAINED CONCRETE FLOOR
106	PROV WAITING	PAV	PAV	GYP/PT	GYP/PT	GL	GYP/PT	EXP2	ALT: STAINED CONCRETE FLOOR
107	TESTING VEST	CPT	VNL	GYP/PT	GYP/PT	GYP/PT	GYP/PT	ACT2	
108	TESTING CENTER	CPT	VNL	GYP/PT	GYP/PT	GYP/PT	GYP/PT	ACT2	
109	PROV TOILET	CT	CT	GYP/PT/CT	GYP/PT/CT	GYP/PT/CT	GYP/PT/CT	GYP/PT	
110	PROV RECEPTION	CPT	VNL	GYP/PT	GYP/PT	GYP/PT	GYP/PT	EXP2	
111	PROV OPEN OFFICE	CPT	VNL	GYP/PT	GYP/PT	GYP/PT	GYP/PT	ACT2	
112	PROV CONF	CPT	VNL	GYP/PT	GYP/PT	GL	GYP/PT	EXP2	
113	PROV OFFICE #4	CPT	VNL	GYP/PT	GYP/PT	GL	GYP/PT	EXP2 / ACT2	
114	PROV OFFICE #3	CPT	VNL	GYP/PT	GYP/PT	GL	GYP/PT	EXP2 / ACT2	
115	PROV OFFICE #2	CPT	VNL	GYP/PT	GYP/PT	GL	GYP/PT	EXP2 / ACT2	

LEGEND

ACT 1 - ACOUSTICAL CEILING TILE
ACTC2 - ACOUSTICAL CEILING TILE CLOUD
ACT3 - ACOUSTICAL CEILING TILE (MOISTURE RESISTANT) - SEE SPECS
CMU - CONCRETE MASONRY UNIT
CONC - CONCRETE
CPT - CARPET
CT - CERAMIC TILE
EXP1 - EXPOSED CEILING
EXP2 - EXPOSED CEILING PAINTED - SEE SPECS
EPX - EPOXY
GL - GLAZING
GYP - GYPSUM WALL BOARD
PAV - PAVERS
PT - PAINT
RUB - RUBBER
SC - SEALED CONCRETE
SV - SHEET VINYL
VCT - VINYL COMPOSITION TILE
VNL - VINYL

Figure 9 Finish schedule.

Elevation drawings and sections are also drawn for landings, stairways, and backfilled retaining walls. The section drawings detail construction techniques or materials. For instance, a stairway will have details, sections, and elevation drawings with information on tread, landing, and handrail construction. The amount of detail will depend on the complexity of the stairway and on the various building code specifications.

Longitudinal and transverse wall sections show construction features that are expanded on in the structural plans. These sections show the following:

- Relationships of all wall features from the footings through the roof
- Footing and foundation placement in relation to other elevations
- Exterior material symbols and notations
- Framework type and placement

As with elevation drawings, the sections provide an overall view of the proposed structure rather than the detail required for construction. Detail notations or callouts will refer directly to the structural drawing sheets. Most callouts will refer to details on roofing **beams** and trusses, foundations, framing features, or other structural components.

1.4.0 Structural Drawings

Structural drawings provide a view of the structural members of the building and how they will support and transmit those loads to the ground. Structural drawings are numbered sequentially and designated by the letter S. Structural drawings are typically located after the architectural drawings in a set of drawings.

A structural engineer prepares structural drawings. They must calculate the forces on the building and the load that each structural member must withstand. The structural support information includes the foundation, size, and reinforcing requirements; the structural frame type and size of each member; and details on all connections required. The structural drawings usually include the following:

- Foundation plans
- Framing plans for floors and roofs
- Support details
- Schedules

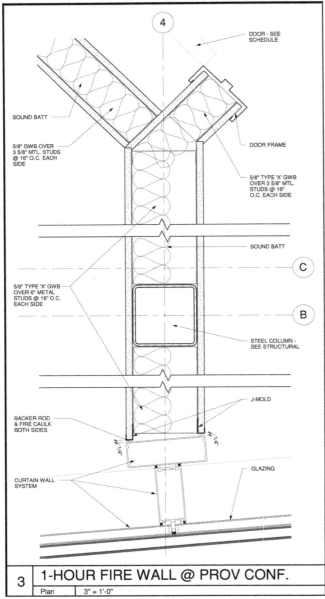

Figure 10 Detail drawing.

- Notes to describe construction and code requirements

Structural support for a commercial building may be steel framing, precast concrete structural members, or cast-in-place concrete. Unlike residences, most modern commercial buildings do not have wood frames.

Structural drawings can stand alone for craftspersons such as framers and erectors. The structural drawings show the main building members and how they relate to the interior and exterior finishes. They do not include information that is unnecessary at the structural stage of construction.

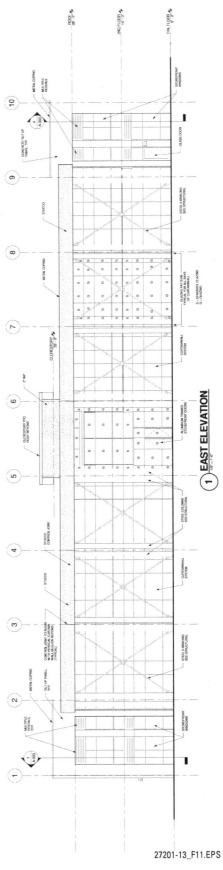

Figure 11 Elevation drawing.

Structural drawings start with the foundation plans. Foundation plans are followed by ground floor or first-floor plans, upper-floor plans, and the roof plan. Only information essential to the structural systems is shown. For example, a second-floor structural plan would show the steel or concrete framing and the configuration and spacing of the loadbearing members. The walls, ceilings, or floors would not be shown.

Structural drawings (*Figure 12*) provide detailed information on the structural features of a building. This includes information on the loadbearing design and materials, such as masonry, reinforced concrete, steel framing, or oversize timber. The structural drawings include plan views, sections, details, schedules, and notes. They provide information on the size and placement of loadbearing members. Structural drawings also show how the members are connected to each other and to other parts of the structure.

Typical structural drawings include a foundation plan, floor framing plans, and a roof framing plan (*Figure 13*). The plan view will have sections, details, schedules, and notes located in any available space on the drawing sheet. Each plan view should have a north directional arrow to maintain a consistent orientation. Plan views are typically drawn to the scale of $\frac{1}{8}$" = 1'–0" or $\frac{1}{4}$" = 1'–0"; sections are $\frac{1}{2}$" or $\frac{3}{4}$" = 1'–0"; details are 1" or $1\frac{1}{2}$" = 1'–0".

Each plan is referenced to a grid identifying the placement of columns and footings. The grid is identical on all plans and includes dimensions. In the structural drawings, a callout sequence number or mark will identify and show the placement of columns on the grid. Some project plans do not use a grid, but use a callout sequence marking. In either case, the identification marks will be referenced to schedules or notes. For example, in *Figure 14* the notation B-2 may be referencing a footing, pier, or column on the appropriate schedule.

Structural drawings show the type of framing and loadbearing for the building. For example, if the floor, roof beams, or trusses place their weight directly on the wall materials, the structure has loadbearing walls. These walls are constructed of materials with high compressive strength such as concrete masonry units (CMUs) or cast-in-place concrete. The walls support their own weight as well as that of the various floor and roof members. The plans for a loadbearing wall will show no structural beams or columns along the wall. This type of construction is typical in smaller commercial buildings with spans of about 40 feet and height of no more than three stories.

In reinforced concrete construction, the loadbearing members usually include reinforced concrete footings, foundations, piers, and columns. Exterior walls are typically nonbearing masonry curtain or panel walls. Structural drawings show the size, type, and placement of reinforcing materials and of various jointing techniques. For instance, sections may provide information for the placement of reinforcing bar (rebar) or welded wire reinforcement, while details may show the types of saddles, chairs, stirrups, or joints to use. A rebar notation will usually include the bar size and the bar spacing. As shown in the schedule in *Figure 15*, footing F6.0 measures 6'–0" wide by 6'–0" long and is 1'–2" thick. It is reinforced with six #6 rebar each way at the bottom.

High-rise buildings are typically steel frame construction with nonbearing masonry curtain or panel walls. The framework for the entire structure is formed by bolting or welding various steel members together. The loads from floors and roofing are transferred to beams and **girders** and down columns to the footings. When the joints are bolted together, the schedules will designate the number, size, and material requirements for the bolts. *Figure 16* shows the most common shapes for steel frame elements and lists the typical plan designations.

Timber frame construction is still used for many commercial buildings. The beams usually have 6" nominal dimensions. They may be solid or laminated. The drawings and schedules will show manufacturer's designations and notations for connecting hardware. Metal parts, such as strap hangers, brackets, base plates, and lag screws, are listed on the schedule. They are used to connect the wood to concrete or steel support.

The standard structural steel notation shown on a construction drawing lists the type or shape of the beam, the depth of the web, and the weight per foot. For example, the notation "W18 × 77" refers to a wide-flange shape with a nominal depth of 18 inches and a weight of 77 pounds per linear

---

The Green Environment

The construction industry is changing. In this new era, the green environment is an important consideration. As a construction craftworker, you must understand how your daily activities at work and at home affect the green environment. With this knowledge, you can make smart choices to reduce your impact.

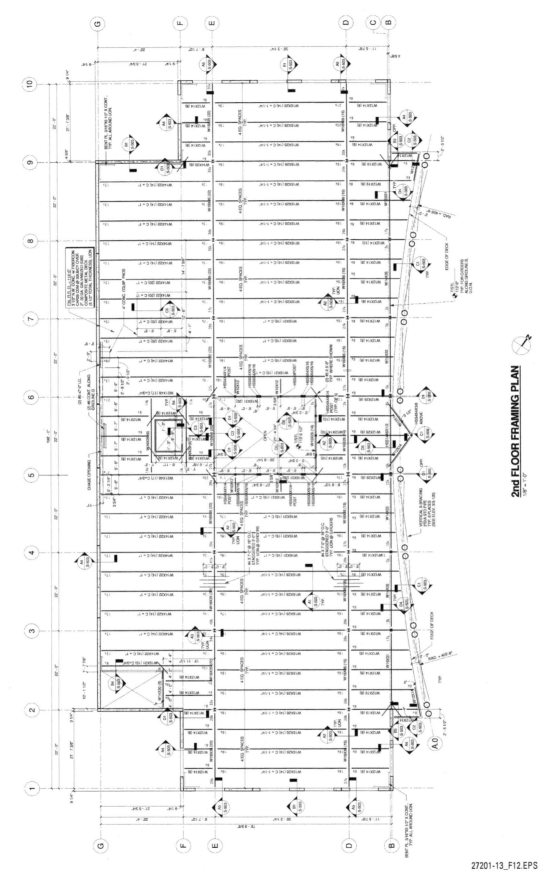

Figure 12 Structural drawing.

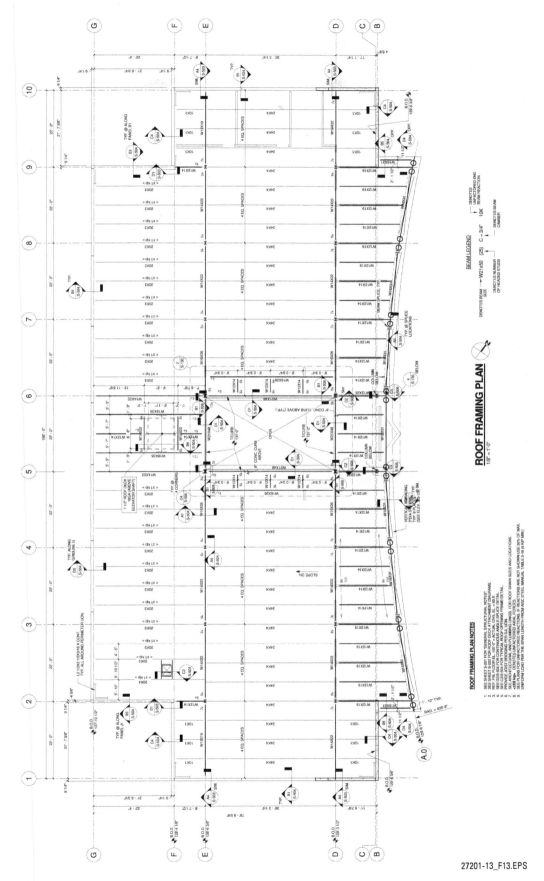

Figure 13 Roof framing plan.

Module 27201 — Commercial Drawings 17

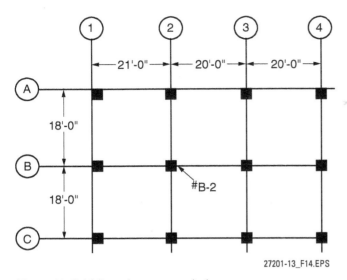

Figure 14 Grid lines for structural plan.

foot. The length of the beam is found on the plan view or the shop drawings. The fabricator will cut these beams to the specified length, label them, and predrill connection holes.

1.4.1 Foundation Plans

Commercial buildings that carry heavy loads receive a great deal of design attention at the foundation and footing level. Soil sampling, laboratory tests, and a professional engineering (PE) analysis determine the foundation type and size. The foundation size is determined by the load to be carried. The type of foundation is determined by a combination of load and soil capacity and characteristics. Foundations are categorized as shallow, intermediate, or deep.

Shallow foundations are set to a depth just below the frost line or slightly lower to reach soil with adequate bearing capacity. Shallow foundations take these forms:

- A continuous reinforced concrete footing around the entire building perimeter, carrying wall loads directly
- Isolated reinforced concrete footings located under loadbearing columns
- A concrete slab that is placed in a single operation and can carry wall loads directly, or through columns, or both
- Grade beams of reinforced concrete set below grade level and supported by other foundation elements

Intermediate foundations are set to a depth generally not exceeding 15' to 20'. Intermediate foundations take these forms:

- Mat foundations are large, heavily reinforced concrete mats under the complete building area; sometimes called raft foundations.
- Drilled piers are concrete or reinforced concrete piers formed by placing concrete in deep columnar holes drilled in the Earth; these are designed to carry column loads or grade beams. In recent years, piers have been made of other time- or cost-effective materials, such as compacted sand or rock.

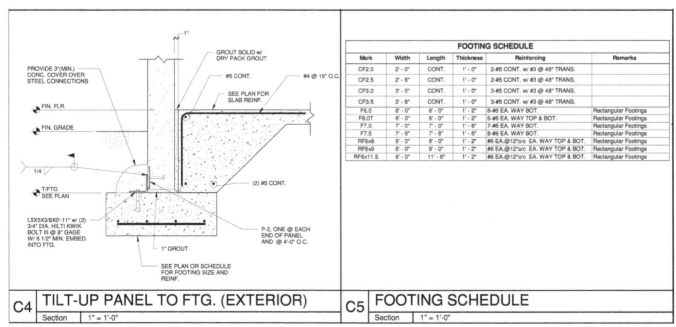

Figure 15 Footing/pier drawing and schedule.

Descriptive Name	Shape	Identifying Symbol	Typical Designation height/wt/ft in lb	Nominal Size height width
WIDE-FLANGE SHAPES	I	W	W21 × 132	21 × 13
MISCELLANEOUS SHAPES	I	M	M8 × 65	8 × 2¼
AMERICAN STANDARD BEAMS	I	S	S8 × 23	8 × 4
AMERICAN STANDARD CHANNELS	[	C	C6 × 13	6 × 2
MISCELLANEOUS CHANNELS	[	MC	MC8 × 20	8 × 3
ANGLES EQUAL LEGS	L	L	L 6 × 6 × ½	6 × 6
ANGLES UNEQUAL LEGS	L	L	L 8 × 6 × ½	8 × 6
STRUCTURAL TEES (cut from wide flange)	T	WT	WT12 × 73	9
STRUCTURAL TEES (cut from am. std. beams)	T	ST	ST9 × 35	9

Figure 16 Structural steel notations.

Deep foundations are set to a depth over 20' to 30'. Deep foundations are used where surface soil is not adequate for building loads. Steel or concrete pilings are driven into the ground until they reach a strong, stable soil layer, or until they generate enough frictional resistance to compensate for the building load. Piles are typically capped with concrete that supports column loads or grade beams.

More detailed information is needed if the foundation goes well below the Earth's surface. However, any foundation plan should provide the following information:

- Plan views for footings, piers, and/or columns with notations on position and size
- Schedules with dimensional notations, shapes, reinforcing, and construction requirements
- Sections showing the smaller construction details of footings, columns, connections, and callouts, which refer to the schedules

1.4.2 Framing Plans

The structural engineer draws a framing plan or diagram for the roof and for each floor level that will be framed. On these drawings, the exterior walls or bearing walls are often drawn in lightly while heavier lines represent the framing. The resulting plan looks very much like a graph or diagram, as shown in *Figure 17*.

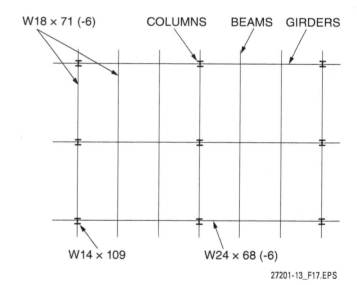

Figure 17 Structural steel framing diagram.

Check the Legend

In order to avoid mistakes in reading the drawings, be sure you understand the symbols and abbreviations used on every drawing set. Symbols and abbreviations may vary widely from one drawing set to another.

Looking at this diagram, or at any framing plan, you should see the following:

- Notations identifying beams, *joists*, and girders by size, shape, and material
- Column, pier, and support locations and their relationship to joists or framing
- Notes or callouts identifying corresponding sections or detail drawings

You may also see details for locations of stairs, recesses, and chimney placements. Details show additional unique framing around these areas. The dimensions on the drawings are center-line dimensions, not actual member dimensions. The member size must be less than the center-line dimensions to allow for construction tolerances.

The columns on a framing plan are shown from the top. Lines running between columns are beams. Beams fasten directly to the columns. Joists fasten between beams or between beams and walls. To keep long spans from swaying or twisting in the center, bridging or support members are placed between joists. All of these members will either be referenced to a schedule or to notes directly written on the plans.

Structural drawings include sheets of details, schedules, and notes with the framing plans. They help craftworkers to understand and follow the specifications for the structure. They provide information on the following areas:

- Reinforcing information for all areas where rebar or welded wire reinforcement will be used
- Information for each type of connection made in framing members
- Bearing-plate information detailing the features of all members that will bear directly on other members
- Information for positioning ties, stirrups, or saddles
- Placement and construction information for any unique features that cannot be adequately described with a drawing

The details also identify load limits, test strength, fastener types, and uniform specifications, which must be applied where specific information is not given.

There is a great deal of information on the framing plans. It is easier to read such plans by isolating the separate bays or spans between columns. Read the details for that area before moving to other areas of the plan.

1.4.3 Shop Drawings

Shop drawings are detail-oriented supplemental drawings that describe the fabrication and erection of certain elements of a structure. Shop drawings are approved during the submittal process and then become part of the overall scope of the legal construction documents.

One type of shop drawing is a rebar drawing, which details the types and locations of reinforcement bar (*Figure 18*). Using the structural drawings as a guide, the rebar detailer prepares drawings that indicate the size, location, spacing, and other information needed to develop bar lists. (These drawings are also used by ironworkers when installing the rebar.) A bar list contains quantities, sizes, grading, lengths, and bending dimensions of the rebar.

The rebar fabricator uses the bar list to cut, bend, and tag rebar. Ironworkers use the bar lists to check shipments, sort rebar, and place rebar in forms.

Each bundle of rebar contains a tag that shows the quantity, size, weight, and bend configuration of a specific bundle of bars. This tag is developed by the fabricator. In many cases, the tag is computer generated from data that went into the development of the bar list. The tag also shows the location in which the bars are to be placed. Bars are usually sorted into bundles of the same type and building location. However, this may vary, depending on the quantity and sizes of the bars.

1.5.0 Mechanical, Electrical, and Plumbing Drawings

For commercial structures, nonstructural systems are detailed in separate drawings. Mechanical drawings show the different mechanical systems of the building such as the heating, ventilating, and air conditioning (HVAC) system. Plumbing systems and fire protection systems are shown on separate drawings. These drawings have a prefix of M, P, and FP, respectively. Plan views are commonly used for these drawings as they offer the best illustration of the location and configuration of the work. The drawings serve as a diagram of the system layouts.

Electrical drawings are labeled with page numbers beginning with the letter E. They contain information on the electrical service requirements for the building and all other electrical aspects of a building.

1.5.1 Mechanical Drawings

Mechanical drawings provide information about the HVAC systems. The work required to install these systems in commercial structures is typically performed by trades or firms that specialize in the particular craft.

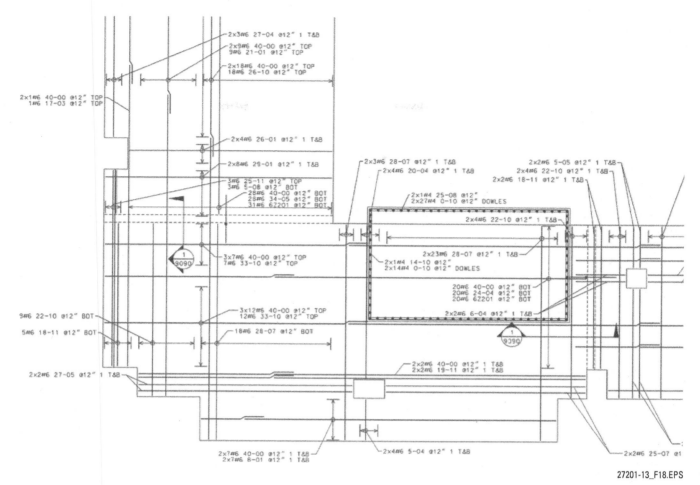

Figure 18 Rebar sizes and placement.

While carpenters generally do not get involved with working on these systems, there are many situations where carpenters are required to make passage for or work around mechanical items. One example would be the through-wall ductwork for a forced-air HVAC system. Because of these types of requirements, you must carefully review all the mechanical drawings for information on dimensions and measurements. Locate services entering the building and passing through interior walls, and coordinate with the HVAC workers.

Mechanical drawings that contain HVAC systems should be carefully studied for the layout, locations, and sizes of ductwork and fin tube radiation baseboard. Special mechanical drawings that illustrate the components of boiler and rooftop- or grade-mounted HVAC units may also be included.

HVAC drawings typically contain schedules that identify the different types of HVAC equipment. As appropriate, the drawings include detail drawings describing the installation of the HVAC equipment. Depending on the nature of the project, these drawings may include a refrigeration-piping schematic, chilled-water coil and hot-water coil piping schematics, and piping runs for other HVAC equipment.

A basic understanding of an air-handling system is a primary requirement for reading an HVAC drawing. In a forced-air system, fans move the heated or cooled air through ductwork into the working areas, offices, and public spaces. Air return ducts collect air from these areas and channel it back through the system or to the outside atmosphere.

The ductwork for these systems is usually one of the following types:

- Individual duct systems
- Trunk duct systems
- Crawl space plenum systems

The first two delivery systems may be used either from the floor or ceiling levels. The ductwork begins at the central unit and moves out to each area of workspace. The plans may specify circular or rectangular ducts. Circular ducts slip together and bend to form angles. Rectangular ducts fit between joists and studs. In the individual duct system, many small ducts may reach from the central unit to each room, where a grilled register delivers the air. In a trunk duct system, large main ducts run the length of the structure and smaller ducts branch off to each room register.

In some forced-air systems, particularly in buildings with a crawl space, the heating unit is designed to heat the exterior walls and the floor above the crawl space. A crawl space plenum system heats the entire crawl space and the conditioned air rises through floor registers. This system requires no ductwork to transfer the warm air since it is transferred through the structural frame.

To gain the most information from HVAC plans, look for the following:

- The direction of the central fan
- Duct supply lines; these deliver the heated or cooled (conditioned) air to all parts of the structure
- Return ducts; some HVAC systems will have return ducts to return air to the central unit for more heating or cooling

Figure 19 shows a typical mechanical drawing for an HVAC system. This involves laying out all the piping, controllers, and air handlers to scale on a basic floor plan. Most of the details that are normally on a floor plan have been removed so that the details of the HVAC system can be seen.

A large amount of information is required for mechanical work. There is limited space on the drawing to show the piping, valves, and connections. Special symbols are used for clarity. *Appendix C* contains some common HVAC symbols.

Detail drawings may be used on HVAC drawings. Unlike the detail drawings shown on architectural drawings, these detail drawings are typically not drawn to scale. Usually, they are drawn as an elevation drawing or perspective view. They show details about the configuration of the equipment.

1.5.2 Electrical Drawings

Electrical drawings identify the layout of the electrical distribution system, the lighting requirements, and the telecommunications and computer connections. Electrical drawings are labeled with page numbers beginning with the letter E. Depending on the complexity of the project, electrical drawings may include the following:

- Site plan for electrical service requirements; for job-site safety, particular attention should be paid to overhead power lines and underground utilities in proximity to the project
- Floor plans for the outlet and switch locations and the branch circuit requirements
- Schematics of the branch circuits
- Location of all outlets, switches, and fixtures
- Lighting plans
- Emergency power and lighting systems
- Power plan
- Life safety systems
- Any backup power generation facilities
- Notes and details to describe other parts of the electrical system

> **WARNING!** For job-site safety, particular attention should be paid to overhead power lines and underground utilities in proximity to the project. Underground utilities must be located before excavation begins. Whether it's a residential or commercial construction project, a locate request must be submitted at least 48 hours (not counting weekends and/or legal holidays) prior to excavation.

Like mechanical drawings, electrical drawings use plan views to show system layouts. Detail drawings and schedules provide clarification. One drawing may include power, lighting, and telecommunications layouts. In more complex structures the systems are shown separately. There are many different symbols for electrical connections and fixtures. Commonly used symbols are shown in *Appendix D*. Special symbols for components such as power supplies, security systems, and circuit boards are usually designated by the manufacturer.

Although you do not need to understand the details of electrical circuitry to read an electrical plan, you should understand the basic design of the system. All systems begin at the service source. Like water and natural gas, it begins with a meter installed by the utility company. From the meter, service moves through a main cutoff switch to the service entry panel. The panel separates service into branch circuits. These circuits are protected from overload by circuit breakers with various capacities. The branches then feed out to specific fixtures and outlets throughout the building.

Electrical systems may be very basic for structures like general warehouses and office buildings. They can also be very complicated for research laboratories and hospitals. Some complex facilities have backup generators and parallel wiring systems.

The power plan illustrates the power requirements of the structure. It shows the panels, receptacles, and the circuitry of power-utilizing

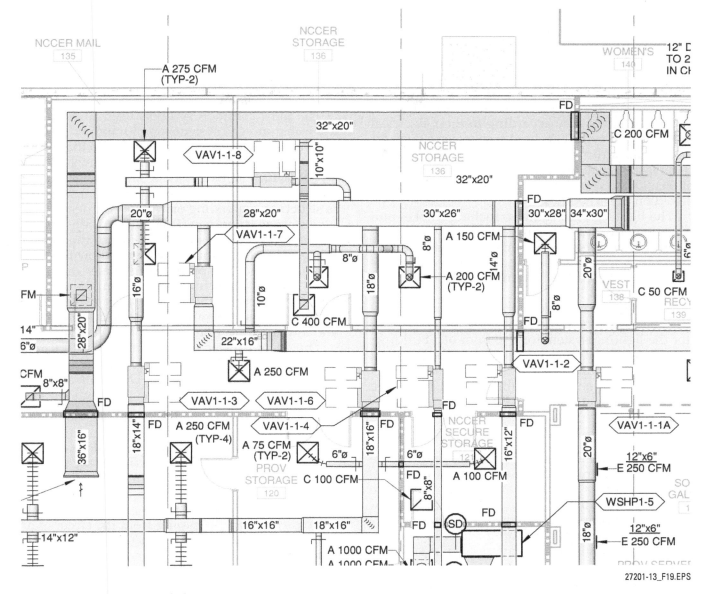

Figure 19 HVAC mechanical plan.

equipment. Some buildings need multiple panels. The power plan would then include panel schedules. Panel schedules list the circuits in the panel, the individual power required for each panel, and the total power requirement for the system. The schedules total the power requirements to help the electrical contractor size the panel. The power company sizes the overall service requirements.

Lighting the New York Skyline

The first searchlight on top of the Empire State building heralded the election of Franklin D. Roosevelt in 1932. A series of floodlights were installed in 1964 to illuminate the top 30 floors of the building. Today, the color of the lights is changed to mark various events. Yellow traditionally marks the opening day of the US Open tennis tournament, but it also marked the day "The Simpsons Movie" was released. Red, white, and blue always honor Independence Day.

The building is lit from the 72nd floor to the base of the TV antenna by 204 metal halide lamps and 310 fluorescent lamps. In 1984, a color-changing apparatus was added in the uppermost mooring mast. There are 880 vertical and 220 horizontal fluorescent lights. The colors can be changed with the flick of a switch.

The lighting plan locates the various lighting fixtures in the building. It is complemented by a fixture schedule that lists the types of light fixtures to be used. The fixture schedule is organized by number or letter. These refer to the manufacturer and model, the wattage of the lamps, voltage, and any special remarks concerning the fixture. Outlets vary according to their electrical capacity and type. Different symbols are used for outlets according to their voltage requirements. This is also true of most other types of electrical items. Details such as type, size, and voltage are listed on the schedules.

The lighting plan can also include smoke- and fire-detection equipment, emergency lighting, cable TV, and telephone outlets. Buildings with more complex electrical systems may include separate plans for fire prevention and telecommunications systems.

Figure 20 contains two drawings for electrical installations. *Figure 20A* shows the electrical plan for the lighting system for one floor of a building. It mainly consists of overhead fluorescent lighting controlled by wall switches. *Figure 20B* is the riser diagram for the main control panel. It shows the layout of the main panel and its connections.

1.5.3 Plumbing Drawings

All buildings that will be occupied require some type of plumbing. Warehouses have simple restrooms, whereas hospitals and restaurants have sophisticated plumbing systems. As with fire protection, plumbing work requires special knowledge and training. Plumbing work is usually performed by a licensed craftsperson. A separate permit and inspection process are required.

Plumbing drawings are considered part of the mechanical plans. However, they are usually placed in their own set of drawings for clarity. Unless the building is very basic, placing them on the same sheets as the HVAC drawings would cause confusion. Plumbing drawings usually include the following:

- Site plan for water supply and sewage disposal systems
- Floor plans for the fire system, including hydrant connections and sprinkler systems
- Floor plans for the water supply system and fixture location
- Floor plans for the waste disposal system
- Riser diagrams to describe the vertical piping features
- Floor plans and riser diagrams for the gas lines

The most common symbols used to designate the water and gas systems on plumbing drawings are shown in *Appendix E*. In addition to these symbols, there will be specific graphic symbols for the layout of such items as toilets, sinks, water heaters, and sump pumps. These will vary from structure to structure, depending on the specific design.

Plumbing drawings (*Figure 21*) usually appear as plan views and as isometric drawings in the form of riser diagrams. Plan views show the horizontal distances or piping runs. Riser diagrams show vertical pipes in the walls. Plumbing drawings for water systems usually show two separate systems: a water source or distribution system, and a waste collection and disposal system.

The incoming water systems operate under pressure. The distribution system shows the hot- and cold-water supply lines. Distribution plans usually include several of the following:

- Piping type and size; the piping types include copper, plastic, brass, iron, or steel piping; sizes usually appear as nominal dimensions approximating the inside diameter (ID) of the pipe
- Pipe fitting at joints; this includes couplings (straight-run joints), elbows (45- or 90-degree bends), tees, and valves
- Cold water supply piping from the street main to the service meter and on to the various fixtures
- Hot water supply piping locations from the water heater to the various fixtures
- A legend identifying cold- and hot-water supply lines

The waste system works using gravity and must be vented. The waste system is known as drain, waste, and vent, or DWV, piping. Reading waste disposal plans, you can expect to see a combination of several notations. You will usually find the following information on a plumbing drawing for a waste disposal system:

The Empire State Building

The Empire State Building has 102 stories. It has 70 miles of water pipe that provides water to tanks at various levels. The highest tank is on the 101st floor. There are two public restrooms on each floor and a number of private bathrooms. There are, however, no water fountains.

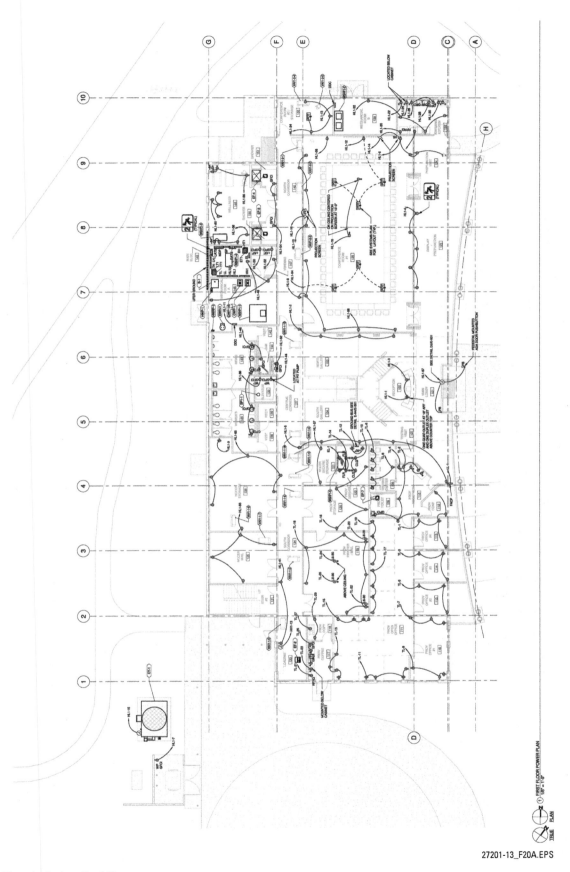

Figure 20 Electrical plan (1 of 2).

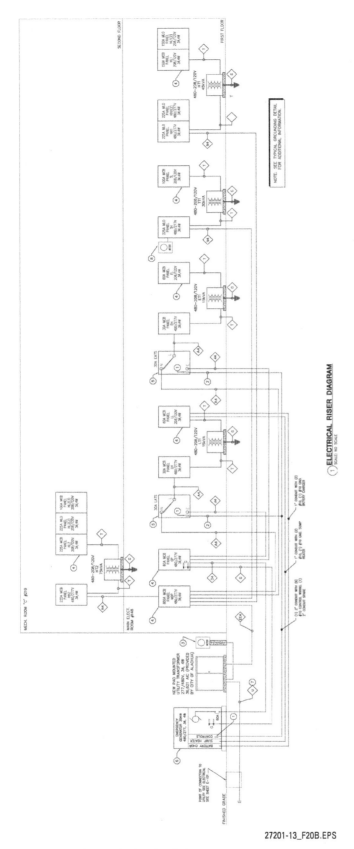

Figure 20 Electrical plan (2 of 2).

- Waste disposal line size
- Location of fixture branches (horizontal disposal pipes)
- Stack locations; vertical disposal pipes are called waste stacks if they carry toilet waste, soil stacks if they carry waste materials other than toilet waste, and vent stacks if they extend through the roof to release gases within the stacks
- Locations for drains, sewer lines, cleanouts, traps, meters, and valves
- Slopes of the fixture branches; because waste disposal occurs using the force of gravity, the slopes of the fixture branches must be specified (the usual slope is approximately ¼" per foot away from the fixtures)

Piping for distribution of natural gas within the structure is often considered part of the plumbing work. The utility company installs a meter where natural gas enters the building. The piping starts at the meter. Natural gas is distributed to the gas-fueled appliances within the building such as boilers, furnaces, water heaters, and rooftop HVAC units.

Typically, black steel pipe with threaded fittings is used to distribute natural gas. The fittings are similar to those of other piping systems. They include elbows, bends, unions, and tees. Valves to control the flow of gas within the pipe are typically brass. Gas systems may require openings through masonry walls to accommodate piping, valves, and regulators. Through-wall piping must not be rigidly connected to the wall due to the different rates of expansion and contraction between the piping and the masonry.

Isometric Drawings

Isometric means equal measurement. A designer uses the true dimension of an object to construct the drawing. An isometric drawing shows a three-dimensional view of where the pipes should be installed. The piping may be drawn to scale or to dimension, or both. In an isometric drawing, vertical pipes are drawn vertically on the sketch, and horizontal pipes are drawn at an angle to the vertical lines.

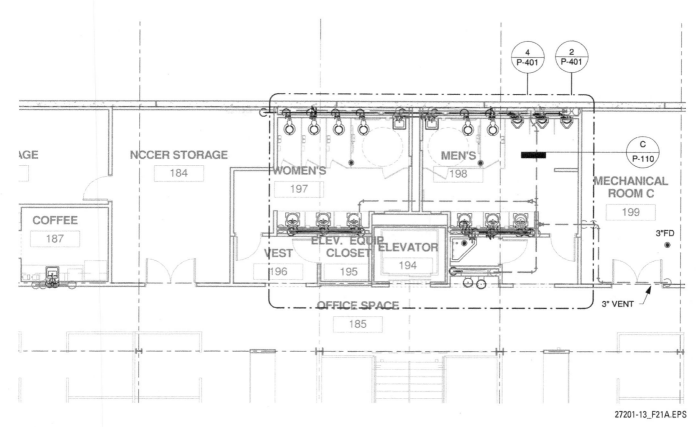

Figure 21 Sanitary plumbing plan (1 of 2).

1.6.0 Comparing Drawings from Different Disciplines

Coordination drawings are created by the individual contractors for each trade in order to prevent a conflict in the installation of materials and equipment. Coordination drawings are produced prior to finalizing shop drawings, cut lists, and other drawings and before the installation begins. Development of these drawings evolves through a series of review and coordination meetings held by the various contractors.

Some contracts require coordination drawings, while others only recommend them. In the case where one contractor elects to make coordination drawings and another does not, the contractor who made the drawings may be given the installation right-of-way by the presiding authority. As a result, the other contractor may have to bear the expense of removing and reinstalling equipment if the equipment was installed in a space designated for use by the contractor who produced the coordination drawings.

Today, most construction drawings are generated using a computer-aided design (CAD) system and architectural software. Some CAD software is capable of building information modeling (BIM). As shown in *Figure 22*, BIM allows designers to prepare three-dimensional (3-D) digital models of an entire structure and all its components. BIM allows people to virtually walk or even fly through a building before it has been built, in order to see how it will look when it has been completed. This allows designers to identify problems and conflicts with various components, known as clash detection, in time to correct them before construction has begun or is completed. BIM also allows designers to estimate the costs of constructing the building and even the cost of facilities maintenance over time.

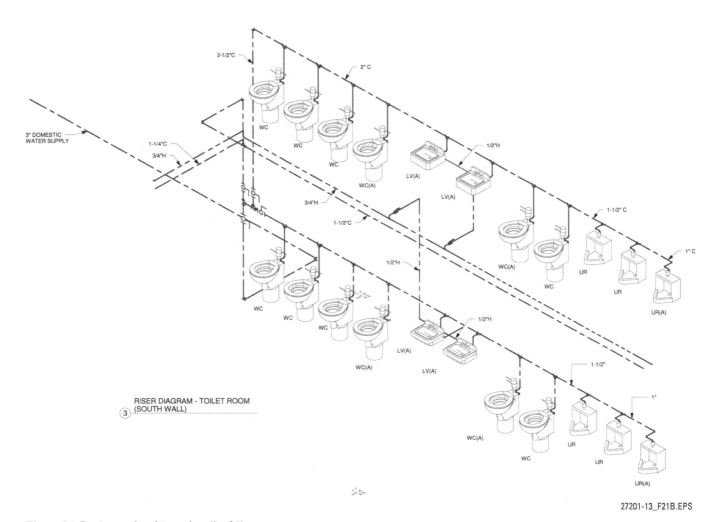

Figure 21 Sanitary plumbing plan (2 of 2).

Figure 22 Building information modeling.

1.0.0 Section Review

1. Compared to residential drawings, the drawings for a commercial project are _____.
   a. extremely similar
   b. smaller
   c. less complex
   d. more complex

2. The site plan is the most common form of _____.
   a. architectural drawing
   b. civil drawing
   c. structural drawing
   d. mechanical drawing

3. The core drawings in any plan set are the _____.
   a. architectural drawings
   b. finish drawings
   c. site drawings
   d. sectional drawings

4. Which of the following types of structural support is *not* typically used in commercial buildings?
   a. steel framing
   b. cast-in-place concrete
   c. wood framing
   d. precast concrete structural members

5. If provided, piping for distribution of natural gas is usually included in the _____.
   a. HVAC plans
   b. plumbing plans
   c. structural plans
   d. mechanical plans

6. The type of drawings created by the individual contractors for each trade in order to prevent a conflict in the installation of their materials and equipment is called a _____.
   a. detail drawing
   b. floor plan
   c. coordination drawing
   d. site plan

SECTION TWO

2.0.0 SPECIFICATIONS

Objective

Define the use of specifications and how they are referenced.
  a. Describe the format of specifications.
  b. Explain how specifications are written.

The specifications for a building or project are the written descriptions of work and duties required of the owner, architect, and consulting engineer. Together with the construction drawings, the specifications form the basis of the contract requirements for the construction of the building or project. Those who use the construction drawings and specifications must always be alert to discrepancies between the construction drawings and the specifications. These are some situations where discrepancies may occur:

- Architects or engineers use standard or prototype specifications and attempt to apply them without any modification to specific construction drawings.
- Previously prepared standard drawings are changed or amended by reference in the specifications only, and the drawings themselves are not changed.
- Items are duplicated in both the drawings and specifications, but an item is subsequently amended in one and overlooked in the other contract document.

Legally, specifications take precedence over drawings. However, the person in charge of the project has the responsibility to determine whether the drawings or the specifications take precedence in a given situation.

2.1.0 Format of Specifications

For convenience in writing, speed in estimating, and ease of reference, the most suitable organization of the specifications is a series of sections dealing with the construction requirements, products, and activities. They should be easily understandable by the different trades. Those people who use the specifications must be able to find all information needed without spending too much time searching for it.

The most commonly used specification format in North America is *MasterFormat*™. *MasterFormat*™ was developed jointly by the Construction Specifications Institute (CSI) and Construction Specifications Canada (CSC). *MasterFormat*™ is divided into four major groupings and 49 divisions with some divisions reserved for future expansion (*Figure 23*).

MasterFormat™ is organized by a six-digit number. The first two digits represent the division number. The next two digits represent subsections of the division and the two remaining digits represent the third-level sub-subsection numbers. A fourth level, if required, is a decimal and number added to the end of the last two digits. For example, the number 132013.04 represents division 13, subsection 20, sub-subsection 13, and sub-sub-subsection 04.

2.2.0 How Specifications Are Written

Writing accurate and complete specifications for building construction is a serious responsibility for those who design the buildings because the specifications, combined with the construction drawings, govern practically all important decisions made over the duration of the project. Compiling and writing specifications is not a simple task, even for those who have had considerable experience in preparing such documents.

MasterFormat™ History

For many years prior to 2004, the organization of construction specifications and supplier's catalogs were based on a standard with 16 sections, known as divisions. The divisions and their subsections were individually identified by a five-digit numbering system. The first two digits represented the division number and the next three individual numbers represented successively lower levels of breakdown. For example, the number 13213 represented division 13, subsection 2, sub-subsection 1, and sub-sub-subsection 3.

DIVISIONS NUMBERS AND TITLES

PROCUREMENT AND CONTRACTING REQUIREMENTS GROUP	
Division 00	Procurement and Contracting Requirements

SPECIFICATION GROUP

GENERAL REQUIREMENTS SUBGROUP

Division 01	General Requirements

FACILITY CONSTRUCTION SUBGROUP

Division 02	Existing Conditions
Division 03	Concrete
Division 04	Masonry
Division 05	Metals
Division 06	Wood, Plastics, and Composites
Division 07	Thermal and Moisture Protection
Division 08	Openings
Division 09	Finishes
Division 10	Specialties
Division 11	Equipment
Division 12	Furnishings
Division 13	Special Construction
Division 14	Conveying Equipment
Division 15	*Reserved*
Division 16	*Reserved*
Division 17	*Reserved*
Division 18	*Reserved*
Division 19	*Reserved*

Figure 23 Example of the *MasterFormat*™ 2012 organization.

A set of written specifications for a single project will usually contain thousands of products, parts, and components, and the methods of installing them, all of which must be covered in either the drawings and/or specifications. No one can memorize all of the necessary items required to describe accurately the various areas of construction. One must rely upon reference materials such as manufacturer's data, catalogs, checklists, and high-quality specifications.

Specifications

Written specifications supplement the related construction drawings in that they contain details not shown on the drawings. Specifications define and clarify the scope of the job. They describe the specific types and characteristics of the components that are to be used on the job and the methods for installing some of them. Many components are identified specifically by the manufacturer's model and part numbers. This type of information is used to purchase the various items of hardware needed to accomplish the installation in accordance with the contractual requirements.

2.2.1 Qualifications

Before bidding on a project, a contractor must be qualified to perform the work. A contractor's safety rating is a primary factor in determining whether the contractor is qualified for a project. A low experience modification rate (EMR) or Occupational Safety and Health Administration (OSHA) incidence rate is required to be qualified to bid on a project. An EMR of 1.0 or above usually disqualifies a contractor from bidding on a project. In addition, contractors bid on projects that fall within their area of expertise. For example, contractors with experience in high-rise construction would likely bid on a 50-story structure.

Your Impact on Your Company's EMR

The EMR is based on how safely workers perform their duties on the job site. A larger number of recordable safety incidents results in a higher EMR and decreases the number of construction projects that a contractor can bid on.

2.0.0 Section Review

1. Specifications follow a standard format _____.
   a. as outlined in the building codes
   b. for convenience in writing and ease of reference
   c. because it is required by law
   d. developed by ANSI

2. Methods to be used when installing specific components are found in the _____.
   a. structural drawings
   b. project schedules
   c. architectural drawings
   d. specifications

Summary

Commercial and industrial construction makes up a large part of the construction market. These structures are usually more difficult to build than residential buildings. They also involve many code, safety, and environmental requirements.

In order to build commercial structures, drawings must be properly interpreted. The basic principles such as scaling and dimensioning apply to both residential and commercial drawings. However, commercial drawings are much more detailed. They can contain additional structural, mechanical, and electrical systems. These require close study and review. Understanding the plan will help keep construction activities well organized and true to the design.

A typical commercial plan set includes civil, life safety, architectural, structural, mechanical, electrical, plumbing, and fire protection drawings. There is a letter designation for each section. Each section may include more than one sheet, depending on the size and complexity of the project. Architectural drawings include the floor plans, wall sections, door and window details, and schedules. Structural drawings include foundation plans, framing plans, support details, schedules, and notes. Mechanical drawings include HVAC plans in both plan view and isometric view. Plumbing drawings and mechanical drawings are similar in format, but the plumbing drawings provide both a plan view and an isometric view of the freshwater system and the wastewater system. Electrical drawings usually include a power plan and a lighting plan.

When reading commercial plans, follow a step-by-step process to avoid confusion and see all the details involved. Begin by carefully reviewing all drawings to gain a general impression of the shape, size, and appearance of the structure. Then, read the project specifications to pick up details not found on the drawings. Begin correlating the floor plans with the exterior elevation drawings. Look at the wall sections to determine the wall types and construction details.

Review the structural plans to determine the foundation requirements and the type of structural system. Carefully consider all items that may require special construction procedures on the architectural and structural drawings. Review all interior elevation drawings. Try to get a clear picture of what the interior of the building will look like. Review the finish schedule and the mechanical, electrical, and plumbing (MEP) plans. In particular, look for work items to be performed by carpenters or features that will affect the work of carpenters. If some part of a drawing is not clear, check with your supervisor. There may be a logical explanation, or there may be a mistake. The architect should clarify and resolve all conflicts in writing.

Specifications provide written instructions for the owner, architect, and engineer. A set of written specifications for a large commercial project will usually contain thousands of products, parts, components, and the methods of installing them. A standard format has been voluntarily adopted for commercial construction throughout North America. Manufacturers and suppliers key their catalogs and data to this standard format. The standard format makes understanding the specifications much easier. However, discrepancies can arise if the standard specifications are not modified to match the drawings of a particular project.

Review Questions

1. In some sets of commercial drawings, the site plans are called _____.
   a. civil drawings
   b. commercial drawings
   c. layout drawings
   d. elevation drawings

2. Architectural drawings are identified with the letter _____.
   a. E
   b. A
   c. M
   d. P

3. When reading commercial plans, you can avoid missing details by using a _____.
   a. step-by-step process
   b. schedule
   c. local code
   d. plan handbook

4. Unlike residential plans, commercial site plans usually include _____.
   a. survey data
   b. paving details
   c. geographic data
   d. elevations

5. The typical interval between contour lines on a site plan is _____.
   a. 1 foot
   b. 2 feet
   c. 5 feet
   d. 10 feet

6. An enlargement of an area on a floor plan is called a(n) _____.
   a. elevation
   b. schedule
   c. section
   d. detail

7. Stairways will usually be shown on elevation drawings and _____.
   a. sections
   b. grids
   c. materials
   d. schedules

8. Drilled piers are designed to carry _____.
   a. footings
   b. grade beams
   c. mats
   d. girders

9. Information on the quantities, sizes, grading, lengths, and bending dimensions of all the rebar for a project is given in the _____.
   a. structural drawing
   b. placing drawing
   c. bar list
   d. bar invoice

10. The structural engineer draws a framing plan for _____.
    a. each floor level that will be framed and the roof
    b. the plumbing, electrical, and HVAC systems
    c. landscaping and grading
    d. the foundation and exterior walls

11. It is easier to read framing plans by _____.
    a. starting at the bottom and working upwards
    b. starting at the top and working downwards
    c. isolating the separate bays or spans between columns
    d. reading the foundation plans first

12. Mechanical drawings for HVAC systems are typically shown in a(n) _____.
    a. section view
    b. plan view
    c. elevation view
    d. cutaway view

13. A plumbing riser diagram shows _____.
    a. vertical pipes in the wall
    b. horizontal distance or pipe runs
    c. the hot- and cold-water supply lines
    d. the couplings, elbows, tees, and valves needed

14. Due to different rates of expansion and contraction between the piping and the masonry, through-wall piping must _____.
    a. not be rigidly connected to the wall
    b. be firmly cemented
    c. be properly sized
    d. be thoroughly insulated

15. The most commonly used specification is developed by Construction Specifications Canada and the _____.
    a. Brick Industry Association (BIA)
    b. American National Standards Institute (ANSI)
    c. Construction Specifications Institute (CSI)
    d. International Code Council (ICC)

Trade Terms Quiz

Fill in the blank with the correct term that you learned from your study of this module.

1. The overall shape of a building site is shown on _____.
2. _____ are large structural support beams, usually located on the building's perimeter.
3. Supported by walls or by columns and girders, _____ are loadbearing horizontal framing members.
4. A financial obligation is referred to as a _____.
5. The types and sizes of specified construction materials, such as doors, are listed in table form on a _____.
6. A _____ is a reference point of known elevation on the construction site.
7. Drawings made at a larger scale to better show specific features are _____.
8. Horizontal wood or steel members that rest on beams and support floor sheathing are called _____.
9. _____ are markings or tags on drawings that direct the user to other documents, such as schedules.
10. Drawings that show an object with all three faces equally inclined to the picture plane are described as _____.
11. The layout, components, and connections of a piping system are shown on a _____.
12. _____ connect points of the same elevation on a site drawing.
13. Views of a structure's front, rear, or side are shown on _____.
14. The _____ is the lowest point of a pipe through which liquid flows.
15. An object is shown as seen from directly above in a _____.
16. The _____ show locations of trees, bushes, and other proposed plantings on the site.
17. Horizontal structural members are supported by vertical structural members known as _____.
18. Surveyor's data is recorded in the _____.

Trade Terms

Beams
Benchmark
Callouts
Civil drawings
Columns
Contour lines
Detail drawings
Elevation drawings
Field notes
Girders
Invert
Isometric drawings
Joists
Landscape drawings
Liability
Plan view
Riser diagram
Schedule

Trade Terms Introduced in This Module

Beams: Loadbearing horizontal framing members supported by walls or columns and girders.

Benchmark: A known elevation on the site used as a reference point during construction.

Callouts: Markings or identifying tags describing parts of a drawing; callouts may refer to detail drawings, schedules, or other drawings.

Civil drawings: A drawing that shows the overall shape of the building within the confines of the site. Also referred to as site plans.

Columns: Vertical structural members that support the load of other members.

Contour lines: Imaginary lines on a civil drawing that connect points of the same elevation. Contour lines normally never cross one another.

Detail drawings: Drawings shown at a larger scale in order to show specific features or connections.

Elevation drawings: Drawings showing a view from the front, rear, or side of a structure.

Field notes: A permanent record of field measurement data and related information.

Girders: Large steel or wood beams supporting a building, usually around a perimeter.

Invert: The lowest point of a pipe through which liquid flows.

Isometric drawings: Three-dimensional drawings in which the object is tilted so that all three faces are equally inclined to the picture plane.

Joists: Horizontal wood or steel members supported by beams; joists support floor sheathing.

Landscape drawings: A drawing that shows proposed plantings and other landscape features.

Liability: An obligation or responsibility, typically financial.

Plan view: A drawing that represents a view looking down on an object.

Riser diagram: Type of drawing that depicts the layout, components, and connections of a piping system.

Schedule: Tables that describe and specify the various types and sizes of construction materials used in a building. Door schedules, window schedules, and finish schedules are the most common types.

Appendix A

MATERIAL SYMBOLS

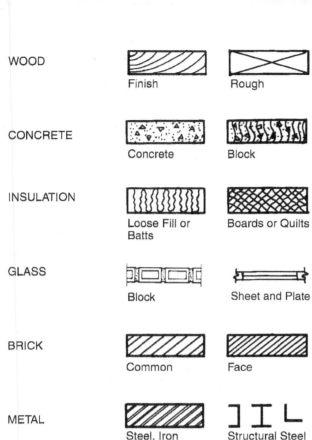

GENERAL PLAN SYMBOLS

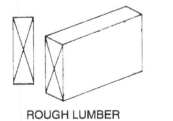
ROUGH LUMBER

METAL

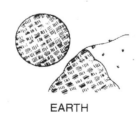
EARTH

CONCRETE

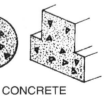

SECTION VIEW SYMBOLS

Appendix B

Topographic Symbols

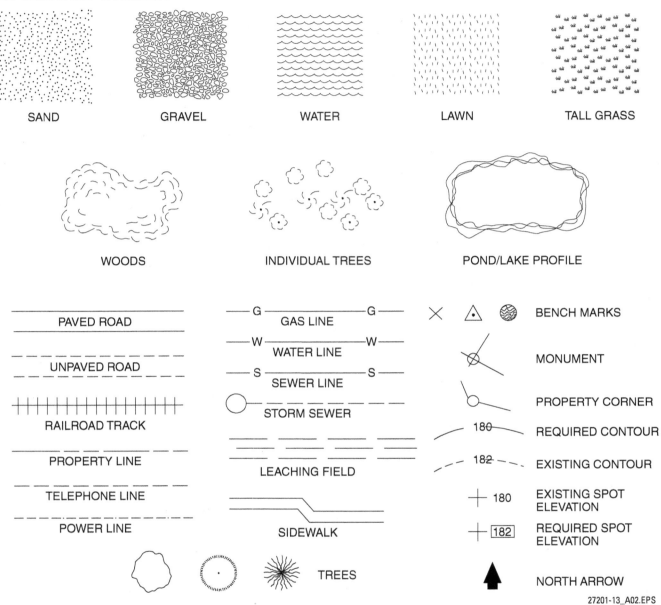

Appendix C

HVAC Symbols

⟷⊠⟶	CEILING DIFFUSER (ARROWS INDICATE DIRECTION OF AIR FLOW)
⤳ ▢	RETURN AIR GRILLE
⊠	SUPPLY DUCT UP
⊠	SUPPLY DUCT DOWN
▢	RETURN DUCT UP
◺	RETURN DUCT DOWN
6"ɸCD / 200¢	NECK SIZE/AIR DEVICE CFM
(T)	THERMOSTAT
	SQUARE TO ROUND TRANSITION
	PARALLEL BLADE DAMPER
	FIRE DAMPER (WALL) (FLOOR)
	AIRFOIL BLADE TURNING VANES
	AIR EXTRACTOR
ɸ	DIAMETER
¢	CFM (CUBIC FEET PER MINUTE)
RA	RETURN AIR
OSA	OUTSIDE AIR
CD	CONDENSATE DRAIN

Appendix D

ELECTRICAL SYMBOLS

GENERAL OUTLETS

Junction Box, Ceiling	Ⓙ
Fan, Ceiling	Ⓕ
Recessed Incandescent, Wall	─Ⓡ
Surface Incandescent, Ceiling	○
Surface or Pendant Single Fluorescent Fixture	▭

SWITCH OUTLETS

Single-Pole Switch	S
Double-Pole Switch	S_2
Three-Way Switch	S_3
Four-Way Switch	S_4
Key-Operated Switch	S_K
Switch w/Pilot	S_P
Low-Voltage Switch	S_L
Door Switch	S_D
Momentary Contact Switch	S_{MC}
Weatherproof Switch	S_{WP}
Fused Switch	S_F
Circuit Breaker Switch	S_{CB}

RECEPTACLE OUTLETS

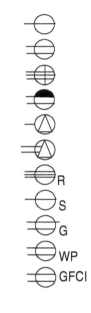

Single Receptacle	
Duplex Receptacle	
Triplex Receptacle	
Split-Wired Duplex Recep.	
Single Special Purpose Recep.	
Duplex Special Purpose Recep.	
Range Receptacle	R
Switch & Single Receptacle	S
Grounded Duplex Receptacle	G
Duplex Weatherproof Receptacle	WP
GFCI	GFCI

AUXILIARY SYSTEMS

Telephone Jack	◁
Meter	Ⓜ
Vacuum Outlet	▽
Electric Door Opener	D
Chime	CH
Pushbutton (Doorbell)	•
Bell and Buzzer Combination	
Kitchen Ventilating Fan	
Lighting Panel	■
Power Panel	
Television Outlet	TV

Appendix E

Plumbing Symbols

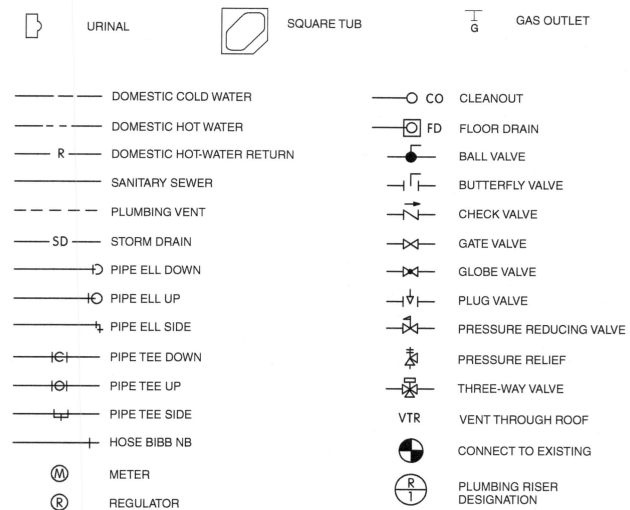

Additional Resources

This module presents thorough resources for task training. The following resource material is suggested for further study.

Architectural Graphic Standards, Eighth Edition. 1988. American Institute of Architects. New York: John Wiley & Sons.

Basics for Builders: Plan Reading & Material Takeoff. 1994. Wayne J. DelPico. Kingston, MA: R.S. Means Company, Inc.

Figure Credits

Provided by The Haskell Company, CO01, Figure 22

King's Material, Inc., Figure 1a

Tilt-Up Concrete Association, Figure 1b

Black & Veatch, Figure 18

The Numbers and Titles used in this textbook are from MasterFormat™ 2012, published by The Construction Specifications Institute (CSI) and Construction Specifications Canada (CSC), and are used with permission from CSI. For those interested in a more in-dept explanation of MasterFormat™ 2012 and its use in the construction industry visit **www.csinet.org/masterformat** or contact: The Construction Specifications Institute (CSI) 99 Canal Center Plaza, Suite 300 Alexandria, VA 22314 800-689-2900; 703-684-0300 **http://www.cisnet.org**, Figure 23

Dumond Chemicals, Inc., SA01

Tom Crane, Photographer, The Athenaeum of Philadelphia, 2000, SA02

Section Review Answer Key

Answer	Section Reference	Objective
Section One		
1. d	1.1.0	1a
2. b	1.2.0	1b
3. a	1.3.0	1c
4. c	1.4.0	1d
5. b	1.5.3	1e
6. c	1.6.0	1f
Section Two		
1. b	2.1.0	2a
2. d	2.2.0	2b

This page is intentionally left blank.

NCCER CURRICULA — USER UPDATE

NCCER makes every effort to keep its textbooks up-to-date and free of technical errors. We appreciate your help in this process. If you find an error, a typographical mistake, or an inaccuracy in NCCER's curricula, please fill out this form (or a photocopy), or complete the online form at **www.nccer.org/olf**. Be sure to include the exact module ID number, page number, a detailed description, and your recommended correction. Your input will be brought to the attention of the Authoring Team. Thank you for your assistance.

Instructors – If you have an idea for improving this textbook, or have found that additional materials were necessary to teach this module effectively, please let us know so that we may present your suggestions to the Authoring Team.

NCCER Product Development and Revision
13614 Progress Blvd., Alachua, FL 32615

Email: curriculum@nccer.org
Online: www.nccer.org/olf

❏ Trainee Guide ❏ Lesson Plans ❏ Exam ❏ PowerPoints Other _____

Craft / Level: _____ Copyright Date: _____

Module ID Number / Title: _____

Section Number(s): _____

Description: _____

_____

_____

_____

_____

Recommended Correction: _____

_____

_____

_____

_____

Your Name: _____

Address: _____

_____

Email: _____ Phone: _____

This page is intentionally left blank.

Cold-Formed Steel Framing

Overview

This module describes the uses and installation of cold-formed steel framing. In commercial and multifamily residential construction, it is common to use steel framing materials in place of wood studs to frame walls and partitions. It is also becoming more common in single-family residential construction as the price of lumber rises. Steel framing, however, requires a carpenter to master tools and joining techniques different from those required in wood framing construction. In order to work a variety of steel framing projects, from installing loadbearing steel assemblies to nonstructural walls, you will need to become familiar with these materials and methods.

Module 27205

Trainees with successful module completions may be eligible for credentialing through NCCER's National Registry. To learn more, go to **www.nccer.org** or contact us at **1.888.622.3720**. Our website has information on the latest product releases and training, as well as online versions of our *Cornerstone* magazine and Pearson's product catalog.

Your feedback is welcome. You may email your comments to **curriculum@nccer.org**, send general comments and inquiries to **info@nccer.org**, or fill in the User Update form at the back of this module.

This information is general in nature and intended for training purposes only. Actual performance of activities described in this manual requires compliance with all applicable operating, service, maintenance, and safety procedures under the direction of qualified personnel. References in this manual to patented or proprietary devices do not constitute a recommendation of their use.

Copyright © 2013 by NCCER, Alachua, FL 32615, and published by Pearson, New York, NY 10013. All rights reserved. Printed in the United States of America. This publication is protected by Copyright, and permission should be obtained from NCCER prior to any prohibited reproduction, storage in a retrieval system, or transmission in any form or by any means, electronic, mechanical, photocopying, recording, or likewise. To obtain permission(s) to use material from this work, please submit a written request to NCCER Product Development, 13614 Progress Blvd., Alachua, FL 32615.

27205 V5

From *Carpentry, Trainee Guide*. NCCER.
Copyright © 2013 by NCCER. Published by Pearson. All rights reserved.

27205
Cold-Formed Steel Framing

Objectives

When you have completed this module, you will be able to do the following:

1. Identify the tools and components of cold-formed steel framing systems and their safe use.
    a. Identify the safety guidelines that should be followed when working with cold-formed steel.
    b. Identify steel framing materials.
    c. List the steel framing tools and fasteners.
    d. Explain how to perform a material takeoff for a steel frame project.
2. Identify the steps to layout and install a steel stud wall.
    a. Describe the basic steel construction methods.
    b. Explain how to frame nonstructural steel walls.
    c. Explain how to frame structural steel walls.
3. Identify other steel framing applications.
    a. Explain how steel framing members are used in floor and roof construction.
    b. Explain how steel framing members are used in ceiling construction.

Performance Tasks

Under the supervision of your instructor, you should be able to do the following:

1. Estimate the amount of materials to complete an instructor-specified steel framing project.
2. Lay out a steel stud wall with openings to include bracing and blocking.
3. Demonstrate the ability to build headers (back-to-back, box, and L-header).

Trade Terms

Blocking	Gauge	Plenum	Stiffening lip
Clip angle	Header	Powder load	Track
Cold-formed steel	Knurled	Racking	Web
C-shape	Lateral	Rim track	
Curtain wall	Mil	Roof rafter	
Diaphragm	Panelization	Shear wall	

Code Note

Codes vary among jurisdictions. Because of the variations in code, consult the applicable code whenever regulations are in question. Referring to an incorrect set of codes can cause as much trouble as failing to reference codes altogether. Obtain, review, and familiarize yourself with your local adopted code.

Industry-Recognized Credentials

If you're training through an NCCER-accredited sponsor, you may be eligible for credentials from NCCER's Registry. The ID number for this module is 27205. Note that this module may have been used in other NCCER curricula and may apply to other level completions. Contact NCCER's Registry at 888.622.3720 or go to **www.nccer.org** for more information.

Contents

Topics to be presented in this module include:

1.0.0 Materials and Tools 1
   1.1.0 Cold-Formed Steel Framing Safety Guidelines 2
   1.2.0 Steel Framing Materials... 2
      1.2.1 Identification of Framing Materials..4
      1.2.2 Furring 4
      1.2.3 Slip Connections 5
   1.3.0 Steel Framing Tools and Fasteners 5
      1.3.1 Cutting Tools..7
      1.3.2 Hole-Cutting Tools..8
      1.3.3 Screws 9
      1.3.4 Pins 10
      1.3.5 Powder-Actuated Fasteners and Powder Loads..................... 10
      1.3.6 Clinching 11
      1.3.7 Bolts and Anchors 12
   1.4.0 Material Takeoff 12
      1.4.1 Joists and Joist Headers 12
      1.4.2 Bottom and Top Tracks 12
      1.4.3 Studs 13
2.0.0 Layout and Installation of Steel Framed Walls.................................... 14
   2.1.0 Basic Steel Construction Methods 14
      2.1.1 In-Line Framing... 14
      2.1.2 Web Stiffeners 14
      2.1.3 Web Holes and Patches 15
      2.1.4 Built-Up Shapes 16
      2.1.5 Bridging, Bracing, and Blocking 16
      2.1.6 Thermal Considerations 16
      2.1.7 Protecting Piping and Wiring 16
   2.2.0 Framing Nonstructural (Nonbearing) Steel Walls............................. 17
      2.2.1 Steel Curtain Walls... 18
      2.2.2 Construction Methods for Steel-Framed Curtain Walls.................. 19
      2.2.3 Bracing for Curtain Walls 20
      2.2.4 Finishing for Curtain Walls.. 20
      2.2.5 Radius (Curved) Walls 20
      2.2.6 Other Nonstructural Wall Assemblies...................................... 20
   2.3.0 Framing Structural Steel Walls 21
      2.3.1 Layout.. 21
      2.3.2 Wall Assembly 22
      2.3.3 Wall Installation 22
      2.3.4 Backing 24
      2.3.5 Shear Walls 24
      2.3.6 Header Assembly 27
      2.3.7 Jambs 29
      2.3.8 Sills 29

3.0.0 Other Steel Framing Applications 31
    3.1.0 Steel Floor and Roof Assemblies.. 31
    3.2.0 Ceiling Systems.. 32
Appendix Common Terms Used in Cold-Formed Steel Framing Work 37

Figures and Tables

Figure 1   Basic components of a steel framing system 3
Figure 2   Steel product label 3
Figure 3   Types of furring channels.. 5
Figure 4   Slip connectors 5
Figure 5   Slip connector application 6
Figure 6   Standard screwgun 6
Figure 7   Screwgun with collated feed 6
Figure 8   Powder-actuated tool 7
Figure 9   Hammer drill 7
Figure 10 Locking C-clamps.. 7
Figure 11 Chop saw 8
Figure 12 Swivel-head shears 8
Figure 13 Aviation snips 8
Figure 14 Self-drilling and self-piercing screws 9
Figure 15 Screw head shapes for typical self-drilling screws 10
Figure 16 Proper pin installation...11
Figure 17 Knurled shank pin 11
Figure 18 Drive pins and threaded studs 11
Figure 19 Powder loads 11
Figure 20 Pneumatic clinching tool 12
Figure 21 Example of structural wall system 14
Figure 22 In-line framing detail 15
Figure 23 Web stiffener... 16
Figure 24 Standard stud punchouts 16
Figure 25 Floor opening formed with built-up headers 17
Figure 26 Proper protection of wiring 18
Figure 27 Slip track... 18
Figure 28 Example of a radius track 20
Figure 29 Track splicing 22
Figure 30 Stud seating 23
Figure 31 Balloon framing 23
Figure 32 Lateral strapping for stud walls.. 25
Figure 33 Bridging with cold-rolled channel (CRC) 26
Figure 34 Example of stud bracing 26
Figure 35 Temporary bracing 26
Figure 36 X-bracing... 26
Figure 37 Proprietary shear wall 26
Figure 38 Wood structural sheathing 26
Figure 39 Attachment of X-bracing.. 27
Figure 40 Loadbearing header 28

Figures and Tables

Figure 41 L-header 28
Figure 42 Completed rough opening 28
Figure 43 Stitch-welded jamb 29
Figure 44 Standard floor joists 31
Figure 45 Example of complex framing using steel trusses 31
Figure 46 Ceiling framework 32

Table 1 Minimum Base Steel Thickness of Cold-Formed Steel Members ... 4

Section One

1.0.0 Materials and Tools

Objective

Identify the tools and components of cold-formed steel framing systems and their safe use.

a. Identify the safety guidelines that should be followed when working with cold-formed steel.
b. Identify steel framing materials.
c. List the steel framing tools and fasteners.
d. Explain how to perform a material takeoff for a steel frame project.

Performance Task 1

Estimate the amount of materials to complete an instructor-specified steel framing project.

Trade Terms

Cold-formed steel: Sheet steel or strip steel that is manufactured by press braking of blanks sheared from sheets or cut lengths of coils or plates, or by continuous roll forming of cold- or hot-rolled coils of sheet steel.

C-shape: Cold-formed steel shape used for structural and nonstructural framing members consisting of a web, two flanges, and two lips.

Gauge: Standard measure of metal thickness. Gauge numbers are only an approximation of thickness.

Header: Horizontal structural framing member used over floor, roof, or wall openings to transfer loads around the opening to supporting structural members.

Knurled: A series of small ridges used to provide a better gripping surface on metal and plastic.

Mil: Unit of measurement equal to $\frac{1}{1000}"$.

Panelization: The process of assembling steel-framed walls, joists, or trusses before they are installed in a structure.

Powder load: Metal casing that contains a powderized propellant and is crimped shut on the end and sealed.

Stiffening lip: Part of a C-shape framing member that extends perpendicular from the flange as a stiffening element. Also called an *edge stiffener*.

Track: A framing member consisting of only a web and two flanges. Track web depth measurements are taken to the inside of the flanges.

Web: Portion of a framing member that connects the flanges.

Designers and builders have long recognized steel for its strength, durability, and functionality. They, along with an increasing number of architects, have recognized steel's important environmental attributes as well. Cold-formed steel framing helps with industry efforts to promote sustainable construction.

By virtue of its material characteristics and properties, steel offers the following significant advantages for building construction:

- Steel studs and joists are strong, lightweight, and made from uniform-quality material. Steel has the highest strength-to-weight ratio of any building material.
- Steel framing members are dimensionally stable and will not warp, crack, rot, or split.
- Steel is uniform and provides a flat surface for sheathing or other material attachment.
- Steel framing does not contribute combustible material to feed a fire. The subsequent lower combustion rating results in lower insurance costs for builders and owners.
- Steel can be engineered to meet the strongest wind and seismic ratings specified by building codes.

> **NOTE**
> The *Appendix* contains a list of terms commonly used in conjunction with cold-formed steel framing.

Going Green

Steel Recycling

Industry-wide, steel has an average of 67 percent recycled content, making it the world's most recycled material. In North America alone, millions of tons of steel are recycled or exported for recycling per year. By comparison, the wood framing used in a majority of residential construction places heavy demands on timber resources. Typically, one acre of forest is used in the construction of a 2,000-square-foot home.

1.1.0 Cold-Formed Steel Framing Safety Guidelines

Working with cold-formed steel presents several different safety issues than working with wood framing members. Cold-formed steel framing safety guidelines are as follows:

- Wear appropriate gloves when handling steel framing members. The gloves should be cut resistant to prevent penetration by sharp edges of the framing members.
- Since metal does not absorb moisture (like wood), steel framing members may become slippery when wet. Use caution when handling wet steel framing members.
- Ensure that proper personal protective equipment (PPE) is used, including hearing protection, respiratory protection, and full-face shield, when cutting steel framing members with a cutoff saw. When metal is cut with a cutoff saw, a loud noise may be emitted and flying metal fragments may be produced. In addition, fumes are emitted from the zinc coating applied to cold-formed steel framing members, sometimes causing zinc chills. Also, since metal fragments are produced when cutting steel, a full-face shield must be worn in addition to the standard eye protection such as wraparound safety glasses.
- The edges of steel framing members may be sharp. Avoid dropping members or placing heavy loads of steel framing members on electrical cords as the steel may cut through the cord and create an electrical hazard.

1.2.0 Steel Framing Materials

Framing components include steel studs, steel joists, and steel roof trusses. Accessories such as the clips, web stiffeners, resilient channels, fastening devices, and anchors are required for complete and proper installation of members. Some manufacturers offer specialized products that enable builders to shorten construction times for complicated curves, arches, and other unique architectural features. The vertical and horizontal framing members serve as structural loadbearing components for a large number of low-rise and high-rise structures.

> **WARNING!**
> Cutting cold-formed steel framing members with an abrasive saw generates zinc fumes since the metal becomes very hot when being cut. Proper respiratory protection must be worn when cutting framing members with an abrasive saw, to prevent the inhalation of the fumes.

Steel framing materials have several advantages over wood framing, such as lower costs, faster assembly and installation, and design flexibility. The components are also noncombustible, lightweight, corrosion resistant, and fabricated to fit together easily. Manufacturing tolerances for cold-formed steel members are governed by ASTM standards. These standards govern length, web depth, flare, crown, bow, and twist. Cold-formed steel components are also coated against corrosion according to ASTM standards. Hot-dipped zinc galvanizing is the most effective coating method. Depending on the thickness of zinc applied to the steel and the environment in which the steel is placed, zinc coatings can protect the steel for more than 1,000 years.

Steel framing is also compatible with all types of surfacing materials. There are a variety of loadbearing and nonbearing systems, but *Figure 1* shows the basic components of a steel framing system. To meet custom material requirements, studs, track, and joist material can be cut within

North American Codes and Standards

The use of cold-formed steel (CFS) members in building construction began around 1850. In North America, however, steel members were not widely used until 1946, when the American Iron and Steel Institute (AISI) Specification was first published. This design standard was primarily based on research sponsored by AISI at Cornell University. Subsequent revisions to the document reflected technical developments, and ultimately led to the publishing of the North American Specification for the Design of Cold-Formed Steel Structural Members. AISI, along with the American National Standards Institute (ANSI), the American Society for Testing and Materials (ASTM) International, the International Code Council (ICC), and the Steel Stud Manufacturers Association (SSMA) govern the design, manufacturing, and use of cold-formed steel and framing. All framing members carry a product identification to comply with the minimum sheet steel thickness, coating designation, minimum yield strength, and manufacturer's name.

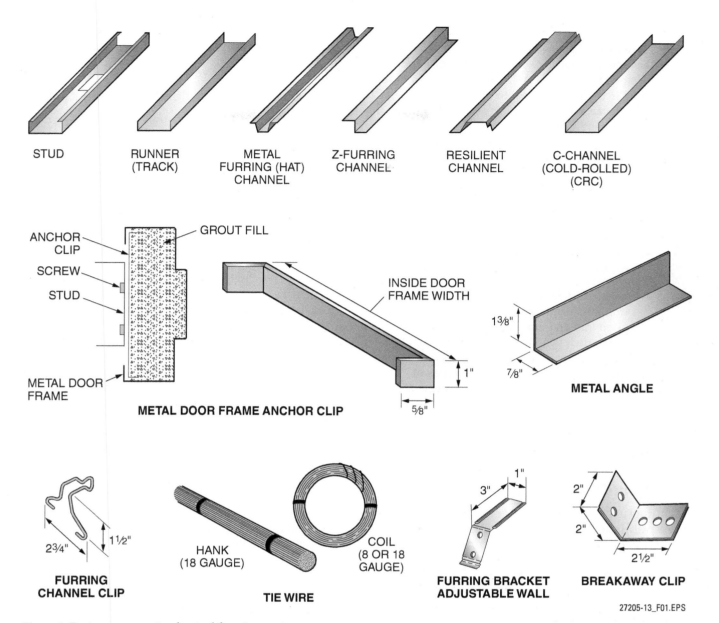

Figure 1 Basic components of a steel framing system.

⅛" of specification. Length is restricted only by the mode of physical transportation—typically 40' for containers and flatbed trucks. Custom ordering allows for less field cutting, and less labor and waste on the job site.

Steel framing members that are not marked with the required identification should not be used. The identification may be etched, stamped, or labeled every 96" along the center line of the **web** (*Figure 2*). Missing identification can result in project delays until materials are identified in accordance with appropriate standards.

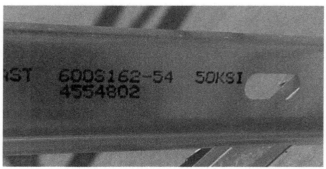

Figure 2 Steel product label.

> **CAUTION**
>
> Modern steel production relies on two technologies: basic oxygen furnace (BOF) or electric arc furnace (EAF). The BOF process uses 25 to 35 percent old steel to make new steel. It produces sheet for products where the main requirement is drawability; that is, products like automotive fenders, steel framing members, refrigerator enclosures, and packaging. The EAF process uses 95 to 100 percent old steel to make new steel. It is primarily used to manufacture products where strength is critical, such as structural beams, steel plates, and reinforcement bars. No matter which process is used, the resulting steel product has a minimum of 25 percent recycled content. Industry-wide, the recycled content of steel averages 67 percent.
>
> During the production process, molten steel is poured into an ingot mold or a continuous caster, where it solidifies into large rectangular shapes known as slabs. The slabs are then passed through a machine with a series of rolls that reduce the steel into thin sheets at desired thickness, strength, and other physical properties. The sheets are sent through a hot-dipped galvanizing process, and are then rolled into coils that weigh approximately 13 tons.

1.2.1 Identification of Framing Materials

Manufacturers of various steel framing products have published widely varying codes for identifying steel components. To eliminate this confusion, the Steel Stud Manufacturers Association (SSMA), representing more than 80 percent of all US steel framing production, developed a universal designator system. The S-T-U-F system has become the industry standard for identifying the most commonly used framing members. The designator consists of a four-part identification code. The first part is a three- or four-digit number indicating the member web depth in $1/100$" (outside-to-outside dimension). The second part is a single letter indicating the type of member:

S = Stud or joist section with stiffening lips
T = Track section
U = Channel section
F = Furring channel section

The third part of the designator is the flange width expressed in $1/100$". The final part of the designator is the mil thickness expressed in $1/1000$".

As an example, the 800S162-54 designator indicates the following:

800 = 8" member web depth (800 × $1/100$" = 8")

S = Stud or joist section with stiffening lips

162 = $1\frac{5}{8}$" flange width (162 × $1/100$" = 1.62" rounded to $1\frac{5}{8}$")

54 = 0.054"
(54 × $1/1000$" = 0.054") base steel thickness

Base steel thickness is either referenced in gauge or mils. *Table 1* lists common base steel thicknesses and equivalent gauge identifiers.

1.2.2 Furring

Design requirements may specify that gypsum board materials be separated from the stud using steel furring channels to provide sound isolation. In this case, the gypsum board is usually furred out using resilient channel (*Figure 3*). Resilient furring channels are used over both metal and wood framing to provide a sound-absorbent spring mounting for gypsum board and should be attached as specified by the manufacturer. The channels not only improve sound insulation, but they help isolate the gypsum board from structural movement, minimizing the possibil-

Table 1 Minimum Base Steel Thickness of Cold-Formed Steel Members

Designation (Thickness in mils)	Minimum Base Steel Thickness Inches (mm)[1]	Reference Gauge Number[2]
18	0.0179 (0.455)	25
27	0.0269 (0.683)	22
30	0.0296 (0.752)	20 – Drywall[3]
33	0.0329 (0.836)	20 – Structural[3]
43	0.0428 (1.09)	18
54	0.0538 (1.37)	16
68	0.0677 (1.72)	14
97	0.0966 (2.45)	12
118	0.1180 (3.00)	10

[1] Design thickness shall be the minimum base steel thickness divided by 0.95.
[2] Gauge thickness is a method of specifying sheet and strip thickness. Gauge numbers are a rough approximation of steel thickness and shall not be used to order, design, or specify any sheet or strip steel product.
[3] Historically, 20-gauge material has been furnished in two different thicknesses for structural and drywall (nonstructural) applications.

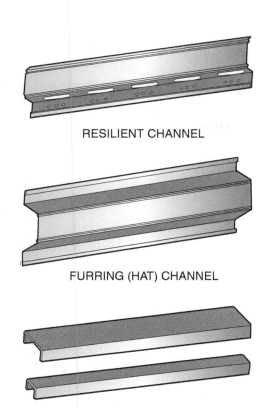

Figure 3 Types of furring channels.

Figure 4 Slip connectors.

ity of cracking. Resilient furring channels may also be used for application of gypsum boards over masonry and concrete walls. In wood frame construction, gypsum board can be screwed to resilient metal furring channels to provide a higher degree of sound control. Furring has additional applications:

- Provide additional space for insulation.
- Allow out-of-plane walls or walls of different thickness to match and have a smooth surface.
- Provide additional space to conceal fixtures or structural elements within a wall.

1.2.3 Slip Connections

Slip connectors (*Figure 4*) are devices that allow for the vertical movement of a structure without imposing additional loads on cold-formed steel framing or other wall components. These connectors are used where a structural system other than steel framing is needed to carry the loads of upper floors and the roof down to the foundation. Loads on the upper portions of a structure cause it to deflect, which in turn may induce loading in walls and wall components that are not designed to carry these loads. Slip connectors are designed to allow for this movement without creating unmanageable stress on components.

Slip connectors are generally located at the top of a wall panel, where it meets the underside of a structural element, such as a floor slab or beam. Under gravity, seismic, or wind loads, this upper portion of the structure may deflect up or down. The connector is designed to allow this movement, restrain the wall system from out-of-plane movement, and prevent any additional axial loading on the stud.

Slip connections are also useful at locations where a wall system is continuous, bypassing intermediate floors (*Figure 5*). Where this occurs, a slip connection extends from the side of the structure and supports the wall components laterally. Connections are also installed at roof bypasses, where the wall system extends past a roof structure to form a parapet or high wall. At this location, the slip connection must permit movement of the structure either up, due to wind uplift, or down, due to gravity loads.

1.3.0 Steel Framing Tools and Fasteners

Steel framing requires more use of power tools than wood framing, so be sure to practice power

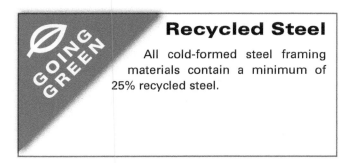

Recycled Steel

All cold-formed steel framing materials contain a minimum of 25% recycled steel.

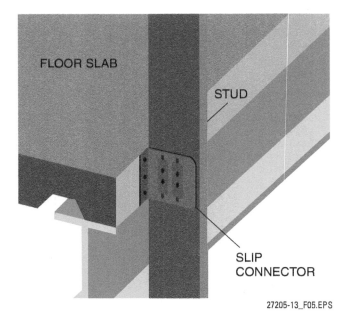

Figure 5 Slip connector application.

Figure 7 Screwgun with collated feed.

tool safety measures at all times, including grounding of all electrical tools. Some of the common tools used in steel framing are as follows:

- *Screwgun* – The screwgun (*Figure 6*) drives screws to connect steel members and attach sheathing material such as oriented strand board (OSB) and gypsum board to steel. Also called a power screwdriver, it increases installation speed and efficiency. Screwguns are available as battery-operated models or have power cords. The preferred screwgun for steel-to-steel connections should have an adjustable clutch and torque setting, with a speed range of 0 to 2,500 revolutions per minute (rpm). A drywall screwgun is recommended for attaching OSB or gypsum board to steel. Feed attachments are also available for screwguns (*Figure 7*). They use collated strips of screws that automatically feed at the end of a bit tip.

> **NOTE**
> A power screwdriver is not just a drill with a screwdriver bit. Power screwdrivers typically have an adjustable depth control to prevent overdriving the screws. Many power screwdrivers have clutch mechanisms that disengage when the screw has been driven to a preset depth. Some power screwdrivers are designed to perform specific fastening jobs, such as fastening gypsum board to walls and ceilings.

> **CAUTION**
> If the screwgun is running too fast, the tip of the screw may burn out before it penetrates the steel. Once the screw is properly seated, the screwgun automatically stops spinning the screw, preventing the screw from stripping.

- *Stud driver* – Two types of stud drivers are used to drive fasteners into concrete slabs, structural steel, and foundations: powder-actuated tools (*Figure 8*) and hammer-drills (*Figure 9*). The holding strength of the fastener in concrete depends on the compressive strength of the concrete, shank diameter and depth of penetration of the fastener, as well as spacing and edge conditions. Headed or threaded drive pins are available with **knurled** shanks to increase holding power in structural steel material. Fasteners loaded in tension may require washers to prevent the fastener from pulling through the steel track. Carpenters must be properly trained in the use of these powder-actuated tools (PATs) and must have an operator's certificate before operating them. The operator must wear a hard hat, heavy boots, and eye, ear, and

Figure 6 Standard screwgun.

Figure 8 Powder-actuated tool.

face protection. Treat a powder-actuated tool as if it were a loaded gun. Before you handle it, determine whether it is loaded with a powder charge, a fastener, or both. Point it away from yourself and away from others at all times. Be sure to recognize when the tool is and is not in locked position.

> **WARNING!**
>
> Powder-actuated fastening tools are to be used only by trained operators in accordance with the operator's manual. Operators must take precautions to protect both themselves and others when using powder-actuated tools:
>
> - Operate the tool as directed by the manufacturer's instructions and use it only for the fastening jobs for which it was designed.
> - To prevent injury or death, make sure that the drive pin cannot penetrate completely through the material into which it is being driven.
> - To prevent a ricochet hazard, make sure the recommended shield is in place on the nose of the tool.
> - Powder-actuated tools must be tested each day before loading.
> - Fasteners should not be driven into very hard or brittle materials, such as cast iron, glazed tile, and face brick.
>
> Other safety considerations are proper use and disposal of loads used in powder-actuated tools.

> **NOTE**
>
> As a courtesy, notify building occupants prior to the use of powder-actuated tools. Discharge of the tool may sometimes be mistaken for gunfire.

- *Locking C-clamps and bar clamps* – These tools come in a variety of sizes and are used to hold steel members together during fastening.

Figure 9 Hammer drill.

C-clamps (*Figure 10*) prevent separation, also known as screw jacking, when the first layer of steel climbs the threads of the screw. Bar clamps are used to hold steel wall members together until permanent fastenings are applied. They are also used to hold headers in place until they are fitted into the top track. All clamps used in steel framing should have regular tips on ends without pads to reach around the steel flanges.

1.3.1 Cutting Tools

Since many steel framing components are prefabricated, the amount of steel cutting required in the field is minimal. However, if field cutting needs to be performed, the following are some common tools and methods:

Figure 10 Locking C-clamps.

> **NOTE:** Some contract documents prohibit the use of certain cutting tools. Always check installation specifications before using any one method.

- *Chop saw* – Chop saws are most commonly used for field cutting, using an abrasive blade to cut quickly through the steel (*Figure 11*). While chop saws are very effective for square cuts and for cutting bundled studs, they are very noisy and emit hot flying metal filings. Be aware that the edge produced by a chop saw is very rough, with sharp burrs left on the steel. There are a number of circular saws on the market that can be used with abrasive blades to cut steel studs. Some of them have special guards to catch flying metal chips, which can be a hazard when using a high-speed saw to cut steel.
- *Swivel-head shears* – Shears are manufactured in both electric- and battery-operated models. *Figure 12* shows typical swivel-head shears. They are used to cut steel studs, runners, sheet metal, and cold-formed steel up to 20-gauge thick. They are portable and make smooth cuts with no abrasive edges. The cutting edge rotates 180 degrees for overhead and side work. Some drawbacks to shears are that they may be difficult to use in cutting tight radii on **C-shapes**, and blades are expensive to replace.
- *Circular saw/dry-cut metal-cutting saw* – Newer types of metal-cutting saws are becoming available with blades made of aluminum oxide that produce a smooth edge. One manufacturer produces a dry-cut saw that collects the metal filings right in the tool.
- *Aviation snips* – Aviation snips (*Figure 13*) are hand tools that can cut cold-formed steel up to 18-gauge thick. They are useful when cutting and coping steel, for snipping flanges, and for making small cuts. Some brands of snips are color coded for left, right, and straight. Always wear gloves and safety goggles when using snips.

1.3.2 Hole-Cutting Tools

Most openings in steel studs are prepunched at the factory. However, it may be necessary to punch additional openings for electrical or tele-

Figure 11 Chop saw.

Figure 12 Swivel-head shears.

Figure 13 Aviation snips.

communications cabling that may be installed later. Holes in webs of studs, joists, and tracks must be in conformance with an approved design or a recognized design standard.

- *Hole punch/steel stud punch* – For small holes, up to approximately 1" to 1½" in diameter, a hole punch can be used for steel members up to 20-gauge thick. Grommets should be inserted to protect wiring from sharp edges. They may also be used to isolate copper or other dissimilar metals from the steel.
- *Hole saw* – For larger holes up to 6" in diameter, and through material thicknesses greater than 20 gauge, hole saws and bits are recommended. The saws are used on a drill motor to cut through the steel. Bits, while more costly, tend to cut through the steel faster.

> **WARNING!** Field-cut holes must follow the requirements in the *Standard for Cold-Formed Steel—General Provisions*. If holes are not cut properly, the integrity of the structural steel members may be compromised.

1.3.3 Screws

The right fastener for any steel framing project will make a significant difference in the structural integrity and quality of your work. Consider the loads to be transmitted through the connection, as well as the thickness, strength, and configuration of the materials to be joined. Fasteners include screws, pins, clinches, and welds, as well as anchor bolts, rivets, powder-actuated fasteners, and expansion bolts. For applications requiring increased stability or when working with steel thicker than 18 gauge, the components should be joined by shielded metal arc welding to provide additional strength. Always follow manufacturer recommendations for fastening systems during all phases of cold-formed steel construction projects.

Screws are the most common type of steel framing fasteners. Since pilot holes are not drilled in steel framing members, screws used to secure the members must be able to make their own holes before they engage. Screws used for steel framing are installed with screwguns, and are available in a variety of head styles to fit a wide range of structural and cosmetic requirements. Screws have three distinct thread thicknesses: coarse (threads spaced farthest apart), medium, and fine (threads spaced closest together). Screws used in steel framing generally have coarse threads for optimum cutting. Thicker steel, such as 12 gauge, requires the use of a fine-threaded screw. Screws are generally finished with zinc or cadmium plating to withstand environmental impacts and provide corrosion resistance. The size and the type of screw needed will depend on the thickness of the sheathing material and steel. Become familiar with the different types of screws to choose from. If the wrong screw is used, the proper connection may not be made, or the screw may break off or not penetrate the steel.

> **CAUTION** When securing screws, overdrilling should be avoided, as it will reduce the effectiveness of the connection and hurt the structural integrity of the sheathing.

There are two main types of steel framing screws: self-drilling and self-piercing (*Figure 14*). Self-drilling screws are designed to drill through layers of steel before any of the screw threads engage. Self-piercing screws have sharp points that can typically penetrate 20- to 25-gauge material with ease. They are commonly used to attached plywood and gypsum board to thinner layers of steel. Code and standard requirements state that for all connections, screws must extend through the steel a minimum of three exposed threads. For most steel-to-steel connections, ½" or ¾" screws are acceptable. When applying plywood, gypsum board, or rigid foam insulation, proper screw length is determined by adding the measured thickness of all materials, and then providing an extra ⅜" allowance for the exposed threads.

Screw heads are available in many different forms. The head locks the screw into place and

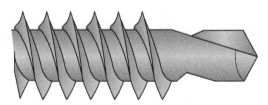
SELF-DRILLING SCREW

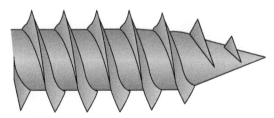
SELF-PIERCING SCREW

Figure 14 Self-drilling and self-piercing screws.

prevents it from sinking through the layers of material. The head also contains the drive type (type of bit tip) to apply the screw. The head profile is selected to avoid interference with other building components.

Common self-drilling screws are shown in *Figure 15* and include the following:

- *Gypsum board (bugle-head) screw* – These screws are used to attach gypsum board to steel framing members. No. 6 sharp-point bugle-head drywall screws were designed for this application. They should be used with a depth-setting nosepiece to avoid tearing the gypsum board or protective paper. For steel thicker than 20 gauge, use a self-drilling bugle-head drywall screw.
- *Flat-head screw* – This screw is used for wood flooring and facings. It is designed to countersink and seat flush without causing splintering or splitting.
- *Wafer-head screw* – This screw is larger than the flat-head screw and is used to secure soft materials to steel framing members. The large head provides a greater bearing surface, and seats flush to achieve a clean, finished appearance.
- *Hex washer-head screw* – The most popular head style for steel-to-steel connections with no sheathing or gypsum board is the hex washer. The washer face provides a surface for the driver socket, ensuring good stability in driving. If sheathing and gypsum board are applied over the fastener, the head style must have a very thin profile to prevent blowouts or bumps at the screw locations.
- *Trim-head screw* – Small trim-head screws are preferred when installing baseboard and other trim. The small head penetrates the trim, leaving a tiny hole easily filled with putty.
- *Oval-head screw* – This screw spans clearance holes and has a low-profile appearance for attaching accessory items to steel stud walls.
- *Pan-head screw* – This screw is used to fasten studs to runner tracks and to connect steel bridging, strapping, or furring channels to studs or joists.
- *Winged screw* – Winged screws are used only to attach plywood to metal framing members.

1.3.4 Pins

Pneumatic or cordless nailers are commonly used in steel framing to attach plywood, OSB, or siding to walls and roofs. Some contractors use only pins, while others use screws around the perimeter and pins in the field of the board.

Plywood or OSB sheathing material may be applied using pins (*Figure 16*). To be effective, the plywood must be held tightly against the steel member before the pin is driven because firing the pin does not tighten the plywood against the steel. Roof sheathing is more easily installed with pins because the carpenter usually stands on the plywood, keeping it tight to the steel. An alternative method is to tack the plywood to the steel with screws along the perimeter first and then pin the field.

There have been advancements in fastener technology to include pins for steel-to-steel connections as well. Connections are accomplished by driving a pin with a knurled shank (*Figure 17*) into the layers of steel.

1.3.5 Powder-Actuated Fasteners and Powder Loads

Powder-actuated fasteners are available in a variety of diameters and lengths, and include drive pins and threaded studs (*Figure 18*). Drive pins are used for permanent installations. Threaded studs are used for applications where the material or equipment fastened to the concrete or steel base material is to be moved. Most drive pins and threaded studs have a plastic fluted washer that is used to center the pin or stud in the PAT during installation.

Figure 15 Screw head shapes for typical self-drilling screws.

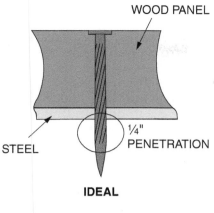

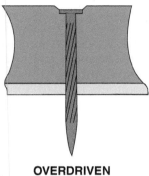

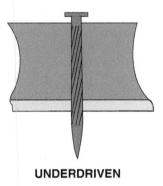

Figure 16 Proper pin installation.

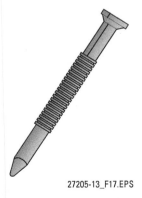

Figure 17 Knurled shank pin.

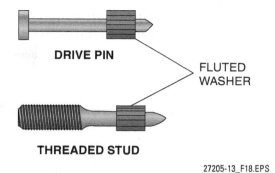

Figure 18 Drive pins and threaded studs.

> **WARNING!**
> Powder-actuated fasteners can be used only by personnel trained and certified in the use of the specific tool. You must carry your certification card with you when using the tool, and the card must cover the tool you are using.
> Be sure to select the proper powder load for the job in accordance with the manufacturer's instructions. If in doubt, check with your supervisor.

Powder loads provide the propellant for powder-actuated tools. Powder loads resemble shell casings used in conventional firearms. PATs have a .22-, .27-, or .32-caliber bore diameter and require a load of the same caliber. The casings for .22- and .27-caliber loads are made from brass and .32-caliber casings are made from nickel. Powder loads range in strength from #1 (lowest velocity) to #6 (high velocity). Color coding is used for ease of identification of powder loads (*Figure 19*).

1.3.6 Clinching

Clinching requires no screws or pins, only a pneumatic tool that press-joins pieces of steel together. The clinching tool, in effect, creates a rivet

Load Color	Powder Level	
Gray	1	Low Power
Brown	2	
Green	3	
Yellow	4	
Red	5	
Purple	6	High Power

Figure 19 Powder loads.

in the metal, as shown in *Figure 20*. Manufacturers of clinching tools provide test reports that verify the strength of the press joining. Clinching systems work well in panelization environments, and are less permanent than welding. If studs are fastened incorrectly with a clinch connection, the stud may be popped out with a screwdriver or drilled out.

1.3.7 Bolts and Anchors

Bolts and anchors are also commonly used to fasten cold-formed steel framing to masonry, concrete, and other steel components. Except for some proprietary anchors, predrilling of holes is necessary. Bolts require the installation of a washer, and must meet or exceed the requirements of *ASTM A307*. Expansion anchors are commonly used for connections to concrete or masonry, and require information from the manufacturer to determine the capacity and spacing requirements.

CLINCHING TOOL

CROSS SECTION OF CLINCHED STEEL

27205-13_F20.EPS

Figure 20 Pneumatic clinching tool.

1.4.0 Material Takeoff

For commercial construction projects, the material quantities are typically calculated by an estimator and ordered to be delivered to the job site. For residential construction projects, a carpenter may need to order building materials and supplies. Therefore, an understanding of basic material quantities is needed to communicate the proper information to the supply house.

For construction projects, a material takeoff is needed to properly determine the amount of materials required. Special structural elements such as floor openings, cantilevers, and partition supports that affect material requirements should be considered. The process begins by checking the specifications for the types and dimensions of materials to be used. It also requires that the prints be checked or scaled to determine the dimensions of the various components needed. Once the types and dimensions of material are determined, a material takeoff can then be performed.

1.4.1 Joists and Joist Headers

To determine the number of floor joists in a frame, divide the length of the building by the joist spacing and add one joist for the end and one joist for each partition that runs parallel to the joists. In a 32' × 64' building with a girder running lengthwise down the center and with no partitions, the number of joists is 49 [(64' × 12" per foot) ÷ 16" OC = 48 + 1 = 49, where OC is on center]. Because there are two rows of joists (one on each side of the girder), the total number of joists needed is 98 (2 × 49 = 98). Each of these joists would be about 18' long. The amount of material needed for the header joists is 128 lineal feet (2 × 64' = 128').

1.4.2 Bottom and Top Tracks

To determine the amount of material needed for top and bottom tracks, use the following procedure:

Step 1 Determine the length of the walls in feet. Multiply the length by 2 to account for the top and bottom tracks.

Step 2 Divide the result by 10 to determine the number of 10' pieces of stock. Round up to the next full number, and allow for waste.

1.4.3 Studs

To determine the number of studs needed, use the following procedure:

Step 1 Determine the length in feet of all the walls.

Step 2 The general industry standard is to allow one stud for each foot of wall length, even when you are framing 16" on center. This should cover any additional studs that are needed for openings, corners, and partition tees.

For the 32' × 64' building noted previously, there are 192' of exterior walls. You will need one stud for every foot of wall, or 192 studs.

1.0.0 Section Review

1. When handling steel framing members, you should wear specific gloves to protect against cuts.
   a. True
   b. False

2. The universal designator system for steel framing materials is called the _____.
   a. FMIS system
   b. S-T-U-F system
   c. SSID system
   d. STMF system

3. To attach OSB sheathing to steel framing, you would use a _____.
   a. drywall screwgun
   b. hammer drill
   c. screwdriver
   d. powder-actuated tool

4. The number of cold-formed steel studs required for a building measuring 50' × 75' is _____.
   a. 50
   b. 75
   c. 125
   d. 250

Section Two

2.0.0 LAYOUT AND INSTALLATION OF STEEL-FRAMED WALLS

Objective

Identify the steps to layout and install a steel stud wall.
a. Describe the basic steel construction methods.
b. Explain how to frame nonstructural steel walls.
c. Explain how to frame structural steel walls.

Performance Tasks 2 and 3

Lay out a steel-framed wall with openings, including bracing and blocking.

Demonstrate the ability to build headers (back-to-back, box, and L-header).

Trade Terms

Blocking: C-shaped track, brake shape, or flat strap attached to structural members or sheathing panels to transfer shear forces.

Clip angle: An L-shaped piece of steel (normally with a 90-degree bend), typically used for connections.

Curtain wall: A light, nonbearing exterior wall attached to the concrete or steel structure of the building.

Diaphragm: A floor, ceiling, or roof assembly designed to resist in-plane forces such as wind or seismic loads.

Lateral: Running side to side; horizontal.

Plenum: An enclosed space, such as the space between a suspended ceiling and an overhead deck, which is used as a return for heating, ventilating, and air conditioning (HVAC) systems.

Racking: Being forced out of plumb by wind or seismic forces.

Roof rafter: A horizontal or sloped structural framing member that supports roof loads.

Shear wall: A wall designed to resist lateral forces such as those caused by earthquakes or wind.

Historically, cold-formed steel framing has been widely accepted for use in nonstructural applications or partition walls. Today, it is commonly used for floor, structural wall, and roof assemblies as well. This section covers steel framing as it applies to nonstructural and structural walls used in residential and commercial structures (*Figure 21*).

2.1.0 Basic Steel Construction Methods

Requirements and guidelines apply to cold-formed steel construction as for all steel members within a building. They include in-line framing, web stiffeners, standard web holes and patches, and proper stud seating. Steel frame assemblies also require certain types of bracing for stability.

2.1.1 In-Line Framing

In-line framing, or direct alignment, is the preferred and most common framing method for providing a direct load path for transfer from studs to joists, through the framing system, to the ground. In this method, cold-formed steel framing members are aligned vertically so that the center line of the joist web is within ¾" of the center line of the structural stud member below, or the center line of the stud web is within ¾" of the center line of the web joist below (*Figure 22*).

Figure 21 Example of structural wall system.

2.1.2 Web Stiffeners

Web stiffeners, or bearing stiffeners (*Figure 23*), are C-shapes or tracks that are used to prevent joists from crippling at the point where the load transfers from a stud into the floor joist under structural walls. C-shapes provide additional rigidity as they are formed with a web, two flanges, and two stiffening lips. At a minimum, the thickness of a stiffener is the same as the floor joist, and the length of the stiffener is the depth of the joist minus ⅜". Stiffeners are installed across the joist depth of the web and on either side of the web. The stiffener is fastened to the web with either three or four No. 8 screws. Three fasteners are used when the screws are installed in a single row, and four when installed with one fastener in each corner.

2.1.3 Web Holes and Patches

Depending upon the application, framing members may be designed and manufactured with standard punchouts—web holes made during manufacturing (*Figure 24*). Holes in webs of studs, joists, and tracks must conform to an approved design. Always refer to the manufacturer recom-

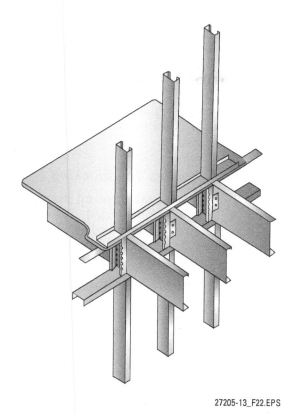

Figure 22 In-line framing detail.

Punch Openings in Steel Studs

Steel studs are manufactured with prepunched openings for wiring, conduit, and piping. A metal stud punch is available for punching additional openings, if needed. This tool works with studs up to 20 gauge. When electrical or telecommunications cabling is routed through the openings, grommets like the one shown should be inserted in the openings to protect the cabling from sharp edges. Other types of devices are made to eliminate conduit rattle.

Never punch a hole in a structural member without considering the structural requirements of the member. Consult the approved standard.

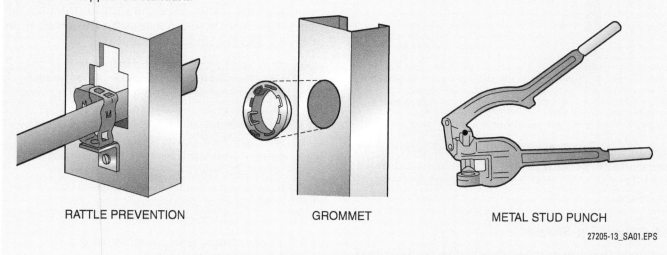

Module 27205   Cold-Formed Steel Framing   15

Figure 23 Web stiffener.

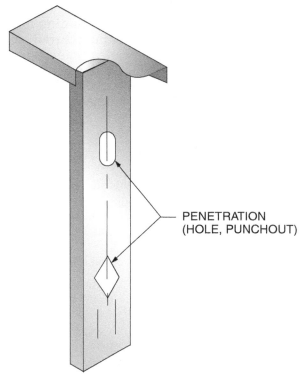

Figure 24 Standard stud punchouts.

mendations regarding web hole and patch size. Proprietary products are also produced with much larger holes that have been engineered and approved.

2.1.4 Built-Up Shapes

C-shapes that are commonly used in cold-formed steel framing provide only a minimal resistance to twisting. Built-up shapes, such as a nested stud and track assembly, provide the increased stiffness of a closed section as well as flat surfaces for attaching finish materials. Built-up shapes are commonly used for door and window jambs, headers (*Figure 25*), beams, and posts. The strength of built-up shapes is determined by the properties of the individual members, as well as the method of fastening.

2.1.5 Bridging, Bracing, and Blocking

The strength of individual framing members is a function of the bracing provided. This bracing restrains the member from moving laterally or twisting, and may be in the form of any one or a combination of three common methods:

- Cold-rolled channels placed through the punchouts
- Steel strapping attached to the flanges with periodic solid blocking and/or X-bracing
- Sheathing attached to the flanges

Overall, system bracing anchors the steel members and provides stability to the entire structure.

2.1.6 Thermal Considerations

Products used in construction may require some additional insulation to meet energy codes. The *Thermal Design and Code Compliance for Cold-Formed Steel Walls*, published by the Steel Framing Alliance, provides designers and contractors with guidance on thermal design of buildings that use cold-formed steel framing members. The thermal performance of a steel-framed structure may also be improved by the batt and other insulating materials within the wall cavity. Some thermal regions will require insulation foam board on the exterior of the frame. Designs should consider the effects of moisture when assessing the application of cavity and continuous insulation.

> **NOTE**
> Use care when using pressure- or fire-treated wood products with cold-formed steel framing, as accelerated corrosion may result. It is preferable not to use pressure-treated wood with steel framing, but if you do, specify a less corrosive pressure treatment, such as sodium borate. Always separate the steel framing material from the pressure-treated wood with a nonabsorbent closed-cell sill seal.

2.1.7 Protecting Piping and Wiring

Copper piping is commonly run through the punchouts of wall studs, creating the potential for direct contact with the galvanized steel and

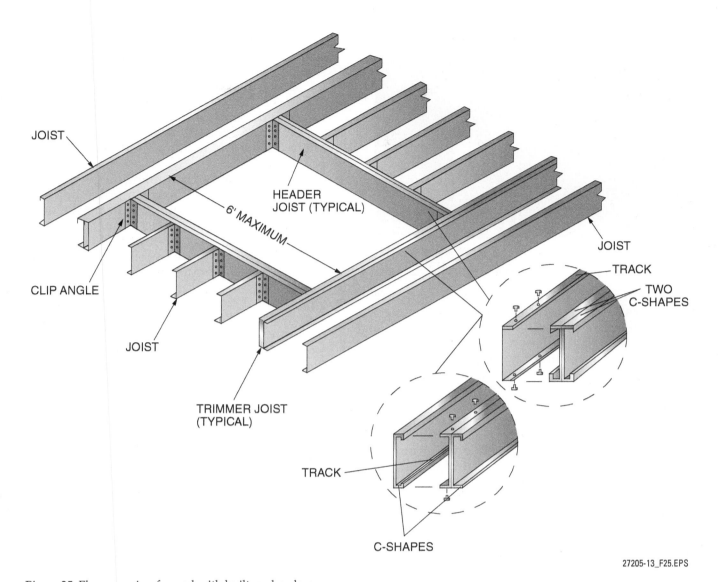

Figure 25 Floor opening formed with built-up headers.

producing a galvanic reaction that will compromise the strength and performance of both the steel and copper. This contact of dissimilar metals must be avoided by installing nonconductive grommets, plastic bushings, or other materials designed to separate the metals.

Plastic piping, on the other hand, does not require protection from contact with steel framing members. Even so, consideration should be given to installing nonmetallic brackets that will isolate the pipe from the hole in order to prevent noise and the potential for an incision in the pipe.

Ensure that wiring sheathed with a nonmetallic coating is separated from the sharp edges typically found in punchouts of wall studs and joists (*Figure 26*). The *National Electrical Code*® (*NEC*®) states that nonmetallic sheathed cable must be protected by bushings or grommets securely fastened in the opening prior to the installation of the cable. Cable that follows the length of the framing member also needs to be secured against movement. This is commonly accomplished by the use of tie-downs such as nylon cable or zipper ties. These are attached to the studs at intervals as required by local building codes.

2.2.0 Framing Nonstructural (Nonbearing) Steel Walls

Framing members used in interior systems may be nonbearing, or may be designed as part of the structural system. This section primarily discusses nonstructural walls that function as space partitions within the exterior walls of a building.

Nonstructural walls are made up of studs, track, and accessories. The primary differences are the characteristics of the materials and the application of connectors and accessories.

Figure 26 Proper protection of wiring.

Nonstructural framing members typically have a base steel thickness of 25, 22, or 20 gauge, compared with a minimum thickness of 20 gauge for structural studs. In addition, the minimum stud flange dimension for nonstructural framing members is 1¼", and the minimum stiffening lip dimension is ⅛", compared with 1⅝" flange and ½" lip for structural studs. Also, nonstructural members typically have a G40 galvanized coating weight, compared with G60 or higher for structural studs. The rules for nonstructural framing are different from those for structural walls, and may be found in the gypsum specification rather than the building codes.

2.2.1 Steel Curtain Walls

Since their introduction in the early 1900s, metal and glass curtain wall systems have become very popular in the architectural design of modern structures. Unlike interior partitions, wind-bearing curtain walls resist loads from exterior wind pressures that, in some cases, exceed 60 pounds per square foot. Cold-formed steel curtain walls are made up of various components:

- *Angles* – Clip angles and continuous angles are used to connect framing members within the curtain wall system.
- *Clip angle* – A steel angle, generally 3" to 12" long, which makes the transition between a framing member and the component supporting it. These angles are used to connect two framing members.
- *Continuous angle* – A steel angle that makes the transition between a stud curtain wall and the primary frame. The angle is typically of hot-rolled thickness (³⁄₁₆" to ⅜"); however, thinner-gauged materials can be used if the span and load requirements are relatively small.
- *Diagonal brace (or kicker)* – A sloping brace used to provide lateral support to a curtain wall assembly. When installed horizontally, this brace is referred to as a strut.
- *Embed* – A hot-rolled steel plate or angle, reinforced with shear studs or steel rebar, which is cast into a concrete floor or beam. Embeds allow for the welded attachment of steel supports.
- *Girts* – Horizontal structural members that support wall panels and are primarily subject to bending under horizontal loads, such as wind load.
- *Slide clip* – A connection device that permits deflection of the primary frame to which a stud attaches, while it braces the stud against lateral forces.
- *Slip track* – A track section used in for-fill curtain wall applications (*Figure 27*). Slip tracks accommodate vertical movements of a primary frame (normally ¼" to ¾"), while bracing the wall against lateral forces. Slip tracks may also be specified at the top of interior gypsum board partitions.

The term *curtain wall* is used to distinguish this system from loadbearing framing. Loadbearing framing requires that the wall members carry the weight of the structure above. With curtain wall framing, the structure is usually already in place, and the wall framing is filled in between the floor

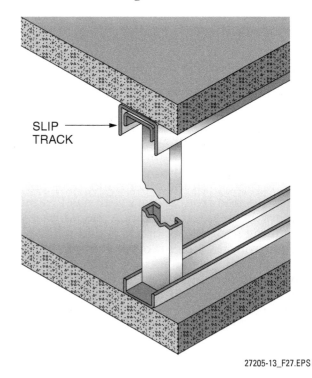

Figure 27 Slip track.

slabs. The stud-to-track gap distance for curtain wall is no more than ¼", unless it is otherwise specified in an approved design. This is different from loadbearing construction, which permits only a ⅛" gap. The only gravity loads that curtain wall framing typically carries are the weight of any cladding or finish materials attached to it.

> **CAUTION**
> For curtain wall installation, the use of components formed from steel measuring less than 20 gauge should be avoided.

2.2.2 Construction Methods for Steel-Framed Curtain Walls

There are four main methods for building assemblies in curtain wall construction:

- *Infill* – This method describes applications where studs are only one story tall, spanning from floor to floor of a structure. Infill framing requires less stud material. Connections are often easier to make, since low-cost powder-actuated fasteners can be used in many applications. In addition, the spans are often shorter than bypass conditions, so thinner steel or wider spacing can be used. The disadvantage of infill framing is that more track material is needed, and wall sections can be difficult to panelize.
- *Panelization* – This method can be used if field measurements are made after the floor systems are in place, or if a slip connector or telescoping stud system is used. When structural steel is used for main framing, the spandrel beam can get in the way of framing. In that case, framing must be attached to the bottom of the beam. If framing takes place outside the spandrel beam, it can cause difficulty in supporting insulation.
- *Bypass framing (balloon framing)* – This method allows a single stud to be used for framing two or more floors, and it requires less track material. The multiple spans can reduce moment stresses in members, allowing for more widely spaced or thinner framing members. In addition, this method makes it easier to prepanelize large sections, including multistory panels. On the other hand, some connections can be more difficult to make, such as bracing and support connections behind columns and spandrel beams. Depending on the condition, double connections may need to be made. This requires clips or slip connectors attached first to the structure, and then again to the stud. This last issue can be mitigated somewhat by the use of connectors that friction-fit inside the stud, thus reducing connection time. Finally, if slab or other structural elements extend into the stud cavity, they may need to be chipped away, or stud framing may need to be altered to correctly install the bypass framing.
- *Stacked wall framing* – This method permits bypass framing, while isolating slip connections to one- or two-story segments. As a result, multistory panels and prefinished panels can be fabricated and installed, including insulation. However, in order to prevent water infiltration, this approach requires special detailing at slip connections and exterior finishes. Because the entire panel weight goes to fixed connections, usually at every other floor, these fixed connections need to be strengthened to carry the added dead load of the taller panels. Also, some connections are more difficult to reach, such as bracing and support connections behind columns and spandrel beams.

In multistory construction, movement may occur at the floor below a curtain wall system, causing the entire wall to move down. In this case, the connector at the top of the wall must have sufficient capacity to allow the wall system to move down without creating tension on wall components.

Other than slip connections, the most typical connections used in metal-stud curtain wall design include the following:

- *Base connections* – Stud to track, track to concrete or steel deck.
- *Head or sill to jamb connections* – Connections at the top and bottom of window or door openings.
- *Continuous angle connections* – Field-welded connection usually found at spandrel framing.
- *Clip angle connections* – Typically at nonslipped bypass or spandrel framing connections and designed to carry both lateral and self-weight forces.
- *Outrigger clips* – Short lengths of angles designed as an axially loaded strut to carry stud lateral reactions back to the structure.
- *Wind girt connections* – Usually at taller spandrel framing cases when the use of diagonal braces is impractical. These connections typically are not required to carry any gravity loads.
- *Diagonal braces or kickers (as opposed to X-bracing for shear walls)* – Stud diagonals providing for a bottom spandrel stud reaction back to the structure.
- *Stud-to-stud connections* – Either lapped or track-to-track. On occasion, these connections

require movement allowance, which can be accommodated by a slip track or slip-pin detail.
- *Knee-wall base* – This is a moment connection at the bottom of a knee or stub wall, usually either a freestanding parapet or a long segment of wall under a continuous or ribbon window condition.

2.2.3 Bracing for Curtain Walls

C-shapes are the most typical cold-formed steel curtain wall member. Due to its shape and geometric properties, a C-shape tends to rotate under lateral load. Unbraced studs may also move out of plane. This is known as torsional-flexural buckling. Mechanical bridging and/or the sheathing materials can restrain the flanges and prevent this buckling. When using discreet bracing rather than sheathing, decreasing the bridging spacing typically increases the member capacities; increasing the spacing will decrease capacities.

In a typical steel-framed exterior curtain wall, member deflections are the primary serviceability issue. Deflection limits are most often determined by the architectural finishes, and are set forth in the project specifications.

2.2.4 Finishing for Curtain Walls

A wide range of finish systems may be applied to curtain wall frames, including the following:

- Brick veneer
- Split-faced block veneer
- Tile or thin-cast brick
- Exterior insulation finish systems (EIFS)
- Glass fiber-reinforced concrete (GFRC)
- Metal panel
- Modified portland cement (stucco)
- Fiber-cement board or siding
- Dimensional stone, such as granite or limestone
- Wood siding and many other finish systems

2.2.5 Radius (Curved) Walls

Some partitions that might be difficult to construct with wood framing members can easily be built with steel framing members. A radius wall is an example. Note that a radius wall is circular in shape. Constructing a radius wall with wood framing members would take many hours. Using steel framing members, a radius wall could be built in less than an hour. A plywood template may be required for more complex radius walls.

The bend can be achieved using several methods. Curved walls can be framed out of steel using curved track for partitions or exterior loadbearing walls. The track can be bent at the job site by slitting the flanges. Ordering curved track from specialty companies also reduces the time needed to construct a radius wall.

Track may be ordered curved to a specified radius. Some specialty companies use power equipment to bend the track without slitting the flanges. Others manufacturers produce a flexible track that can be ordered and formed on the job site based on the desired effect. This provides a clean, neatly bent track to an exact radius. Wall track is bent around the flanges (*Figure 28*).

2.2.6 Other Nonstructural Wall Assemblies

- *Chase walls* – Chase walls are commonly used in commercial and residential construction to conceal plumbing and other utilities. For oversized utilities or acoustical requirements, two separate walls with a void in between may be framed in advance. Cross bracing between the two walls may be required.
- *Fire-rated assemblies* – Building codes require that certain partitions provide fire-rated separation of interior spaces, with a specific rating as to how long the partition or assembly will last under the design load before being penetrated by a fire. Approved assemblies must be constructed precisely as described in the directories published by the rating agencies.
- *Area separation walls* – Although this term sounds redundant, it refers to a specific group of rated walls designated by code requirements, which are designed to permit structural failure on one side of the wall while still providing fire protection.
- *Head-of-wall conditions* – Some building codes require that wall assemblies accommodate hor-

Figure 28 Example of a radius track.

izontal and vertical movement while still providing a fire-resistive barrier. Proprietary and nonproprietary devices and assemblies have been designed and tested to meet code requirements.

- *Shaft wall systems* – These systems are special types of rated wall systems that may be constructed from one side only. Higher sound and fire ratings may be achieved by attaching additional layers of gypsum board to the outer or inner face. A 1-inch-core board is specifically made for this purpose. Horizontal shaft wall may also be used for soffits or plenums where ratings are required.

2.3.0 Framing Structural Steel Walls

Structural walls support the weight of the building and protect occupants from wind loads and other forces of nature. A basic, cold-formed structural steel wall includes structural studs, tracks, fasteners, bracing, and bridging. Headers for window and door openings may be constructed using structural studs, joists, or proprietary cold-formed shapes. In addition, shear walls are typically framed within the stud wall assembly.

Structural steel wall framing members have a material base steel thickness of 20 to 10 gauge, with a minimum metallic zinc coating of G60. Typically, structural steel studs are produced in sizes from 2½" to 8", with flanges ranging from 1⅜" to 2½" or more. The size of the stiffening lips depends on the flange size. Structural track is sized to accommodate the web depth of wall studs, and the flanges are typically sized 1¼" to 3". The selection of a particular member will depend on its intended use. A number of proprietary and nonproprietary accessories are also available for structural wall assembly, including cold-formed channel for bridging, as well as clips, angles, and straps.

2.3.1 Layout

It is important to lay out the wall studs accurately to align with other structural assemblies such as roof and joist framing. Steel framing is typically spaced at 24" on center to take full advantage of the framing member.

Place the top and bottom track members on the straight edge of the panel table. They should be arranged with the webs next to each other, clamped together temporarily. They should fit tightly against the edge of the end stop (the straight edge at the end of the wall). Mark the layout of the wall studs on the flanges of the top and bottom tracks, starting with a wall stud at the end of the wall. Use highly visible ink, such as a black felt-tip marker. Place a line at the web location and an X on the side of the line to indicate the stud flanges.

Mark the next stud location to match the first truss or roof rafter location from the end wall. Continue marking every 24" (or 16", depending on the layout) for the full length of the wall. Where the exterior corner walls intersect, the wall that runs to the edge of the foundation has an extra stud. This stud is 3" from the end and acts as a backer to screw to the shorter intersecting wall.

Next, identify the rough openings in the walls. Check the architectural drawings to find door and window locations. If the dimensions are not provided, scale them off the drawing. Mark the location for the center of the openings on the top and bottom tracks. Use a red felt-tip marker to distinguish these marks from the layout marks. Check the door and window sizes on the drawings and verify rough openings with actual window sizes. Add 12" to the width of the window openings to allow for two jack studs on each side of the header with room to spare.

Using a tape measure, center the dimensions over the red marks on the track. Mark each end of the tape measure. This shows the location of the webs of the king studs. Put the X on the side of the mark away from the window. The webs of the king studs will be on the rough opening side. Remember to mark both the top and bottom tracks with 12" added to the rough window openings. This dimension is standardized to simplify header ordering and assembly. Two king studs may not be required at every opening, and framers may vary the length of the headers. However, standardizing header lengths helps to simplify cut lists.

All loadbearing studs must be aligned with the trusses, joists, or rafters above or below the wall. Because these walls carry loads, it is important to fit each stud tightly in the track member to allow the stud to properly carry the axial (downward) load from above. The top and bottom wall tracks must be of equal or greater thickness than the studs. Panelized walls may be constructed on a concrete slab, floor deck, or panel table. For all systems of wall construction, make sure that the surface the wall is built on is level. Walls must not be out of plumb more than ⅛" for every 10'.

Walls may be framed in two ways. Full-length walls may be framed up to 40' depending on available assistance at the job site. Be aware that longer walls also tend to twist; they could get damaged if there is not a big-enough crew to keep the wall straight. Shorter sections need more plumbing and alignment, but they are easier to assemble and may be built and spliced together. The track will also need to be spliced (*Figure 29*).

2.3.2 Wall Assembly

Before beginning assembly of a structural wall, ensure that the foundation or the bearing surface is free of all defects. The bearing surface should be uniform, with a maximum 1/8" gap between it and the track.

After layout is complete, separate the top and bottom tracks, and install a wall stud at each end of the wall between the top and bottom tracks. Clamp the stud flanges to the track flanges with locking C-clamps at each end. Tap the track on one end with a hammer to seat the studs as tightly as possible in the track. Fitting each stud tightly and perpendicular to the track keeps the wall straight. Screw one low-profile No. 8 screw through the flange of the track into the flange of the stud on either side of the stud. If an elevated panel table is used, the framer may be able to install the screw (from underneath) on the other flange as well. If not, once all the studs and headers are in place, and all screws are installed on one side, flip the wall over to install the screws on the other side.

When framing for a gypsum board covering, the framer often only installs screws on one side of the wall. Check local codes. However, for load-bearing walls, screws must be installed on each flange on both sides of the wall to keep the studs from twisting and to provide the proper connection for in-line framing. Continue twisting the studs into the track. Install the studs all the same way, with the open side of the C-shape facing the same direction and toward the start of the layout. Align the punchouts in the studs to provide straight runs for bracing if needed. The studs should be aligned so that they all face the same direction on parallel loadbearing walls.

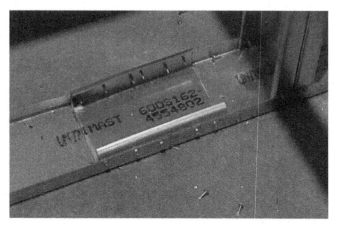

Figure 29 Track splicing.

> **CAUTION**
> Do not patch or reinforce errors in holes or punchouts yourself. If a structural member is damaged, a design professional should be consulted for any corrective action.

> **WARNING!**
> Field modifications to framing members cannot be attempted without the approval of the design professional. Damaged materials should never be installed, because they may compromise structural integrity.

Install the king studs at the rough openings with the hard side of the stud facing the rough opening and punchouts aligned. Do not install studs at the markings between the king studs. These markings will be for the cripple studs when the rough openings are framed. Continue down the length of the wall until all the studs are twisted and screwed in place. Do not remove the wall panel from the table until the headers and rough openings are completed.

During wall assembly, it is important to seat the studs into the track as tightly as possible. Studs must be seated with a gap of no more than 1/8" to ensure that the building loads are transferred through the studs, and not through the fastener (*Figure 30*).

Structural walls may be constructed using any of the following methods:

- *Stick building* – Walls are framed on the job site, one stud at a time. Walls can be built on a flat surface of the site, such as a concrete slab.
- *Panelization* – This reduces construction time while improving the efficiency and quality of the steel framing assembly.
- *Pre-engineered method* – This assembly typically increases the size and spacing of structural steel members. In some cases, the spacing can be as much as 4' or 6' on center.

Screws are most commonly used to attach track to cold-formed steel floor or roof assemblies, but welding may be used when specified. Powder-actuated fasteners may be required when the track connects to structural steel framing members or concrete. Expansion bolts are used at jamb locations and corners, while expansion anchors are typically used at shear wall locations.

2.3.3 Wall Installation

If anchor bolts are used in the foundation, measure their locations and place holes in the bottom track of the wall panels so that the walls will fit

ACCEPTABLE

UNACCEPTABLE

27205-13_F30.EPS

Figure 30 Stud seating.

over the bolts. If strap anchors or other kinds of anchors are used, this will not be necessary. Place temporary bracing material near the foundation to prepare for the wall to be raised. Any stud material at the job site, preferably 12' long, may be used for temporary bracing. Caulk the concrete foundation with weatherproof caulking material and use foam closed-cell sill sealer beneath the track.

Move the wall panel and set the bottom track on the foundation. Position it over the anchor bolt locations and tilt the wall up. Leaving the wall tilting slightly outward, clamp the temporary brace material to the wall studs in two or three locations (every 8' to 12' along the wall), depending on the length of the wall. Make sure the braces do not lap past the inside face of the wall. Before removing the clamps, secure a No. 10 hex-head screw through the brace into the stud. Install a brace every 8' to 12' along the wall. Secure the bottom of the brace with a stake driven into the ground or other solid surface. Screw through the stud into the stake to hold it in place. Repeat this process with all wall panels until all loadbearing walls are standing.

Some builders choose to frame walls in place, especially for one- or two-person crews. In this case, the track should be cut for the full length of the wall. Mark the top and bottom track for layout, and anchor the bottom track in place, securing the studs in each end of the track. Position the top track at the ends and with intermediate studs. Use a string line and level to position the remaining studs for the wall. Install headers, X-bracing, or plywood with the wall standing.

A balloon frame can be used for some structures (*Figure 31*). This is categorized as a pre-engineered structure. The carpenter and the engineer must work closely together to size the members and develop details for balloon framing.

27205-13_F31.EPS

Figure 31 Balloon framing.

> **WARNING!**
> Before tilting up a wall assembly, ensure that the braces do not hang past the wall edges. Snagging on the braces is very dangerous and could result in injury.

Proper bracing is required to prevent a cold-formed steel framing member from twisting and buckling. In wall construction, there are several common methods for bracing. They all have different applications and purposes. *Figures 32* through *35* are examples of typical types of bracing.

Intermediate stud bracing – Intermediate stud bracing is used when gypsum board or structural sheathing is not applied to both sides of loadbearing walls, such as garage walls (*Figure 34*).

Temporary bracing – There are two types of basic temporary bracing: one for panelized walls and one for installed walls. If a straight wall is constructed on a panel table, installing plywood or temporary bracing prevents racking when the wall is removed from the table. Before the wall is taken off the table, check for squareness by diagonally measuring the panel. Adjust if necessary. Lay extra studs or truss material across the wall diagonally, and screw the bracing to the wall studs, especially at door openings where the bottom track is weak. Leave the bracing on the wall until the wall is installed and permanently braced. These precautions will help provide straight walls ready for installation.

Installation of steel-framed walls requires adequate temporary bracing in order to resist loads during construction until permanent bracing can be installed (*Figure 35*).

2.3.4 Backing

Wall-mounted fixtures, appliances, and chair rails will require backing for proper support. Solid wood backing installed between the studs or light-gauge sheet metal should be fastened to the studs before the wall finish material is applied. Interior elevation drawings or separate backing drawings typically contain information regarding the proper placement of backing.

2.3.5 Shear Walls

Lateral (shear) loads are typically resisted by a system of interconnected shear walls and floor and roof/ceiling diaphragms that work together to transfer applied wind and seismic loads to the foundation of the structure. These diaphragms are called shear walls and can be created by properly attaching wood structural sheathing, steel decking, X-bracing (*Figure 36*), or other materials to the floor, wall, ceiling or roof framing. In addition to field-fabricated shear wall systems, there are now several proprietary, high-strength pre-engineered systems that may help shorten construction time (*Figure 37*).

The wall design for any steel-framed structure should provide for the placement of shear walls. In multistory construction, proper alignment (stacking and load path) is also required to ensure there is adequate shear transfer between roof or floor diaphragms and shear walls, or other assemblies such as bar joists, long-span deck, and wood framing. Shear walls are usually connected

Advantages of Steel Framing

Metal framing offers a great deal of flexibility in building design. Fire ratings are easier to achieve because, unlike wood, steel framing is noncombustible. In addition, steel framing can be produced in lengths not readily available with wood.

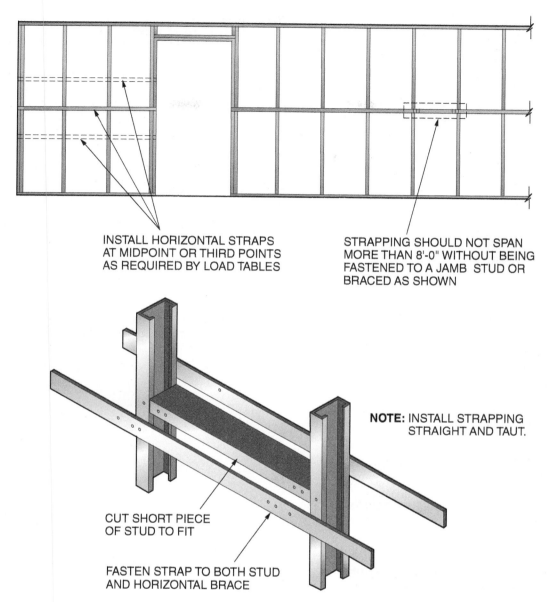

Figure 32 Lateral strapping for stud walls.

to the foundation using proprietary or engineered connectors and hold-downs.

The most common ways of applying shear-wall bracing are structural sheathing and X-bracing. Structural sheathing, such as Type II plywood or OSB, may be adequate to keep the wall from racking, depending on the design, as long as there are not excessive openings in the wall (*Figure 38*). In order for structural sheathing to be effective, it should be installed with the long dimension parallel to the stud framing (vertical orientation). The plywood or OSB may be secured to the wall while panelizing, or after the wall is plumb and level. X-bracing is another way to obtain shear strength when structural sheathing is not used. X-braces are diagonal steel straps attached to the walls with screws or welded connections (*Figure 39*). This bracing must be designed by an engineer, and the straps must be inspected for the correct number of fasteners. Do not tighten the straps until the walls are plumbed and aligned.

> **NOTE:** All concealed cavities such as headers in exterior walls must be pre-insulated before they are installed.

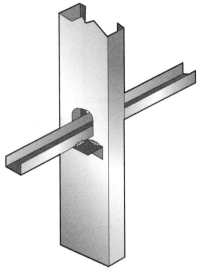

Figure 33 Bridging with cold-rolled channel (CRC).

Figure 36 X-bracing.

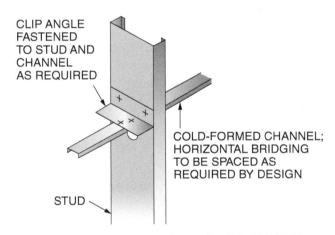

Figure 34 Example of stud bracing.

Figure 37 Proprietary shear wall.

Figure 35 Temporary bracing.

Figure 38 Wood structural sheathing.

NOTES:
- Install strapping as close to 45° as possible.
- Place straps on both sides of stud wall in order to prevent eccentric loading.
- Check for increased axial load applied to wall studs due to tension in straps. Double studs will typically be required at strap ends.
- Install wall straps straight and taut.

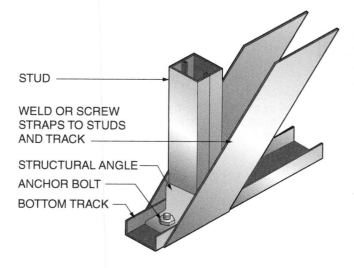

DIAGONAL STRAPPING FOR SHEAR WALLS

GUSSET PLATE X-BRACING

Figure 39 Attachment of X-bracing.

2.3.6 Header Assembly

Loadbearing headers are typically boxed, unpunched C-shapes that are capped on the top and bottom with track sections (*Figure 40*). They must be engineered for bending and shear strength, along with web crippling (crushing) at the locations of the loads from above. The two types of headers that are most commonly built from standard C-shapes are the box and back-to-back headers. A third type is a steel angle in the shape of an *L*, also called an L-header. Prefabricated headers may also be ordered from manufacturers. Selection of the header type depends on loads and applications. L-headers are being used more frequently in low-rise and multifamily construction because they use fewer fasteners and less material.

To install the header, loosen one of the king studs at the top of the rough opening, and install the header, with the open end up, by inserting it into the top track. This is usually a tight fit. Clamp the header at one end with the bar clamp. Apply pressure, working from one side of the header to the other using the clamp to tightly push the header into the top track. Make sure that the header is fitted tightly into the track before screwing it in place. Reposition the king studs that were loosened, and screw them back into place.

Back-to-back headers are formed by placing two C-shapes with the webs of the members touching each other. They are positioned in the top track of the wall and finished just like a box header.

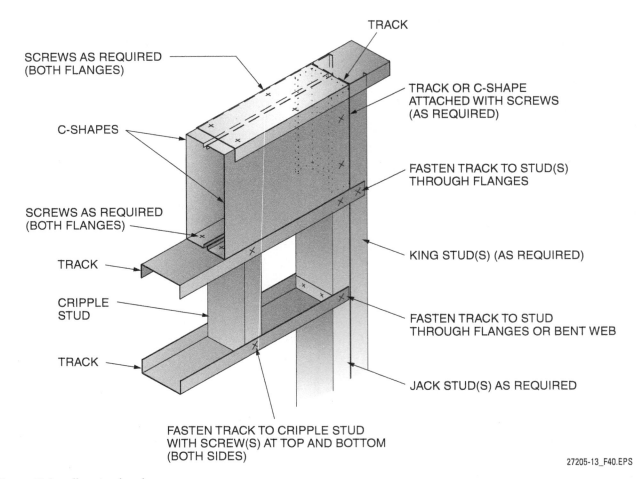

Figure 40 Loadbearing header.

The L-header (*Figure 41*) consists of one or two angle pieces that fit over the top track. This saves labor because there is no special fabricating, and the number of screws is reduced. The L-shape itself spans the opening for the header.

Once the header is installed in the top track, the remainder of the rough opening may be framed (*Figure 42*).

NOTE: After header assembly, the window opening on the drawing should be rechecked to verify the position and height of the window in the wall.

Figure 41 L-header.

Figure 42 Completed rough opening.

Mark the jack (trimmer) studs for the top of the window opening. Another mark should be made on the trimmer studs at the bottom of the window. Cut two track pieces to fit between the jack studs to make up the head and sill pieces. These pieces should be cut 2" longer so the flanges can be clipped 1" on each side.

Clip the flanges and bend the web down toward the flanges. Set the head and sill pieces in the opening, keeping the hard side of the track toward the opening. Screw the tab at the flange of each piece into the jack studs with one No. 8 screw at each tab. The tabs on the webs of the header track should also be screwed into the trimmer studs with two No. 8 screws.

The cripple studs should be cut to fit between the header track and head pieces, as well as between the sill piece and bottom track. The cripples should maintain the spacing layout (16" or 24" on center) for ease in installing gypsum board and sheathing. Screw the cripple studs into place with a No. 8 screw at each track flange on both sides.

Door openings do not require a bottom sill. However, the bottom track should run continuously at the bottom of the door to hold the wall together temporarily. The track can be cut out after the wall is plumb, level, and permanently braced.

2.3.7 Jambs

Loadbearing jambs require a minimum of two studs on each side of the framed opening (one trimmer plus one king), with more if required by the building code or by engineering analysis. The studs must be fastened together to act as one member, either by capping with track and screw fastenings or by stitch welding (*Figure 43*).

2.3.8 Sills

Sill members are usually single-track sections that are the same width as the wall stud used to frame an opening. They are clipped and screwed to the jamb studs. In cases where the allowable lateral load is exceeded, sills must be constructed with multiple track sections.

Figure 43 Stitch-welded jamb.

Fire-Rated Construction

A fire rating represents the ability of a wall, floor, roof, or structural member to withstand fire for a period of time, ranging from one hour to four hours. A rating of two hours, for example, means that a structural member such as a wall, floor, or roof would not allow flame or hot gases to pass through it for two hours. The fire rating also refers to the ability of a structure or material to withstand the force of water sprayed from a hose. The fire rating of a frame wall is based on its construction materials. Fire-rated walls are built in different fashions, depending on their application.

2.0.0 Section Review

1. In in-line framing, framing members are aligned vertically so the center line of the joist web and center line of the structural stud is within _____.

    a. ¾"
    b. 4"
    c. 8"
    d. 36"

2. When assembling a structural steel-framed wall, the maximum allowable gap between stud and track is _____.

    a. ⅛"
    b. ¼"
    c. ½"
    d. ¾"

3. In wall framing, the normal spacing for steel studs is _____.

    a. 16" on center
    b. 16" apart
    c. 24" apart
    d. 24" on center

Section Three

3.0.0 OTHER STEEL FRAMING APPLICATIONS

Objective

Identify other steel framing applications.
  a. Explain how steel framing members are used in floor and roof construction.
  b. Explain how steel framing members are used in ceiling construction.

Trade Terms

Rim track: A horizontal structural member that is connected to the end of a floor joist.

While walls are the most common elements to be constructed with cold-formed steel framing members, other building elements may also be steel-framed. These elements include floors, roofs, and ceilings.

3.1.0 Steel Floor and Roof Assemblies

Cold-formed steel floor assemblies typically use standard C-shape floor joists (*Figure 44*), proprietary floor joists, pre-engineered steel floor trusses, rim track, web stiffeners, clip angles, hold-down anchors, and fasteners. They can be installed on crawl spaces and stem walls, as well as directly to interior structural walls. They are similar to conventional framing and use single- or multiple-span installation techniques.

Recent years have seen a dramatic rise in the use of cold-formed steel framing members in roof assemblies. They allow for easy and standardized assembly and are durable and noncombustible. In addition to the standard C-shape member, scores of proprietary shapes, fabrication methods, and installation requirements are available from truss manufacturers nationwide. *Figure 45* shows an example of custom-designed steel trusses.

Figure 44 Standard floor joists.

Figure 45 Example of complex framing using steel trusses.

3.2.0 Ceiling Systems

Ceiling systems typically consist of a suspended drywall grid with gypsum board. An alternative method is to use furring and cold-rolled channel to provide a rigid framework for suspended gypsum board and other ceiling assemblies, with the steel members suspended (*Figure 46*). The furring channel is commonly clipped or wire tied, perpendicular to the underside of the U-channel at appropriate intervals for attaching gypsum board with screws.

This furring channel can also be screw-attached to other structural steel members, and can be attached directly to the bottoms of bar joists. In the latter case, the furring is installed perpendicular to the joists and wire-tied at appropriate intervals. Note that wire ties are required for fire-rated and multilayer assemblies.

Figure 46 Ceiling framework.

3.0.0 Section Review

1. Steel roof trusses are _____.
   a. seldom used today
   b. durable and noncombustible
   c. difficult to erect
   d. available only as standard C-shape members

2. In a ceiling system using steel members, gypsum board is attached with _____.
   a. bolts
   b. wire ties
   c. drive pins
   d. screws

SUMMARY

Cold-formed steel framing is more widespread in the construction industry than it was just a few years ago. Steel will not warp, crack, rot, or split. It is immune to termites and will not add fuel to a fire. It is a perfect fit for noncombustible construction and environmentally conscious building, and is more cost effective than lumber. In a wide variety of residential and commercial buildings, cold-formed steel is being used not just for partition walls, but as the main structural component. While the layout of cold-formed steel framing members may be very similar to conventional wood framing, there are important differences in components and techniques to remember. The studs, fasteners, and various accessories are designed to best take advantage of the light weight and design flexibility of steel.

Review Questions

1. When a stud has a designator of 800S162-54, the web depth of the stud is _____.
   a. 0.8"
   b. 8"
   c. 80"
   d. 800"

2. In the S-T-U-F system, the letter U represents _____.
   a. channel sections
   b. track sections
   c. studs with stiffening lips
   d. furring channel sections

3. The organization that created the universal designator system for steel components is the _____.
   a. American Iron and Steel Institute
   b. OSHA
   c. United Steel Alliance
   d. Steel Stud Manufacturers Association

4. Devices that allow for the vertical movement of a structure without imposing additional loads on cold-formed steel framing are called _____.
   a. slotted studs
   b. adjustors
   c. slip connectors
   d. bypass sections

5. The maximum speed of screwguns for steel-to-steel screw connections is _____.
   a. 1,500 rpm
   b. 2,000 rpm
   c. 2,500 rpm
   d. 3,000 rpm

6. A tool that can cut steel components of up to 20 gauge in thickness and leave no abrasive edges is a _____.
   a. chop saw
   b. snip
   c. C-clamp
   d. swivel-head shear

7. When selecting a screw to attach gypsum board to steel framing parts, the screw length should be the thickness of all the material plus _____.
   a. ⅛"
   b. ¼"
   c. ⅜"
   d. ½"

8. A common caliber for powder-actuated tools is _____.
   a. 18 caliber
   b. 27 caliber
   c. 38 caliber
   d. 45 caliber

9. When using bolts to fasten steel framing to concrete, you must install _____.
   a. washers
   b. spacers
   c. caulking
   d. cotter pins

10. The maximum center-line-to-center-line tolerance for in-line framing of cold-formed steel is _____.
    a. ⅛"
    b. ¼"
    c. ½"
    d. ¾"

11. A steel framing component that has a web, two flanges, and two lips is called a _____.
    a. header
    b. C-shape
    c. Z-furring channel
    d. hat channel

12. Structural stud sizes typically range from _____.
    a. 1" to 6"
    b. 1⅝" to 6"
    c. 2" to 5"
    d. 2½" to 8"

13. For a structural wall, the maximum allowable gap between the foundation and the bottom track is _____.

    a. 1/16"
    b. 1/8"
    c. 1/4"
    d. 3/8"

14. Diagonal steel straps attached to the walls are used to _____.

    a. brace shear walls
    b. temporarily brace panelized walls
    c. provide structural sheathing
    d. support headers

15. Furring channels in a ceiling system can be attached to the bar joists, and must be _____.

    a. welded in place
    b. parallel to the joists
    c. adjusted with slip connectors
    d. perpendicular to the joists

Trade Terms Quiz

Fill in the blank with the correct term that you learned from your study of this module.

1. A framing member with a U-shaped cross section is known as a _____.
2. A tool with a _____ handle has small ridges to provide a good gripping surface.
3. A lightweight nonbearing exterior wall, called a _____, is attached to the concrete or steel structure of a building.
4. _____ is a C-shaped track or other components used to transfer shear forces.
5. A structural framing member that may be sloped or horizontal, a _____ supports roof loads.
6. A _____ transfers loads around a window or door opening to supporting structural members.
7. Press braking or continuous roll forming is used to manufacture _____.
8. A(n) _____ is an enclosed space used as an air return channel in heating, ventilating, and air conditioning (HVAC) systems.
9. A horizontal structural member, the _____ is connected to the end of a floor joist.
10. The flanges on a steel framing member are connected by the _____.
11. A(n) _____ is a piece of steel, usually bent at a 90-degree angle, that is used to make connections.
12. _____ is an out-of-plumb condition caused by wind or seismic pressure on a wall.
13. Bracing prevents side-to-side, or _____, movement of wall framing.
14. A basic steel framing member that has a web, two flanges, and two lips is called a _____.
15. A wall that is designed to resist wind or other lateral forces is called a _____.
16. The _____ is a measure of material thickness equal to $1/1000$ inch.
17. A _____ is also known as an edge stiffener.
18. Designed to resist shear forces such as wind, the _____ is an assembly used in a floor, ceiling, or roof.
19. The practice of assembling wall sections, trusses, and other components before placing them in a structure is referred to as _____.
20. The sealed metal casing containing propellant used in a powder-actuated tool is known as a _____.
21. The thickness of steel framing members is typically expressed in _____.

Trade Terms

Blocking
Clip angle
Cold-formed steel
C-shape
Curtain wall
Diaphragm
Gauge

Header
Knurled
Lateral
Mil
Panelization
Plenum
Powder load

Racking
Rim track
Roof rafter
Shear wall
Stiffening lip
Track
Web

Trade Terms Introduced in This Module

Blocking: C-shaped track, brake shape, or flat strap attached to structural members or sheathing panels to transfer shear forces.

Clip angle: An L-shaped piece of steel (normally with a 90-degree bend), typically used for connections.

Cold-formed steel: Sheet steel or strip steel that is manufactured by press braking of blanks sheared from sheets or cut lengths of coils or plates, or by continuous roll forming of cold- or hot-rolled coils of sheet steel.

C-shape: Cold-formed steel shape used for structural and nonstructural framing members consisting of a web, two flanges, and two lips.

Curtain wall: A light, nonbearing exterior wall attached to the concrete or steel structure of the building.

Diaphragm: A floor, ceiling, or roof assembly designed to resist in-plane forces such as wind or seismic loads.

Gauge: Standard measure of metal thickness. Gauge numbers are only an approximation of thickness.

Header: Horizontal structural framing member used over floor, roof, or wall openings to transfer loads around the opening to supporting structural members.

Knurled: A series of small ridges used to provide a better gripping surface on metal and plastic.

Lateral: Running side to side; horizontal.

Mil: Unit of measurement equal to $\frac{1}{1000}$".

Panelization: The process of assembling steel-framed walls, joists, or trusses before they are installed in a structure.

Plenum: A enclosed space, such as the space between a suspended ceiling and an overhead deck, which is used as a return for heating, cooling, and ventilation (HVAC) systems.

Powder load: Metal casing that contains a powderized propellant and is crimped shut on the end and sealed.

Racking: Being forced out of plumb by wind or seismic forces.

Rim track: A horizontal structural member that is connected to the end of a floor joist.

Roof rafter: A horizontal or sloped structural framing member that supports roof loads.

Shear wall: A wall designed to resist lateral forces such as those caused by earthquakes or wind.

Stiffening lip: Part of a C-shape framing member that extends perpendicular from the flange as a stiffening element. Also called an *edge stiffener*.

Track: A framing member consisting of only a web and two flanges. Track web depth measurements are taken to the inside of the flanges.

Web: Portion of a framing member that connects the flanges.

Appendix

Common Terms Used in Cold-Formed Steel Framing Work

AISC: American Institute of Steel Construction.

AISI: American Iron and Steel Institute.

Base steel thickness: The thickness of bare steel exclusive of all coatings.

Bracing: Structural elements that are installed to provide restraint or support (or both) to other framing members so that the complete assembly forms a stable structure.

Ceiling joist: A horizontal structural framing member that supports ceiling components and may be subject to attic loads.

Cold-formed sheet steel: Sheet steel or strip steel that is manufactured by press braking of blanks sheared from sheets or cut lengths of coils or plates, or by continuous roll forming of cold- or hot-rolled coils of sheet steel. Both forming operations are performed at ambient room temperature, that is, without any addition of heat such as would be required for hot forming.

Cold-formed steel: *See cold-formed sheet steel.*

Component assembly: A fabricated assemblage of cold-formed steel structural members that is manufactured by the component manufacturer, which may also include structural steel framing, sheathing, insulation, or other products.

Component design drawing: The written, graphic, and pictorial definition of an individual component assembly, which includes engineering design data.

Component designer: The individual or organization responsible for the engineering design of component assemblies.

Component manufacturer: The individual or organization responsible for the manufacturing of component assemblies for the project.

Component placement diagram: The illustration supplied by the component manufacturer identifying the location assumed for each of the component assemblies, which references each individually designated component design drawing.

Cripple stud: A stud that is placed between a header and a window or door head track, a header and a wall top track, or a window sill and a bottom track to provide a backing to attach finishing and sheathing material.

Design thickness: The steel thickness used in design that is equal to the minimum base steel thickness divided by 0.95.

Edge stiffener: That part of a C-shape framing member that extends perpendicular from the flange as a stiffening element.

Erection drawings: *See installation drawings.*

Erector: See *installer*.

Flange: That portion of the C-shape framing member or track that is perpendicular to the web.

Floor joist: A horizontal structural framing member that supports floor loads and superimposed vertical loads.

Framing contractor: See *installer*.

Framing material: Steel products, including but not limited to structural members and prefabricated structural assemblies, ordered expressly for the requirements of the project.

General contractor: See *installer*.

Harsh environments: Coastal areas where additional corrosion protection may be necessary.

In-line framing: Framing method where all vertical and horizontal load-carrying members are aligned.

Installation drawings: Field installation drawings that show the location and installation of the cold-formed steel structural framing.

Installer: Party responsible for the installation of cold-formed steel products.

Jack stud: A stud that does not span the full height of the wall and provides bearing for headers. Also called a *trimmer stud*.

King stud: A stud, adjacent to a jack stud, that spans the full height of the wall and supports vertical and lateral loads.

Lip: See *edge stiffener*.

Material supplier: An individual or entity responsible for furnishing framing materials for the project.

Nonstructural member: A member in a steel framed assembly that is limited to a transverse load of not more than 10 lb/ft² (480 Pa); a superimposed axial load, exclusive of sheathing materials, of not more than 100 lb/ft (1,460 N/m); or a superimposed axial load of not more than 200 lb (890 N).

Punchout: A hole made during the manufacturing process in the web of a steel framing member.

Shop drawings: Drawings for the production of individual component assemblies for the project.

Span: The clear horizontal distance between bearing supports.

Standard cold-formed steel structural shapes: Cold-formed steel structural members that meet the requirements of the SSMA *Product Technical Guide*.

Strap: Flat or coil sheet steel material typically used for bracing and blocking that transfers loads by tension and/or shear.

Structural engineer-of-record: The design professional who is responsible for sealing the contract documents, which indicates that the structural engineer-of-record has performed or supervised the analysis, design, and document preparation for the structure and has knowledge of the requirements for the load-carrying structural system.

Structural member: A floor joist, rim track, structural stud, wall track in a structural wall, ceiling joist, roof rafter, header, or other member that is designed or intended to carry loads.

Structural stud: A stud in an exterior wall or an interior stud that supports superimposed vertical loads and may transfer lateral loads, including full-height wall studs, king studs, jack studs, and cripple studs.

Stud: A vertical framing member in a wall system or assembly.

Trimmer: See *jack stud*.

Truss: A coplanar system of structural members joined together at their ends, usually to construct a series of triangles that form a stable beam-like framework.

Yield strength: A characteristic of the basic strength of the steel material defined as the highest unit stress that the material can endure before permanent deformation occurs, as measured by a tensile test in accordance with *ASTM A370, Standard Test Methods and Definitions for Mechanical Testing of Steel Products*.

Additional Resources

This module presents thorough resources for task training. The following resource material is suggested for further study.

American Iron and Steel Institute (AISI). Provides a variety of resources related to the use of cold-formed steel for construction applications, **http://www.steel.org**

ASTM A370, Standard Test Methods and Definitions for Mechanical Testing of Steel Products. 2012. ASTM International.

National Electrical Code®. 2011. National Fire Protection Association.

National Fire Protection Association (NFPA). Works to reduce the worldwide burden of fire and other hazards by providing and advocating codes and standards, research, training, and education, **http://www.nfpa.org**

Product Technical Guide. 2012. Steel Stud Manufacturers Association.

Standard for Cold-Formed Steel—General Provisions. 2001. American Iron and Steel Institute.

Steel Framing Alliance (SFA). An advocate of cold-formed steel structures, **http://www.steelframing.org**

Steel Stud Manufacturers Association (SSMA). A manufacturers' trade group that promotes the use of steel framing members in structures, **http://www.ssma.com**

Thermal Design and Code Compliance for Cold-Formed Steel Walls. 2008. Steel Framing Alliance.

Figure Credits

Steel Framing Alliance, Figure 2, Figure 4, Figure 10, Figure 11, Figure 14, Figure 16, Figure 17, Figure 20, Figure 21, Figure 22, Figure 23, Figure 24, Figure 25, Figure 26, Figure 27, Figure 28, Figure 29, Figure 30, Figure 31, Figure 34, Figure 35, Figure 36, Figure 37, Figure 38, Figure 39 right, Figure 41, Figure 42, Figure 43, Figure 44, Figure 46, Table 1

Senco Products, Inc., Figure 7

Simpson Strong-Tie Company, Inc., Figure 8

Klein Tools, Inc., Figure 13

Chicago Metallic Corporation, Figure 33, Figure 39 left

Erico, Inc., SA01

Section Review Answer Key

Answer	Section Reference	Objective
Section One		
1.a	1.1.0	1a
2.b	1.2.1	1b
3.a	1.3.0	1c
4.d	1.4.3	1d
Section Two		
1.a	2.1.1	2a
2.a	2.2.1	2b
3.d	2.3.1	2c
Section Three		
1.b	3.1.0	3a
2.d	3.2.0	3b

NCCER CURRICULA — USER UPDATE

NCCER makes every effort to keep its textbooks up-to-date and free of technical errors. We appreciate your help in this process. If you find an error, a typographical mistake, or an inaccuracy in NCCER's curricula, please fill out this form (or a photocopy), or complete the online form at **www.nccer.org/olf**. Be sure to include the exact module ID number, page number, a detailed description, and your recommended correction. Your input will be brought to the attention of the Authoring Team. Thank you for your assistance.

Instructors – If you have an idea for improving this textbook, or have found that additional materials were necessary to teach this module effectively, please let us know so that we may present your suggestions to the Authoring Team.

NCCER Product Development and Revision
13614 Progress Blvd., Alachua, FL 32615

Email: curriculum@nccer.org
Online: www.nccer.org/olf

❏ Trainee Guide   ❏ Lesson Plans   ❏ Exam   ❏ PowerPoints   Other _____

Craft / Level: _____   Copyright Date: _____

Module ID Number / Title: _____

Section Number(s): _____

Description: _____

_____

_____

_____

_____

Recommended Correction: _____

_____

_____

_____

_____

Your Name: _____

Address: _____

_____

Email: _____   Phone: _____

This page is intentionally left blank.

Exterior Finishing

OVERVIEW

Like roofing materials, there is a wide variety of finishing materials used on the exteriors of homes and commercial buildings. They include wood, brick, vinyl, metal, and fiber-cement board. Each type of material has its own preparation requirements and installation practices. Exterior finishing materials are not installed just to make the building attractive; they serve to protect the building from the elements. Finish material that is not installed per the manufacturer's instructions can also result in voiding the warranty. Another element of exterior finish is the closing of roof overhangs with soffit and fascia.

Module 27204

Trainees with successful module completions may be eligible for credentialing through NCCER's National Registry. To learn more, go to **www.nccer.org** or contact us at **1.888.622.3720**. Our website has information on the latest product releases and training, as well as online versions of our *Cornerstone* magazine and Pearson's product catalog.

Your feedback is welcome. You may email your comments to **curriculum@nccer.org**, send general comments and inquiries to **info@nccer.org**, or fill in the User Update form at the back of this module.

This information is general in nature and intended for training purposes only. Actual performance of activities described in this manual requires compliance with all applicable operating, service, maintenance, and safety procedures under the direction of qualified personnel. References in this manual to patented or proprietary devices do not constitute a recommendation of their use.

Copyright © 2013 by NCCER, Alachua, FL 32615, and published by Pearson, New York, NY 10013. All rights reserved. Printed in the United States of America. This publication is protected by Copyright, and permission should be obtained from NCCER prior to any prohibited reproduction, storage in a retrieval system, or transmission in any form or by any means, electronic, mechanical, photocopying, recording, or likewise. To obtain permission(s) to use material from this work, please submit a written request to NCCER Product Development, 13614 Progress Blvd., Alachua, FL 32615.

27204 V5

From *Carpentry, Trainee Guide*. NCCER.
Copyright © 2013 by NCCER. Published by Pearson. All rights reserved.

27204
EXTERIOR FINISHING

Objectives

When you have completed this module, you will be able to do the following:

1. Describe the safety hazards when working with exterior finish materials.
   a. Identify safety hazards that are present when working at elevations.
   b. Describe safety hazards when working with hand and power tools, equipment, and exterior finishing materials.
2. Describe the various types and applications of exterior finish materials.
   a. Identify the types of wood siding.
   b. Identify vinyl and metal siding materials and components.
   c. List applications for fiber-cement siding.
   d. Discuss the types of veneer finishes.
   e. List specialty exterior finishes.
   f. Explain the purpose of flashing.
3. Explain how to install exterior finish materials.
   a. Describe surface preparation that must be performed prior to installing exterior finish materials.
   b. Discuss the types of furring and insulation that might be applied to exterior walls.
   c. Explain how to establish a straight reference line.
   d. Describe how to install wood siding.
   e. Describe how to install vinyl and metal siding.
   f. Describe how to install fiber-cement siding.
   g. Explain how to install cornices.
4. Describe the estimating procedure for exterior finish projects.
   a. Explain how to perform a takeoff on panel and board siding.

Performance Tasks

Under the supervision of your instructor, you should be able to do the following:

1. Install three of the most common siding types in your area.
2. Estimate the amount of lap or panel siding required for a structure.

Trade Terms

Board-and-batten siding	Fascia	Rabbet
Brown coat	Finish coat	Rake
Building paper	Frieze board	Safety data sheet (SDS)
Cornice	Ledger	Scratch coat
Course	Lookout	Soffit
Duty rating	Louver	Veneer
Eave	Plancier	Vent

Code Note

Codes vary among jurisdictions. Because of the variations in code, consult the applicable code whenever regulations are in question. Referring to an incorrect set of codes can cause as much trouble as failing to reference codes altogether. Obtain, review, and familiarize yourself with your local adopted code.

Industry-Recognized Credentials

If you're training through an NCCER-accredited sponsor, you may be eligible for credentials from NCCER's Registry. The ID number for this module is 27204. Note that this module may have been used in other NCCER curricula and may apply to other level completions. Contact NCCER's Registry at 888.622.3720 or go to **www.nccer.org** for more information.

Contents

Topics to be presented in this module include:

1.0.0 Exterior Finishing Safety 1
  1.1.0 Working at Elevations 1
    1.1.1 Ladders 1
    1.1.2 Scaffolds 2
    1.1.3 Aerial Lifts 2
  1.2.0 Tools, Equipment, Materials, and Related Safety 3
    1.2.1 Hand and Power Tools 3
    1.2.2 Equipment 4
    1.2.3 Material Safety 5
2.0.0 Types and Applications of Exterior Finishing Materials 6
  2.1.0 Wood Siding 6
    2.1.1 Beveled Siding 10
    2.1.2 Board-and-Batten Siding 10
    2.1.3 Reverse Batten (Board-on-Batten) Siding 10
    2.1.4 Board-on-Board Siding 11
    2.1.5 Tongue-and-Groove Siding 12
    2.1.6 Shiplap Siding 13
    2.1.7 Shingle Siding or Shakes 13
    2.1.8 Panelized Shakes/Shingles 14
    2.1.9 Plywood Siding 14
  2.2.0 Vinyl and Metal Siding 14
    2.2.1 Vinyl and Metal Siding Materials and Components 15
  2.3.0 Fiber-Cement Siding 16
  2.4.0 Veneer Finishes 16
    2.4.1 Stucco (Cement) Finishes 17
    2.4.2 Brick and Stone Veneer 18
  2.5.0 Specialty Finishes 19
  2.6.0 Flashing 20
3.0.0 Installing Exterior Finishing Materials 23
  3.1.0 Preparing Surfaces 23
  3.2.0 Furring and Insulation Techniques 23
    3.2.1 Aluminum Foil Underlayment 24
    3.2.2 Window and Door Buildout 24
    3.2.3 Undersill Furring 24
    3.2.4 Undereave Furring 24
  3.3.0 Establishing a Straight Reference Line 25
  3.4.0 Installing Wood Siding 25
    3.4.1 Installing Beveled Siding 25
    3.4.2 Installing Board-and-Batten Siding 28
    3.4.3 Installing Tongue-and-Groove and Shiplap Siding 30
    3.4.4 Installing Shakes and Shingles 31
    3.4.5 Installing Plywood Siding 33

3.5.0 Installing Vinyl and Metal Siding 33
   3.5.1 Installing Inside Corner Posts 34
   3.5.2 Installing Outside Corner Posts 34
   3.5.3 Installing a Starter Strip 34
   3.5.4 Installing Window and Door Trim 35
   3.5.5 Installing Gable End Trim 37
   3.5.6 Cutting Procedures 37
   3.5.7 Installing Siding and Corner Cap 38
   3.5.8 Installing Siding around Windows and Doors 40
   3.5.9 Installing Siding at Gable Ends 41
   3.5.10 Installing Siding under Eaves 43
   3.5.11 Caulking and Cleanup 43
3.6.0 Installing Fiber-Cement Siding 44
3.7.0 Installing Cornices 44
   3.7.1 Aluminum or Vinyl Fascia and Soffits 49
4.0.0 Estimate Exterior Finish Materials 54
4.1.0 Estimating Panel and Board Siding 54
   4.1.1 Estimating Board-and-Batten Siding 54
   4.1.2 Estimating Board-on-Board Siding 55
   4.1.3 Estimating Nails for Siding 55

Figures and Tables

Figure 1 Aerial lifts 2
Figure 2 Siding installation tools 4
Figure 3 Power shears 4
Figure 4 Portable brake 5
Figure 5 Common wood siding styles (1 of 2) 8
Figure 5 Common wood siding styles (2 of 2) 9
Figure 6 Commonly used nails 10
Figure 7 Wood or metal inside corner strips 10
Figure 8 Beveled siding 11
Figure 9 Board-and-batten siding 11
Figure 10 Reverse batten siding 11
Figure 11 Board-on-board siding 12
Figure 12 Applying flat-grain board 12
Figure 13 Tongue-and-groove siding 12
Figure 14 Styles of shiplap siding 13
Figure 15 Wood shingles 14
Figure 16 Surface textures and designs of common plywood siding 16
Figure 17 Typical vinyl and metal siding materials 17
Figure 18 Typical vinyl and metal siding installation components 18
Figure 19 Stucco section 18
Figure 20 Stucco lock 18
Figure 21 Typical DEFS water management 19
Figure 22 Typical flashing installation 21
Figure 23 Cutting sill extensions 23

Figure 24 Furring for vertical siding 24
Figure 25 Undersill furring 24
Figure 26 Undereave furring 25
Figure 27 Straight reference line 25
Figure 28 Installation of plain beveled siding 26
Figure 29 Siding junctures at gable ends 27
Figure 30 Example of incorrect nailing 27
Figure 31 Using a siding gauge 28
Figure 32 Installing siding around windows 29
Figure 33 Metal corner caps 29
Figure 34 Shingle and wood lap siding corner caps 29
Figure 35 Mitered corner 29
Figure 36 Outside corner using corner boards 30
Figure 37 Wall starting board 30
Figure 38 Approaching an installed window 31
Figure 39 Leaving an installed window 31
Figure 40 Installing wood shingles 32
Figure 41 Double coursing a shingle wall 32
Figure 42 Plywood joint suggestions 34
Figure 43 Inside corner post 35
Figure 44 Correct nailing of flanges 35
Figure 45 Installing a starter or J-channel strip 35
Figure 46 Installing aluminum window trim 36
Figure 47 Boxing in sill ends 36
Figure 48 Cutting J-channel 37
Figure 49 J-channel 37
Figure 50 Installing a piece of flashing 37
Figure 51 Gable end trim 37
Figure 52 Using tin snips 38
Figure 53 Reopening a lock after cutting 38
Figure 54 Overlapping panels 39
Figure 55 Sequence of installation 39
Figure 56 Proper staggering of joints 40
Figure 57 Inserting a backer tab for metal siding 40
Figure 58 Leaving room for corner caps 40
Figure 59 Cutting a panel around a window 41
Figure 60 Fitting panels at door and window sills 41
Figure 61 Using a scrap piece to measure clearance 42
Figure 62 Using furring with J-channel 42
Figure 63 Completing the installation at a window top 42
Figure 64 Installing siding at gable ends 43
Figure 65 Trim extends the length of the wall 43
Figure 66 Installing siding under eaves 43
Figure 67 Typical fiber-cement lap siding installation details (1 of 2) 45
Figure 67 Typical fiber-cement lap siding installation details (2 of 2) 46
Figure 68 Typical fiber-cement panel siding installation details 47
Figure 69 Box cornice 48
Figure 70 Types of wood cornice trim molding 48

Figures and Tables

Figure 71 Closed cornice 48
Figure 72 Typical aluminum/vinyl soffit and trim installation 49
Figure 73 Typical aluminum/vinyl soffit and trim materials 51
Figure 74 Nail spacing chart for a plywood closed soffit 51
Figure 75 Determining the ledger position 51
Figure 76 Marking lookout locations on a ledger 52
Figure 77 Determining the length of corner lookouts 52
Figure 78 Installing the soffit and fascia boards 52
Figure 79 Rake soffit cut to wall line 52
Figure 80 Boxing in the cornice return 53

Table 1 Ladder Duty Ratings 2
Table 2 Waste Allowances 54
Table 3 Board-and-Batten Lumber Requirements 54
Table 4 Board-on-Board Lumber Requirements 55

SECTION ONE

1.0.0 EXTERIOR FINISHING SAFETY

Objective

Describe the safety hazards when working with exterior finish materials.
  a. Identify safety hazards that are present when working at elevations.
  b. Describe safety hazards when working with hand and power tools, equipment, and exterior finishing materials.

Trade Terms

Duty rating: Load capacity of a ladder.

Fascia: The exterior finish member of a cornice on which the rain gutter is usually hung.

Safety data sheet (SDS): A document that must accompany any hazardous substance. The SDS identifies the substance and gives the exposure limits, the physical and chemical characteristics, the kind of hazard it poses, precautions for safe handling and use, and specific control measures.

Soffit: The underside of a roof overhang.

Unlike studs, joists, and other framing members, exterior finishing materials are visible to the owner and other building inhabitants. Therefore, it is important to ensure that the finishing materials are not marred or otherwise damaged during storage and installation. Even though the structural integrity of a structure may be sound, an owner may perceive construction to be less-than-satisfactory if the exterior finishing materials are not properly installed in good condition.

The usual work clothing may be worn, keeping in mind that a loose shirt and pants cuffs are a hazard. No jewelry should be worn. Work shoes are probably the most important item of clothing. The shoes should be of the safety type; provide good support; and have a thick, nonskid sole for standing on scaffolds and ladders. Installing exterior finishing materials also presents safety hazards that may not be encountered in other carpentry work. A good carpenter always remembers to put safety first in all situations. Follow all applicable Occupational Safety and Health Administration (OSHA) standards, as well as local and national building codes. A job hazard analysis and fall protection work plan is necessary in order to perform exterior finishing.

When installing exterior finish materials, the weather conditions are often overlooked, and this can be fatal. When working at elevations, surface wind can create a hazardous situation. When carrying large pieces of exterior siding, soffit, or fascia boards, always point the edge into the wind. If wind strikes the flat surface of these items, it could carry you over the edge of the work area and cause serious injuries. Strong winds can also carry away unsecured ladders and scaffolding, causing severe injury to persons working below. Be aware of the danger of snow, ice, and rain. These conditions create slippery surfaces.

1.1.0 Working at Elevations

Falls from elevated surfaces are one of the leading causes of fatalities among construction workers. When installing exterior finishing materials, a carpenter is commonly required to work above ground level on ladders, scaffolds, aerial lifts, or other elevated work platforms. OSHA *Subpart M* requires fall protection for platforms or other work surfaces with unprotected sides or edges that are 6' or higher than the ground or level below it. However, some state OSHA regulations and company policies may require fall protection for heights less than 6'.

1.1.1 Ladders

Ladders are used to install exterior finishing materials and accessories at elevations. Any time work is performed above ground level, there is a risk of accidents. Reduce the risk by carefully inspecting a ladder before use and by using it properly. Check the rungs and rails for cracks or other damage. Do not use the ladder if there is any damage. OSHA requires regular inspections of ladders and an inspection just prior to use. General guidelines for the safe use of ladders are as follows:

- All ladders must be inspected and approved by a competent person.
- A worker must be trained in the use of a ladder.
- Do not stand on or above the second step from the top of a stepladder and fourth rung from the top of an extension ladder.
- Wear shoes with nonslip soles.
- Do not climb a closed stepladder.
- Ensure a ladder is on firm footing.
- Do not climb on the back of a ladder; it is not designed to support a person.

- Do not climb a ladder if you are not physically or mentally up to the task.
- Keep your body centered on the ladder when climbing.
- Climb a ladder while facing it. Always maintain three points of contact with the ladder at all times—two feet and one hand, or two hands and one foot.
- Do not exceed the duty rating of the ladder. See *Table 1*.
- No more than one person is permitted on a ladder at a given time.
- Place an extension ladder at a 75½-degree angle. The bottom of an extension ladder should be set back 1 foot for each 4 feet of ladder length. Tie off the ladder when it is in the proper position.
- Always look up for overhead hazards before climbing a ladder. Climbing near power lines must be avoided.

A complete discussion of ladders and ladder safety is included in the *Core Curriculum*.

1.1.2 Scaffolds

Scaffolds provide safe elevated work platforms for people and materials. Scaffolds are designed to comply with high safety standards, but normal wear and tear or accidentally placing too much weight on scaffolds can weaken them and make them unsafe.

The aluminum adjustable scaffold systems are widely used to provide a working platform when installing exterior finishing materials on large buildings. With these systems, the distance from the building facade remains the same from the bottom to the top. Exact specifications on spacing dimensions, planking, permissible heights and loads, and other details are provided in OSHA regulations.

Inspect all scaffolds before using them. Check for bent, broken, or badly rusted tubes. Check for loose connections. Correct these issues before using the scaffold.

Table 1 Ladder Duty Ratings

Duty Ratings	Load Capacities
Type IAA	375 lb, extra heavy duty/professional use
Type IA	300 lb, extra heavy duty/professional use
Type I	250 lb, heavy duty/industrial use
Type II	225 lb, medium duty/commercial use
Type III	200 lb, light duty/household use

All scaffolds must be placed on a firm footing and leveled. Per OSHA, if a scaffold is more than 10' high, it must be equipped with top rails, midrails, and toeboards, or use personal fall arrest systems. All connections must be pinned to prevent slipping. Some company safety rules may require fall protection at heights above 4 feet.

> **CAUTION**
> Only a competent person—per OSHA definition—has the authority to supervise setting up, moving, and disassembling scaffolds. Only a competent person can approve the use of scaffolds on the job site.

A complete discussion of scaffolds and scaffold safety is included in the *Core Curriculum*.

1.1.3 Aerial Lifts

Aerial lifts are used to raise and lower workers to and from elevated locations. There are two main types of lifts: boom lifts and scissor lifts. Both types are available in various models. Some models can be used for installing exterior finish materials. *Figure 1* shows two types of commonly used aerial lifts.

Boom lifts have a single arm that extends a work platform/enclosure capable of holding one or two workers. Some models have a jointed (articulated) arm that allows the work platform to

BOOM-SUPPORTED WORK PLATFORM (BOOM LIFT)

SELF-PROPELLED ELEVATING WORK PLATFORM (SCISSOR LIFT)

Figure 1 Aerial lifts.

be positioned both horizontally and vertically. Scissor lifts raise a work enclosure vertically by means of crisscrossed supports.

Most models of aerial lifts are self-propelled, allowing workers to move the platform as work is performed. The power to move these lifts is provided by several means, including electric motors; gasoline, propane, or diesel engines; and hydraulic motors.

Only trained and authorized workers may use an aerial lift. Safe operation requires the operator to understand all limitations and warnings, operating procedures, and operator requirements for maintenance of the aerial lift. All startup and shutdown procedures must follow manufacturers' instructions.

Safety precautions unique to aerial lifts must also be followed. Remember that the other safety precautions previously discussed in this module also apply to aerial lifts. Each manufacturer provides specific safety precautions in the operator's manual that is available with the equipment. Specifically, OSHA *CFR (Code of Federal Regulations) 1926.453* defines and governs the use of aerial lifts as expressed in the following precautions:

- Avoid using the lift outdoors in stormy weather or in strong winds. Know the lift's wind limitations.
- Prevent people from walking beneath the work area of the platform.
- Use a personal fall arrest system as required for the type of lift being used. Use approved anchorage points.
- Do not use an aerial lift on uneven ground.
- Lower the lift and lock it into place before moving the equipment. Also, lower the lift, shut off the engine, set the parking brake, and remove the key before leaving it unattended.
- Stand firmly on the floor of the basket or platform. Do not lean over the guardrails of the platform, and never stand on the guardrails. Do not sit or climb on the edge of the basket or use planks, ladders, or other devices to gain additional height.

1.2.0 Tools, Equipment, Materials, and Related Safety

Many common tools and equipment are used during the installation of exterior finish materials. In addition, a few unique tools are used solely for the purpose of installing exterior finish materials.

1.2.1 Hand and Power Tools

The installation of exterior finish materials should be accomplished with the proper tools (*Figure 2*). Power staplers, power saws, power or hand sheet-metal tin snips, utility knives, and hammers are the most common tools used, and each has its own potential dangers. In most cases, work will be done from the surface of a scaffold or the rung of a ladder, so greater caution is required. Other common tools are shown in *Figure 2*.

A carbide-tipped power shears (*Figure 3*) may be used to cut fiber-cement siding. The use of power shears minimizes the amount of toxic silica dust when cutting the material. A power saw using a fine-toothed, carbide-tipped or dry-diamond circular saw blade, or a score-and-snap knife with a tungsten-carbide tip may also be used to cut fiber-cement siding.

Scaffold Systems

Besides sectional, freestanding, manufactured scaffolds that can be assembled in any number of configurations, continuously adjustable aluminum scaffolding systems are available for cornice and siding working heights up to 50'. Typically, an OSHA-recognized system of this type consists of aluminum poles and a standing platform assembly with a safety railing/workbench and safety net. The standing platform can be raised and lowered as an assembly on the supporting poles to obtain the optimum working height. Most systems may be joined both vertically and horizontally for security as well as portability.

27204-13_SA01.EPS

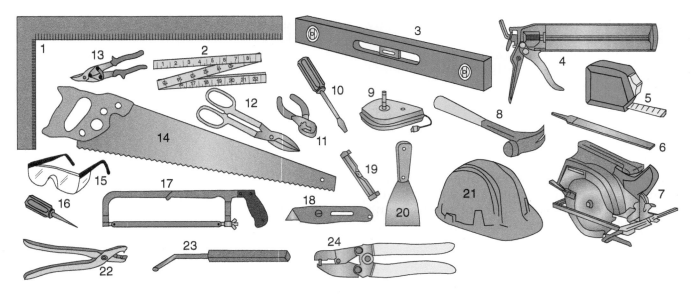

1. Framing square
2. Folding rule
3. 2' – 6' level
4. Caulking gun
5. Tape measure
6. Fine-tooth file
7. Circular saw (optional)
8. Hammer
9. Chalkline
10. Screwdriver
11. Pliers
12. Tin snips (duckbill type) or power hand shears
13. Aviation snips
14. Crosscut saw
15. Safety glasses
16. Steel awl
17. Hacksaw with fine-tooth blade (24 teeth per inch)
18. Utility knife
19. Line level
20. 3" putty knife
21. Hard hat
22. Snap lock punch
23. Vinyl-siding unlocking tool
24. Nail hole punch

Figure 2 Siding installation tools.

ELECTRIC (OR PNEUMATIC) HAND SHEAR

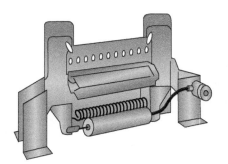

PNEUMATIC SHEAR (LAP SIDING ONLY)

Figure 3 Power shears.

WARNING! Proper respiratory protection must be worn when drilling, cutting, or sanding fiber-cement siding, to avoid the inhalation of the toxic silica dust that can cause the fatal lung disease called silicosis.

1.2.2 Equipment

The following equipment is required when applying metal or vinyl siding:

- *Portable brake* – For job-site bending of custom trim sections such as fascia trim, window casing, and sill trim, a portable brake is extremely useful (*Figure 4*). Utilizing white or colored coil stock, precision bending, including multiple bends, can be accomplished. These machines are lightweight and can be carried to the job site and set in place. Various sizes and brake styles are available. As shown in *Figure 4*, some are equipped with a lengthwise rolling cutter to allow sizing the trim stock to the desired width.

> **WARNING!**
> Exercise extreme care when working with a portable brake. A portable brake has many pinch points, which may result in injury to the user.

- *Cutting table* – A cutting table allows a standard circular saw to be mounted in a carrier and held away from the work to avoid damaging the siding. This table can be used for measuring and crosscutting, as well as for making miters and bevels. The table is constructed of lightweight aluminum and can be easily set up on the job site by one worker.

1.2.3 Material Safety

Some of the most common siding materials are wood, vinyl, metal, fiber-cement, stone, brick, stucco, and vinyl. Each type of material has its own inherent safety issues that should be considered prior to working with the material. General safety precautions similar to those discussed in the *Core Curriculum* should be followed.

Safety precautions must be observed when cutting certain siding materials, including western red cedar, stucco, masonry coatings, treated lumber, and fiber-cement siding. The dust resulting from cutting or mixing such products can be hazardous to inhale or may be an allergen. Safety data sheets (SDSs) furnished by the manufacturers of siding materials must be consulted for any applicable hazards before cutting siding products.

An SDS must accompany every shipment of hazardous substance and must be available at the job site. SDSs are used to properly manage, use, and dispose of hazardous materials safely. Information found on an SDS includes the following:

Figure 4 Portable brake.

- Identity of the substance
- Exposure limits
- Physical and chemical characteristics of the substance
- Type of hazard the substance presents
- Precautions for safe use and handling
- Reactivity of a substance
- Specific control measures
- Emergency first-aid procedures
- Manufacturer contact information

Additional information on SDSs is found in the *Core Curriculum*.

1.0.0 Section Review

1. Fall protection for platforms or work surfaces higher than 6 feet is required by OSHA Subpart _____.
   a. A
   b. F
   c. M
   d. T

2. Information on the use, management, and disposal of a hazardous substance is contained in the _____.
   a. SDS
   b. owner's manual
   c. HSSS
   d. hazard warning document

SECTION TWO

2.0.0 TYPES AND APPLICATIONS OF EXTERIOR FINISHING MATERIALS

Objective

Describe the various types and applications of exterior finish materials.
a. Identify the types of wood siding.
b. Identify vinyl and metal siding materials and components.
c. List applications for fiber-cement siding.
d. Discuss the types of veneer finishes.
e. List specialty exterior finishes.
f. Explain the purpose of flashing.

Trade Terms

Board-and-batten siding: A type of vertical siding consisting of wide boards with the joint covered by narrow strips known as battens.

Brown coat: A coat of plaster with a rough face on which a finish coat will be placed.

Building paper: A heavy paper used for construction work. It assists in weatherproofing the walls and prevents wind infiltration. Building paper is made of various materials and is not a vapor barrier.

Cornice: The construction under the eaves where the roof and side walls meet.

Course: One row of brick, block, or siding as it is placed in the wall.

Finish coat: The final coat of plaster or paint.

Frieze board: A horizontal finish member connecting the top of the sidewall, usually abutting the soffit. Its bottom edge usually serves as a termination point for various types of siding materials.

Rabbet: A groove cut in the edge of a board so as to receive another board.

Rake: The slope or pitch of the cornice that parallels the roof rafters on the gable end.

Scratch coat: The first coat of cement plaster consisting of a fine aggregate that is applied through a diamond mesh reinforcement or on a masonry surface.

Veneer: A brick face applied to the surface of a frame structure.

The primary purpose of any exterior finishing is to provide protection from the elements. Some of the most common siding materials are wood, metal, vinyl, fiber-cement, stone, brick, and stucco. Wood, because of its availability and workability, is the most widely used.

Before the installation of any exterior finishing material, the base material to which the finishing material will be fastened must be made weather resistant. House wrap is usually installed over the plywood or oriented strand board (OSB) exterior sheathing. For house wrap, a 6" to 12" lap at every horizontal joint is required. At corners, a 3' lap from both sides is usually recommended. All window and door openings must be wrapped with the material to prevent water penetration.

After the sheathing and house wrap are installed, all boxed cornices, rake sections, windows, and exterior door frames are installed. Exterior window and door trim, if not part of the assembly, should be installed in the same way as interior trim, which is described in Module Ten, *Window, Door, Floor, and Ceiling Trim*. After any cornices are installed and finished, the siding is applied, and the roof drainage system gutters and downspouts are selected and installed.

Most of the common cornices and wall finishes, along with their installation methods, are covered in Section 3.7.0 of this module. Be sure to check the manufacturers' instructions and local building codes, which may require different or more specific construction/installation methods.

A wide variety of exterior finishing materials are used for residential and commercial buildings, including wood, vinyl, and fiber-cement siding. Veneer and specialty exterior finish materials, such as exterior insulation and finish systems (EIFS), are commonly used on commercial buildings. Depending on the type of material, some materials are installed horizontally, while others are installed vertically.

2.1.0 Wood Siding

The woods used most often for siding are western red cedar (WRC), bald cypress, Douglas fir, western hemlock, western larch, ponderosa pine, red pine, southern white pine, sugar pine, and redwood. These woods are shaped into many different siding styles. *Figure 5* is a summary of the most common styles, sizes, and nailing patterns.

Building Wrap

One of the more popular building wraps is a spun-bonded olefin material such as ProWrap® or Tyvek®. This material is airtight but breathes to allow water vapor to pass from inside a structure to outside. It is very tough and resists tearing and liquid water penetration. Many local codes do not require building wrap, but it is necessary for a wind- and weather-resistant seal. All horizontal and vertical joints must be sealed with a tape approved for that purpose by the manufacturer. Proper installation and taping is especially important to prevent the wrap from being damaged by high winds before the exterior finishing material is completed.

> **WARNING!**
> The dust from western red cedar is an allergen and can cause respiratory ailments including asthma and rhinitis. It can also cause eye irritation and skin disorders, including dermatitis, itching, and rashes. Avoid inhaling the dust or getting it on your skin or in your eyes.

Siding, casing, box, finish, ring-shank, or spiral-shank stainless steel or steel nails with hot-dipped galvanized or noncorrosive coatings (*Figure 6*) are commonly used to apply wood siding. The size will vary from 6d to 10d. The siding nail is considered the best nail for wood siding except in high-wind areas. Then, a spiral-shank nail is required.

Pneumatic nailers can be used to install wood siding. However, a flush-mount attachment should be used or the pressure of the tool should be decreased to prevent overdriving the nails. Depending on the type and thickness of the siding, pneumatic nailers may cause excessive splitting of the siding.

Finishing Projects

Before beginning any exterior finishing project, ensure that you check local codes and manufacturers' instructions for important information that can affect the installation of materials. Such information can include whether aluminum siding must be grounded, if metal drip caps must be installed on cornices, the size of rain gutters, and so forth.

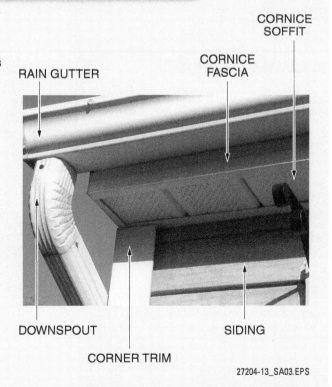

SIDING PATTERNS	SIZES (THICKNESS AND WIDTH)	NAILING 6" AND NARROWER	NAILING 8" AND WIDER
PLAIN BEVELED OR BUNGALOW 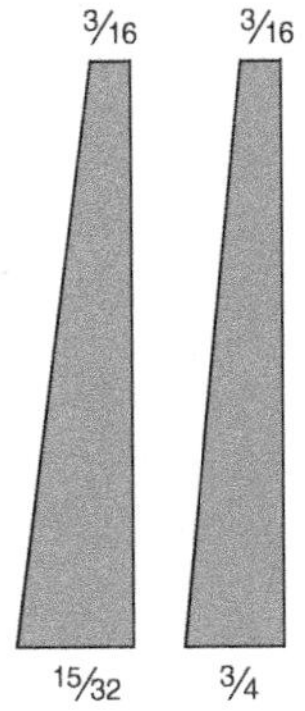 — Bungalow (colonial) is slightly thicker than plain beveled. Either can be used with the smooth or saw-faced surface exposed. Patterns provide a traditional-style appearance. Recommend a 1" overlap. Do not nail through overlapping pieces. Horizontal applications only. Cedar bevel is also available in $7/8$" × 10, 12. (3/16, 3/16, 15/32, 3/4)	$1/2 \times 4$ $1/2 \times 5$ $1/2 \times 6$ $5/8 \times 8$ $5/8 \times 10$ $3/4 \times 6$ $3/4 \times 8$ $3/4 \times 10$	PLAIN 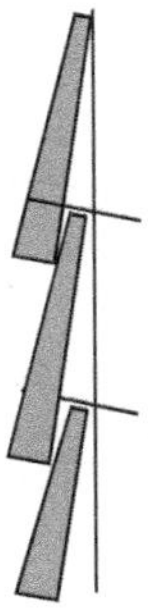 — Recommend 1" overlap. One siding or box nail per bearing, just above the 1" overlap.	PLAIN 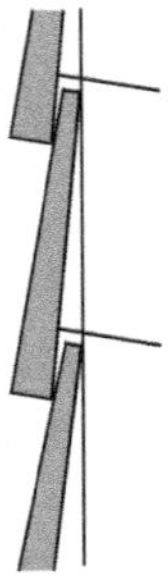 — Recommend 1" overlap. One siding or box nail per bearing, just above the 1" overlap.
DOLLY VARDEN 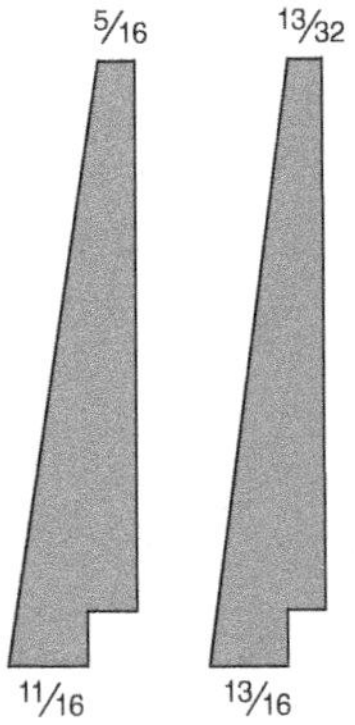 — Dolly Varden is thicker than bevel and has a rabbeted edge. Surface smooth or saw textured. Provides a traditional-style appearance. Allows for $1/2$" overlap, including an approximate $1/8$" gap. Do not nail through overlapping pieces. Horizontal applications only. Cedar Dolly Varden is also available in $7/8$" × 10, 12. (5/16, 13/32, 11/16, 13/16)	STANDARD DOLLY VARDEN $3/4 \times 6$ $3/4 \times 8$ $3/4 \times 10$ THICK DOLLY VARDEN 1×6 1×8 1×10 1×12	RABBETED EDGE 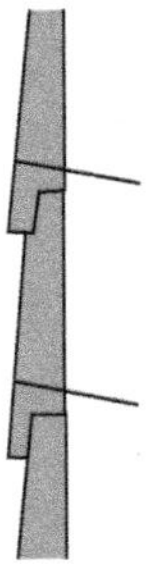 — Allows for $1/2$" overlap. One siding or box nail per bearing, 1" up from bottom edge.	RABBETED EDGE 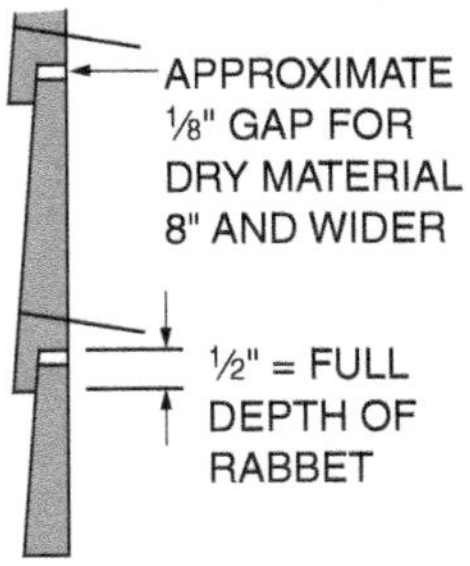 — APPROXIMATE $1/8$" GAP FOR DRY MATERIAL 8" AND WIDER. $1/2$" = FULL DEPTH OF RABBET. Allows for $1/2$" overlap. One siding or box nail per bearing, 1" up from bottom edge.
TONGUE-AND-GROOVE (T&G) — T&G siding is available in a variety of patterns. T&G lends itself to different effects aesthetically. Refer to Western Wood Products Association (WWPA) "standard patterns" (G-16) for pattern profiles. Sizes given are for plain T&G. Do not nail through overlapping pieces. Vertical, diagonal, or horizontal applications.	1×4 1×6 1×8 1×10 NOTE: T&G PATTERNS MAY BE ORDERED WITH $1/4$", $3/8$", OR $7/8$" TONGUES. FOR WIDER WIDTHS, SPECIFY THE LONGER TONGUE AND PATTERN.	PLAIN  — Use one casing nail per bearing to blind nail.	PLAIN  — Use two siding or box nails 3" – 4" apart to face-nail.

Figure 5 Common wood siding styles (1 of 2).

SIDING PATTERNS	SIZES (THICKNESS AND WIDTH)	NAILING	
		6" AND NARROWER	8" AND WIDER
DROP 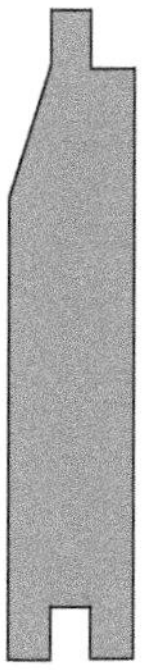 Drop siding is available in 13 patterns, in smooth, rough, and saw-textured surfaces. Some are T&G (as shown), others are shiplapped. Refer to Western Wood Products Association (WWPA) "standard patterns" (G-16) for pattern profiles with dimensions. A variety of looks can be achieved with different patterns. Do not nail through overlapping pieces. Horizontal or vertical applications.	¾ × 6 ¾ × 8 ¾ × 10	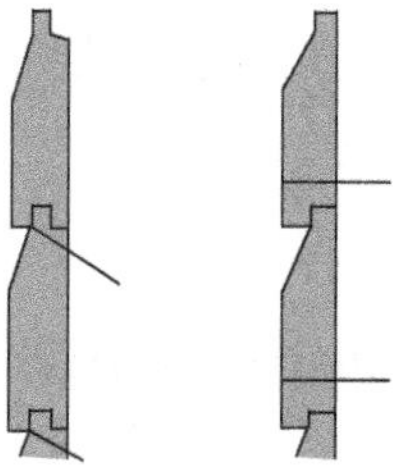 T&G PATTERN / SHIPLAP PATTERN Use casing nails to blind nail T&G patterns, one nail per bearing. Use siding or box nails to face-nail shiplap patterns 1" up from bottom edge.	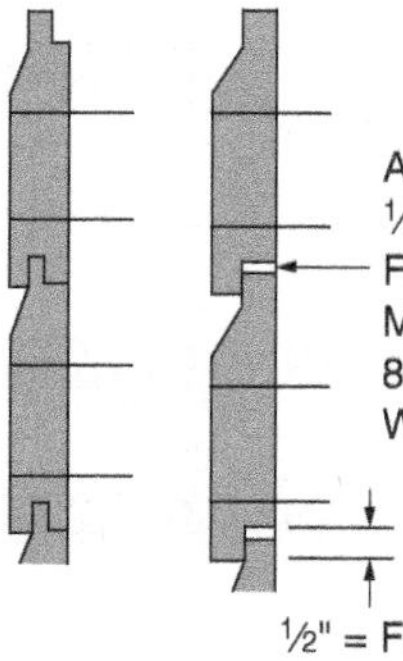 T&G PATTERN / SHIPLAP PATTERN APPROX. ⅛" GAP FOR DRY MATERIAL 8" AND WIDER ½" = FULL DEPTH OF RABBET Use two siding or box nails 3" – 4" apart to face-nail starting 1" up from bottom edge.
CHANNEL RUSTIC Channel rustic has a ½" overlap (including an approximate ⅛" gap) and a 1" to 1¼" channel when installed. The profile allows for maximum dimensional change without adversely affecting appearance in climates of highly variable moisture levels between seasons. Available smooth, rough, or saw textured. Do not nail through overlapping pieces. Vertical, diagonal, or horizontal applications.	¾ × 6 ¾ × 8 ¾ × 10	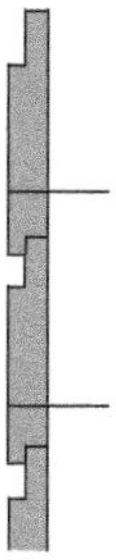 Use one siding or box nail to face-nail once per bearing, 1" up from bottom edge.	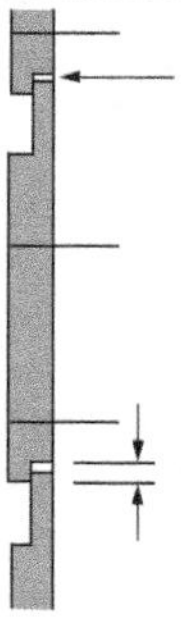 APPROXIMATE ⅛" GAP FOR DRY MATERIAL 8" AND WIDER ½" = FULL DEPTH OF RABBET Use two siding or box nails 3" – 4" apart to face-nail starting 1" up from bottom edge.
LOG CABIN 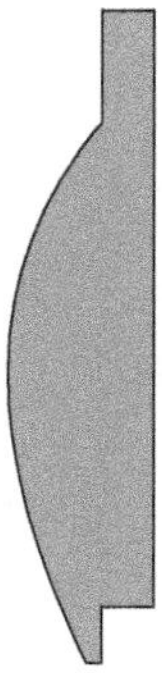 Log cabin siding is 1½" thick at the thickest point. Ideally suited to informal buildings in rustic settings. The pattern may be milled from appearance grades (commons) or dimensional grades (2× material). Allows for ½" overlap, including an approximate ⅛" gap. Do not nail through overlapping pieces. Vertical or horizontal applications.	1½ × 6 1½ × 8 1½ × 10 1½ × 12	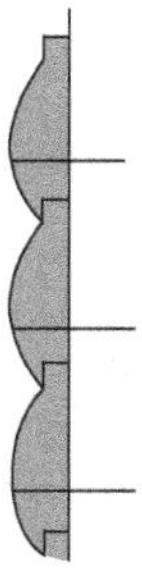 Use one siding or box nail to face-nail once per bearing, 1½" up from bottom edge.	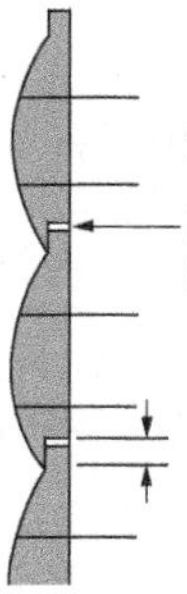 APPROXIMATE ⅛" GAP FOR DRY MATERIAL 8" AND WIDER ½" = FULL DEPTH OF RABBET Use two siding or box nails 3" – 4" apart, per bearing to face-nail starting 1½" up from bottom edge.

Figure 5 Common wood siding styles (2 of 2).

Figure 6 Commonly used nails.

After the flashing is installed and before applying the siding, inside corner strips are usually installed at all inside corners of the structure, as shown in *Figure 7*. In some more costly projects, inside and outside corner pieces are not used and the siding is mitered to fit.

2.1.1 Beveled Siding

Beveled siding is a pattern most often associated with traditional architecture, but it can also be used with success in contemporary structures. Beveled siding is available in plain or bungalow and rabbeted (Dolly Varden) styles (see *Figure 8*). Plain beveled siding produces a strong shadow line. Rabbeted beveled siding provides a somewhat snugger lap and can be installed faster, with a greater coverage than beveled siding. The surfaced side is normally used for painted finishes and the rough side for natural finishes and a more informal look.

2.1.2 Board-and-Batten Siding

Board-and-batten siding is an attractive, versatile, squared-edge siding that is widely used and accepted by architects and contractors throughout the building industry (see *Figure 9*).

Board-and-batten siding is easy to apply and is weathertight. Since it is surfaced on four sides (S4S), it does not require expensive millwork. All of these factors contribute to making it an economical and practical vertical siding.

There are many variations of board-and-batten siding, but the most widely used is the vertical placement of wide boards, with the joints covered by narrow battens. There are a number of different sizes and textures of lumber used.

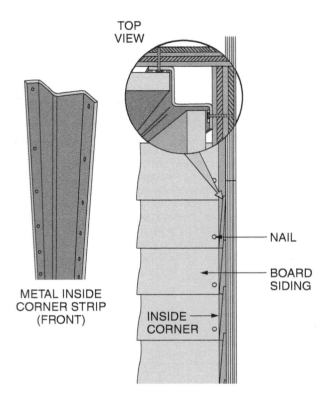

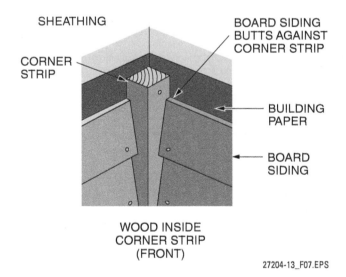

Figure 7 Wood or metal inside corner strips.

2.1.3 Reverse Batten (Board-on-Batten) Siding

Reverse batten (*Figure 10*) is also an attractive vertical siding, giving the building a very sharp, well-defined, deep vertical shadow line. This play of narrow shadow and wide surface creates the illusion that the boards on the surface are free floating. This method can be especially attractive when using rough-sawn boards.

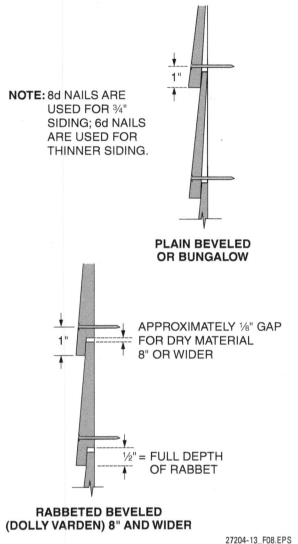

Figure 8 Beveled siding.

2.1.4 Board-on-Board Siding

Board-on-board siding is another type of vertical siding. Not only does this method create a vertical shadow line, but it allows the architect or builder to maintain a uniformity in the width of the material used (see *Figure 11*).

Simulated Board-and-Batten Siding

A board-and-batten or reverse-batten effect can be obtained on an exterior plywood sheathing by covering the vertical seams with a board or batten and then spacing additional boards or battens between the seams for the desired effect. Sometimes a complete board-and-batten system is applied over a wood sheathing.

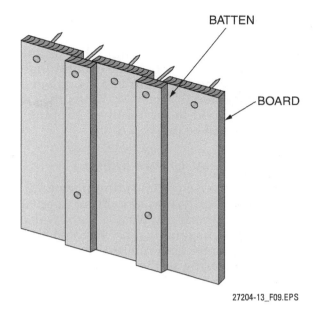

Figure 9 Board-and-batten siding.

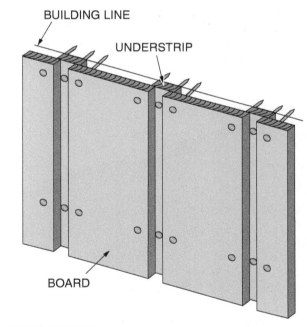

REVERSE BATTEN:
Drive one 8d nail per bearing through the center of the understrip and two 10d nails per bearing through the outer boards.

Figure 10 Reverse batten siding.

For board-on-board siding, apply the underboards first, spacing them to allow a 1½" overlap by the outer boards at both edges. Use standard nailing for underboards, with one 8d nail per bearing. The outer boards must be nailed twice per bearing to ensure proper fastening. Nails having some free length do not hold the outer boards so rigidly as to cause splitting if there is movement from humidity changes. Drive 10d siding nails so that the shanks clear the edges of the underboards by approximately ¼". This provides sufficient bearing for nailing, while allowing clearance for the underboards to expand slightly.

When applying flat-grain boards, orient each board as indicated in *Figure 12*. When the crown surface is exposed to the weather, cupping and grain raising will be prevented.

2.1.5 Tongue-and-Groove Siding

Tongue-and-groove (T&G) siding (*Figure 13*) can be applied vertically, horizontally, or at an angle and provides a perfectly weathertight wall. It is often installed diagonally. The diagonal application of siding creates an interesting exterior pattern and is pleasing to the eye. As shown, T&G siding is generally available in several styles.

On the exterior elevations of the construction drawings, the architect will indicate the location and the application angle. The most common angle is 45 degrees, but make sure of this by checking the plans very carefully. In vertically applied siding, do not use boards larger than 1 × 4 or 1 × 6, because using wider boards will cause problems.

Tongue-and-groove drop siding (*Figure 13*) is normally applied only horizontally or vertically. Horizontal T&G drop siding is more water resistant than plain T&G because the top joint is protected by the overhang of the board above, making water penetration of the joint improbable. Like plain T&G siding, T&G drop siding can be blind-nailed.

The advantage that T&G siding has over shiplap siding is that 6" or narrower boards can be blind-nailed, while shiplap siding cannot. Both types of T&G siding are self-aligning, so they take practically no effort to apply after the first piece is set in the correct position.

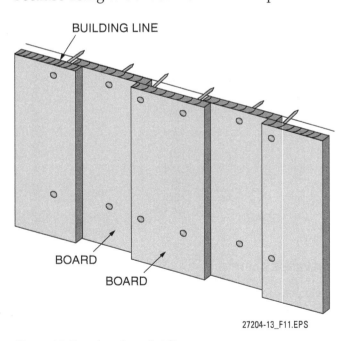

Figure 11 Board-on-board siding.

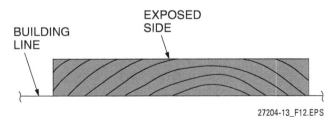

Figure 12 Applying flat-grain board.

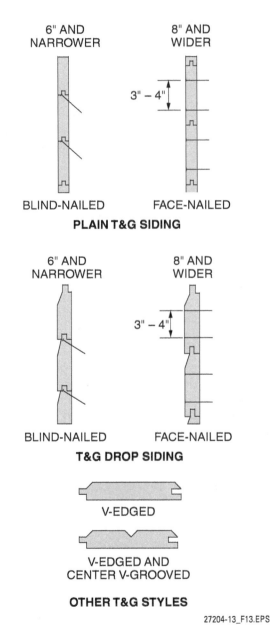

Figure 13 Tongue-and-groove siding.

Architectural Accents

Vertical or diagonal applications of T&G V-edged or V-grooved and shiplap-style siding are often used as architectural accents on interior or external wall surfaces.

2.1.6 Shiplap Siding

Plain shiplap siding can be installed vertically, horizontally, or diagonally. In addition to the plain shiplap patterns, it is generally available in four other patterns, with the most common being the V-edged. Plain shiplap installs with a flush edge. This tends to minimize the direction of the courses and instead accentuates the texture and grain of the wood. On the other hand, V-edged shiplap creates a definite shadow line and indicates the direction of the courses (*Figure 14*). Other styles include drop, channel rustic, and log cabin. They can be used either horizontally or vertically.

Any style shiplap siding that is 6" or narrower can be face-nailed 1" from the bottom with one nail per bearing. Siding that is wider than 8" should be face-nailed with two nails per bearing. The general rule is that the nails should be long enough to penetrate at least 1½" into the studs, or the studs and wood sheathing combined.

Use 8d nails for 1" siding and 6d nails for thinner stock. Nails should be spaced 1½" from the edge of the overlap and 2" from the edge of the underlap for 8" boards. Nail other widths proportionately.

2.1.7 Shingle Siding or Shakes

The use of shingle siding or shakes as a sidewall finish results in a very attractive, rustic, architecturally interesting siding. Red cedar or cypress is normally used to make the shingles or shakes. They are very durable because of their decay resistance. What is referred to as the normal wood shingle is sawn by machine and is manufactured in 16", 18", and 24' vertical lengths. The widths are random. The shingles are tapered, with a butt thickness of ⅜" to ¾" (*Figure 15*). The wood shake is hand-split from a log and is available in taper-split form. Because they are hand split, wood shakes are generally more expensive than sawn shingles, but this price difference is compensated for by their beautiful rustic appearance when applied.

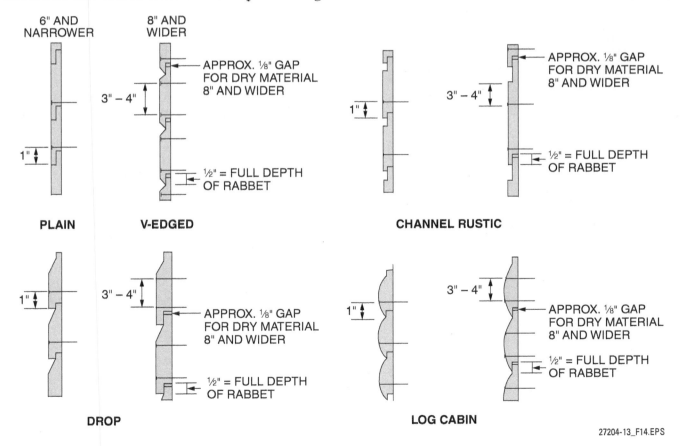

Figure 14 Styles of shiplap siding.

Mating T&G Siding

In some cases, the tongue and groove of T&G siding may have to be forced together to obtain a uniform mating of each course of siding. If necessary, use a hammer and a scrap block of siding on the course being installed to force it onto the tongue of the preceding course. If a board is slightly warped and does not mate evenly along its entire length, secure the board with nails at one or both ends up to the point that the warp begins. Set additional nails in the siding beyond the point of warp. Then, drive a flat, broad chisel into the underlayment-nailing surface with its beveled edge against the siding. Use the chisel as a lever to force the siding into position and then nail the siding into place. Repeat along the length of the board as necessary until the board is seated and secured. Make sure to maintain any inside-groove spacing that may be required.

2.1.8 Panelized Shakes/Shingles

In most cases, panelized shakes/shingles are simply shakes glued or stapled to a backer board of plywood. They are available in widths of 4' and 8'. The panels are available in natural wood and are prefinished in several basic colors of stain. The panels can be applied rapidly, using spiral-shank nails colored to match the panels. The manufacturers' installation instructions should be followed.

2.1.9 Plywood Siding

The use of plywood as an exterior finishing material has been rapidly growing among architects and builders because of the speed of installation and the reliability of the waterproof adhesives being used. The beauty and diversity of the available surfaces have also added to its growing popularity. Because of its strength, plywood is nailed directly to the studs, eliminating the need for sheathing. This is another advantage of plywood siding, because it saves not only the cost of the plywood sheathing, but also the cost of its installation.

Plywood siding is available in thicknesses of ⅜", ½", ⅝", and ¾"; however, ⅝" is the most commonly used. Some of the common textures and designs of plywood siding are shown in *Figure 16*.

Manufacturers generally recommend 7d or 9d siding nails or box nails made of hot-dipped galvanized, aluminum, or stainless steel. Ring-shank and box nails are the types usually specified. The nailing pattern is 6" on center (OC) on the edges and 12" OC in the field. Staples can also be used as long as they are the appropriate size and type.

2.2.0 Vinyl and Metal Siding

Vinyl and metal siding are applied in new construction as well as over existing finishes for remodeling work. Both vinyl and metal siding are manufac-

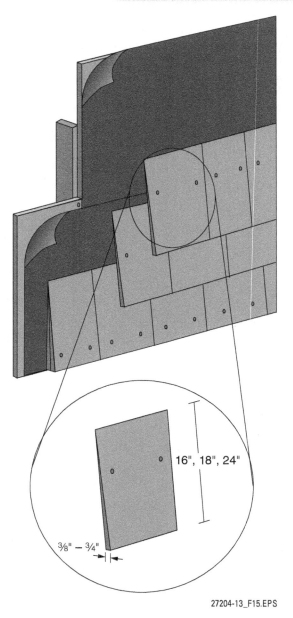

Figure 15 Wood shingles.

tured to look like beveled siding and are available in many colors and finishes. Inside corner posts, door and window trim, individual corner pieces, starter strips, and butt supports are also available. Metal siding and trim are usually supplied with a baked-on or plastic finish. Many manufacturers offer siding with a rigid insulating backing board. This makes the siding less susceptible to exterior damage and also increases its rigidity.

2.2.1 Vinyl and Metal Siding Materials and Components

Figure 17 shows some of the styles of horizontal siding materials currently available. *Figure 18* shows a variety of installation components used with vinyl or metal siding. Horizontal metal siding is usually limited to single- or double-lap styles.

The major advantages of vinyl siding are its low cost, ease of handling and installation, and resistance to denting. It is also colorfast in sunlight, waterproof, decay resistant, and termite proof. Its major disadvantage is that its resistance to impact damage is very low in cold temperatures. If installed improperly, it will break during expansion or contraction over wide temperature variations.

Metal siding resists damage from temperature extremes and is decay resistant and termite proof. However, unlike vinyl, it is susceptible to salt-spray and impact damage. Metal siding is also more difficult to handle and install.

In addition to the components previously listed, the following materials may be required when installing vinyl or metal siding:

- Building wrap or aluminum breather foil
- Touch-up paint in colors to match the siding (for kitchen fans, service cables, etc.)
- Caulking (preferably a butyl caulk)
- Aluminum, plain-shank, or spiral-shank nails (1½" for general use; 2" for re-siding; 2½" or more to nail insulated siding into soffit sheathing; 1" to 1½" trim nails colored to match siding)

A minimum penetration of ¾" (excluding the point of the nail) into solid lumber is required for nailing to be effective with plain-shank nails.

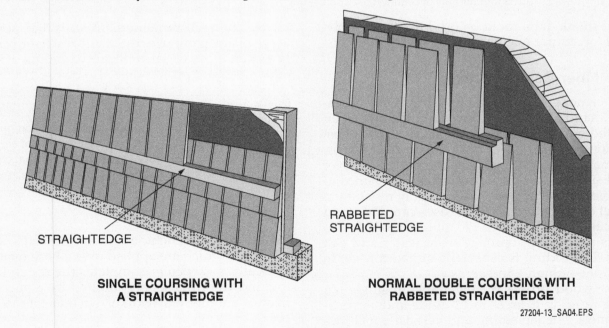

Straightedge Shake/Shingle Application

For single coursing, use straightedges tacked at the butt line to rest shingles on for spacing selection. For double coursing, use straightedges with a rabbeted edge so that the outer course is about ¼" below the inner course. Sort the shakes/shingles for proper seam overlap and lay them butt down on the straightedge. Then, nail the shakes/shingles to the wall. For a ribbon-style double coursing, a reversed straightedge with a deeper rabbet can be used to shift the outer shake/shingle up so that about 1" to 1½" of the lower part of the inner shake/shingle is exposed. With any method, use a shingling hatchet to trim and fit the edges if necessary. Butt ends are not trimmed. If rebutted and rejoined shakes/shingles are used, no trimming should be necessary.

SINGLE COURSING WITH A STRAIGHTEDGE

NORMAL DOUBLE COURSING WITH RABBETED STRAIGHTEDGE

27204-13_SA04.EPS

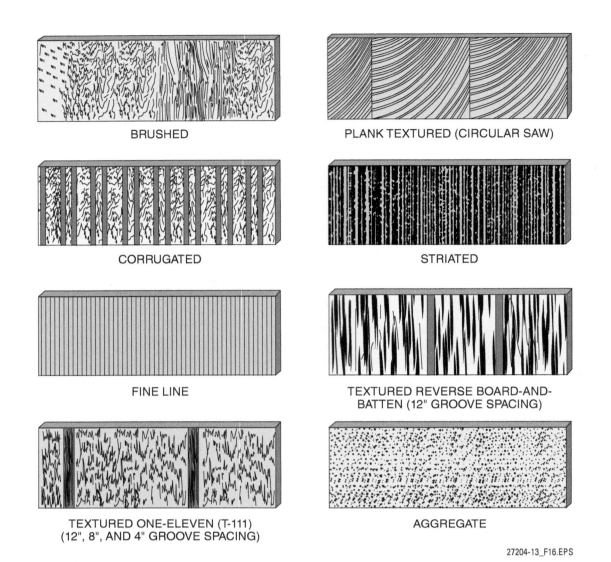

Figure 16 Surface textures and designs of common plywood siding.

Screw-shank nails could be used through ½" plywood for similar effectiveness.

2.3.0 Fiber-Cement Siding

Fiber-cement siding is made using portland cement, sand, fiberglass and/or cellulose fiber, selected additives, and water. It is usually pressure-formed and heat-cured into planks or panels. The major advantage of fiber-cement siding over wood siding is that it is decay resistant, noncombustible, and termite proof. It is highly resistant to impact damage and in some cases can withstand hurricane-force winds of 130 mph or more. It also resists permanent damage from water and salt spray. This siding is especially suited for use in fire-prone or high-wind areas.

Like wood siding, it is available in single-lap siding ranging from 6" to 12" wide and as vertical panels. The lap siding and vertical panels are available with a number of different surface patterns. The recommended finish is 100 percent acrylic latex paint over an alkali-resistant primer; however, gloss or satin oil/alkyd paints over an alkali-resistant primer may also be used.

> **WARNING!** Because dry material will be drilled, cut, and/or abraded, proper respiratory protection must be used when working with fiber-cement siding to avoid inhalation of toxic silica dust that can cause a fatal lung disease called silicosis.

2.4.0 Veneer Finishes

Veneer finishes include stucco and brick/stone materials, which are applied over a base material. Typically, a carpenter completes the rough framing and the finishes are applied by a specialty firm or another craftworker, such as a brick mason.

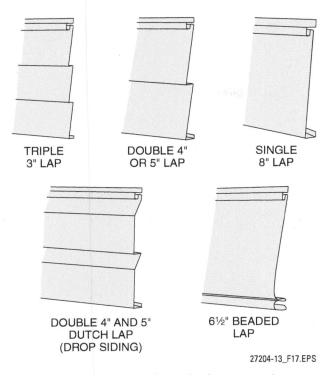

Figure 17 Typical vinyl and metal siding materials.

2.4.1 Stucco (Cement) Finishes

Stucco is a durable cement-based coating for exterior walls that is normally applied by painters or masons.

When applying stucco over frame walls, use wood sheathing, exterior gypsum, or cement board as a base material. **Building paper** is placed over the sheathing and 2 × 8 or 4 × 8 panels of metal reinforcement mesh, called diamond mesh (1⅜" × ⅜" openings), are placed over that. The diamond mesh is applied with special fasteners that hold it away from the wall slightly so that it will become embedded in the first coat of plaster.

> **NOTE**
> When applying stucco to concrete block walls, the diamond mesh is not required, but all other requirements apply.

Three coats of cement plaster must be applied (*Figure 19*). The first coat is called the **scratch coat**, the second is the **brown coat**, and the last is called the **finish coat**. The first two coats can be troweled on with a rough finish. The finish coat can be applied rough or smooth, as noted in the specifications. In many cases, these coats are applied using texture sprayers. Based on Portland Cement Association recommendations, the total thickness of the three coats should be approximately 1".

Control joints must be placed in accordance with the manufacturer's specifications.

When applying wood trim next to stucco, such as the **frieze board** or half timbers used to simulate Tudor architecture, the trim should have a rabbet along the back joining edge, called a stucco lock (*Figure 20*). The stucco lock prevents water penetration around the stucco at the juncture of the stucco and the wood trim.

Fiber-Cement Siding Styles

A number of architectural styles of fiber-cement siding are available and can be used to obtain different effects. Planks are available as smooth or wood grained, and panels are available as plain, stucco, and vertical wood grained.

HARDIPANEL® SMOOTH VERTICAL SIDING

HARDIPANEL® STUCCO VERTICAL SIDING

HARDIPANEL® SIERRA-8 VERTICAL SIDING

HARDIPANEL® SIERRA-4 VERTICAL SIDING

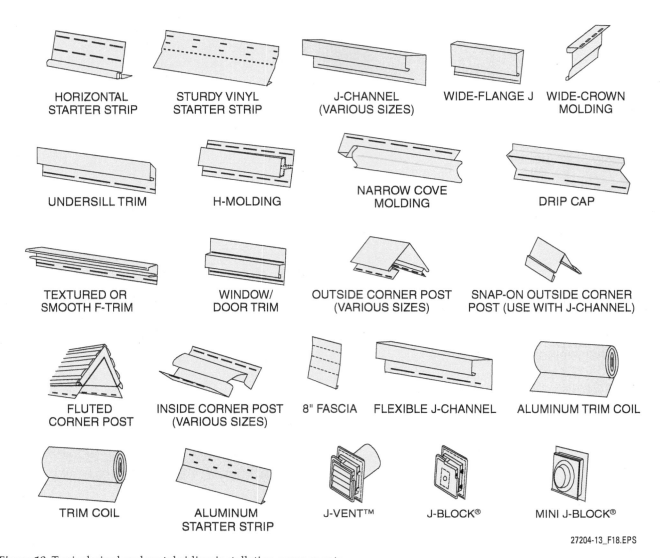

Figure 18 Typical vinyl and metal siding installation components.

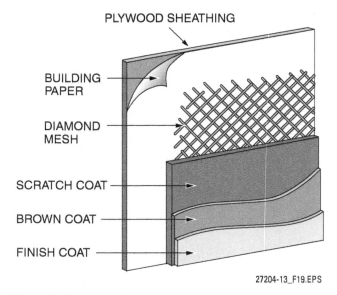

Figure 19 Stucco section.

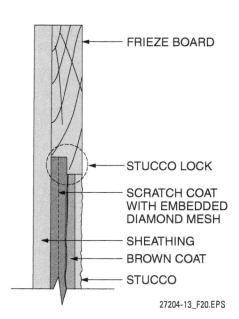

Figure 20 Stucco lock.

Stucco Drying Time

Stucco must be allowed to dry between coats. The amount of time required depends on the environmental conditions that exist at the time the coats are applied. Refer to the manufacturer's instructions for the recommended drying/curing times.

2.4.2 Brick and Stone Veneer

The most durable materials that can be used for exterior wall finishes are brick, synthetic stone, and natural stone. Brick and synthetic stone are factory produced, while stone is a natural material. All three materials are attractive and require little or no maintenance. Brick is made from a clay mixture and is hardened by baking it in a kiln. The nominal size of a brick is 2¼" × 3¾" × 8". This size may vary slightly due to the hardening processes in the kiln. Synthetic stone is made from a cement mixture. Natural stone is either found on the surface or is mined from an open-pit quarry.

2.5.0 Specialty Finishes

Direct-applied exterior finish systems (DEFS) and exterior insulation and finish systems (EIFS) are designed as water-managed systems (*Figure 21*). In appearance, they are similar to traditional stucco or masonry finishes, but they employ different types and applications of material.

Water-managed systems are usually defined as wall cladding systems that:

- Provide specific drainage methods for intruding water that penetrates beyond the cladding
- Provide protection for water-sensitive construction elements
- Are applied to water-durable or water-resistant substrates that can tolerate exposure to water

In most cases, the substrate is a fiber-cement or a fiberglass-coated and treated gypsum panel that can be used over a wood sheathing with underlayment or nonstructural sheathing.

Veneer Moisture Barrier

A building wrap is usually applied over the sheathing. The wrap should overlap the top of the flashing used at the bottom of the wall for water drainage.

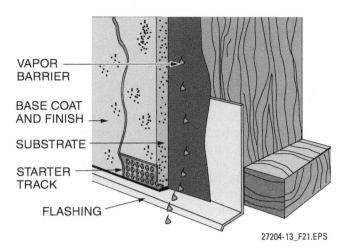

Figure 21 Typical DEFS water management.

All water-managed systems for framed construction or masonry construction have very specific means and methods for flashing and directing incidental water that enters around or through windows, doors, and other openings. For this reason, and to achieve satisfactory performance as well as warranty protection, the manufacturer's instructions must be rigorously followed when installing the components of these systems. In addition, a vapor barrier is usually required on the interior side of exterior walls.

Water-managed DEFS and EIFS wall claddings are almost identical, except that in an EIFS, insulating boards, usually made of expanded polystyrene (EPS), are fastened to the substrate surface and a mesh reinforcement is bonded to the insulation board under the base coat. The insulation boards can vary from 1" to 4" in thickness, depending on the insulation value required.

In addition to the normal finish for DEFS, several masonry facings can also be used. These include ceramic tile set in a latex-fortified grout on top of a latex-fortified mortar base coat and bond coat or exposed aggregate set in an epoxy base coat.

DEFS and EIFS wall-cladding surface finishes are noncombustible, making them ideal for use in fire-prone areas. With two layers of substrate, as well as fire-rated interior insulation and two layers of fire-rated interior drywall, these systems are fire-rated for up to two hours. Like fiber-cement siding, these systems are decay resistant, termite proof, and can withstand high winds.

Normally, the base coat and surface finishes for these systems are applied by painters or masons using texture sprayers. Finishes involving thin brick, tile, or aggregate are usually applied by masons. Carpenters normally install the flashing, water barrier, substrate(s), seam tape, insulation, and/or fiber mesh.

The following are typical general application characteristics/guidelines for these systems:

- When applied to standard wood studs or light-gauge steel studs at 16" OC, the walls can withstand wind loads of 40 pounds per square foot (psf). Greater wind loads are allowable with the use of larger studs or closer stud spacing.
- These systems can be used for ceilings, soffits, curtain walls, bearing walls, panelization elements, and privacy fences.
- The final surface finish materials can be tinted any color.
- These systems cannot be used as sill finishes.
- The substrate may not be used as structural sheathing. Racking resistance must be accomplished by separate bracing.
- All windows, doors, and other openings must be properly flashed.
- DEFS on steel framing must be laterally braced. To enhance crack resistance over steel framing, mesh reinforcement is required over the entire substrate to reinforce the base coat.
- Multiple layering of the substrate can be used to achieve various architectural effects such as banding on various levels of a building.
- The construction drawings and manufacturers' specific instructions must be rigidly followed.

2.6.0 Flashing

Before the exterior finishing materials are installed, flashing must be installed around all openings. The primary purpose of flashing is to prevent water that may penetrate the finishing materials from eventually entering the exterior walls and causing decay, water damage of interior surfaces, or mold and mildew. If water does penetrate the siding, flashing is designed to channel it back out again, thus avoiding any water damage (*Figure 22*).

Flashing usually consists of galvanized sheet metal, aluminum, or a synthetic material; however, on rare occasions, copper and stainless steel may be used. Normally, aluminum is not used for flashing masonry because of corrosion problems.

When brick or stone veneer is used for a frame building, flashing is installed at the base of the sheathing and above door and window openings to channel the water to the outside through weepholes. Frame construction at the water table also requires that flashing be used, as illustrated in *Figure 22*.

Some metal-covered or vinyl-covered windows and doors are manufactured with a flashing flange at the tops and sides, and do not require separately installed flashing; however, it is usually a good practice to install flashing as a precaution.

DEFS/EIFS Codes

Always check state and local building codes before using DEFS or EIFS cladding. Some states and/or localities have prohibited its use on residential structures due to deficiencies in earlier versions of the water-management systems, compounded by faulty installation that resulted in severe damage. Some manufacturers offer improved versions of DEFS and EIFS cladding systems that are designed to alleviate these problems. Make sure that any system used is the latest version offered by the manufacturer. Also, make sure that the manufacturer's instructions for installation are rigorously followed.

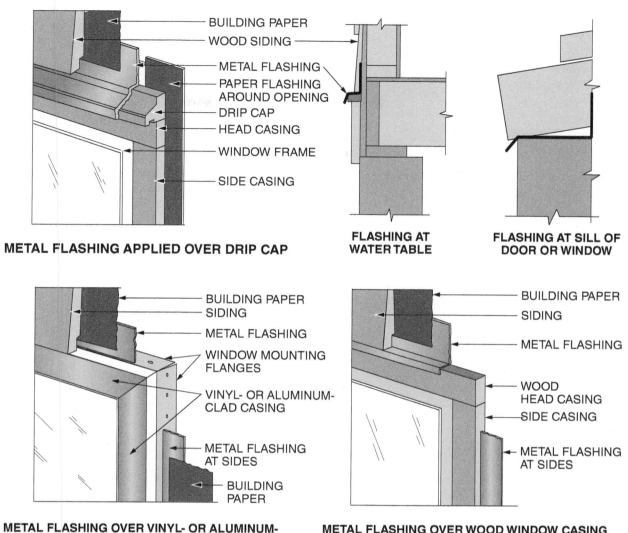

Figure 22 Typical flashing installation.

2.0.0 Section Review

1. In high-wind areas, the best nail for wood siding is the _____.
   a. casing nail
   b. box nail
   c. spiral-shank nail
   d. siding nail

2. Susceptibility to salt-spray damage is a disadvantage of _____.
   a. red cedar siding
   b. metal siding
   c. vinyl siding
   d. fiber-cement siding

3. When cutting fiber-cement siding, proper respiratory protection must be worn to prevent inhaling toxic _____.
   a. fumes
   b. asbestos fibers
   c. silica dust
   d. cement particles

4. When applying stucco, the three coats should have a total thickness of _____.
   a. ⅜"
   b. ¾"
   c. 1"
   d. 1½"

5. EIFS and DEFS are designed as _____.
   a. waterproofing systems
   b. environmental control systems
   c. hydrostatic diversion systems
   d. water-managed systems

6. On brick or stone veneer walls, flashing is used to channel water to the outside through _____.
   a. sole plate vents
   b. weepholes
   c. drip grooves
   d. moisture wicking

Section Three

3.0.0 Installing Exterior Finishing Materials

Objective

Explain how to install exterior finish materials.
a. Describe surface preparation that must be performed prior to installing exterior finish materials.
b. Discuss the types of furring and insulation that might be applied to exterior walls.
c. Explain how to establish a straight reference line.
d. Describe how to install wood siding.
e. Describe how to install vinyl and metal siding.
f. Describe how to install fiber-cement siding.
g. Explain how to install cornices.

Performance Task 1

Install three of the most common siding types in your area.

Trade Terms

Eave: The lower part of a roof, which projects over the side wall.

Ledger: A board to which the lookouts are attached and which is placed against the outside wall of the structure. It is also used as a nailing edge for the soffit material.

Lookout: A member used to support the overhanging portion of a roof.

Louver: A slatted opening used for ventilation, usually in a gable end or a soffit.

Plancier: Similar to a soffit, but the member is usually fastened to the underside of a rafter rather than the lookout.

Vent: A small opening to allow the passage of air.

Exterior finishing materials must be properly installed to resist penetration from wind and moisture infiltration and to provide a pleasing finished appearance. Many times, home and building owners evaluate the quality of the construction for an entire building solely on the appearance of the exterior finishing materials. Therefore, it is important that the exterior finishing materials are carefully installed.

3.1.0 Preparing Surfaces

The quality of the finished job depends on good preparation of the work surface, especially for remodeling work. Keep the following points in mind when preparing the surface for exterior finishing materials:

- Check for low places in the plane of the wall and build out (shim out) if required.
- Prepare the entire building a few courses at a time. Securely nail all loose boards and loose wood trim. Replace any rotted boards.
- Scrape away old paint buildup, old caulk, and hardened putty, especially around windows and doors where it might interfere with the positioning of new trim. New caulk should be applied to prevent air infiltration.
- Remove downspouts and other items that will interfere with the installation of new siding.
- Tie shrubbery and trees back from the base of the building to avoid damage.
- Window sill extensions may be cut off so that the J-channel can be installed flush with the window casing (*Figure 23*). However, if the building owner wishes to maintain the original window design, coil stock can be custom-formed around the sill instead of cutting away the sill extensions.

3.2.0 Furring and Insulation Techniques

Furring strips may be required in order to provide a smooth, even base for nailing on the new siding. Normally, ⅜"-thick wood lath strips are used over wood construction and 1 × 3 strips are used over brick and masonry. Furring is not usually necessary in new construction, but

Figure 23 Cutting sill extensions.

older homes often have uneven walls, and furring out low spots or shimming can help to prevent the siding from appearing wavy. If possible, it is preferable to remove any old exterior siding down to the sheathing to avoid furring. If extensive furring is used under vinyl siding, a backer board may be required for reinforcement of the siding.

For horizontal siding, the furring should be installed vertically at 16" OC. The air space at the base of the siding should be closed off with strips applied horizontally. Window, door, gable, and eave trim may have to be built out to match the thickness of the wall furring.

The furring for vertical siding is essentially the same as for horizontal siding, except the wood strips are securely nailed horizontally into structural lumber on 16" to 24" centers. When using 1 × 3 furring, be sure to check what effect, if any, the additional thickness will have on the use of trim (see *Figure 24*).

3.2.1 Aluminum Foil Underlayment

Aluminum reflector foil, if used, is a good insulator and can be used advantageously as an underlayment to siding. It may be stapled directly to the existing wall or over ¾" furring strips to provide an additional air space and better insulation. Reflector foil for remodeling projects must be the perforated or breather type to allow for the passage of water vapor.

Foil should be installed with the shiny side facing the air space (outward with no furring; inward if applied over furring). Foil is generally available in 36"- and 48"-wide rolls. Nail or staple the foil just before applying the siding.

When applying foil over furring, be careful not to let the foil collapse into the air space. Place the foil as close as possible to openings and around corners where air leaks are likely to occur. Overlap the side and end joints by 1" to 2".

3.2.2 Window and Door Buildout

Some trim buildout at windows and doors may be required to maintain the original appearance of the house when using furring strips or underlayment board. This is particularly true when the strips or underlayment board are more than ½" thick. Thicker furring and underlayment generally provide added insulation value and are usually a good investment for the homeowner, particularly if the home is under-insulated.

3.2.3 Undersill Furring

Building out below the window sill is often required in order to maintain the correct slope angle if a siding panel needs to be cut to less than full height. The exact thickness required will be apparent when the siding courses have progressed up the wall and reached this point (*Figure 25*).

3.2.4 Undereave Furring

For the same reason, furring is usually required to maintain the correct slope angle if the last panel needs to be cut to less than full height. The exact thickness required will be apparent when the siding courses have progressed up the wall and reached this point (see *Figure 26*).

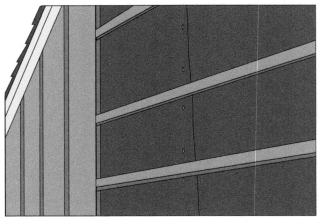

Figure 24 Furring for vertical siding.

Figure 25 Undersill furring.

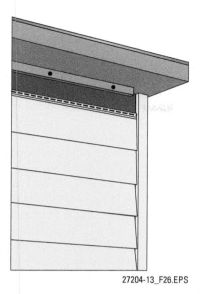

Figure 26 Undereave furring.

3.3.0 Establishing a Straight Reference Line

The key element in a successful siding installation is establishing a straight reference line upon which to start the first course of siding. The suggested procedure is to measure equal distances downward from the eaves. This ensures that the siding appears parallel with the eaves, soffits, and windows, regardless of any actual settling of the house from true level (see *Figure 27*).

To establish a reference line, find the lowest corner of the house. Partly drive a nail about 10" above the lowest corner, or high enough to clear the height of a full siding panel. Stretch a taut chalkline from this corner to a similar nail installed at another corner. Reset this line based upon measuring down equal distances from points on the eaves. Repeat this procedure on all sides of the house until the chalklines meet at all corners.

Before snapping the chalklines, check for straightness. Be alert to sag in the middle of the line, particularly if a line is more than 20' long. If preferred, lines may be left in place while installing the starter strip, as long as they are checked periodically for excess sag.

If the house is level, an alternative is to use a water level or a laser level to set the chalkline approximately 2" (or the width of the starter strip) from the lowest point of the old siding and locate the top of the starter strip at that line. Take the level reading at the corners and centers of the chalkline for best results. The water level can be used for measurements up to 100', and is accurate to ±1⁄16" at 50' feet

3.4.0 Installing Wood Siding

Different installation procedures are used for the various types of wood siding. Installation procedures for the most common type of wood siding are included in this section.

3.4.1 Installing Beveled Siding

Before installing beveled siding, make a story pole that is as long as the height of the wall. Draw a line about 1" below the top of the foundation, but at least 6" above grade level. Follow these guidelines to install the siding:

- Use a divider set at the siding exposure to mark each course of siding on the story pole, then transfer the marks to the wall at the corners, windows, and doors. Make sure the story pole is plumb before marking the wall.

> **NOTE**
> The exposure may be adjusted, as required, so that the spacing to the top comes out even, as long as a minimum overlap of 1" is observed. If possible, also adjust spacing so single pieces of siding will run continuously above and/or below the majority of windows or other wall openings without requiring notching or small slivers of siding above or below the opening.

- It is a good idea to snap a chalkline through the siding location marks on long sides of the building, so that each piece of siding may be nailed to a mark. This will ensure that the siding will have a straight, neat appearance, devoid of waves and sags. If desired, a spacing gauge can be used to place siding, as shown in *Figure 28*. However, a chalkline should be used as a reference.

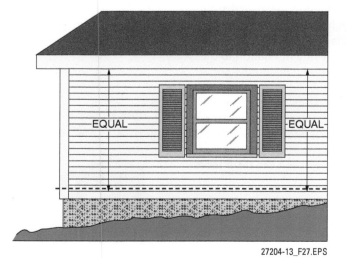

Figure 27 Straight reference line.

CAUTION: Before installing siding, make sure that the sheathing and siding are dry. If excessive moisture is present during the application, later drying and shrinking may cause end gaps and stress warping of the siding.

- Start by placing the bottom course of siding around the perimeter of the building.

NOTE: When using plain beveled siding, place a furring strip behind the bottom edge of the starting course. The strip should be approximately the thickness of the top 1" of the siding. This will allow the siding to project the same distance that the next course of siding will project above it (refer to *Figure 28*). Also, it is important that the lowest edge of the siding be at least 6" above the ground level. The high humidity and free water often present at the base of a foundation because of landscaping can cause finish difficulties and structural problems. It is also very important that the end grain of siding butt joints and the bottom edge of the first course of siding be given a water-repellent treatment.

- When placing the remainder of the siding, do not allow the butt joints to align vertically. Tight-fitting butt joints can be obtained by cutting the siding about 1/16" longer than the measurement. Bow the piece slightly to position the ends, then snap into place.
- If the gable ends will use a different type of finish siding, provide a siding juncture as shown in *Figure 29*.

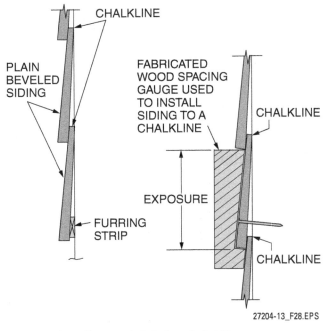

Figure 28 Installation of plain beveled siding.

Reference Lines

For horizontal siding, it is critical that the horizontal lines of vinyl or metal siding are parallel to the soffit/roof line of a structure even if the structure or soffit/roof line is out of level. This is necessary to prevent the horizontal lines of the siding from being at an angle to the soffit/roof line, which would exaggerate the out-of-level structural error. For vinyl or metal siding installation, establishing the initial reference line is even more important than with wood lap siding because the spacing of the courses up to the soffits is not generally adjustable. The position of the line in relation to the soffit/roof lines must be checked and rechecked to make sure that it is equidistant at all points.

On long walls or gable ends of a structure, a laser level can be used to set the lower edge of the siding at intermediate points even if the soffit/roof lines are out of level. This is accomplished by setting the laser level to an out-of-level position (in the plane parallel to the wall) that allows the transit crosshairs to sweep across the same story pole location mark at both outside corners of the wall. However, the laser level must be level in the plane perpendicular to the wall.

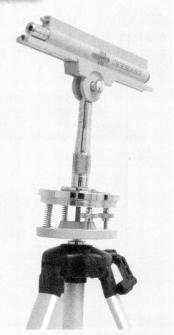

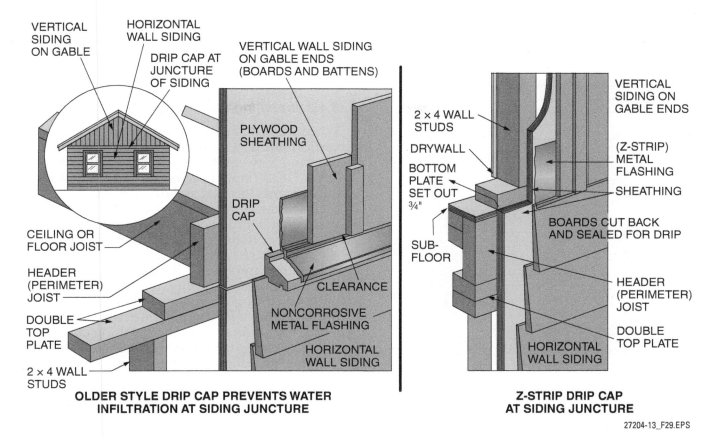

Figure 29 Siding junctures at gable ends.

Beveled siding is face-nailed with one siding nail per bearing (8d for ¾" siding and 6d for thinner siding), so that the shank of the nail clears the tip of the undercourse. Allow ⅛" above the tip for expansion. Lap beveled siding by at least 1" and rabbeted beveled siding by ½". Use aluminum nails, hot-dipped galvanized nails, or stainless steel nails. Do not nail through the overlap (*Figure 30*).

As mentioned previously, plain or bungalow beveled siding must have a minimum lap of at least 1". However, it may be larger to allow for spacing adjustment.

- Use a siding gauge, also known as a preacher, when installing siding to fit between a corner of a wall and a window, two window casings, or a door and window casing (*Figure 31*).

Siding Reference Marks

It may be desirable to set nails at the siding reference marks so that a line can be attached to them for siding alignment. They can also be used for a chalkline if that is the method used to mark a reference line for siding alignment.

When fitting siding around windows, it may be necessary to notch the siding to fit (*Figure 32*). Flashing must be installed at the top of window and door casings to prevent water from infiltrating the structure.

The three ways to finish corners for beveled siding are corner caps (*Figures 33* and *34*), mitered corners (*Figure 35*), and corner boards (*Figure 36*).

Figure 30 Example of incorrect nailing.

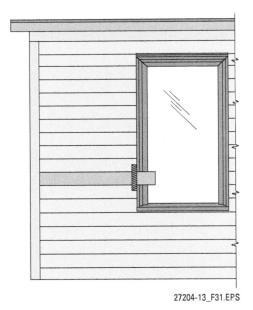

Figure 31 Using a siding gauge.

Mitered corners (*Figure 35*) provide an attractive way to finish corners, but because of the additional labor involved, this method is normally used only on more expensive structures. The angle shown in *Figure 35* is 47 degrees instead of the conventional 45 degrees used for most miter cuts. This prevents gapping at the corners when the siding dries. When applying the siding, force the mitered corners together firmly and nail the mitered ends to the sheathing, not to each other.

Corner boards (*Figure 36*) are thicker than the siding projection and are nailed at all outside corners of the building. The beveled siding simply butts snugly to these boards.

3.4.2 Installing Board-and-Batten Siding

When framing an exterior wall that is to receive vertical siding, it is necessary to install horizontal blocking between the studs, from top to bottom at 24" OC. When applying the siding, space the underboards ½" apart and drive the nails midway between the edges at each bearing. A major advantage of board-and-batten construction is that with proper nailing, the boards are free to move slightly with changes in moisture content, but are held snugly in place by the battens. To allow for this movement capability, only one nail should be used through the center of a board at each bearing. The nails should penetrate 1½" into the studs, the studs and wood sheathing combined, or the blocking. If this depth of penetration is not possible, use ring- or spiral-shank nails for their increased holding power. One 8d nail is nailed midway between the edges of the underboard at each bearing.

With the underboards in place, fasten the battens using 10d nails. These should overlap each edge of the board underneath by at least 1". The nails should be driven directly through the center of the batten so that the shank passes between the underboards.

Siding Gauge

To make a siding gauge, sometimes called a preacher, select a piece of ⅜" or ½" hardwood long enough to accommodate the width of the siding used plus 2¼" and proceed as follows:

Step 1 Center the siding on the block of hardwood (A).
Step 2 Lay out the width of the siding plus ¼" for clearance (B).
Step 3 Lay out the thickness of the siding plus ⅝" (C).
Step 4 Allow 1" around all of the inside cuts (D).
Step 5 Bevel the corners as shown at (E).

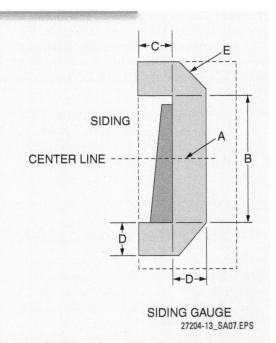

SIDING GAUGE

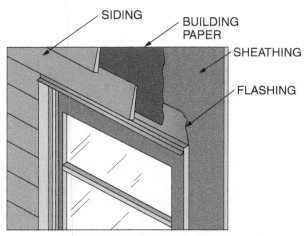

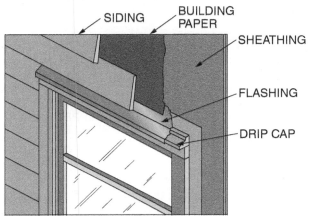

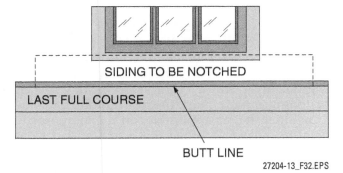

Figure 32 Installing siding around windows.

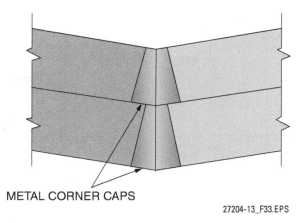

Figure 33 Metal corner caps.

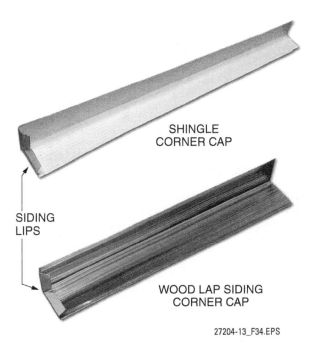

Figure 34 Shingle and wood lap siding corner caps.

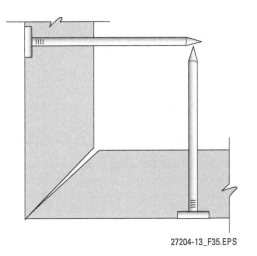

Figure 35 Mitered corner.

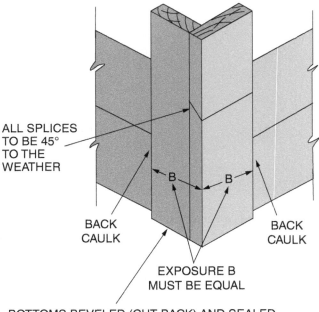

Figure 36 Outside corner using corner boards.

3.4.3 Installing Tongue-and-Groove and Shiplap Siding

Vertical T&G and shiplap siding are installed in similar manners. The following installation hints can be used when installing 6" to 8" (or less) vertical siding:

- Install a wall starting board. Plumb the tongue edge of the board with the grooved edge beyond an outside corner (*Figure 37*) or against an inside corner. Temporarily tack the board in place.
- Mark the board along the length of the corner. For an outside corner, mark along the back side of the board flush with the corner. For an inside corner, mark along the face of the board using a spacer 6" or 7" long and wide enough to extend beyond the groove depth of the board.
- Rip off the grooved edge to the mark and slightly back-bevel the edge.

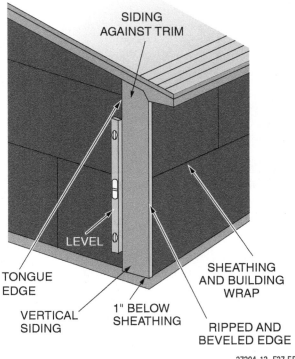

Figure 37 Wall starting board.

- Position the board at the corner with the ripped edge flush with the corner.
- Install the board and recheck the plumb of the tongue edge, then face-nail the board at the corner. Blind-nail the tongue edge.

When approaching an installed window or door, temporarily install and tack a full-width board to the wall just before the window (or door), as shown in *Figure 38*. Using a 6" or 7" length of scrap siding of the same width with just the tongue cut off, mark the top, bottom, and side of the opening, as applicable, on the temporarily installed board. Remove the board and cut out the marked opening. In the same location as the temporary board, permanently install and nail another board with exactly the same width. Then,

Installing Corner Caps

Corner caps are usually used to cover the outside corners of wood siding. They are available in various lengths for different siding widths and are usually installed as each course of siding is applied. Some caps can be installed after all courses are applied. Each course of siding is cut off flush or slightly back from the sheathing on the adjoining wall at each corner and nailed to the sheathing. The lips of the corner cap are tapped up under the course of siding and the cap is nailed to the sheathing above the lower edge of the next course of siding. Always make sure that the cap and siding panels are flush before nailing the cap.

Corner Flashing

As an added precaution in areas of the country subject to wind-driven rain, vertical flashing is sometimes placed around all outside corners for mitered siding or siding that uses corner boards. The flashing should be wide enough so that it extends 3" to 4" on each side of the corner.

install and face-nail the cutout board at the opening edge. Blind-nail the tongue edge.

When leaving an installed window or door, temporarily install and tack two scrap siding pieces (one piece for a door) to the wall as spacers (*Figure 39*). The piece(s) must be wide enough to extend beyond the opening at positions above and below the opening (above for a door). Temporarily install and tack a board of the same width, cut to full length, against the scrap piece(s) with the tongue of the scrap piece(s) inserted. For a door, plumb the tongue edge before tacking. Using a 6" or 7" length of scrap siding of the same width with the tongue cut off, mark the top, bottom, and side of the opening, as applicable, on the temporarily installed board. Remove the board and cut out the marked opening. Remove the scrap spacer(s). Then, install and face-nail the cutout board at the opening edge. Blind-nail the tongue edge.

The wall ending must be planned to prevent ending with a narrow sliver of siding. Stop several feet short of the end of the wall and space off the remaining distance to determine the width of the last board. If random widths are available, use them to allow a reasonably wide ending board; otherwise, rip and regroove several boards to achieve the same effect. Install the boards up to the last board. Then, temporarily install the last board and mark the back side of the board flush with the corner. Rip the board and permanently install it by face-nailing at the corner.

3.4.4 Installing Shakes and Shingles

Before applying wood shingles or shakes, check the local building codes or with the building in-

Furring Strips

If Styrofoam™ or wood fiber insulation (⅝" or more) is used on a wall, some plans require the use of horizontal furring strips to provide an adequate surface for nailing the vertical boards. The furring strips can be nailed on 16" or 24" centers.

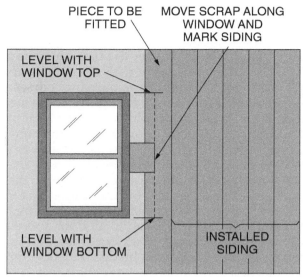

Figure 38 Approaching an installed window.

spector in your area and make sure that the fire codes will permit their use.

Solid nailing or stapling is a must for wood shingles or shakes applied to the exterior sidewalls. The base should consist of plywood, tongue-and-groove, or shiplap sheathing. The sheathing or furring strips should first be covered with building paper, insulation board, and/or house wrap. The two basic methods of shingle sidewall application are single course and double course.

In single-course application, the shingles are applied as in roof construction, but greater weather exposures are permitted. The maximum recommended weather exposures with single-course wall construction are 8½" for 18" lengths and 11½" for 24" lengths. Shingle walls will have

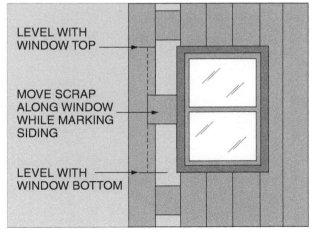

Figure 39 Leaving an installed window.

two plies of shingles at every point, whereas shingle roofs will have three-ply construction.

The first course of shingles may be doubled at the bottom (*Figure 40*). After the first course is applied, lay out the story pole with all of the courses indicated on it. Make sure you have the courses arranged to line up as closely as possible to the top of all door and window openings. Sometimes it may be necessary to change the exposure slightly on the course at the door and window heads and at the window sills so that they will line up. If such an adjustment is necessary, make sure it is slight so that it is not noticeable when viewing the other exposures.

Using a story pole, transfer those markings to the ends of the building and to all door and window openings. Snap a chalkline at long runs. A furring strip or 1 × 2 straightedge may be temporarily tacked to the building at each course so that the shingle butts may be placed on it for alignment before they are nailed. To form closed joints on outside corners of the sidewalls, shingles in adjoining courses may be alternately overlapped and edge shaved to a close fit.

Another method is to miter the two adjoining shakes in each course, but because of the time consumed in doing this, it is usually too expensive. Inside corners may be woven in alternate overlaps or may be closed by nailing a 1 × 1 square molding or corner board in the corner before the shingles are applied.

The nailing for single coursing is accomplished by using 3d (1¼") corrosion-resistant nails, such as hot-dipped zinc, aluminum, or stainless steel. Only two nails are used per shingle, and each is placed approximately ¾" from the side edge of the shingle and approximately 1" above the butt line of the next course. Drive the nails flush, but not so hard that the head crushes the wood.

The double-course method of wood shingle sidewall application (*Figure 41*) is much the same as the single-course method, with a few exceptions. Double coursing allows for the application of extended weather exposure shingles over coursing-grade shingles. Double coursing also provides deep, intense, bold shadow lines. When double coursed, a shingle wall should be tripled at the foundation line by using a double undercourse.

The double-course nailing requires that the outer-course shingle be secured with two 5d (1¾") small-head, corrosion-resistant nails, driven 1" to 2" above the butts, approximately ¾" in from each side. Additional nails are driven about 4" apart across the face of the shingle in a straight line.

The outside corners should be constructed with an alternate overlap of shingles between successive courses. Inside courses may be mitered or woven over a metal flashing, or they may be

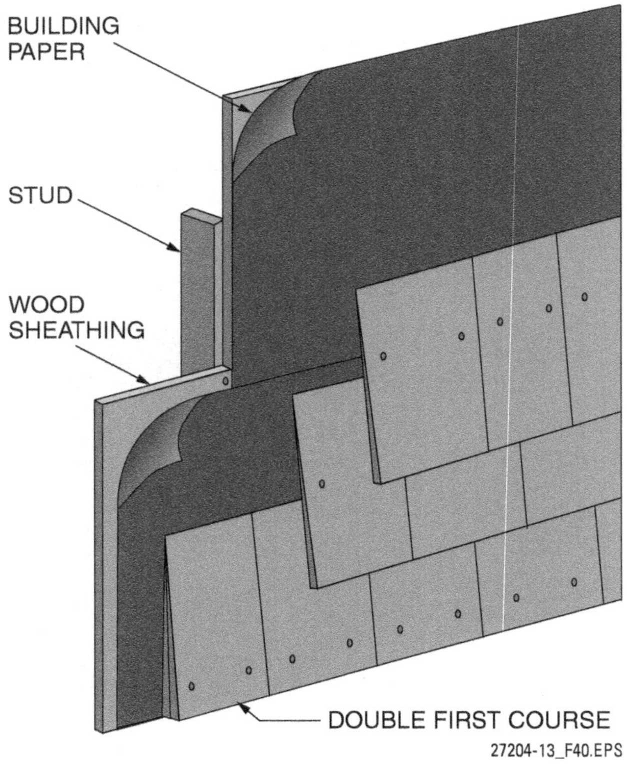

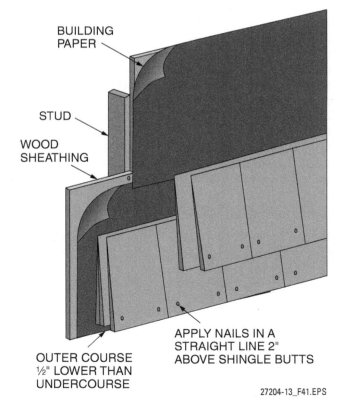

Figure 40 Installing wood shingles.

Figure 41 Double coursing a shingle wall.

made by nailing an S4S (surfaced four sides) 1½" or 2" square strip in the corner, after which the shingles of each course are fastened to the strip.

3.4.5 Installing Plywood Siding

The spacing of wall studs or nailing supports for the ⅜" plywood is a maximum of 16" OC, but thicker plywood (½", ⅝", and ¾") permits a spacing of 24" OC. Blocking is usually required at all horizontal joints (subject to local building codes). Some of the joint suggestions are indicated in *Figure 42*.

Be sure to refer to the manufacturer's technical specifications prior to installing the siding.

To apply lap plywood siding, follow these suggestions:

- If sheathing is not used, place horizontal blocking at 4'-0" centers.
- Install building wrap between the siding and the studs.
- Use a starter strip that is the same thickness as the siding for the first course.
- Coat the edges of the siding with a primer or a water-repellent finish before application.
- Vertical joints should be staggered. These joints must be centered over studs with a tapered wedge at least 1⅝" wide behind the joint.
- Use 8d noncorrosive siding or box nails. Insert one nail at each stud on the bottom edge of the siding.

At all vertical joints, nail 4" OC for siding 12" wide or less. Nail 8" OC for siding 16" wide or more. All nails should be placed ¼" back from the edge of the plywood. Set and putty all casing nails. Box nails are driven flush.

3.5.0 Installing Vinyl and Metal Siding

Prior to the installation of vinyl or metal siding, trim accessories are installed to accept the ends of the siding. Inside corner posts, outside corner posts, starter strips, window and door trim, and gable-end trim pieces are some of these trim accessories.

Shake/Shingle Architectural Accents

Like board-and-batten siding, special styles of shakes and shingles, known as fancy-butt shingles or shakes, are sometimes used on gables as accents similar to those used on Victorian-style homes. In other cases, uniformly or randomly spaced staggered-length shakes/shingles are used to enhance a rustic appearance.

STAGGERED-LENGTH SHAKES/SHINGLES

FANCY-BUTT SHINGLE ACCENT

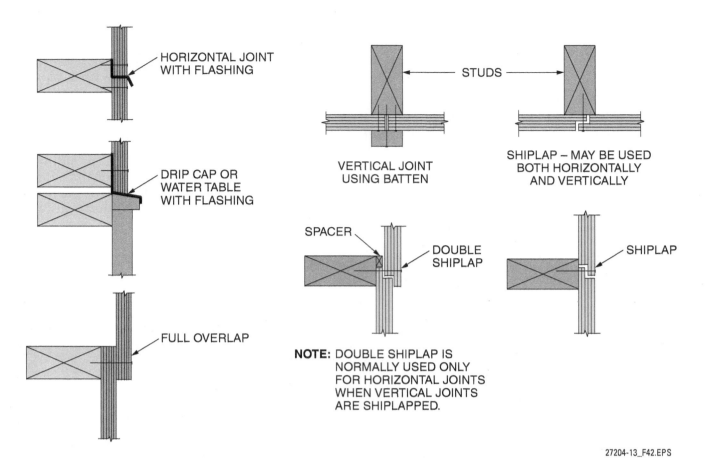

Figure 42 Plywood joint suggestions.

3.5.1 Installing Inside Corner Posts

The inside corner posts are installed before the siding is hung. Depending on the type of siding (insulated or noninsulated), deeper or narrower posts may be required. The post is set in the corner full length, reaching from ¼" below the bottom of the starter strip up to ¼" from the eave or gable trim. Nail the upper slot at the top of the slot, then nail approximately 8" to 12" on both flanges with aluminum nails in the center of the slots. Make sure the post is set straight and true. The flange should be nailed securely to the adjoining wall, but do not overdrive the nails so as to cause distortion. If a short section is required, use a hacksaw to cut it. If a long section is required, the posts should be overlapped, with the upper piece outside.

The siding is later butted into the corner and then nailed into place, allowing approximately a 1/16" to ¼" space between the post and the siding for expansion purposes (*Figure 43*).

3.5.2 Installing Outside Corner Posts

If used, the outside corner post produces a trim appearance and will accommodate the greatest variety of siding types. Most outside corner posts are designed to be installed before the siding is hung, in a manner similar to the inside corner post. If desired, old corner posts may sometimes be removed. Set a full-length piece over the existing corner, running from ¼" below the bottom of the starter strip to ¼" from the underside of the eave. If a long corner post is needed, overlap the corner post sections, with the upper piece outside.

Nail the uppermost slot at the top of the slot, then nail approximately every 8" to 12" with aluminum nails on both flanges in the center of the slots. Make sure the flanges are securely nailed (*Figure 44*), but avoid distortion caused by overdriving nails. Use a hacksaw to cut short sections, if required. If insulated siding is being used, wider corner posts are needed.

3.5.3 Installing a Starter Strip

Using the chalkline previously established as a reference line, take equal-distance measurements, as shown in *Figure 45*, and install the starter strip or J-channel strip all the way around the bottom of the building depending on the material at the base of the building. If insulated siding is

Figure 43 Inside corner post.

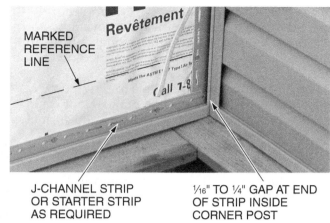

Figure 45 Installing a starter or J-channel strip.

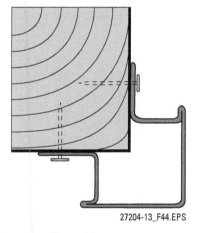

Figure 44 Correct nailing of flanges.

used, the starter strip should be furred out to a distance equal to the thickness of the backer. It is very important that the starter strip be straight and meet accurately at all corners because it will determine the line of all siding panels installed. Where hollows occur in the old wall surface, shim out behind the starter strip to prevent a wavy appearance in the finished siding.

When using individual corner caps, install the starter strip up to the edge of the house corner. Use aluminum nails spaced not more than 8" apart to fasten the starter strip. Nail the starter strip as low as possible. Be careful not to bend or distort the strip by overdriving the nails. The strip should not be nailed tight. Cutting lengths of starter strip is best accomplished with tin snips. Butt the sections together.

Starter strips may not work in all situations. For example, other accessory items, such as J-channels or all-purpose trim, may work better in starting the siding course over garage doors and porches or above brick. These situations must be handled on an individual basis as they occur.

3.5.4 Installing Window and Door Trim

For a superior job in remodeling work, old window sills and casings can be covered with aluminum coil stock that is bent to fit on the job site. The advantage is freedom from maintenance.

Sometimes, window and door casings need to be built out to retain the original appearance of the house or to improve the appearance. To do this, use appropriate lengths and thicknesses of

Hanging Vinyl and Metal Siding

Siding will expand when heated and contract when cooled. The expansion will amount to approximately 1/8" per 10' length for a 100°F temperature change. An allowance for this expansion or contraction should be made when installing siding. If the siding is installed in hot weather, the product is already warm and at least partly expanded; therefore, less room will be required to allow for temperature expansion.

Vinyl and metal siding installation is commonly referred to as hanging siding since nails should not be driven tightly against the siding surface. Rather, the siding is hung from the nails to allow expansion and contraction of the siding.

good-quality lumber and nail them securely to the existing window casings. Remove any storm windows before covering the casings with aluminum coil stock sections custom-formed on the job site.

Forming aluminum sections to fit window casings is done using a portable brake. Door casings are handled in a similar manner. *Figure 46* shows the installation of aluminum window trim.

If there is a step in the wood sill, it can best be covered by bending two separate sill cover pieces with interlocking flanges, as shown in *Figure 47*. By using tin snips and bending flanges on the job, the old sill ends can be boxed in to provide a neat appearance and to prevent water penetration.

J-channel is used around windows and doors to receive the ends of the siding. Side J-channel members are cut longer than the height of the window or door and notched at the top. Notch the top J-channel member at a 45-degree angle and bend the tab down to provide flashing over the side members (*Figure 48*). Caulk should be used behind J-channel members to prevent water infiltration between the window and the channel. See *Figure 49*.

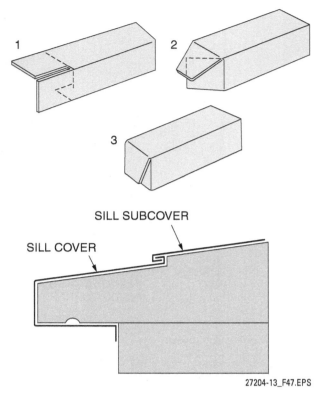

Figure 47 Boxing in sill ends.

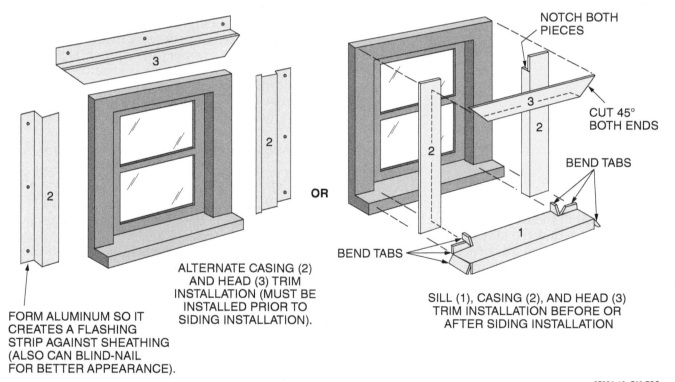

Figure 46 Installing aluminum window trim.

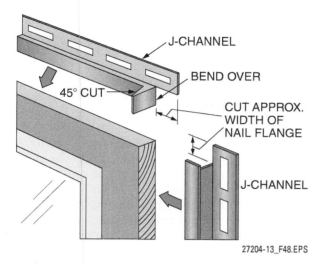

Figure 48 Cutting J-channel.

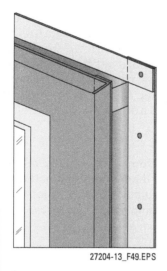

Figure 49 J-channel.

To provide protection against water infiltration, a flashing piece, which is cut from coil stock or a precut piece of step flashing, is slipped under the base of the side J-channel members. It should be positioned so that it overlaps the top lock of the panel below, as shown in *Figure 50*.

3.5.5 Installing Gable End Trim

Before applying the siding, J-channels should be installed to receive the siding at the gable ends, as shown in *Figure 51*. Where the left and right sections meet at the gable peak, allow one of the sections to butt into the peak, with the other section overlapping it. A miter cut is made on the face flange of this piece to provide a better appearance. All old paint buildup should be removed before installing J-channels. Nail the J-channels every 12" using aluminum nails.

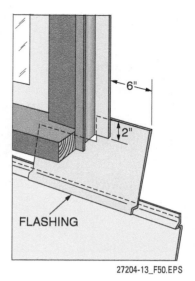

Figure 50 Installing a piece of flashing.

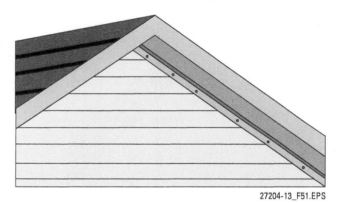

Figure 51 Gable end trim.

3.5.6 Cutting Procedures

For precision cutting, a power saw is the most convenient tool to use. Cutting one panel at a time is recommended. A special jig that will keep the saw base clear of the work is preferred in order to prevent damaging panels. For vinyl, reverse a fine-tooth blade to produce a smoother cut. For aluminum or steel, use a minimum 10-tooth aluminum cutting blade or an abrasive blade. A bar of soap may be rubbed on the blade to produce a smoother cut on the siding panel and prolong blade life. Feed the saw through the work slowly to prevent flutter against the blade. Safety goggles must be worn at all times while operating a power saw.

Individual panels can be cut with tin snips. Start by drawing a line across the panel using a square. Begin cutting at the top lock (*Figure 52*) and continue toward the bottom of the panel. For metal panels, break the panel across the butt edge and snip through the bottom lock. For metal siding, use a screwdriver to reopen the lock, which may become flattened by the tin snips (*Figure 53*).

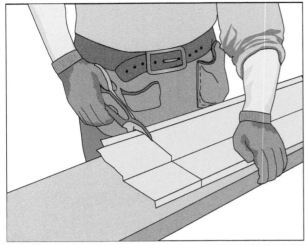

Figure 52 Using tin snips.

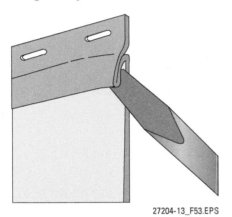

Figure 53 Reopening a lock after cutting.

Aviation snips are sometimes used to cut the top and bottom locks, and a utility knife is used to score and break the face of the panel and to cut vinyl panels. For straight cuts, the best choice is duckbill snips. For aluminum, a heavy score is made on the panel and the piece is bent back and forth until it snaps cleanly along the score line.

On window cutouts, a utility knife and tin snips may be used for both vinyl and aluminum. Use duckbill tin snips to cut accessories such as all-purpose trim, J-channel, and starter strips. Use a hacksaw to cut accessories such as corner posts.

3.5.7 Installing Siding and Corner Cap

For metal siding, check with local building codes to see if they require that the first course of siding be grounded to reduce the danger from lightning. Extra care must be taken when applying the first course of siding because it establishes the base for all other courses. Apply a panel by hooking the bottom lock of the panel into the interlock bead of the starter strip. Make sure the lock is engaged.

Do not force it, which might cause distortion of the panel and result in a warped shadow line. Double-check for continuous locking along the panel before proceeding further. Also, check carefully for proper alignment at the corners.

At the corner posts, slide the panel into the recess first, then exert upward pressure to lock the panel into place along its entire length. Allow clearance for expansion, as necessary. If individual corner caps are being used, keep the panels back from the corner edges (¾" for noninsulated siding and ¼" for insulated siding) to allow for later fitting of the individual corners. Panels must be hung with aluminum nails through the center of the factory-slotted holes every 16" to 24" along their entire lengths or as specified by the manufacturer. Nails must be driven into sound lumber, such as ¾" penetration into house framing with plain-shank nails or through ½" plywood with screw-shank nails. Nails or screws should be set about ¹⁄₁₆" to ⅛" away from the panel. If they are set tight, the panels may warp, bend, or crack due to expansion and contraction. On low spots, fasten the panel on both sides of the low spot and allow the panel to float over the low spot.

> **CAUTION**
> Do not force the panels up or down when nailing them into position. The panels should not be under vertical tension or compression after being nailed into place.

> **NOTE**
> In some cases, prevailing wind direction must also be considered when determining the direction of the siding seam laps. With some vinyl siding, winds in excess of 50 to 60 mph entering under the siding through the seams can tear off multiple siding panels. To prevent this, the seams may have to be lapped in the direction of the prevailing wind in high-wind areas. In addition, the maximum nailing distance must not be exceeded.

On the sides of the building, start at the rear corner and work toward the front so that the lapping will be away from the front and less noticeable. On the front of the building, start at the corners and work toward the entrance door for the same reason. For best appearance when lapping, the factory-cut ends of the panels should cover the field-cut ends.

Metal panels should overlap each other by about ½". A maximum of ⅝" and a minimum of ⅜" is a good rule of thumb. Vinyl manufacturers usually recommend a 1" overlap with a double-size nailing flange cut. Thermal expansion requirements need to be considered in panel overlaps. Cut away the top lock strip on the overlapped panel by twice the amount of the intended overlap (*Figure 54*).

Avoid panel lengths of 24" or less and make sure that the factory-cut ends are always on top of the field-cut ends. The job should start at the rear of the house and work toward the front, as shown in *Figure 55*.

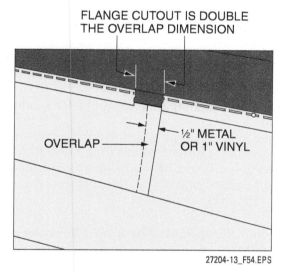

Figure 54 Overlapping panels.

Figure 55 Sequence of installation.

For the best appearance, the staggering of joints should be planned before the installation (*Figure 56*). Avoid installing siding in a set pattern. A set pattern may be more labor- and cost-effective, but results in a poor overall appearance. It is best to plan the job so that any two joints in line vertically will be separated by at least two courses. At a bare minimum, separate panel overlaps on the next course by at least two feet. Joints should be avoided on panels directly above and below windows. Shorter pieces that develop as work proceeds can be used for smaller areas around windows and doors.

A High-Wind-Load Vinyl Siding

One manufacturer offers a flexible-hem vinyl siding that is designed to be installed under vertical tension and nailed tight. This allows the panel to float over low spots and move to accommodate expansion and contraction. When fastened with staples, its specifications indicate that it will withstand winds of up to 235 mph.

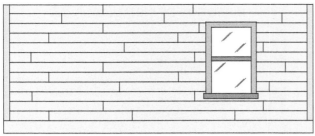

Figure 56 Proper staggering of joints.

Backer tabs are used with 8" horizontal non-insulated aluminum siding only. They ensure rigidity, evenness of installation, and tight endlaps. They are used at all panel overlaps and behind panels entering corners. After the panel has been locked into place, slip the backer tab behind the panel with the flat side facing out (*Figure 57*). The backer tab should be directly behind and even with the edge of the first panel of the overlap. Nail the backer tab into place.

Individual corner caps, if used, may be used for 8" horizontal aluminum lap siding instead of outside corner posts. The siding courses on adjoining walls must meet evenly at the corners. To allow room for the cap, install the siding with ¾" clearance from the corner (¼" clearance for insulated siding). Refer to *Figure 58*.

Complete one wall first. On the adjacent wall, install one course of siding, line the course up, and install the corner cap. Each corner cap must be fitted and installed before the next course of siding is installed. A jig can be constructed to facilitate the alignment, or a special tool may be purchased for this purpose.

Install the cap by slipping the bottom flanges of the corner cap under the butt of each siding panel. Use slight, steady pressure to press the cap into place. If necessary, insert a putty knife between the panel locks, prying slightly outward to allow room for the flanges to slip in. Gentle tapping

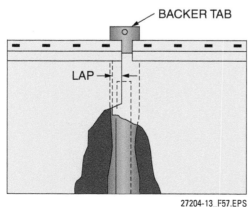

Figure 57 Inserting a backer tab for metal siding.

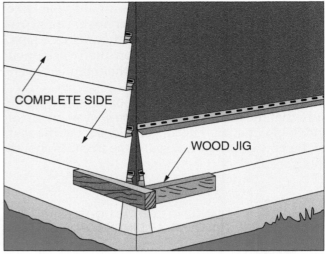

Figure 58 Leaving room for corner caps.

with a rubber mallet and wood block can also be helpful.

When the cap is in position, secure it with 2" or 2½" nails, or nails that are long enough for ¾" penetration into solid wood or sheathing. Nail through at least one of the prepunched nail holes in the top of the corner cap.

3.5.8 Installing Siding Around Windows and Doors

As the siding courses reach a window, a panel will probably need to be cut narrower to fit the space under the window opening. Plan this course of siding so that the panel will extend on both sides of the opening. Hold the panel in place to mark for the vertical cuts. Use a small piece of scrap siding as a template, placing it next to the window and locking it into the panel below (*Figure 59*). Make a mark on this piece ¼" below the sill height to allow clearance for all-purpose trim. Do the same on the other side of the window, because windows are not always absolutely level.

The vertical cuts are made from the top edge of the panel with duckbill snips, aviation snips, or a power saw. For aluminum, the lengthwise (horizontal) cut is scored once with the utility knife

---

Aluminum Siding Corner Caps

Aluminum siding corner caps are similar to those used for wood lap siding corners. Before nailing the corner caps, always make sure that the cap and siding panels are flush.

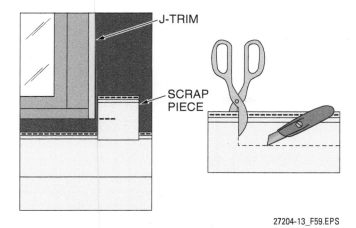

Figure 59 Cutting a panel around a window.

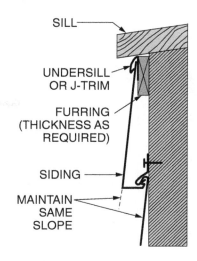

and bent back and forth until the unwanted piece breaks off. For vinyl, the horizontal cut is made with a utility knife. For steel, the horizontal cut is made with tin snips.

The raw edge of the panel should be trimmed with all-purpose trim for the exact width of the sill. First, determine if furring is required behind the cut edge to maintain the slope angle with adjacent panels. Nail the correct thickness of furring under the sill and install all-purpose trim over it with aluminum nails, close up under the sill for a tight fit (*Figure 60*). On vinyl panels, use a snap-lock punch to place raised ears on the raw edge of the panel. Slide the panel upward so as to engage the undersill or J-channel, the J-channels on the window sides, and the lock of the panels below.

Fitting panels over door and window openings is almost the same as making undersill cutouts, except that the clearances for fitting the panel are different. The cut panel on top of the opening needs more room to move down to engage the interlock of the siding panel below on either side of the window. Mark a scrap piece template without allowing clearance and then make saw cuts ¼" to ⅜" deeper than the mark (*Figure 61*). This will provide the necessary interlock clearance.

Check the need for furring over the top of the window or door in order to maintain the slope angle and install it, if required (*Figure 62*). Make sure the furring is pressure treated and is spaced off the bottom of the J-channel.

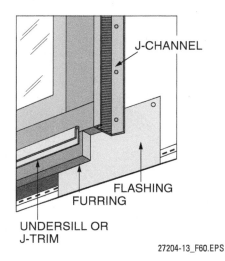

Figure 60 Fitting panels at door and window sills.

Cut a piece of all-purpose trim the same width as the raw edge of the cut panel and slip it over this edge of the panel before installing it. Drop the panel into position, engaging the interlocks on the siding panels below. The all-purpose trim can now be pushed downward to close any gap at the juncture with the J-channel. Refer to *Figure 63*.

3.5.9 Installing Siding at Gable Ends

When installing siding on gables, diagonal cuts will have to be made on some of the panels. To make a pattern of cutting panels to fit the gable

Pressure-Treated Furring and Furring Substitutes

If used, always make sure that any wood furring for vinyl or metal siding is pressure treated to resist insects and decay. In place of pressure-treated furring in some installations, J-channel is used under the sill. Then undersill trim installed under the siding provides the furring spacing and a panel locking point. At the top of the opening, installing undersill trim under the siding provides the furring spacing and panel locking point.

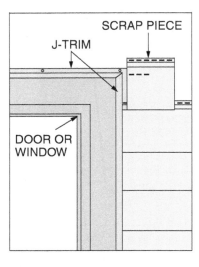

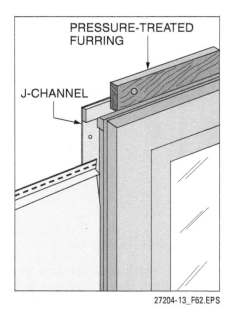

Figure 62 Using furring with J-channel.

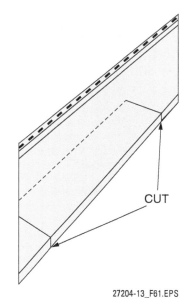

Figure 61 Using a scrap piece to measure clearance.

slope, use two short pieces of siding as templates (*Figure 64*). Interlock one of these pieces into the panel below. Hold the second piece against the J-channel trim on the gable slope. Along the edge of this second piece, scribe a line diagonally across the interlock end panel and cut along this line with tin snips or a power saw.

> **NOTE**
> Always check a cutting pattern after each course for accuracy. This is necessary because roof slopes are not always straight.

This cut panel is a pattern that can be used to transfer cutting marks to each successive course along the gable slope. All roof slopes can be handled in the same manner as gable end slopes.

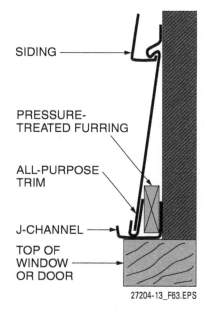

Figure 63 Completing the installation at a window top.

Slip the angled end of the panel into the J-channel previously installed along the gable end. Lock the butt into the interlock of the panel below. Remember to allow for expansion or contraction where required. If necessary, face-nail with 1¼" (or longer) painted-head aluminum nails in the apex of the last panel at the gable peak. Touch-up enamel in matching siding colors can also be used for exposed nail heads.

Do not cover existing louvers or vents. Attic ventilation is necessary in summer to reduce temperatures and in winter to prevent the accumulation of moisture.

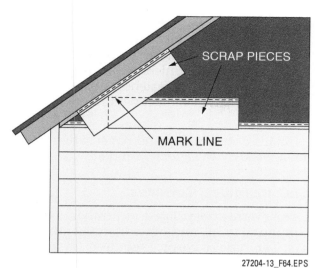

Figure 64 Installing siding at gable ends.

3.5.10 Installing Siding Under Eaves

The last panel course under the eaves will almost always have to be cut lengthwise to fit in the remaining space. Usually, furring will be needed under this last panel to maintain the correct slope angle. Determine the proper furring thickness and install it. Nail all-purpose trim or J-channel to the furring with aluminum nails. The trim should be cut long enough to extend the length of the wall (*Figure 65*).

To determine the width of the cut required, measure from the bottom of the top lock to the eave, subtract ¼", and mark the panel for cutting. Take measurements at several points along the eaves to ensure accuracy. Score the panel with the utility knife and bend it until it snaps. For vinyl panels, use a snaplock punch to place raised ears (16" or 24" apart) along the top cut edge so that it will lock into the J-channel.

For aluminum siding, apply gutter seal to the nail flange of the all-purpose trim. Slide the final panel into the trim. Engage the interlock of the panel below.

On metal panels, the lock may be flattened slightly using a hammer and a 2' or 3' piece of lumber before the final panel is installed, so the panel will grip more securely. Press the panel into the gutter seal adhesive. With this technique, face nails will not be required. Refer to *Figure 66*.

3.5.11 Caulking and Cleanup

In general, caulk is applied around doors, windows, and gables where the siding meets wood or metal, except where accessories are used to make caulking unnecessary. Caulk is also needed where siding or siding accessories meet brick or stone around chimneys and walls. Caulking around faucets, meter boxes, and other panel cutouts must be done neatly.

It is important to get a deep caulk bead that is ¼" minimum in depth, not just a wide bead. To achieve this, cut the plastic caulking cartridge tip straight across rather than at an angle. Move the gun evenly and apply steady, even pressure on the trigger. A butyl type of caulk is preferred as it has greater flexibility. Most producers supply

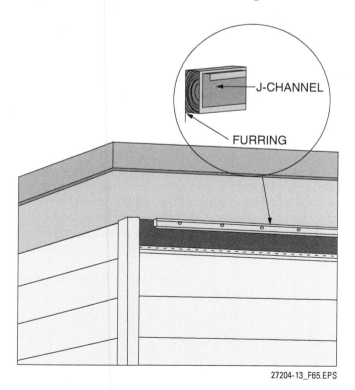

Figure 65 Trim extends the length of the wall.

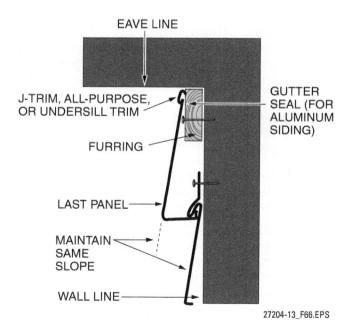

Figure 66 Installing siding under eaves.

caulk in colors to match siding and accessories. Do not depend on caulk to fill gaps more than ⅛" wide, as the expansion or contraction of the siding may cause the caulk to crack.

Reinstall all fixtures, brackets, downspouts, and similar items that were removed. Accessories that were not replaced, such as kitchen-fan outlets or service cables, may be painted to match the new siding color. Most manufacturers have touch-up paint or matching paint formulas, which can be purchased at a local paint store. All scrap pieces, cartons, nails, and other materials should be removed and the job site left neat and clean each day.

3.6.0 Installing Fiber-Cement Siding

To achieve satisfactory performance and for warranty purposes, fiber-cement siding must be installed and finished as specified by the manufacturer. General installation guidelines are as follows:

- Fiber-cement siding may be applied over walls sheathed with wood or insulation board up to 1" thick and with studs spaced not more than 24" OC or over unsheathed walls with studs spaced not more than 16" OC. The lowest edge of the siding should not be in contact with earth or standing water. When cutting, be sure to prime all cut ends with an alkali-resistant primer.
- In accordance with the manufacturer's instructions, moisture-resistant paper or felt may be required under the siding when the siding is applied directly to studs or over wood sheathing.
- Fiber-cement siding may be cut with a power saw using a fine-toothed, carbide-tipped or dry-diamond circular saw blade, electric or pneumatic carbide-tipped power shears, or a score-and-snap knife with a tungsten-carbide tip.

> **WARNING!** Because dry material will be drilled, cut, and/or abraded, proper respiratory protection must be used when working with fiber-cement siding to avoid inhalation of toxic silica dust that can cause a fatal lung disease called silicosis.

- Only galvanized steel, copper, or stainless steel flashing and screws or nails may be used when installing fiber-cement siding. Siding nails may be used, but for maximum wind resistance, use a standard 2" 6d nail or an 8-18 bugle-head screw through the overlap to a stud.
- Galvanized steel with a powder or baked enamel finish or vinyl inside/outside corners and other trim can be used and painted to match the siding finish. Never use aluminum trim components or fasteners because they will corrode when in contact with fiber-cement siding.

Details of typical lap and panel siding installation are shown in *Figures 67* and *68*.

3.7.0 Installing Cornices

Cornices are constructed of lumber, as well as aluminum, vinyl, and other manufactured products. The type of cornice required for a particular structure is shown on the wall sections of the construction drawings. The two general types of cornices are the closed cornice and box cornice.

Of the two types of cornices, box cornices are the most commonly used. If a ceiling is installed, the roof area above any ceiling cannot be ventilated unless vertical vents through the frieze board are provided. A closed cornice is the least desirable type of cornice because it does not allow roof ventilation and provides little protection to the side of the building.

With a box cornice, the rafter overhang is entirely boxed in by the roof covering, the fascia, and a bottom strip called a plancier or soffit. *Figure 69* shows examples of one type of box cornice.

Box cornices use wood, aluminum, vinyl, or exterior gypsum materials for soffits and sometimes for fascia. *Figure 70* shows various types of trim molding used on cornices.

A roof with no rafter overhang normally has a closed cornice (*Figure 71*). This cornice consists of a single strip called a frieze board. The frieze board is beveled on its upper edge to fit close under the overhang of the eaves and rabbeted on its lower edge to overlap the upper edge of the top siding course.

When a wood shingle or shake roof is to be installed, a strip of wood shingles is used to provide a roof overhang several inches beyond the molding. In this instance, the strip can serve as both a drip edge and starter strip for a wood shingle roof or as a support for the starter strip of an asphalt shingle roof. Some codes may require the installation of a metal drip cap on the edge of the strip of shingles. If trim is used, it usually consists of molding installed as shown in *Figure 71*.

A roof with a rafter overhang will have a box cornice if it has a cornice at all.

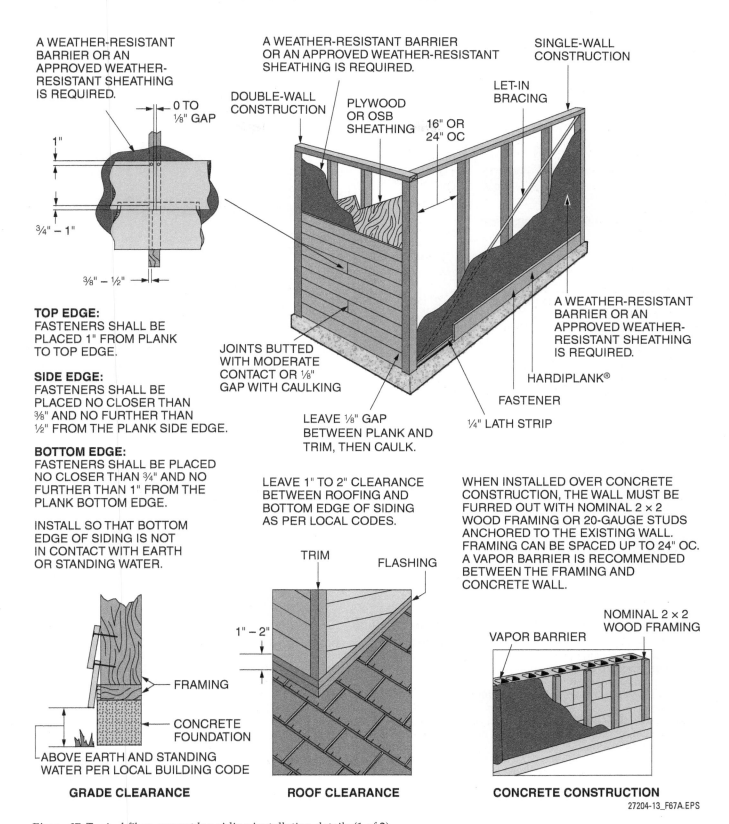

Figure 67 Typical fiber-cement lap siding installation details (1 of 2).

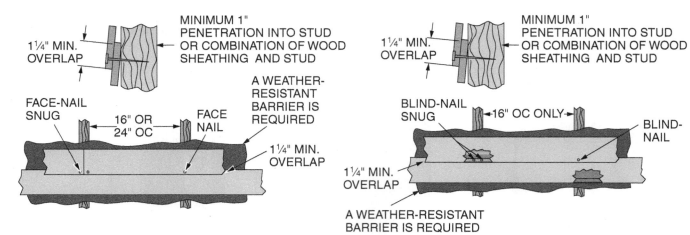

CORROSION-RESISTANT NAILS
(GALVANIZED‡ OR STAINLESS STEEL)
- 6d (0.118" shank × 0.267" HD (head diameter) × 2" long)
- Siding nail (0.089" shank × 0.221" HD × 2" long)†
- Siding nail (0.091" shank × 0.221" HD × 1½" long)*
- ET & F pin (0.100" shank × 0.25" HD × 1½" long)†

CORROSION-RESISTANT SCREWS
- Ribbed bugle-head or equivalent (No. 8–18 × 0.323" HD × 1⅝" long) Screw must penetrate ¼" or 3 threads into metal framing.

CORROSION-RESISTANT NAILS
(GALVANIZED‡ OR STAINLESS STEEL)
- Siding nail (0.089" shank × 0.221" HD × 2" long)†
- 11 gauge roofing nail (0.121" shank × 0.371" HD × 1¼" long)

CORROSION-RESISTANT SCREWS
- Ribbed bugle-head or equivalent (No. 8–18 × 0.375" HD × 1¼" long) Screws must penetrate ¼" or 3 threads into metal framing.

FACE-NAILED

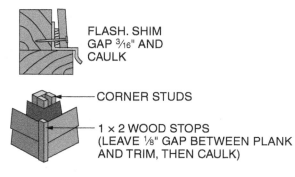

BLIND-NAILED
(NOT APPLICABLE FOR 12" WIDE SIDING)

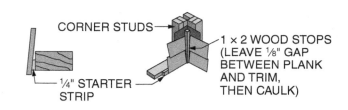

NOTES: **TRIM DETAILS**
* For face-nail application of 9½" wide or less siding to OSB, fasteners are spaced a maximum of 12" OC.
† The use of a siding nail or roofing nail may not be applicable to all installations where greater wind loads or higher exposure categories of wind resistance are required by the local building code. Consult the applicable building code compliance report.
‡ Hot-dipped galvanized nails are recommended.

FASTENING REQUIREMENTS:
- Drive fasteners perpendicular to siding and framing.
- Fastener heads should fit snug against siding (no air space). (Examples 1 & 2)
- Do not underdrive nail heads or drive nails at an angle. (Example 4)
- If nail is countersunk, caulk nail hole and add a nail. (Example 3)

Example 1   SNUG

Example 2   FLUSH

Example 3   COUNTERSUNK, CAULK AND ADD NAIL

Example 4   DO NOT UNDERDRIVE NAILS

Figure 67 Typical fiber-cement lap siding installation details (2 of 2).

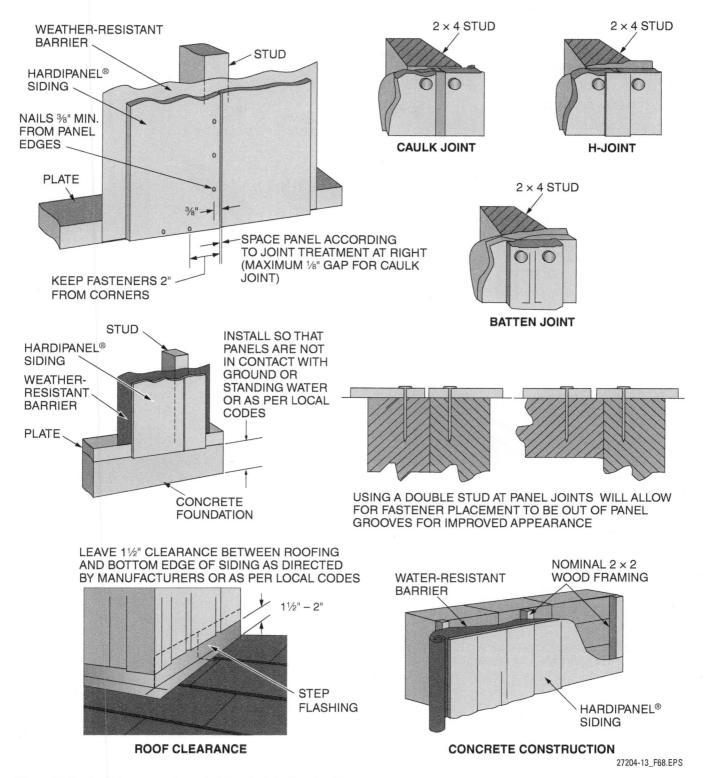

Figure 68 Typical fiber-cement panel siding installation details.

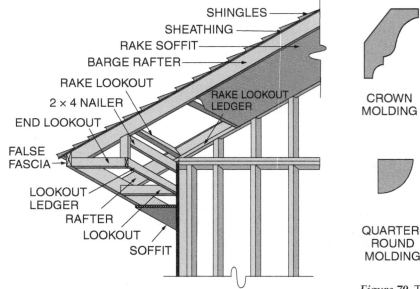

PARTIALLY COMPLETED BOX CORNICE WITH CORNICE RETURN AND RAKE SECTION

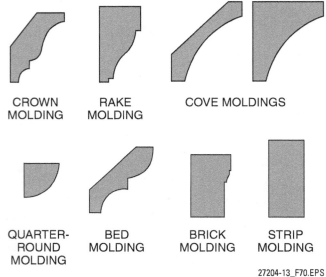

Figure 70 Types of wood cornice trim molding.

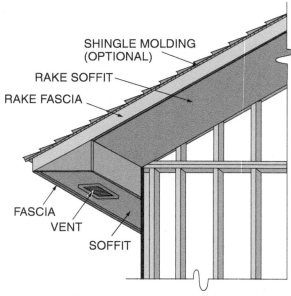

COMPLETED BOX CORNICE WITH CORNICE RETURN AND RAKE SECTION

Figure 69 Box cornice.

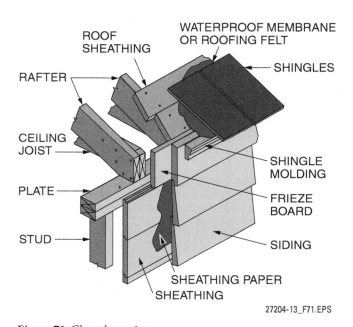

Figure 71 Closed cornice.

3.7.1 Aluminum or Vinyl Fascia and Soffits

Prefinished aluminum and vinyl have been used extensively in cornice construction and have proven to be very satisfactory. *Figure 72* shows a typical installation detail of an aluminum or vinyl fascia eave trim and a soffit. The actual installation will vary according to the manufacturer. *Figure 73* shows typical soffit and trim materials. Each soffit panel piece is cut to the desired width and then slid into place from the end of the cornice. Each panel interlocks with the adjoining pieces. Depending on the materials used and the amount of overhang, lookouts and ledgers may be required for the support of the soffit to prevent sag and wind damage. A lookout is a horizontal member used to support the overhanging portion of a roof.

The fasteners used in wood cornice work normally consist of various sizes and types of nails. Common steel nails and cement-coated box nails are used in areas where they will be completely enclosed and protected from the weather.

Hot-dipped galvanized nails are commonly used to fasten any cornice member exposed to the weather. These nails should be of sufficient length to hold the material in place and may be any of the types commonly manufactured.

Aluminum and stainless steel nails are available for use on exterior trim to eliminate rust streaks. Stainless steel nails are used on the highest-quality work where the added cost of the nails is incidental.

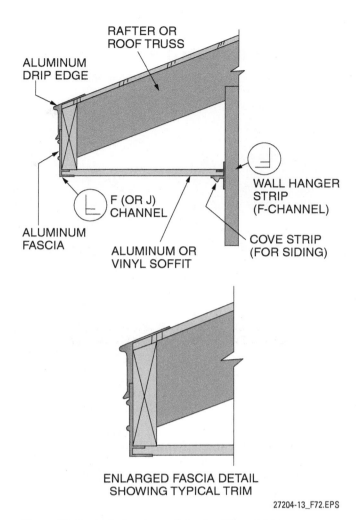

Figure 72 Typical aluminum/vinyl soffit and trim installation.

Cornice Returns

A variety of more complicated cornice returns exists, as shown in these illustrations.

REVERSED CORNICE RETURN

ANGLED CORNICE RETURNS

COLONIAL-MOLDING CORNICE RETURN

27204-13_SA10.EPS

Figure 74 is a nail spacing chart for plywood used as a closed soffit. Many times, the failure of plywood to maintain a level surface after installation is not due to a product failure, but is caused by improper installation.

The tools used in the installation of the cornice and related finish are the same as those used for rough and finish carpentry. Guidelines for constructing a box cornice are as follows:

Step 1 Start by marking the location of the ledger (*Figure 75*), then snap a chalkline to ensure the ledger will be level. Secure the ledger to the wall. The ledger makes it easier to install the lookouts.

Step 2 Use a straight board to mark the lookout positions on the ledger (*Figure 76*).

Step 3 Measure the distance from the face of the ledger to the ends of the rafters to determine the length of the lookouts.

Step 4 Trim the corner lookouts to match the slope of the rafters (*Figure 77*).

Step 5 Attach the lookouts to the ledger, then install the assembly, nailing through the sheathing into studs where possible.

Step 6 Once the lookouts are in place, the soffit and fascia boards can be attached (*Figures 78* and *79*).

Step 7 The cornice enclosure must be cut to fit, and the cornice end piece is grooved to fit inside the cornice enclosure (*Figure 80*).

Stainless Steel Nails

While aluminum nails are rustproof, they may be corroded by certain chemicals in industrial or high vehicular traffic environments. Quality stainless steel nails are not affected by any environment.

Metal Drip Caps and Edges

Some codes require the use of a metal drip edge (sometimes called a drip cap) at the edges of all roofs, including the cornices. Others may permit just the use of a row of wood shingles as a drip edge at the cornice edge of the roof or a row of wood shingles at the cornice in combination with a perimeter metal drip cap. The purpose of a drip edge is to direct rain water away from the cornice fascia.

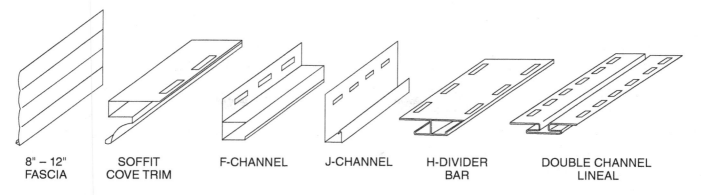

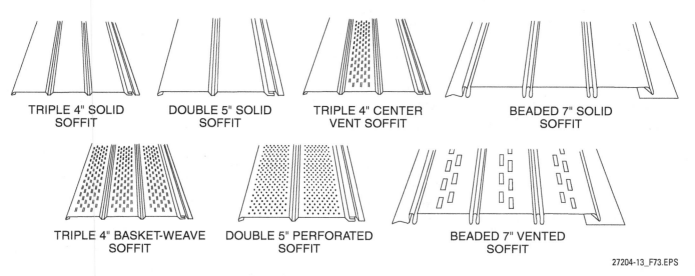

Figure 73 Typical aluminum/vinyl soffit and trim materials.

PLYWOOD THICKNESS	MAXIMUM SPACING SUPPORT	NAIL SIZE	NAIL SPACING EDGE	NAIL SPACING FIELD
3/8"	24"	6d	6"	12"
1/2"	24"	6d	6"	12"
5/8"	48"	8d	6"	12"

Figure 74 Nail spacing chart for a plywood closed soffit.

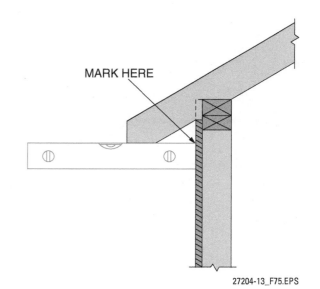

Figure 75 Determining the ledger position.

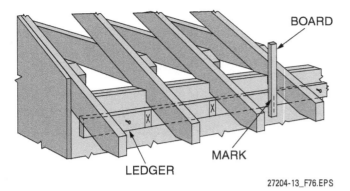

Figure 76 Marking lookout locations on a ledger.

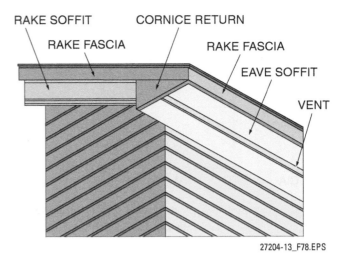

Figure 78 Installing the soffit and fascia boards.

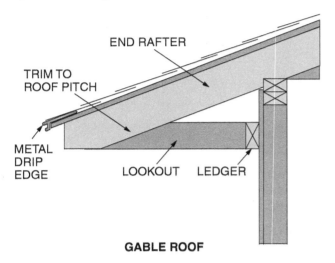

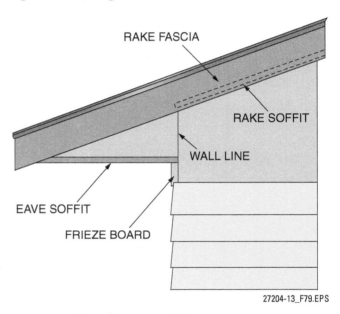

Figure 79 Rake soffit cut to wall line.

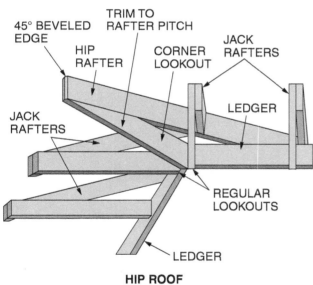

Figure 77 Determining the length of corner lookouts.

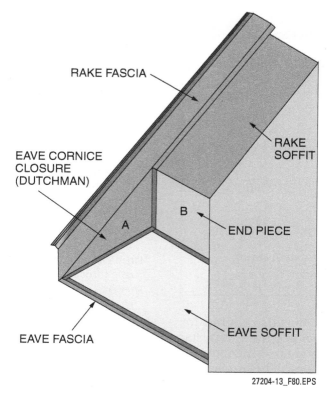

Figure 80 Boxing in the cornice return.

3.0.0 Section Review

1. To prevent rust streaks when installing exterior finishing materials, use _____.
   a. box nails
   b. finish nails
   c. nongalvanized nails
   d. stainless steel nails

2. When getting a surface ready for exterior finishing materials, you should prepare the entire building a few courses at a time.
   a. True
   b. False

3. Furring strips for vertical siding should be installed horizontally on _____.
   a. 12" centers
   b. 16" to 24" centers
   c. 16" centers
   d. 12" to 18" centers

4. To establish a straight reference line for siding installation, find _____.
   a. the lowest corner of the house
   b. the highest peak of the roof
   c. the first wall stud
   d. the concrete foundation's center

5. Another name for a siding gauge is a _____.
   a. preacher
   b. course spacer
   c. story pole
   d. tell-tale

6. Most outside corner posts are installed _____.
   a. after the siding is hung
   b. at the same time as the siding
   c. before the siding is hung
   d. only with TRG siding

7. Fiber-cement siding can be applied to sheathed walls with studs spaced up to _____.
   a. 12" OC
   b. 16" OC
   c. 18" OC
   d. 24" OC

Section Four

4.0.0 Estimate Exterior Finish Materials

Objective

Describe the estimating procedure for exterior finish projects.
 a. Explain how to perform a takeoff on panel and board siding.

Performance Task 2

Estimate the amount of lap or panel siding required for a structure.

The amount of siding materials ordered for a project is based on an accurate material takeoff. In general, the area of the exterior wall surfaces is first determined. This area is divided by the coverage of the siding to be applied.

4.1.0 Estimating Panel and Board Siding

To estimate the amount of siding material required for a project, proceed as follows:

Step 1 Determine the total area of the structure to be covered by adding up the areas of all walls and gables using the following formulas for the area of a rectangle or triangle:

Rectangular area = width × height

Triangular area = $\frac{\text{width} \times \text{height}}{2}$

Step 2 Subtract the total area of all openings.

Step 3 For wood board-type siding, add the waste percentages for the size and lap, as listed in *Table 2*. For metal or vinyl siding, add 10 percent.

> **NOTE:** Except for large openings such as a garage or sliding doors, omit Step 2 for 2 × 8 or 4 × 8 sheet material.

Step 4 To determine the number of squares of board-type siding required, divide the total area plus the waste percentage by 100 sq ft. Round up to the next whole square.

Table 2 Waste Allowances

Siding	Size and Lap	Percent to Add
Beveled	1 × 4 – ¾"	38
	1 × 5 – ⅞"	45
	1 × 6 – 1"	33
	1 × 8 – 1¼"	33
	1 × 10 – 1½"	29
	1 × 12 – 1½"	23
Drop siding and rustic (shiplapped)	1 × 4	28
	1 × 5	21
	1 × 6	19
	1 × 8	16
Drop siding and rustic (dressed and matched)	1 × 4	23
	1 × 5	18
	1 × 6	16
	1 × 8	14
Triangular areas or diagonal installation*		10

*The 10 percent is in addition to other allowances.

Number of squares = $\frac{\text{Total area} + \text{waste percentage}}{100}$

Step 5 For 2 × 8 or 4 × 8 panel siding, divide the total area by the area of a panel to determine the number of panels required. Round up to the next whole panel.

Number of 2 × 8 panels = $\frac{\text{total area}}{16}$

Number of 4 × 8 panels = $\frac{\text{total area}}{32}$

4.1.1 Estimating Board-and-Batten Siding

To aid the estimator, the lumber requirements to cover one square (100 sq ft) are indicated in *Table 3*.

Table 3 Board-and-Batten Lumber Requirements

Width of Boards	Board Feet
6"	147
8"	139
10"	132

4.1.2 Estimating Board-on-Board Siding

To aid the estimator, the lumber requirements to cover one square (100 square feet) are indicated in *Table 4*.

4.1.3 Estimating Nails for Siding

For beveled siding, approximately 1 lb of 6d nails per 100 square feet of siding or 1½ lb of 8d nails per 100 square feet of siding is needed.

For board-and-batten siding, the amount of nails needed to apply 100 square feet of siding will depend on their size. For 8d nails, 2½ to 3 lb of nails will be required.

Table 4 Board-on-Board Lumber Requirements

Width of Boards	Board Feet/100 ft^2
6"	150
8"	139
10"	129
12"	123

For board-on-board siding, approximately 25 lb of 8d nails per 1,000 board feet (more if using 10d nails) will be required.

4.0.0 Section Review

1. When estimating nail quantities needed for beveled siding, allow 1.5 lb per 100 sq ft for _____.

    a. 6d nails
    b. 8d nails
    c. 10d nails
    d. 16d nails

SUMMARY

You must be aware of the materials and the general methods of installation for a variety of exterior finishes and roof drainage systems. This module covered the installation methods for flashing and insulation, types of cornices and their fabrication/installation, along with descriptions and installation methods for a variety of wood, metal, and vinyl siding. It also discussed other exterior finishes, including stucco, masonry, and various special exterior finish systems. To achieve satisfactory performance and for warranty purposes, when installing any type of exterior finish or related accessory, always refer to the manufacturer's installation instructions.

Review Questions

1. From a safety standpoint, a carpenter's most important piece of clothing is probably _____.
   a. a long-sleeved shirt
   b. work shoes
   c. work gloves
   d. headgear

2. OSHA requires that a scaffold must be equipped with top rails, midrails, and toeboards (or a personal fall arrest system) if it is _____.
   a. more than 6' high
   b. on uneven ground
   c. not anchored to a building
   d. more than 10' high

3. The primary purpose of an exterior finish is to _____.
   a. provide a base for the final finish
   b. prevent entry of water at all openings
   c. provide protection from the elements
   d. allow the use of the most cost-effective interior construction

4. When building wrap is applied to the exterior of a structure, the wrap should be overlapped at the corners by _____.
   a. 2'
   b. 1'
   c. 3'
   d. 6'

5. A plancier is similar to a _____.
   a. soffit
   b. ledger board
   c. frieze board
   d. lookout

6. The horizontal framing member that the soffit is attached to is called a _____.
   a. jack rafter
   b. hip rafter
   c. lookout
   d. cornice return

7. For effective nailing, plain-shank nails must have a minimum penetration into solid lumber of _____.
   a. ½"
   b. ⅝"
   c. ¾"
   d. 1"

8. One of the most common flashing materials is _____.
   a. stainless steel
   b. asphalt building paper
   c. galvanized sheet metal
   d. copper

9. Two common styles of beveled siding are _____.
   a. colonial and log cabin
   b. log cabin and rustic
   c. plain or bungalow and rabbeted (Dolly Varden)
   d. rabbeted and tongue-and-groove

10. Vertical siding where boards overlap boards underneath that are the same size as the overlapping boards is called _____.
    a. board-and-batten
    b. reverse batten
    c. board-on-board
    d. shiplap

11. The type of shiplap siding shown in *Review Question Figure 1* is known as _____.
    a. channel rustic
    b. drop
    c. V-edged
    d. plain

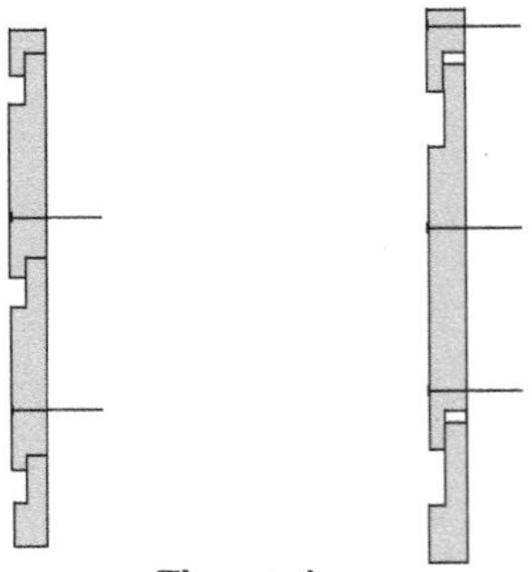

Figure 1

12. The type of wood commonly used to make shingle or shake siding is _____.
    a. red cedar
    b. oak
    c. pine
    d. cherry

13. A plywood siding panel called textured one-eleven (T-111) has a surface pattern that _____.
    a. is striated
    b. is brushed
    c. has widely spaced grooves
    d. has a plank texture

14. When plywood is used as exterior siding, it should be _____.
    a. glued to the sheathing
    b. nailed to the studs
    c. glued to the studs
    d. nailed to the sheathing

15. A finish that is *not* recommended for fiber-cement siding is _____.
    a. 100 percent acrylic latex paint
    b. flat alkyd stain
    c. gloss alkyd paint
    d. satin alkyd paint

16. Single-lap fiber-cement siding is available in widths of _____.
    a. 4"–8"
    b. 6"–12"
    c. 9"–18"
    d. 10"–16"

17. In a stucco finish, water penetration at the juncture of the stucco and wood trim is prevented through the use of a _____.
    a. sheathing
    b. stucco lock
    c. brown coat
    d. finish coat

18. In a water-managed system, a common substrate is _____.
    a. plywood sheathing
    b. brick veneer
    c. asphalt-impregnated felt
    d. fiber-cement panels

19. A ledger is used to simplify the installation of _____.
    a. fascia boards
    b. rafters
    c. frieze boards
    d. lookouts

20. If a siding panel must be cut to less than full height under a window, furring may be needed to _____.
    a. close the gap between panels
    b. provide a nailing surface
    c. maintain the correct slope angle
    d. support the sill

21. A carpenter-made device used to mark the vertical location for each course of beveled siding is called a _____.
    a. siding gauge
    b. story pole
    c. course stick
    d. preacher

22. When butting vinyl siding into the inside corner post, the clearance between the siding and post should be _____.
    a. 1/16" to 1/4"
    b. 1/2" to 3/4"
    c. 3/4" to 1"
    d. 1" to 1 1/2"

23. When using 1/2" plywood for a closed soffit and the soffit is supported every 24" OC, the nail spacing along the edges should be _____.
    a. 1"
    b. 2"
    c. 3"
    d. 6"

24. Which of the following metals will corrode when placed in contact with fiber-cement board siding?
    a. aluminum
    b. galvanized steel
    c. copper
    d. stainless steel

25. For a smooth finish, a fine-toothed, carbide-tipped blade should be used on a power saw when cutting _____.
    a. stucco
    b. fiber-cement siding
    c. insulated metal siding
    d. EIFS

Trade Terms Quiz

Fill in the blank with the correct term that you learned from your study of this module.

1. The side walls and roof of a building meet under the eaves at the _____.
2. The _____ describes a ladder's load capacity.
3. Projecting over the side wall, the _____ is the lower part of a roof.
4. To mate with another board, a groove called a _____ is cut into the edge of a board.
5. A finish coat of plaster is placed over the rough-faced _____.
6. A _____ is a single row of siding, block, or brick on a wall.
7. The top edge of siding materials applied to a sidewall often abut the bottom edge of a _____.
8. Rain gutters are typically hung on the exterior finish member called the _____.
9. Narrow strips cover joints between wider boards in the vertical type of siding known as _____.
10. A brick _____ may be applied to the surface of a frame structure.
11. The slope of a cornice on a gable end, paralleling the roof rafters, is the _____.
12. Information about a hazardous substance is found on its' _____.
13. A _____ is a roof overhang's underside.
14. Usually attached to the underside of a rafter, a _____ is the same as a soffit.
15. The last coat of paint or plaster applied to a wall is known as the _____.
16. Applied directly to masonry or through a fine-mesh diamond reinforcement, the _____ is the first layer of cement plaster.
17. Lookouts are attached to the _____, which is a board attached to the outside wall of a structure.
18. _____ is applied to a structure's walls to prevent wind infiltration.
19. To allow the passage of air, a small opening called a _____ is inserted in a wall, soffit, or roof.
20. Usually located in a gable end or a soffit, a _____ is a slatted ventilation opening.
21. An overhanging portion of a roof is supported by a member called a _____.

Trade Terms

Board-and-batten siding	Course	Frieze board	Rabbet	Soffit
Brown coat	Duty rating	Ledger	Rake	Veneer
Building paper	Eave	Lookout	Safety data sheet (SDS)	Vent
Cornice	Fascia	Louver	Scratch coat	
	Finish coat	Plancier		

Trade Terms Introduced in This Module

Board-and-batten siding: A type of vertical siding consisting of wide boards with the joint covered by narrow strips known as battens.

Brown coat: A coat of plaster with a rough face on which a finish coat will be placed.

Building paper: A heavy paper used for construction work. It assists in weatherproofing the walls and prevents wind infiltration. Building paper is made of various materials and is not a vapor barrier.

Cornice: The construction under the eaves where the roof and side walls meet.

Course: One row of brick, block, or siding as it is placed in the wall.

Duty rating: Load capacity of a ladder.

Eave: The lower part of a roof, which projects over the side wall.

Fascia: The exterior finish member of a cornice on which the rain gutter is usually hung.

Finish coat: The final coat of plaster or paint.

Frieze board: A horizontal finish member connecting the top of the sidewall, usually abutting the soffit. Its bottom edge usually serves as a termination point for various types of siding materials.

Ledger: A board to which the lookouts are attached and which is placed against the outside wall of the structure. It is also used as a nailing edge for the soffit material.

Lookout: A member used to support the overhanging portion of a roof.

Louver: A slatted opening used for ventilation, usually in a gable end or a soffit.

Plancier: Similar to a soffit, but the member is usually fastened to the underside of a rafter rather than the lookout.

Rabbet: A groove cut in the edge of a board so as to receive another board.

Rake: The slope or pitch of the cornice that parallels the roof rafters on the gable end.

Safety data sheet (SDS): A document that must accompany any hazardous substance. The SDS identifies the substance and gives the exposure limits, the physical and chemical characteristics, the kind of hazard it poses, precautions for safe handling and use, and specific control measures.

Scratch coat: The first coat of cement plaster consisting of a fine aggregate that is applied through a diamond mesh reinforcement or on a masonry surface.

Soffit: The underside of a roof overhang.

Veneer: A brick face applied to the surface of a frame structure.

Vent: A small opening to allow the passage of air.

Additional Resources

This module presents thorough resources for task training. The following resource material is suggested for further study.

Vinyl Siding Institute website. **www.vinylsiding.org**
Cedar Shake & Shingle Bureau website. **www.cedarbureau.org**

Figure Credits

JLG Industries, Inc., Figure 1
Milwaukee Electric Tool Corp., Figure 3
TAPCO INTEGRATED TOOL SYSTEMS, Figure 4
James Hardie Building Products, Figure 67A, Figure 67B, Figure 68, SA06

Alum-A-Pole Corporation, SA01
Cummins Industrial Tools, SA07
Copyright 2006 CertainTeed Corporation, used with permission, SA10

Section Review Answer Key

Answer	Section Reference	Objective
Section One		
1. c	1.1.0	1a
2. a	1.2.3	1b
Section Two		
1. c	2.1.0	2a
2. b	2.2.1	2b
3. c	2.3.0	2c
4. c	2.4.1	2d
5. d	2.5.0	2e
6. b	2.6.0	2f
Section Three		
1. d	3.7.1	3g
2. a	3.1.0	3a
3. b	3.2.0	3b
4. a	3.3.0	3c
5. a	3.4.1	3d
6. c	3.5.2	3e
7. d	3.6.0	3f
Section Four		
1. b	4.1.3	4a

This page is intentionally left blank.

NCCER CURRICULA — USER UPDATE

NCCER makes every effort to keep its textbooks up-to-date and free of technical errors. We appreciate your help in this process. If you find an error, a typographical mistake, or an inaccuracy in NCCER's curricula, please fill out this form (or a photocopy), or complete the online form at **www.nccer.org/olf**. Be sure to include the exact module ID number, page number, a detailed description, and your recommended correction. Your input will be brought to the attention of the Authoring Team. Thank you for your assistance.

Instructors – If you have an idea for improving this textbook, or have found that additional materials were necessary to teach this module effectively, please let us know so that we may present your suggestions to the Authoring Team.

NCCER Product Development and Revision
13614 Progress Blvd., Alachua, FL 32615

Email: curriculum@nccer.org
Online: www.nccer.org/olf

❑ Trainee Guide   ❑ Lesson Plans   ❑ Exam   ❑ PowerPoints   Other _____

Craft / Level: _____   Copyright Date: _____

Module ID Number / Title: _____

Section Number(s): _____

Description: _____

_____

_____

_____

_____

Recommended Correction: _____

_____

_____

_____

_____

Your Name: _____

Address: _____

_____

Email: _____   Phone: _____

This page is intentionally left blank.

Thermal and Moisture Protection

Overview

A properly insulated building will be comfortable to live or work in and will be economical to heat and cool. Without insulation, warm air will escape the building in cold weather, causing the heating system to operate constantly. This results in an increased use of energy. In hot weather, lack of insulation will allow warm air to penetrate the building, and the air conditioning system will have to work harder, with the same results. Vapor barriers are also important and they must be used to prevent moisture from penetrating the building. Moisture can cause a variety of additional serious problems. A skilled carpenter will know how to select and install insulating materials and vapor barriers.

Module 27203

Trainees with successful module completions may be eligible for credentialing through NCCER's National Registry. To learn more, go to **www.nccer.org** or contact us at **1.888.622.3720**. Our website has information on the latest product releases and training, as well as online versions of our *Cornerstone* magazine and Pearson's product catalog.

Your feedback is welcome. You may email your comments to **curriculum@nccer.org**, send general comments and inquiries to **info@nccer.org**, or fill in the User Update form at the back of this module.

This information is general in nature and intended for training purposes only. Actual performance of activities described in this manual requires compliance with all applicable operating, service, maintenance, and safety procedures under the direction of qualified personnel. References in this manual to patented or proprietary devices do not constitute a recommendation of their use.

Copyright © 2013 by NCCER, Alachua, FL 32615, and published by Pearson, New York, NY 10013. All rights reserved. Printed in the United States of America. This publication is protected by Copyright, and permission should be obtained from NCCER prior to any prohibited reproduction, storage in a retrieval system, or transmission in any form or by any means, electronic, mechanical, photocopying, recording, or likewise. To obtain permission(s) to use material from this work, please submit a written request to NCCER Product Development, 13614 Progress Blvd., Alachua, FL 32615.

27203 V5

From *Carpentry, Trainee Guide*. NCCER.
Copyright © 2013 by NCCER. Published by Pearson. All rights reserved.

27203
THERMAL AND MOISTURE PROTECTION

Objectives

When you have completed this module, you will be able to do the following:

1. Describe the safety and health hazards when working with insulation.
   a. List the personal protective equipment (PPE) that is required when working with insulation.
   b. Describe how to safely handle insulation.
2. Describe the various types of insulation and their characteristics.
   a. Explain how to determine R-value requirements.
   b. Describe flexible insulation and list its characteristics.
   c. Describe loose-fill insulation and list its characteristics.
   d. Describe rigid or semirigid insulation and list its characteristics.
   e. Describe reflective insulation and list its characteristics.
   f. List miscellaneous types of insulation.
3. Describe the various installation methods for insulation.
   a. Explain how to install flexible insulation.
   b. Explain how to install loose-fill insulation.
   c. Explain how to install rigid or semirigid insulation.
   d. Explain how to install reflective insulation.
4. Identify the requirements for moisture control, waterproofing, and ventilation, and describe the related installation methods.
   a. List various methods to control moisture in a structure.
   b. Identify methods to waterproof a structure.
5. Describe the estimating procedure for thermal and moisture projects.

Performance Tasks

Under the supervision of your instructor, you should be able to do the following:

1. Install blanket insulation in a wall.
2. Install a vapor barrier on a wall.
3. Install selected building wraps.

Trade Terms

Condensation	Diffusion	Permeable	Vapor barrier
Convection	Perm	Permeance	Water stop
Dew point	Permeability	R-value	Water vapor

Code Note

Codes vary among jurisdictions. Because of the variations in code, consult the applicable code whenever regulations are in question. Referring to an incorrect set of codes can cause as much trouble as failing to reference codes altogether. Obtain, review, and familiarize yourself with your local adopted code.

Industry-Recognized Credentials

If you're training through an NCCER-accredited sponsor, you may be eligible for credentials from NCCER's Registry. The ID number for this module is 27203. Note that this module may have been used in other NCCER curricula and may apply to other level completions. Contact NCCER's Registry at 888.622.3720 or go to **www.nccer.org** for more information.

Contents

Topics to be presented in this module include:

1.0.0 Safety Requirements for Insulation Projects 1
    1.1.0 Personal Protective Equipment 1
    1.2.0 Materials Handling 1
2.0.0 Insulation and Its Characteristics 4
    2.1.0 Determining R-value Requirements 4
    2.2.0 Flexible Insulation 8
    2.3.0 Loose-Fill Insulation 9
    2.4.0 Rigid or Semirigid Insulation 9
    2.5.0 Reflective Insulation 13
    2.6.0 Miscellaneous Types of Insulation 14
3.0.0 Installing Insulation 16
    3.1.0 Installing Flexible Insulation 16
    3.2.0 Installing Loose-Fill Insulation 17
    3.3.0 Installing Rigid or Semirigid Insulation 20
4.0.0 Moisture Control, Waterproofing, and Ventilation 22
    4.1.0 Moisture Control 22
        4.1.1 Interior Ventilation 23
        4.1.2 Vapor Barriers 24
        4.1.3 Materials 24
        4.1.4 Installation in Crawl Spaces 25
        4.1.5 Installation in Slabs 26
        4.1.6 Installation in Walls 26
        4.1.7 Installation in Roofs 27
    4.2.0 Waterproofing 27
        4.2.1 Water Stops 27
        4.2.2 Joint Treatment 28
        4.2.3 Vapor Barrier for Cold Storage and Low-Temperature Facilities ... 28
        4.2.4 Air Infiltration Control 28
5.0.0 Estimating Typical Insulation Requirements 33
Appendix Recommended Regional R-Values 39

Figures and Tables

Figure 1   Example of a fiberglass building insulation safety data sheet 2
Figure 2   Typical R-value identification 6
Figure 3   R-values of typical wall construction 9
Figure 4   Climate zones in the United States 10
Figure 5   Flexible blanket insulation 11
Figure 6   Flexible batt insulation 11
Figure 7   Loose-fill insulation 12
Figure 8   Rigid foam board 12
Figure 9   Reflective foil-faced batt insulation 13
Figure 10 Sprayed-in-place insulation 14

Figure 11	Blanket installation without integral vapor seal	17
Figure 12	Blanket installation with integral vapor seal	18
Figure 13	Batt insulation with separate vapor barrier	19
Figure 14	Ceiling insulation at wall and soffit	19
Figure 15	Typical plastic soffit baffle (shown upside down)	19
Figure 16	Perimeter floor insulation	19
Figure 17	Loose-fill insulation	19
Figure 18	Leveling loose-fill insulation	20
Figure 19	Rigid insulation installed under a concrete slab	20
Figure 20	Rigid insulation installed under a slab-and-down footing	20
Figure 21	Effects of insulation and vapor barrier	23
Figure 22	Various methods of roof ventilation	24
Figure 23	Vapor barrier installation	25
Figure 24	Installing insulation batts between ceiling joists with vapor barrier down	25
Figure 25	Vapor barrier installation for crawl spaces	26
Figure 26	Thickened-edge slab vapor barrier installation	26
Figure 27	Surface-mounted vapor barrier on a slab	26
Figure 28	Applying waterproofing material	27
Figure 29	Below-grade waterproofing application	27
Figure 30	Water stops used in joints	28
Figure 31	Building wrap	29
Figure 32	Building wrap accessories	29
Figure 33	Starting a roll of building wrap	30
Figure 34	Top plate detail	30
Figure 35	Cutting and folding wrap at an opening	31
Figure 36	Installing flashing around an opening	31
Figure 37	Installing wrap with a window in place	31
Figure 38	Dividing a plan view into rectangular or square areas	33
Figure A-1	Average low temperatures across the United States	39

Table 1	Insulation Materials	5
Table 2	R-values of Common Materials	7–8
Table 3	Recommended R-Values of Insulation	10
Table 4	Typical Comfort Standards	10
Table 5	Perm Ratings of Various Vapor Retarder Materials	25
Table 6	Typical Insulation Coverage for Various Types of Packaging and R-Values	34
Table A-1	Recommended R-Values of Insulation	39

This page is intentionally left blank.

SECTION ONE

1.0.0 SAFETY REQUIREMENTS FOR INSULATION PROJECTS

Objective

Describe the safety and health hazards when working with insulation.
   a. List the personal protective equipment (PPE) that is required when working with insulation.
   b. Describe how to safely handle insulation.

Insulation projects have many inherent hazards. Some insulation is installed at an elevated location. In addition, hazards, such as cuts or fiberglass strands, caused by tools and materials being used must be considered before beginning an insulation project. Always complete a job hazard analysis (JHA) before an insulation project to identify the potential hazards and recommend action that can be taken to minimize the hazard. In addition to working at elevations, the following other potential hazards may be encountered on an insulation project:

- Cuts that may occur when cutting insulation with utility knives
- Respiratory hazards when installing fiberglass and other types of insulation
- Flying debris as a result of cutting or handling insulation
- Tools falling from a roof, ladder, or other elevated surface onto workers below

1.1.0 Personal Protective Equipment

Proper personal protective equipment (PPE) must be worn when working with and installing insulation. Gloves should be worn to minimize cuts and abrasions, and to protect the skin from irritation. For eye protection, safety glasses and/or goggles must be worn. If the safety glasses do not fit tightly, secure them in place with a lanyard. Respiratory protection, such as a dust mask, should be worn when working with fiberglass insulation to prevent the fibers from being inhaled. Some types of insulation materials may require additional respiratory protection. The *Core Curriculum* provides information regarding the proper use of PPE.

1.2.0 Materials Handling

The OSHA Hazard Communication Standard requires chemical manufacturers, distributors, or importers to provide safety data sheets (SDSs)—formerly known as material safety data sheets (MSDSs)—to communicate the hazards of chemical products (*Figure 1*). Many insulation and roofing materials contain chemicals that may be hazardous if not handled properly. Therefore, SDSs must accompany the insulation or roofing materials when they are delivered to the job site. The SDSs must then be properly stored for reference in case of an emergency. Always know the location of the SDSs on the job site.

SAFETY DATA SHEET

MANUFACTURER Company X
ADDRESS 1234 1st Avenue
Anywhere, NV 54321

PHONE 555.555.5555
CONTACT John Doe, EHS Manager, 555.555.5555 x555
PRODUCT IDENTIFICATION Fiberglass building insulation batts
COMMON NAME Fiberous glass
LAST REVIEWED June 28, 2013

SECTION ONE: COMPOSITION

INGREDIENTS	HAZARD	CAS NO	PERCENTAGE
Fiberous glass	Nuisance dust	065997-17-3	85–98
Ammonia		001309-64-4	0–5
Formaldehyde		025104-55-6	0–5
Kraft product—mineral oil	Mild irritant	678383-87-9	5–10

SECTION TWO: PHYSICAL DATA

Boiling point	N/A	Specific gravity (H_2O) = 1	2.1	Vapor pressure (mm Hg)	N/A
Evaporation rate	N/A	Vapordensity (Air = 1)	N/A	Solubility in water	none
Percent volatile by volume	<1%				

Appearance/odor: Resilient or solid structure containing glass fibers and binding materials. May have a slight odor.

SECTION THREE: FIRE AND EXPLOSION DATA

Flashpoint	N/A	Flammability limits	N/A
Extinguishing media	Foam, dry chemical, water		

SECTION FOUR: REACTIVITY / DECOMPOSITION DATA

Stability	Stable	Incompatibility	None	Hazardous polymerization	Will not occur

SECTION FIVE: HEALTH HAZARDS / PERSONAL PROTECTIVE EQUIPMENT

Primary route of entry	Dusts and fibers from this product may cause mechanical irritation of the nose, throat, and respiratory tract. Use of a 2-strap NIOSH-approved N-95 filtering facepiece is recommended.
Skin contact	May causeirritation. Loose-fitting clothing, gloves, and eye protection are recommended.

SECTION SIX: EMERGENCY / FIRST-AID PROCEDURES

Skin contact	Wash with mild soap and running water. Use a washcloth to help remove fibers. To avoid further irritation, do not rub or scratch affected areas. If irritation persists get medical attention. Never use compressed air to remove fibers from the skin. If fibers are seen penetrating from the skin, the fibers can be removed by applying and removing adhesive tape so that the fibers adhere to the tape and are pulled out of the skin.
Eye contact	Flush with large amounts of water until irritation subsides, as least 15 minutes. See a physician if irritation persists.
Inhalation	Remove to fresh air. Drink water to clear throat and blow nose to evacuate dust. If coughing and irritation develop, call a physician.

Figure 1 Example of a fiberglass building insulation safety data sheet.

1.0.0 Section Review

1. Manufacturers must provide safety data sheets for all roofing or insulation materials that _____.
   a. have sharp edges
   b. are chemically hazardous
   c. are easily shattered
   d. contain fiberglass

2. If safety glasses do not fit tightly, they should be secured with a lanyard.
   a. True
   b. False

Section Two

2.0.0 INSULATION AND ITS CHARACTERISTICS

Objective

Describe the various types of insulation and their characteristics.

a. Explain how to determine R-value requirements.
b. Describe flexible insulation and list its characteristics.
c. Describe loose-fill insulation and list its characteristics.
d. Describe rigid or semirigid insulation and list its characteristics.
e. Describe reflective insulation and list its characteristics.
f. List miscellaneous types of insulation.

Trade Terms

R-value: The resistance to conductive heat flow through a material or gas.

Vapor barrier: A material used to retard the flow of vapor and moisture into walls and prevent condensation within them. The vapor barrier must be located on the warm side of the wall.

Four important considerations for the construction of any building include the following: thermal insulation, moisture control and ventilation, waterproofing, and air infiltration control. This module covers these areas and presents materials and procedures that can be applied to ensure effective installations.

Vapor barriers, also known as vapor retarders or vapor diffusion retarders (VDR), are an important part of moisture control. A vapor barrier is any material that prevents the passage of water. A properly installed vapor barrier will protect ceilings, walls, and floors from moisture originating within a heated space.

Some vapor barrier materials, such as kraft paper, are attached to blanket or batt insulation. They are installed when insulation is installed. Others, such as aluminum foil, may be applied to the back of gypsum drywall during its installation. Polyethylene film used as a vapor barrier is applied over studs and ceiling joists after insulation is installed. Vapor barriers are also installed under slabs, between the gravel cushion and the cast-in-place concrete.

Insulation materials can be divided into four general classifications, as shown in *Table 1*. The materials shown in *Table 1* are used in the manufacture of five basic categories of insulation: flexible, loose-fill, rigid or semirigid, reflective, and miscellaneous.

The R-value of the insulation is marked on the insulation itself or its packaging. See *Figure 2*.

2.1.0 Determining R-value Requirements

A conventionally insulated house can suffer a 55-percent heat loss either by losing heat through frame walls or by way of air infiltration. Because of the need for energy conservation, the building industry has been working to devise new products and methods of installation to prevent heat loss in buildings.

Heat loss can be significantly reduced by using insulation sheathing. There are several brands and types of insulation sheathing. These include Dow Styrofoam™, Celotex Tuff-R™ and Celotex Thermax™ sheathing, expanded polystyrene, fiberboard sheathing, Monsanto Fome-Cor™, and thermo-ply foil-faced paperboard. When selecting a particular insulating sheathing, the R-value and local building codes should be considered.

Installation procedures for insulation sheathing may vary, depending on the type of sheathing selected for use. Insulation sheathing is nonstructural. Adequate corner bracing, such as diagonal 1 × 4 let-in wood bracing, flat or profiled steel bracing, or plywood at corners overlaid with foam sheathing, should be used to comply with local building codes. Dow Styrofoam™ residential sheathing is laminated with a durable plastic film for added resistance to damage and abuse. This sheathing comes with tongue-and-groove edges on ¾" to 1" thicknesses. Celotex foam sheathings (Tuff-R™ and Thermax™) can be easily cut with a utility knife to any shape needed to conform to irregular wall angles or to fit snugly around window or door openings and other projections. Tuff-R™ insulation sheathing is semirigid and can bend around corners to reduce air infiltration.

Siding may be applied directly over the insulation sheathing. Brick, wood, aluminum, and vinyl siding are fastened to the wood frame construction by nailing through the sheathing. Care must be taken when driving nails so that the sheathing is not crushed.

Shakes and shingles can also be applied by installing furring strips or a plywood nailer base over the insulating sheathing. A stucco finish can also be applied over an acceptable lath fastened over the sheathing.

Table 1 Insulation Materials

Classification	Material	Comments
Mineral	Rock	Rock and slag are used to produce wool by grinding and melting the materials and blowing them into a fine mass.
	Slag	
	Glass	
	Vermiculite	
	Perlite	
Natural fiber	Wood	Many vegetable products are processed and formed into various shapes, including blankets and rigid boards.
	Sugarcane	
	Cornstalks	
	Cotton	
	Cork	
	Redwood bark	
	Sawdust or shavings	
	Sheep's wool	
Plastic	Polystyrene	
	Polyurethane	
	Polyisocyanurate	
	Phenolic	
Metal	Foil	Metallic insulating materials are generally applied to rigid boards or papers and used primarily for their reflective value.
	Tin plate	
	Copper	
	Aluminum	

Many other factors must be considered when insulating an exterior wall where exterior finish is to be applied. Windows are available with insulating glass. The insulating glass consists of two or three pieces of glass with an air space between them. The edges are sealed and the air space is then turned into a partial vacuum. All materials in a wall system are rated for various insulation and sound values, and different combinations can achieve large differences in insulation and acoustic characteristics.

Aluminum and plastic siding are available with an insulation board backing. This aids in insulating an exterior wall but is not effective by itself.

Most materials used in construction have some insulating value. Air is an excellent insulator if it is confined to very small spaces and is kept very still. Manufactured insulation material is based on

Insulation

There are many different types of insulation. Each type has specific applications for which it is best suited.

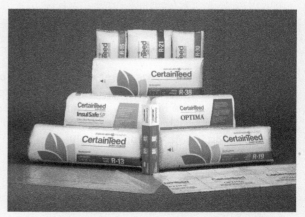

Figure 2 Typical R-value identification.

trapping a large amount of air in a large number of very small spaces to provide resistance to the transfer of heat and sound. Double-pane and triple-pane windows use this method to reduce heat loss.

The amount of insulation in a building directly affects heating and cooling costs. It also affects the value of the building. Some jurisdictions require that a permanent certificate, stating the insulative properties of material used in the structure, be placed on the building's electrical box. When required, this certificate is completed by the builders or designer.

One of the laws of physics states that heat will always flow (or conduct) through any material or gas from a higher-temperature area to a lower-temperature area.

The term *R-value* refers to the resistance to conductive heat flow through a material or gas. R-value is expressed as:

$$R = \frac{1}{k} \text{ or } \frac{1}{C}$$

Where

K = amount of heat in British thermal units (Btus) transferred in one hour through 1 sq ft of a material that is 1" thick and has a temperature difference between its surfaces of 1°F; also called the coefficient of thermal conductivity

C = conductance of a material, regardless of its thickness; the amount of heat in Btus that will flow through a material in one hour per sq ft of surface with 1°F of temperature difference

R = thermal resistance; the reciprocal (opposite) of conductivity or conductance

The higher the R-value, the lower the conductive heat transfer. *Table 2* shows the R-values of a number of common building materials, including some common insulating materials.

The total heat transmission through a wall, roof, or floor of a structure in Btus per sq ft per hour with a 1°F temperature difference is called the total heat transmission or U-value. It is expressed as follows:

$$U = \frac{1}{R_1 + R_2 + \ldots + R_n}$$

Where

$R_1 + R_2 + \ldots + R_n$ represents the sum of the individual R-values for the materials that make up the thickness of the wall, roof, or floor

The lower the U-value, the lower the heat transmission.

The R-values of two typical wall structures are shown in *Figure 3*. While the R-values provide a convenient measure to compare heat loss or gain, the total U-value for a structure is used in the cal-

Air Versus Inert Gas

The air trapped between the panes of double-pane and triple-pane windows makes a good insulator, but there are better alternatives. Inert gases that block more heat than air create a stronger barrier. That's why some window manufacturers fill the space between the panes with argon—an inert, nontoxic, nonflammable gas.

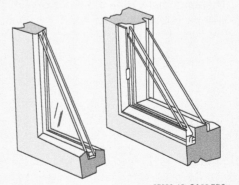

Table 2 (1 of 2) R-values of Common Materials

Material	Thickness	R-Value °F/Sq ft/hr in Btus
Air film and spaces		
Air space bound by ordinary materials	¾"	0.96
Air space bound by ordinary materials	¾" to 4"	0.94
Exterior surface resistance	–	0.17
Interior surface resistance	–	0.68
Masonry units		
Sand and gravel concrete block	4"	0.71
Sand and gravel concrete block	8"	1.11
Sand and gravel concrete block	12"	1.28
Lightweight concrete block	4"	1.50
Lightweight concrete block	8"	2.00
Lightweight concrete block	12"	2.27
Face brick	4"	0.44
Common brick	4"	0.80
Masonry materials		
Concrete, oven-dried sand, and gravel aggregate	1"	0.11
Concrete, undried sand, and gravel aggregate	1"	0.08
Stucco	1"	0.20
General building materials		
Wood sheathing or subfloor	¾"	0.94
Fiberboard sheathing (regular density)	½"	1.32
Fiberboard sheathing (intermediate density)	½"	1.14
Fiberboard sheathing (nail base)	½"	1.14
Plywood	⅜"	0.47
Plywood	½"	0.62
Plywood	¾"	0.93
Bevel-lapped wood siding	½" × 8"	0.81
Bevel-lapped wood siding	¾" × 10"	1.05
Vertical tongue-and-groove (cedar or redwood)	¾"	1.00
Asbestos-cement	¼"	0.21
Gypsum board	⅜"	0.32
Gypsum board	½"	0.45
Interior plywood paneling	¼"	0.31
Building paper (permeable felt)	–	0.06
Plastic film	–	0.00

culations for sizing the structure's heating and cooling equipment.

By doubling the R-value of a wall or roof, the conductive heat loss or gain can theoretically be reduced by half. However, it is important to note that as insulation thicknesses are increased, the heat transmission (U-value) is decreased, but not in a direct relationship. Increases of insulation will continue to decrease heat loss, but at lower and lower percentages. At some point, it is not economically feasible to add more insulation. The same is true for double-, triple-, and quadruple-pane windows. It must also be noted that conductive heat loss or gain does not include heat gains or losses due to air leaks or radiation through windows or other openings.

Increasing energy costs and mandated government energy conservation have resulted in much higher R-value requirements for new construction. While building code and design standards for in-

Table 2 (2 of 2) R-values of Common Materials

Material	Thickness	R-Value °F/Sq ft/hr in Btus
Insulating materials		
Fibrous batts (from rock slag or glass)	2" × 2¾"	7.00
Fibrous batts (from rock slag or glass)	3" × 3½"	11.00
Fibrous batts (from rock slag or glass)	5" × 5½"	19.00
SM brand Styrofoam® plastic foam	1"	5.41
TG brand Styrofoam® plastic foam	1"	5.41
IB brand Styrofoam® plastic foam	1"	4.35
Molded polystyrene beadboard	1"	4.17
Polyurethane foam	1"	5.88
Woods		
Fir, pine, and similar softwoods	¾"	0.94
Fir, pine, and similar softwoods	1½"	1.89
Fir, pine, and similar softwoods	2½"	3.12
Fir, pine, and similar softwoods	3½"	4.35
Maple, oak, and similar hardwoods	1"	0.91

sulation have been traditionally based on average low-temperature zones and charts based on the range of low temperatures expected (see *Appendix*), the requirements are constantly changing. The International Code Council (ICC) now recommends insulation values be based on climate zones, which are determined by local temperature and humidity levels (see *Figure 4* and *Table 3*).

In warm climates of the country, many codes now require almost the same amount of insulation for air conditioning as the cold climates require for heating. In some cases, codes are using comfort standards similar to those shown in *Table 4*. The all-weather standard requires that the insulation provided must be adequate to maintain a desired interior temperature during periods of extreme outside temperatures, both high and low. To meet the moderate standard, the insulation provided must be adequate to maintain a desired temperature during periods of average outside temperature extremes.

In general, insulation must be installed where any exterior surface of a structure is exposed to a thermal difference relative to its internal surface. These areas are:

- Roofs
- Above ceilings
- In exterior walls
- Beneath floors over crawl spaces
- Around the perimeter of concrete floors and around foundations

As you study the information in *Tables 2 and 3*, you will notice that the ceiling insulation has the greatest R-value. Since warm air rises, sufficient ceiling insulation must be installed to prevent the warm air from rising through the ceiling.

2.2.0 Flexible Insulation

Flexible insulation is usually manufactured from fiberglass in blanket form (*Figure 5*) and fiberglass or mineral wool in batt form (*Figure 6*). In some cases, it is manufactured from wood fiber or cotton, and treated for fire, decay, insect, and rodent resistance. The blankets are available in 16" or 24" widths and the batts in 15" or 23" widths. Both are furnished in thicknesses ranging from 1" to 12". The batts are packaged in flat bundles in lengths of 24", 48", or 93", and may be unfaced or faced with asphalt-laminated kraft paper or fire-resistant foil scrim (FSK) with or without nailing flanges. The blankets are furnished in rolls that may be encased in an asphalt-laminated kraft paper or plastic film. In most cases, they have a

Overinsulating

Installing excess insulation wastes money and may cause other problems. If a building is overinsulated and lacks sufficient ventilation and water-barrier protection, moisture can collect inside. This promotes the growth of mold and fungus. It is even possible that cancer-causing radon gas could be trapped in the building, accumulating over time to dangerous levels.

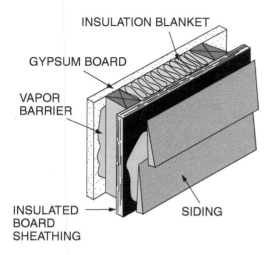

2 × 4 STUD WALL WITH RIGID BOARD:

TYPE	R-VALUE
AIR FILMS*	0.9
¾" WOOD EXTERIOR SIDING	1.0
1" POLYSTYRENE RIGID BOARD	5.0
3½" BATT OR BLANKET INSULATION	11.0
VAPOR BARRIER	0.0
½" GYPSUM BOARD	0.5
	18.4 TOTAL R

2 × 6 INSULATED STUD WALL:

TYPE	R-VALUE
AIR FILMS*	0.9
¾" WOOD EXTERIOR SIDING	1.0
¾" INSULATION BOARD	2.0
5½" INSULATING BLANKET	19.0
VAPOR BARRIER	0.0
½" GYPSUM BOARD	0.5
	23.4 TOTAL R

*Stagnant air film that forms on any surface

Figure 3 R-values of typical wall construction.

facing with nailing flanges. Some blankets are available with an FSK facing. Blankets and batts with kraft or film casing and/or facings are combustible and must not be left exposed in attics, walls, or floors.

Fiberglass batt insulation at a 3½" thickness (standard rating or high rating) may be used on exterior walls between the studs. This has been the normal insulation thickness in the past, because of the use of 3½"-wide studs spaced 16" on center (OC). However, in the northern parts of the country, some builders have been using 2 × 6 studs spaced 16" or 24" OC, which has allowed for an increase in the wall insulation to 5 ½". This thickness is ample insulation for all parts of the United States.

2.3.0 Loose-Fill Insulation

Loose-fill insulation is supplied in bulk form packaged in bags or bales (*Figure 7*). In new construction, it is usually blown or poured and spread over the ceiling joists in unheated attics. In existing construction that was not insulated when it was built, the material can be blown into the walls as well as the attic.

The materials used in loose-fill insulation include rock or glass wool, wood fiber, shredded redwood bark, cork, wood pulp products such as shredded newspaper (cellulose insulation), and vermiculite. All wood products, including paper, must be treated for resistance to fire, decay, insects, and rodents.

Shredded paper absorbs water easily and loses considerable R-value when damp. In addition to wall surface and/or ceiling vapor barriers, it is essential to install a waterproof membrane along the eaves to prevent water leakage.

The R-value of loose-fill insulation depends on proper application of the product. The manufacturer's instructions must be followed to obtain the correct weight per square foot of material as well as the minimum thickness. Before loose insulation is installed, the area of the space to be insulated is calculated (minus adjustments for framing members). Then, the required number of bags or pounds of insulation is determined from the bag-label charts for the desired R-value.

2.4.0 Rigid or Semirigid Insulation

Rigid or semirigid insulation is available in sheet or board form and is generally divided into two groups: structural and nonstructural. It is available in widths up to 4' and lengths up to 12'.

Structural insulating boards come in densities ranging from 15 to 31 pounds per square foot. They are used as sheathing, roof decking, and wallboard. Their primary purpose is structural, while their secondary purpose is insulation. The structural types are usually made of processed wood, cane, or other fibrous vegetable materials.

Nonstructural rigid foam board (*Figure 8*) or semirigid fiberglass insulation is usually a lightweight sheet or board made of fiberglass or

---

Loose-Fill Insulation

A disadvantage of loose-fill insulation is that it tends to settle over time. Insulation may have to be added to refill the cavities formed. Loose-fill insulation should not be covered by materials that could pack or crush it.

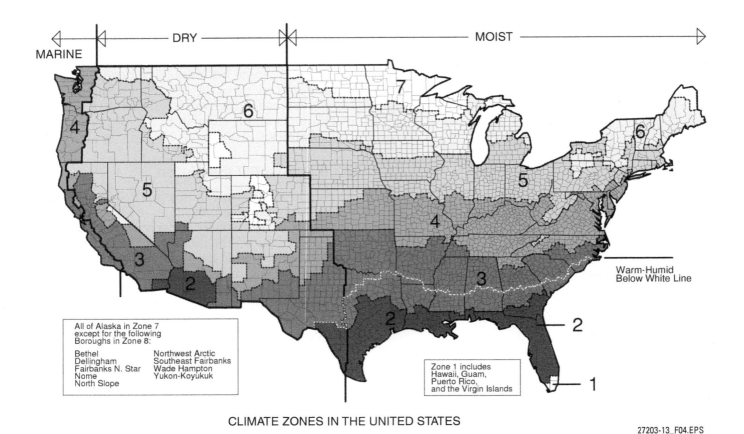

Figure 4 Climate zones in the United States.

Table 3 Recommended R-Values of Insulation

Climate Zone	Floors	Walls[1]	Ceilings
1	13	13	30
2	13	13	30
3	19	13	30
4 except Marine	19	13	38
5 and Marine 4	30[2]	19[2]	38
6	30[2]	19[2]	49
7 and 8	30[2]	21	49

[1] Wood frame walls
[2] Alternative options exist

Table 4 Typical Comfort Standards

Comfort Standard	Insulation Location	Insulation R-Value
All-weather	Walls	R-19
	Ceilings	R-30 to R-38
	Floors	R-19
Moderate	Walls	R-13
	Ceilings	R-26
	Floors	R-13
Minimum	Walls	R-11
	Ceilings	R-19
	Floors	R-11

foamed plastic such as polystyrene, polyurethane, polyisocyanurate, and expanded perlite. Most of these products are waterproof and can be used on the exteriors or interiors of foundations, under the perimeters of concrete slabs, over wall sheathing, and on top of roof decks.

In other cases, the foam board is sealed with an air infiltration film and siding is applied. The foam boards generally range in thickness from 1" to 4" with R-values up to R-30. Because all foam insulation is flammable, it cannot be left exposed; it must be covered with at least ½" of fireproof material.

Some manufacturers also provide rigid foam cores that can be inserted in concrete blocks or used with masonry products to provide additional insulation in concrete block or masonry walls.

Figure 5 Flexible blanket insulation.

Figure 6 Flexible batt insulation.

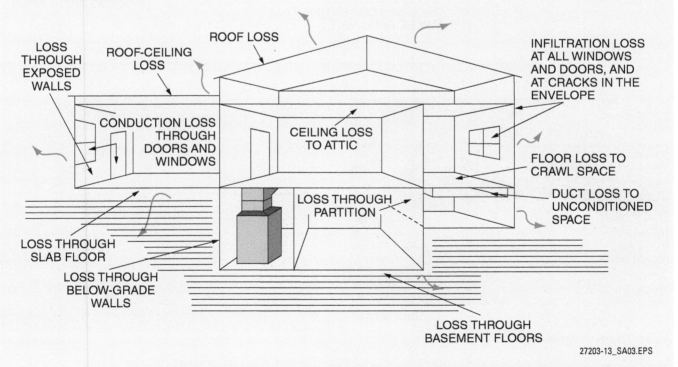

Heat Losses

At relatively cold temperatures, a building loses heat in many ways and through different areas. These include heat lost directly through its walls, ceiling, and roof. Heat also escapes through windows, doors, and gaps or cracks in the structure.

Heat Gains

At relatively warm temperatures, exposed walls and roofs absorb heat from the sun. Heat also enters a building through windows, doors, and gaps or cracks in the structure.

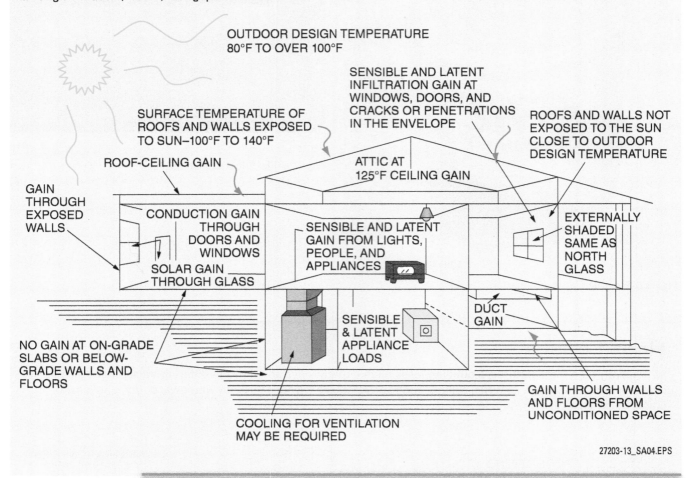

Figure 7 Loose-fill insulation.

Figure 8 Rigid foam board.

Fiberglass Ingredients

The primary ingredients of fiberglass are silica (sand) and recycled (previously melted) glass. The spun glass is held together with a chemical binder. The binder's ingredients include formaldehyde, phenol, and ammonia. The ammonia in the binder sometimes gives fiberglass a strong odor.

While the insulation itself does not burn, the binding material will burn off the glass fibers when the temperature rises high enough (about 350°F). For this reason, fiberglass insulation should not be used in applications that would subject the chemical binder to temperatures approaching its flash point.

The most common rigid and semirigid insulation types include the following:

- *Rigid expanded polystyrene* – This material has an R-value of R-4 per inch of thickness. Water significantly reduces this value because rigid expanded polystyrene is not water resistant. It is not recommended for below-grade insulation. It is the lowest in cost. This material is also called beadboard.
- *Rigid extruded polystyrene* – The R-value of this material is about R-5 per inch of thickness. It is water resistant and can be used below grade. When used in above-grade applications, it is subject to damage by ultraviolet light and must be coated or covered.
- *Rigid polyurethane and polyisocyanurate* – Initially, these boards have R-values up to R-8 per inch. However, over time the R-value drops to between R-6 and R-7 due to escaping gases. This is referred to as aged R-value. These products are subject to damage by ultraviolet light and must be coated or covered.
- *Semirigid fiberglass* – The R-value of this material is about R-4 per inch. Boards of this kind are used on below-grade slabs and walls to provide water drainage as well as insulation. They are also used under membrane roofs as insulation. On below-grade applications, the walls or floors must be waterproofed with a coating or membrane between the wall or floor and the insulation to block water penetration.

2.5.0 Reflective Insulation

Reflective insulation (*Figure 9*) usually consists of multiple outer layers of aluminum foil bonded to inner layers of various materials for strength. The number of reflecting surfaces (not the thickness of the material) determines its insulating value. To be effective, the metal foil must face an open air space that is ¾" or more in depth. In some cases, reflective material is bonded to flexible insulation as the inside surface for both insulation and vapor seal purposes.

Figure 9 Reflective foil-faced batt insulation.

Fiberglass Insulation Safety

Flexible fiberglass is probably the first thing that comes to mind when the average person thinks of insulation. Most of us don't realize that this common material must be handled carefully. The tiny strands of glass in fiberglass insulation can irritate skin, injure eyes, and cause a variety of respiratory problems. While insulation installers must wear protective equipment, their responsibilities don't end there. If debris from the installation is not properly removed or if existing insulation is disturbed, fiberglass particles could spread through the building. Fiberglass that enters an heating, ventilating, and air conditioning (HVAC) system will be carried to all parts of the building. Always use care when handling fiberglass insulation. This protects you as well as the building's current and future occupants.

Insulation Weight

Insulation materials have weight. This weight must be considered when designing a building. One of the advantages of using lightweight insulation board, such as rigid polyurethane and polyisocyanurate, is that it permits greater freedom of design. The lighter the insulation, the less weight loadbearing members of the structure must support.

2.6.0 Miscellaneous Types of Insulation

There are other types of insulation that do not fit the previous four categories. These types are as follows:

- *Foamed-in-place insulation* – This type of insulation can be applied to new or existing construction using special spray equipment. It can be injected between brick veneer and masonry walls; between open studs or joists; and inside concrete blocks, exterior wall cavities, party walls, and piping cavities. The material must be applied by trained and certified contractors.
- *Sprayed-in-place insulation* – Usually, these types of insulation (*Figure 10*) consist of confetti-like or fibrous inorganic material either mixed with an adhesive or sprayed against a wall with an adhesive coating. They are often left exposed for acoustical as well as insulating properties. Like foamed-in-place products, sprayed-in-place insulation should be applied by trained contractors.
- *Lightweight aggregates* – Insulation material consisting of perlite, vermiculite, blast furnace slag, sintered clay products, or cinders is often added to concrete, concrete blocks, or plaster to improve their insulation quality and reduce heat transmission.

> **CAUTION**
> In the 1970s, urea formaldehyde foamed-in-place insulation was injected into many homes. However, due to improper installation, the foam shrank and gave off formaldehyde fumes. As a result, its use was banned in the United States and Canada. Later, it was allowed back on the market in certain areas of the United States. A urethane foam that expands on contact can also be used. It does not have a formaldehyde problem, but it does emit cyanide gas when burned. As a result, it requires fire protection and, like urea formaldehyde, it may also be banned in some areas of the country.

Another foamed-in-place product is a phenol-based synthetic polymer (Tripolymer® made by C.P. Chemical Co.) that is fire resistant and does not drip or create smoke when exposed to high heat. This material does not expand once it leaves the delivery hose of the proprietary application equipment.

Figure 10 Sprayed-in-place insulation.

Reflective Insulation

Reflective insulation, by itself, can only block radiated heat. At relatively hot temperatures, it helps keep buildings cooler by deflecting heat from the sun. At relatively cold temperatures, it can do little to prevent heat from escaping the building.

Sprayed- and Foamed-in-Place Insulation

Sprayed- and foamed-in-place insulation materials are well suited for irregular surfaces. These include walls and ceilings that are curved or that have beams, pipes, or other equipment protruding from them. Foams and sprays can be built-up in layers to the desired insulation thickness.

2.0.0 Section Review

1. Heat loss can be significantly reduced by using _____.
   a. vapor barriers
   b. shingles
   c. insulation sheathing
   d. concrete masonry units

2. Flexible batt insulation (fiberglass or mineral wool) is manufactured in widths of 15" or _____.
   a. 19"
   b. 23"
   c. 28"
   d. 32"

3. The number of bags or weight of loose-fill insulation needed to reach specific R-vaules is found _____.
   a. on bag-label charts
   b. in the Hazard Communication Standard
   c. on the aluminun foil bonding
   d. on the SDS

4. The least expensive type of rigid or semirigid insulation is _____.
   a. rigid polyurethane
   b. rigid extruded polystyrene
   c. semirigid fiberglass
   d. rigid expanded polystyrene

5. Reflective insulation always has a single layer of aluminum foil.
   a. True
   b. False

6. When exposed to fire, some urethane foam insulation can emit _____.
   a. polyurethane fumes
   b. chlorine gas
   c. cyanide gas
   d. formaldehyde fumes

7. To be effective, the metal foil surface of reflective insulation must face an open air space with a depth of at least _____.
   a. ¼"
   b. ½"
   c. ¾"
   d. 1"

SECTION THREE

3.0.0 INSTALLING INSULATION

Objective

Describe the various installation methods for insulation.
a. Explain how to install flexible insulation.
b. Explain how to install loose-fill insulation.
c. Explain how to install rigid or semirigid insulation.

Performance Tasks 1 and 2

Install blanket insulation in a wall.
Install a vapor barrier on a wall.

Before installation, building plans and codes must be checked to determine the R-values and the types of insulation required or permitted for the structure being insulated. Then, the required amount of insulation for the structure must be calculated. Any specific instructions provided by the selected manufacturer must be followed when installing the insulation.

3.1.0 Installing Flexible Insulation

> **WARNING!** Wear proper eye protection, respiratory protection, and gloves when handling and installing insulation.

Use the following procedure when installing typical flexible insulation:

Step 1 For walls, measure the inside cavity height and add 3". From the wall, lay the distance out on the floor and mark it. Unroll blanket insulation or lay batts on the floor. Use two layers or more. At the cut mark, compress the insulation with a board and cut it with a utility knife. On blanket or faced insulation, remove about 1" of insulation from the ends to provide a stapling flange at the top and bottom.

> **NOTE** Refer to manufacturer's recommendations for proper installation.

Step 2 If a separate interior vapor seal will be installed, install blanket or faced insulation so that the stapling flange is fastened to the inside surfaces of the wall studs, top plate, and sole plate (*Figure 11*).
- If the facing of the blanket or batt is the vapor seal, install the stapling flange on the face of the studs and overlap them by at least 1" (*Figure 12*). For faced or blanket insulation, use a power, hand, or hammer stapler to first staple the top flange to the plate.
- Align and staple down the sides.
- Staple the bottom flange to the sole plate. Pull the flanges tight and keep them flat when stapling. Space staples about 12" apart if stapling to the face of the studs; on the sides, space them about 6" apart.
- For unfaced batt insulation, install the batt at the top and bottom first and push it tight against the plates.
- Evenly push the rest of the batt into the cavity (*Figure 13*). For narrow spaces around windows and doors, stuff the spaces with pieces of insulation and cover it with a plastic or tape vapor seal.

> **WARNING!** Exercise caution when installing insulation around electrical outlet boxes and other wall openings or devices. Failure to do so may result in electrocution.

Step 3 Faced or blanket insulation for ceilings or floors is usually installed from the bottom in the same manner as the walls. Unfaced batts can be installed from either the top or the bottom.

- Make sure that ceiling insulation extends over the wall into the soffit area (*Figure 14*). Also make sure soffit baffles (*Figure 15*) are inserted over and cover the ceiling insulation. The baffles should be fastened to the roof deck to hold them in place so that they do not slide down into the soffit and block ventilation. Soffit baffles allow air to enter the attic to properly ventilate it.
- For floors, ensure that the insulation is installed around the perimeter of the floor against the header (*Figure 16*). Floor insulation over a basement is installed with the vapor barrier facing down.

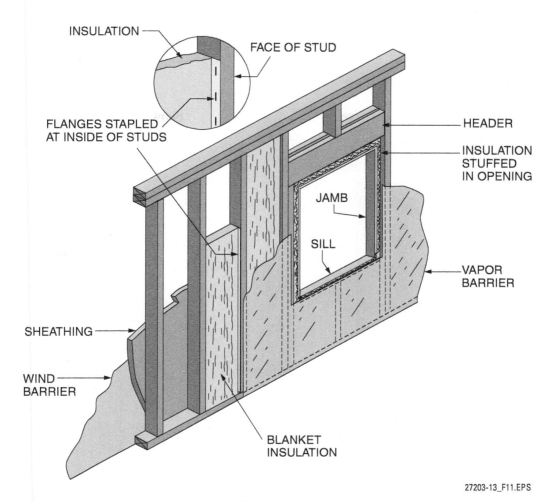

Figure 11 Blanket installation without integral vapor seal.

- Over a crawl space, the vapor barrier faces up. In either case, the insulation can be supported below by a wire mesh (chicken wire), if desired.

3.2.0 Installing Loose-Fill Insulation

For new construction, loose-fill insulation is used primarily for attic insulation. On older construction, it can also be blown into wall cavities through holes drilled at the center and tops of exterior walls. The following steps only cover attic or ceiling installation. Refer to manufacturer's recommendations for proper installation.

> **WARNING!** Wear proper eye protection, respiratory equipment, and gloves when handling and installing insulation.

Step 1 Make sure that the finished ceiling below has been installed. Also, ensure that a separate vapor barrier has been installed between the loose-fill insulation and finished ceiling to prevent moisture penetration of the insulation and to prevent the fine dust from the insulation from penetrating the ceiling in the event of future cracks (*Figure 17*). Make sure that soffit baffles and blocking have been installed to prevent the material from spilling into the soffits.

Cathedral Ceilings

If a cathedral ceiling incorporates gypsum drywall attached to the bottom of the rafters, airflow must be maintained from the soffit to the ridge. Proper ventilation must also be maintained above and below skylights to prevent buildup of heat and moisture. Check your local code for the appropriate methods to use when working with cathedral ceilings.

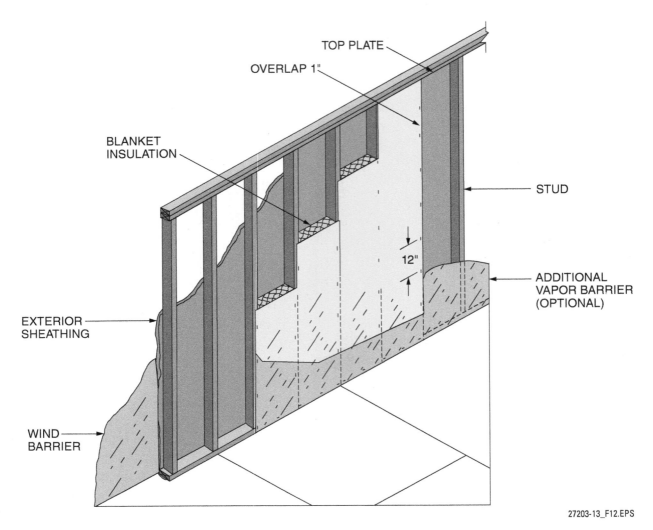

Figure 12 Blanket installation with integral vapor seal.

Step 2 If the final insulation depth will be higher than the ceiling joists, permanently install strike-off boards, as shown in *Figure 18*. Pour the insulation from bags or blow the insulation over the ceiling joists using special equipment. Using a straightedge, tamp the insulation and then level it to the required depth for the R-value desired.

Polystyrene Forms

Structural forms made of polystyrene are sometimes used for residential and light commercial construction. Concrete is poured into the forms, which are left in place to provide insulation for the walls. The forms usually provide sufficient insulation by themselves, but check the local code for these requirements.

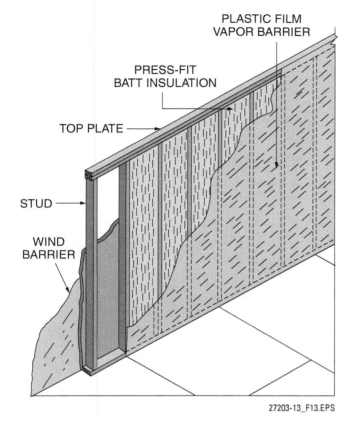

Figure 13 Batt insulation with separate vapor barrier.

Figure 15 Typical plastic soffit baffle (shown upside down).

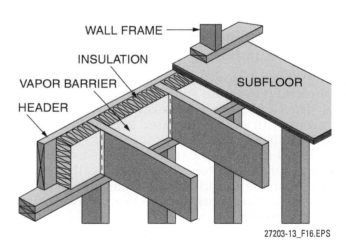

Figure 16 Perimeter floor insulation.

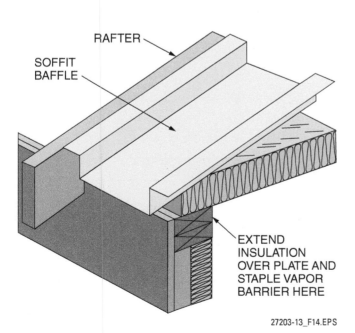

Figure 14 Ceiling insulation at wall and soffit.

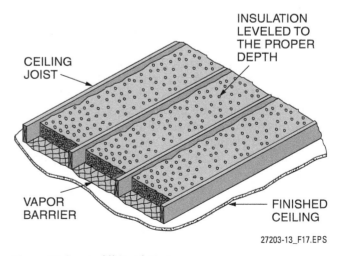

Figure 17 Loose-fill insulation.

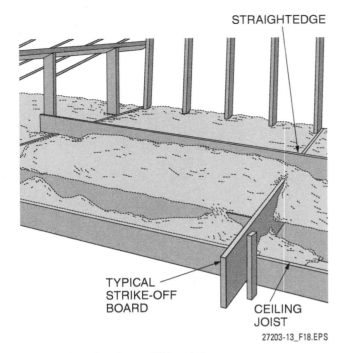

Figure 18 Leveling loose-fill insulation.

3.3.0 Installing Rigid or Semirigid Insulation

Rigid insulation panels can be fastened like sheathing over the studs or wood sheathing of a structure. Nails with large heads/washers or screws with washers are used to prevent crushing the insulation.

Rigid insulation panels may be installed on the exterior of a foundation. Typically, the exterior of the foundation is waterproofed first. Then, the panels are applied over special mastic and secured with concrete nails to hold them in place until the mastic sets. For existing construction, the panels may be installed on the interior of the foundation if the walls are adequately waterproofed.

Figure 19 shows typical methods of installing rigid insulation under surface slabs. Usually, the insulation is only applied around the perimeter of the slab, anywhere from 24" to 36" from the edge of the slab and/or down the inside of the slab footings to below the frost line (*Figure 20*). A vapor barrier should be applied under the slab and over any insulation under the slab. Reflective insulation is installed in the same manner. Refer to manufacturer's recommendations for proper installation.

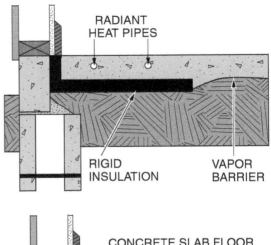
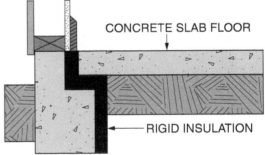
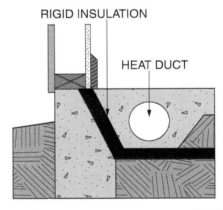

Figure 19 Rigid insulation installed under a concrete slab.

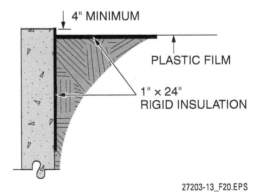

Figure 20 Rigid insulation installed under a slab-and-down footing.

3.0.0 Section Review

1. When installing flexible insulation between studs, staple the side flanges at intervals of _____.
   a. 3"
   b. 6"
   c. 9"
   d. 12"

2. Strike-off boards are sometimes used in the installation of _____.
   a. loose-fill insulation
   b. rigid insulation
   c. flexible insulation
   d. foamed-in-place insulation

3. When installing rigid insulation under a surface slab, the insulation should _____.
   a. be at least 3" in thickness
   b. extend 24" to 36" in from the perimeter of the slab
   c. be placed beneath the entire slab
   d. be installed above a vapor barrier

SECTION FOUR

4.0.0 MOISTURE CONTROL, WATERPROOFING, AND VENTILATION

Objective

Identify the requirements for moisture control, waterproofing, and ventilation, and describe the related installation methods.
  a. List various methods to control moisture in a structure.
  b. Identify methods to waterproof a structure.

Performance Task 3

Install selected building wraps.

Trade Terms

Condensation: The process by which a vapor is converted to a liquid, such as the conversion of the moisture in air to water.

Convection: The movement of heat that either occurs naturally due to temperature differences or is forced by a fan or pump.

Dew point: The temperature at which air becomes oversaturated with moisture and the moisture condenses.

Diffusion: The movement, often contrary to gravity, of molecules of gas in all directions, causing them to intermingle.

Perm: The measure of water vapor permeability. It equals the number of grains squared of water vapor passing through a 1-square-foot piece of material per hour, per inch of mercury difference in vapor pressure.

Permeability: The measure of a material's capacity to allow the passage of liquids or gases.

Permeable: Porous; having small openings that permit liquids or gases to seep through.

Permeance: The ratio of water vapor flow to the vapor pressure difference between two surfaces.

Water stop: Thin sheets of rubber, plastic, or other material inserted in a construction joint to obstruct the seepage of water through the joint.

Water vapor: Water in a vapor (gas) form, especially when below the boiling point and diffused in the atmosphere.

Proper moisture control, waterproofing, and ventilation must be provided to a structure to allow it to breathe while still resisting rain, snow, heavy winds, and other weather elements,

4.1.0 Moisture Control

Water vapor contained in air can readily pass through most building materials used for wall construction. This vapor caused no problem when walls were porous because it could pass from the warm wall to the outside of the building before it could condense into liquid water (*Figure 21*).

When buildings were first constructed with insulation in the walls to cut down on heat loss, moisture in the air passed through the insulation until it reached a point cold enough to cause it to condense. The condensed moisture froze in very cold weather and reduced the efficiency of the insulation. The ice contained within the wall thawed as the weather warmed, and the resulting water in the wall caused studs and sills to decay over time.

For these reasons, it is important to keep cellars, basements, crawl spaces, exterior walls, and attics dry. Moisture in crawl spaces, basements, and attics also encourages destructive insects such as termites, as well as the growth of mold. In the case of crawl spaces, moisture often rises from the ground into the crawl space during periods of heavy rain. To prevent the concentration of this damaging moisture, some precautions must be taken in the original design of the structure:

- The earth must slope down and away about 20' from the structure, carrying surface water away.
- The crawl space should be protected from moisture by a vapor barrier on the ground.
- The foundation walls should be penetrated with vents so that moisture will not be trapped in the crawl space.
- A vapor barrier should be installed between the insulation and the subfloor.

Basements usually have the most trouble with condensation in summer during humid weather. The earth under the concrete basement floor is comparatively cool, causing the floor of the basement to be a cold surface. The hot air is saturated with moisture and condenses when it comes in contact with the cooler surfaces of the floor and walls. This problem is difficult to control. If the surface of the concrete is rough and porous, the moisture will sink in and not cause a wet-

be solved by the use of dehumidification devices during the summer months.

Moisture weeping through the concrete floor is a different problem. In new construction, this is controlled by installing perimeter drainage and a vapor barrier under the concrete slab. When installing polyethylene film as an underslab vapor barrier, be careful not to tear, puncture, or damage the film in any way. Any passageways for moisture will defeat the purpose of the vapor barrier. Prior to pouring the concrete slab, make sure the polyethylene film is placed properly and is free of punctures. Keep all construction debris away from the vapor barrier.

To keep moisture from rising up into the basement, 6" of coarse gravel should be placed over the compacted earth to provide drainage to the perimeter drain before the slab is poured. A polyethylene film should be placed on top of the gravel to keep the concrete from penetrating into the gravel and possibly weakening the slab. In very wet areas or areas with a high water table, floor drainage, in addition to a gravel bed, may also be required.

4.1.1 Interior Ventilation

One of the best ways to reduce or eliminate the chances of moisture damage in attics or in the space between the rafters and the finished roof is through proper ventilation. Ventilation provides a stream of outside air to remove trapped moisture before it is allowed to do any damage. In insulated attics, baffles (blocking strips) are used to keep the insulation material from getting into the vented areas. With the increased use of blown-in insulation in attics, baffles are being required by code in some areas.

The amount of ventilation required varies by climate and building codes. Attics and gable and hip roofs may be ventilated with a variety of louvers and vents. Flat roofs are ventilated with a combination of eave vents and roof stacks (*Figure 22*).

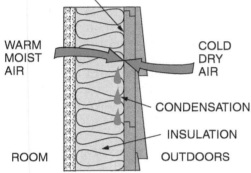
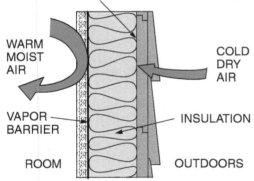

Figure 21 Effects of insulation and vapor barrier.

ness problem. If, however, the floor is dense and smoothly finished, the tightly knit grains of concrete form a vapor barrier of sorts, and the water collects on the slab. This problem can usually

Mold

Moisture accumulating inside a building can damage the structure and promote the growth of mold. While it is not always harmful, this mold may cause allergic reactions or other respiratory problems in some people. Airborne mold spores can also cause infections, primarily in people whose immune systems are compromised.

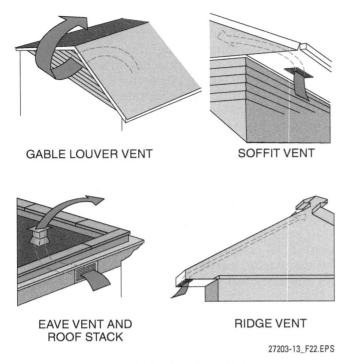

Figure 22 Various methods of roof ventilation.

4.1.2 Vapor Barriers

Vapor barriers are any material or substance that will not permit the passage of water vapor or will do so only at an extremely slow rate. The **permeability** of a substance is a measure of its capacity to allow the passage of liquids or gases. Water vapor permeability is the property of a substance to permit the passage of water vapor and is equal to the **permeance** of a substance that is 1" thick. The measure of water vapor permeability is the **perm**. This equals the number of grains of water vapor passing through a 1-sq-ft piece of material per hour, per inch of mercury difference in vapor pressure. All you really have to remember is that any material that has a perm rating of 1.0 or less is considered a vapor retarder and will not allow the passage of any appreciable or harmful amounts of water vapor. Any material with a rating higher than 1.0 is a breathable material that will permit the passage of water vapor in whatever degree its perm rating indicates. The higher the perm number, the greater the amount of water vapor that will pass through the material in a given time; 0.0 is totally impermeable (*Table 5*).

A properly installed vapor barrier will protect ceilings, walls, and floors from moisture originating within a heated space (*Figure 23*).

An insulated wall will divide two temperature gradients. The area on the inside of the structure will normally be warmer than the air on the outside. The vapor barrier is usually located on the warm side to prevent moisture from moving through the insulation to the cool side and condensing.

Properly designed subroof ventilation is the best weapon for preventing water vapor infiltration into a steeply sloped roof, but is less effective on roofs with low slopes because natural **convection** decreases with diminishing roof height. Moisture dissipation occurs through **diffusion** and wind-induced ventilation.

Normally, the ventilation requirement for a gable roof is 1 sq ft of free air ventilation for every 300 sq ft of ceiling area if a vapor barrier exists under the ceiling. If no vapor barrier is present, the requirement is 1 sq ft for every 150 sq ft of ceiling area. The total requirement must be split evenly between the inlet vents and the outlet vents.

Free air ventilation is the rating of the ventilation devices, taking into account any restrictions caused by screening, louvers, and other devices.

4.1.3 Materials

Common vapor barrier materials include asphalted kraft paper, aluminum foil, and polyethylene film.

Asphalted kraft paper is usually incorporated with blanket or batt insulation. It serves as a

Ice Dams

In colder climates, ice dams can be a problem. Ice dams are formed along the edge of a sloping roof when a building's attic is not properly insulated and ventilated. Heat escaping through the roof melts accumulated snow, forming icicles along the edge of the roof. Over time, water collects under the outer layer of snow and is trapped by the ice. This water backs up under the shingles and penetrates the roof, causing water damage and other problems.

Some homeowners use special snow rakes to remove snow from the roof. This helps prevent ice dams from forming, but it is labor intensive. If used improperly, the rake may actually damage the roof. The best way to prevent ice dams is to properly insulate and ventilate the building's attic.

Ice dams can usually be avoided by installing plenty of insulation and providing ample ventilation in the attic.

Table 5 Perm Ratings of Various Vapor Retarder Materials

Material	Permeance
Aluminum foil (1 mil)	0*
Aluminum foil (0.35 mil)	0.05*
Polyethylene (4 mil)	0.08*
Polyethylene (6 mil)	0.06*
Polyester (1 mil)	0.07*
Saturated and coated roll roofing	0.05**
Reinforced kraft and asphalt-laminated paper	0.3**
Asphalt-saturated and asphalt-coated vapor barrier paper	0.2 to 0.3*
15-lb tarred felt	4.0**
15-lb asphalt felt	1.0**
12.5-lb asphalt	0.5**
22-lb asphalt	0.1*
Built-up membrane (hot-mopped)	0**

* Per *ASTM E96-66, Water Vapor Transmission of Materials in Sheet Form*
** Per *ASTM C355-64, Water Vapor Transmission of Thick Material*

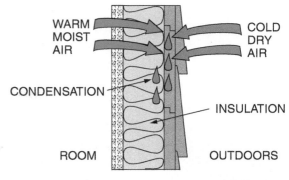
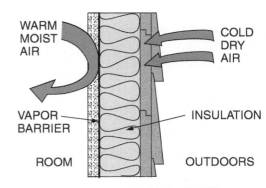

Figure 23 Vapor barrier installation.

means for attaching the insulation to the building framework and as a reasonably good vapor barrier when installed on the warm side of the wall or ceiling (*Figure 24*).

Aluminum foil may be incorporated with blanket or batt insulation in the same manner as kraft paper. It is also applied to the back of gypsum lath and gypsum wallboard where it works as a relatively effective vapor barrier.

Polyethylene film is applied over the studs and ceiling joists after the insulation is installed. When wallboard with polyethylene film or foil backing is used, the insulation will normally be plain batts or blankets that do not have an integral vapor barrier. As a vapor barrier, polyethylene film is stapled over the studs and also covers the window frames. This helps to keep the window frames and sashes clean during application and finishing of the gypsum wallboard. The film should be overlapped 2" to 4" and sealed with special mastic or tape.

4.1.4 Installation in Crawl Spaces

The ground under a ventilated crawl space should be covered with a vapor barrier ground cover to protect the underside of the house from condensation (see *Figure 25*). A vapor barrier should also be installed over the subfloor above the crawl space.

Besides installing vapor barriers, crawl spaces should be properly vented to permit the escape

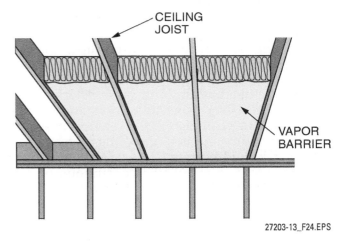

Figure 24 Installing insulation batts between ceiling joists with vapor barrier down.

of moisture. Usually, this is accomplished by the use of a proper number of screened foundation vents installed in the above-grade foundation surrounding the crawl space. The normal requirement is 1 sq ft of free air ventilation for every 160 sq ft of crawl space area when a vapor barrier ground cover is used.

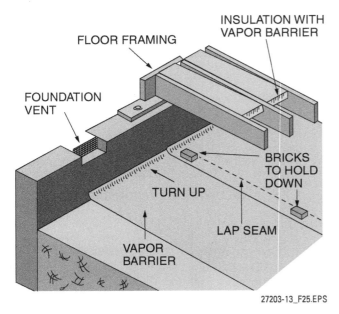

Figure 25 Vapor barrier installation for crawl spaces.

4.1.5 Installation in Slabs

When allowed to proceed unchecked, moisture will migrate from the ground upward through concrete and into the building, where it can cause moisture problems, damage, and higher energy costs. Even though the water table may be several feet below the slab, moisture vapor will migrate up to and through concrete slabs.

Up to 80 percent of the moisture entering a structure does so by migrating from the ground beneath the structure. Moisture vapor passes through concrete more readily than liquid moisture.

Moisture in a building can cause deterioration of interior finishes, especially floors and equipment. Moisture can also add to energy costs by raising humidity and taxing cooling systems that require dehumidification.

Vapor barriers should be continuous under the slab. Great care must be taken not to tear or puncture the barrier. Keep all construction debris away from the barrier location. Vapor barrier installation must be done by qualified contractors.

When used in thickened-edge slab construction, as shown in *Figure 26*, a vapor barrier is placed between the gravel cushion and the poured concrete. The same arrangement is used for other types of slab-on-grade construction.

Figure 27 shows a method of constructing a finished floor over a concrete slab, which affords double protection against moisture. The sealer or waterproofing is placed on the slab itself, and a vapor barrier is suspended above the slab.

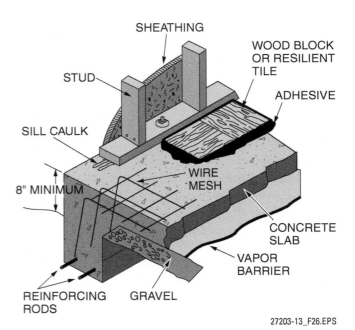

Figure 26 Thickened-edge slab vapor barrier installation.

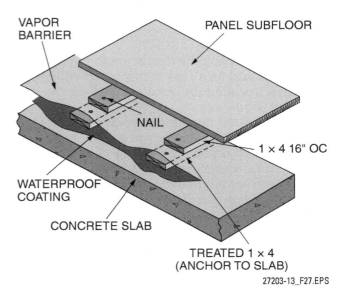

Figure 27 Surface-mounted vapor barrier on a slab.

4.1.6 Installation in Walls

A polyethylene-sheet vapor barrier is easy to apply to frame walls where no integral barrier is provided or where a supplementary barrier is preferred. A flap would normally overlap both floor and ceiling barriers to seal the interior off completely. Adjacent sheets of the film are overlapped 2" to 4" and are sealed with a special mastic or tape.

When vapor barriers are applied to walls, particular attention should be paid to fitting the material around electrical outlet boxes, exhaust fans, light fixtures, registers, and plumbing. Considerable water vapor can escape through the cracks around the equipment, travel from the warm side of the wall to the cold side, and condense on the sheathing or siding. This is especially true if the insulation is poorly fitted at the top and bottom.

4.1.7 Installation in Roofs

A major cause of failure in built-up roofs is condensation of moisture vapor, which rises from inside the building and penetrates the roof deck insulation. When this vapor reaches its dew point, which can occur inside the insulation or at the cool outer surface, it condenses. This results in a reduction or total loss of the thermal efficiency of the insulation, as well as dripping and damage. To prevent this, select a vapor barrier that is both easy to apply and resistant to job-site abuse. Install it on the warm side of the roof deck insulation.

4.2.0 Waterproofing

The single most critical area for waterproofing construction is the below-grade foundation wall. Rising water tables, hydrostatic pressures, structural movement, and groundwater all require a special type of protection. A liquid waterproofing system applied by spray methods ensures the high buildup of film thickness needed to cope with these problems (*Figure 28*).

For all below-grade applications of waterproofing, be sure to fill all cracks, crevices, and grooves. Ensure that the coating is continuous and free from breaks and pinholes.

Carry the coating over the exposed tops and outside edges of the footing (*Figure 29*), forming a cove at the junction of the wall and footing.

Spread the coating around all joints, grooves, and slots, and into all chases, corners, reveals, and soffits. Bring the coating up to the finished grade.

Do not place backfill for 24 to 48 hours after application. Where possible, backfill should be placed within approximately seven days to avoid any unnecessary damage due to construction activities. Take care to place the backfill in a manner that will not rupture or damage the film or cause the coating or membranes to be displaced on the coated surface.

4.2.1 Water Stops

Water stops are thin sheets of rubber, plastic (polyvinyl chloride [PVC]), or other material inserted in a construction joint to obstruct the seep-

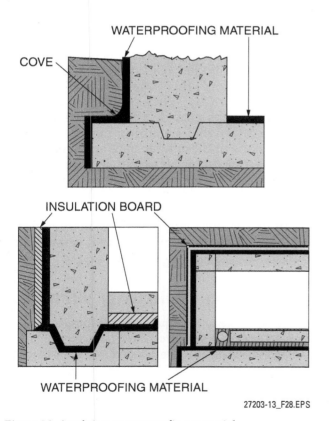

Figure 28 Applying waterproofing material.

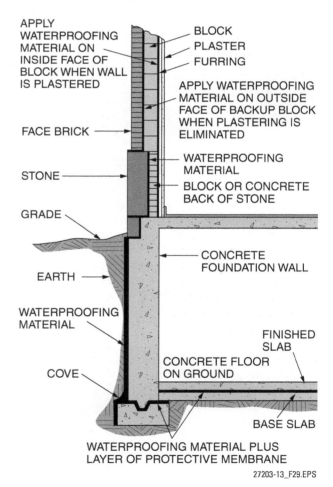

Figure 29 Below-grade waterproofing application.

age of water through the joint. PVC water stops may be used for installations in underpasses, tunnels, tanks, locks, walls, swimming pools, siphons, sewage disposal plants, reservoirs, culverts, sewage treatment plants, channels, drums, filtration plants, foundations, bridges, basements, abutments and decks, mineshafts, aqueducts, retaining walls, and roofs. Refer to *Figure 30* for various applications of water stops.

4.2.2 Joint Treatment

Clean and debris-free joints in structures are critical. They must maintain integrity during movement, yet remain permanently waterproof and airtight. Therefore, it is important to select the proper joint treatment system to avoid problems with moisture penetration at the construction joints.

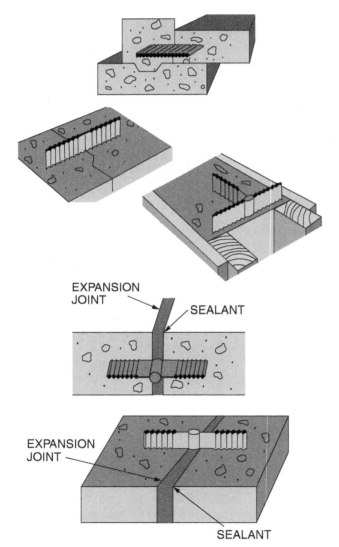

Figure 30 Water stops used in joints.

4.2.3 Vapor Barrier for Cold Storage and Low-Temperature Facilities

Cold-storage vapor barriers are designed for use in areas of extremely low temperatures to halt the migration of damaging moisture vapor into and through the insulation. Most cold-storage vapor barriers consist of two layers of kraft paper, each extrusion coated with black polyethylene; a layer of aluminum foil; two layers of nonasphaltic adhesive; and two layers of high-tensile-strength reinforcing fibers embedded in the adhesive.

4.2.4 Air Infiltration Control

In addition to insulation, the exterior sheathing of a structure should be covered to prevent wind pressure from causing infiltration of outside air into the structure. To achieve maximum energy efficiency in a structure, air infiltration must be strictly controlled.

Traditionally, structures have been covered with water-resistant building paper to help prevent water leakage through the primary barrier (siding) from reaching the structural sheathing or other components of the structure. Additionally, the paper had to be water permeable to allow moisture inside the walls of the structure to pass through and evaporate. To some extent, the paper reduced air infiltration of the structure, especially when board sheathing was used.

For a number of years, products called house wraps or building wraps have been used to replace building paper. These products, under brand names such as Tyvek® or ProWrap®, are easier to apply and perform the same functions as building paper (*Figure 31*). When properly applied and sealed, the wraps provide a nearly airtight structure no matter what sheathing material is used. Most versions of these wraps are an excellent secondary barrier under all siding, including stucco and exterior insulation finish systems (EIFSs).

Many of these products are made of spun, high-density, polyethylene fibers randomly bonded into an extremely tough, durable sheet material. They are usually available in several versions and

---

Securing Building Wraps

While not necessarily required by the local code, a properly secured wrap provides a building with a weather-resistant seal. Make sure the tape used to seal edges and joints in the wrap is approved by the manufacturer. The wrap must be properly installed and secured.

RESIDENTIAL APPLICATION

COMMERCIAL APPLICATION

Figure 31 Building wrap.

SCREWS WITH PLASTIC WASHERS

CONTRACTOR'S TAPE

Figure 32 Building wrap accessories.

weights for residential and light commercial use. Special versions may be available with vertical water channeling permanently pressed into the material for stucco and EIFS. The material is usually furnished in rolls in various sizes from 18" wide to 10' wide and in lengths from 100' to 200'.

Nails with large heads, nails or screws with plastic washers (*Figure 32*), or 1"-wide staples may be used to secure the wrap to wood, plastic, insulating board, or exterior gypsum board. Screws and washers are used for steel construction. Special contractor's tape (*Figure 32*) or sealants compatible with the wrap are used to seal the edges and joints of the wrap.

> **WARNING!**
> Some building wraps are slippery and should not be used in any application where they can be walked on. Because the surface will be slippery, use pump jacks or scaffolds for exterior work above the lower floor. If ladders must be used, extra precautions must be taken to prevent the ladders from sliding on the wrap.

Always refer to the manufacturer's instructions for specific installation information. House or building wrap is generally installed as follows:

Step 1 Using two people and beginning at a corner on one side of the structure, leave 6" to 12" of the wrap extended beyond the corner to be used as an overlap on the adjacent side of the structure (*Figure 33*). Align the roll vertically and unroll it for a short distance. Check that the stud marks on the wrap align with the studs of the structure. Also check that the bottom edge of the wrap extends over and runs along the line of the foundation. Secure the wrap to the corner at 12" to 18" intervals.

Step 2 Unroll the wrap 2 or 3 more feet and ensure that it overlaps and runs along the line of the foundation. Secure the wrap vertically at 12" to 18" intervals on each stud, using the stud marks as a fastening guide. Continue around the structure, covering all openings. If a new roll is started, overlap the end of the previous roll 6" to 12" to align the stud marks of the new roll with the studs of the structure.

Step 3 If the upper parts of the structure require coverage, repeat Steps 1 and 2, starting above the existing wrap. Make sure that the bottom edge of this layer of wrap aligns along the top edge of the lower wrap and overlaps it by 6" to 12".

Step 4 At the top plate, make sure the wrap covers both the lower and upper (double) top plate (*Figure 34*), but leave the flap loose for the time being.

Step 5 At each opening, use one of the following two methods to cut back the wrap.

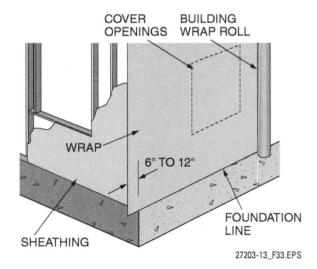

Figure 33 Starting a roll of building wrap.

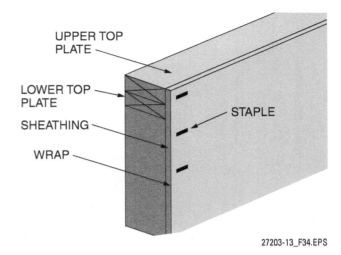

Figure 34 Top plate detail.

> **NOTE**
> Always follow the window or door manufacturer's recommendations for flashing windows or doors.

Method 1—Uninstalled Windows/Doors:
- At the opening, cut the wrap as shown in *Figure 35*. Fold the three flaps around the sides and bottom of the opening and secure every 6". Trim off the excess.
- At the outside, install 6" flashing along the bottom of the opening, then up the sides over the top of the wrap.
- Install head flashing at the top of the opening under the wrap and over the side flashing (*Figure 36*). Tape the flap ends to the head flashing using tape approved by the manufacturer.

Method 2—Windows/Doors with Flanges:
- Create a top flap of the wrap. Insert a head flashing under the flap and over the flange.
- Extend the flashing to the sides about 4" and tape the flap to the head flashing.
- On the remaining sides, trim the wrap to overlap the flange area and tape the edge to the flanges (*Figure 37*).

Step 6 Secure all the bottom edges of the wrap to the foundation with the recommended joint sealer, then fasten the lower edge to the sill. At the top plate, seal the edge to the upper plate with the sealer and fasten the edge to the plate.

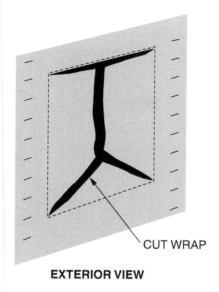

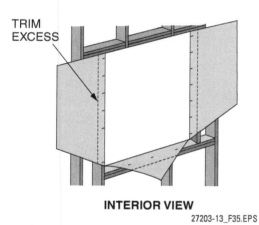

Figure 35 Cutting and folding wrap at an opening.

Step 7 Seal all vertical and horizontal joints in the wrap with the recommended tape.

Step 8 Before applying the siding, repair any damage or tears in the wrap with tape or sealant.

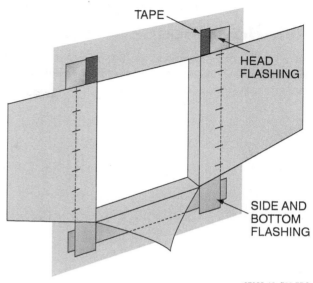

Figure 36 Installing flashing around an opening.

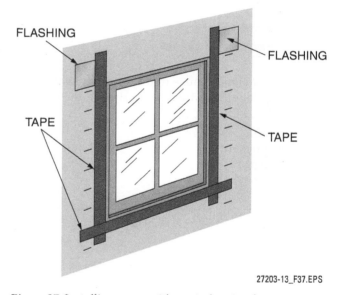

Figure 37 Installing wrap with a window in place.

Window Flashing System

Take special care to prevent moisture from entering the structure around windows and doors. The DuPont® flashing system is designed for this purpose.

27203-13_SA05.EPS

4.0.0 Section Review

1. The area of a house that typically has the most trouble with condensation during the summer is the _____.
   a. attic
   b. kitchen
   c. bathroom
   d. basement

2. The most common material used for water stops is _____.
   a. PVC
   b. aluminum
   c. polyurethane
   d. asphalt-impregnated felt

Section Five

5.0.0 Estimating Typical Insulation Requirements

Objective

Describe the estimating procedure for thermal and moisture projects.

The following is a method of estimating the amount of insulation for the walls, ceilings, and floors of a single-story structure. If no plans are available and the codes specify only a minimum R-value for a structure, refer to the comfort level standards in this module and select a desired comfort level based on the occupancy of the structure. Then, perform the following steps to calculate the amount of required insulation material:

Step 1 Determine the square footage of exterior walls to be insulated:
- Determine from the plans or measure the perimeter length of each exterior wall in feet.
- Add the lengths of all exterior walls to find the total perimeter of the structure.
- Multiply the total perimeter by the ceiling height to find the total square footage of the walls:

  Exterior perimeter (ft) ×
  ceiling height (ft) =
  total sq ft of walls

- Determine from the plans or measure the square footage of each opening in the perimeter walls:

  Height (in) × width (in) =
  opening size (sq in) ÷ 144 sq in/sq ft =
  opening size (sq ft)

- Determine the total square footage of all openings by adding the square footage of each opening.
- Subtract the total square footage of all openings from the total square footage of the walls to find the square footage of insulation required for the walls:

  Total wall area (sq ft) −
  total opening area (sq ft) =
  sq ft of wall insulation

Step 2 Calculate the square footage of ceiling/floor to be insulated. The square footage of a floor will be the same as the ceiling. The square footage of either one can be calculated, and the result can be used for both.
- Divide the ceiling or floor plan into rectangular or square areas, and from the plans or by measurement, determine the length and width of each area (*Figure 38*).
- For each area, multiply the length by the width to determine the square footage:

Length (ft) × width (ft) = sq ft of each area

- Add the square footage of all areas to find the total square footage of insulation required for the ceiling or floor.

Step 3 Add the total insulation square footage required for the walls, the ceiling, and, if required, the floor to determine the total square footage of insulation required for the structure.

Step 4 Divide the square footage of the structure by the coverage per package of insulation for the R-value required. The coverage information will be given in the manufacturer's information for the insulation to be used. See *Table 6* for several examples of package coverage for various R-values and package sizes.

> **NOTE:** The walls, ceiling, and floor of a structure may require different insulation R-values. Check your local code to determine what the requirements are for each installation.

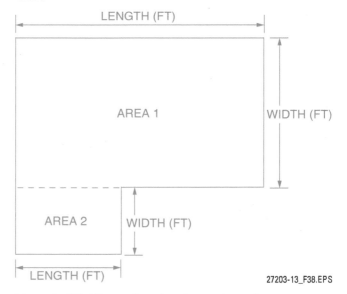

Figure 38 Dividing a plan view into rectangular or square areas.

Table 6 Typical Insulation Coverage for Various Types of Packaging and R-Values

R-Value	Thickness	Width × Length	Square Feet per Package	Pieces per Package
11	3½"	15" × 94"	88	9
13	3⅝"	15" × 94"	88	9
19	6¼"	15" × 94"	49	5
30	9½"	16" × 48"	37	7
38	12"	24" × 48"	48	6

5.0.0 Section Review

1. To find the square footage of a building's walls when calculating insulation needed, multiply the building's perimeter by the ceiling height, then subtract the square footage of openings in the wall.
    a. True
    b. False

SUMMARY

This module presented the materials and procedures that can be applied to ensure that effective insulation, moisture control, ventilation, waterproofing, and air infiltration control are achieved. Carpenters will be involved in the application of these materials and must be thoroughly familiar with the materials and the typical installation techniques covered in this module.

Review Questions

1. The R-value is a measure of the ability of a material to _____.
   a. resist the passage of moisture
   b. resist heat conduction
   c. allow cold air to enter a building
   d. convert water vapor into a liquid

2. The safety data sheets provided with materials that contain hazardous chemicals must be_____.
   a. posted prominently on the job site
   b. properly stored for reference in case of emergency
   c. given to each worker on the job
   d. attached to the building plans

3. A vapor barrier is any material that _____.
   a. is used to cover building walls
   b. prevents the infiltration of air
   c. prevents the passage of water
   d. is used to prevent roof leaks

4. Insulation sheathing can be installed to significantly reduce _____.
   a. heat loss
   b. construction time
   c. the need for bracing
   d. labor costs

5. Resistance to conductive heat flow through a material or gas is measured as an _____.
   a. C-value
   b. F-value
   c. N-value
   d. R-value

6. Manufactured insulation is based on trapping _____.
   a. large amounts of air in a few small spaces
   b. large amounts of air in a large number of very small spaces
   c. large amounts of air in a few large spaces
   d. small amounts of air in a few large spaces

7. For thermal transmission control purposes, insulation does not have to be installed _____.
   a. above ceilings
   b. in exterior walls
   c. in interior walls
   d. beneath floors over crawl spaces

8. Flexible insulation may be faced with fire-resistant foil scrim, which is abbreviated as _____.
   a. FRSI
   b. FSK
   c. FFS
   d. FRFS

9. A loose-fill insulation material that does not need to be treated to resist fire or insect infestation is _____.
   a. wood fiber
   b. shredded paper
   c. glass wool
   d. cork

10. Rigid foam insulation may have insulation values as high as _____.
    a. R-8
    b. R-18
    c. R-22
    d. R-30

11. Rigid and semirigid insulation is available in sheets up to 4' wide and _____.
    a. 4' long
    b. 6' long
    c. 8' long
    d. 12' long

12. The primary purpose of structural insulating boards is _____.
    a. insulation
    b. appearance
    c. structural
    d. waterproofing

13. The insulating value of reflective insulation depends upon _____.
    a. the thickness of the material
    b. the number of reflective surfaces
    c. the presence of a vapor barrier
    d. proper ventilation

14. When cutting flexible insulation to prepare for installation, measure the height of a wall cavity (space between studs), then add _____.
    a. 2"
    b. 3"
    c. 4"
    d. 5"

15. Before applying rigid insulation to the exterior of a foundation, the foundation wall should be _____.
    a. sealed
    b. sprayed with a primer
    c. waterproofed
    d. roughened for good adhesion

16. Rigid insulation is usually installed around the perimeter of a concrete slab, under the slab to a distance from the edge of _____.
    a. 8" to 12"
    b. 12" to 18"
    c. 18" to 24"
    d. 24" to 36"

17. For flat roofs, ventilation is achieved by a combination of eave vents and _____.
    a. roof stacks
    b. parapet louvers
    c. ridge vents
    d. venturi tubes

18. Moisture dissipation from attic spaces occurs through a combination of wind-induced ventilation and _____.
    a. convection
    b. radiation
    c. conduction
    d. diffusion

19. When a vapor barrier is used under the ceiling, proper free air ventilation for a gable roof is defined as 1 sq ft for every _____.
    a. 150 sq ft of attic area
    b. 160 sq ft of attic area
    c. 300 sq ft of attic area
    d. 320 sq ft of attic area

20. A vapor barrier material that is totally impermeable will have a perm rating of _____.
    a. 0.0
    b. 0.5
    c. 1.0
    d. 1.5

21. A vapor barrier is any material with a perm rating of less than _____.
    a. 0.01
    b. 0.05
    c. 0.1
    d. 1.0

22. Asphalted kraft paper is usually incorporated with _____.
    a. semirigid insulation
    b. blanket or batt insulation
    c. reflective insulation
    d. structural foam insulation

23. Of moisture entering a structure, migration of moisture from the ground below the building accounts for up to _____.
    a. 50 percent
    b. 60 percent
    c. 70 percent
    d. 80 percent

24. When waterproofing material is sprayed on exterior below-grade walls, backfilling of the walls should be avoided for _____.
    a. one to two days
    b. two to three days
    c. three to four days
    d. five to six days

25. The vertical seams of building wrap are usually overlapped at each corner of the building by _____.
    a. 2" to 6"
    b. 3" to 4"
    c. 6" to 12"
    d. 12" to 18"

Trade Terms Quiz

Fill in the blank with the correct term that you learned from your study of this module.

1. A _____ is used to prevent condensation within walls by retarding the flow of vapor and moisture into them. It must be located on the wall's warm side.

2. A material that has small openings allowing gases or liquids to seep through is said to be _____.

3. The resistance of a material to conductive heat flow is stated as its' _____.

4. _____ is the temperature at which condensation occurs because the air has become oversaturated with moisture.

5. Heat movement caused by a fan or pump, or occurring naturally because of temperature differences is called _____.

6. The _____ of a material is a measure of its capacity to allow the passage of gases or liquids.

7. Conversion of the moisture in air to water is an example of the process of _____.

8. A _____ is a thin sheet of rubber or other material inserted in a construction joint to obstruct water seepage through the joint.

9. _____ is expressed as the ratio, between two surfaces, of water vapor flow to vapor pressure difference.

10. Water in a gaseous state, such as that diffused in the atmosphere, is known as _____.

11. The _____ is the unit of measure of water vapor permeability.

12. _____ is the movement of gas molecules in all directions, often contrary to gravity.

Trade Terms

Condensation
Convection
Dew point
Diffusion
Perm
Permeability
Permeable
Permeance
R-value
Vapor barrier
Water stop
Water vapor

Trade Terms Introduced in This Module

Condensation: The process by which a vapor is converted to a liquid, such as the conversion of the moisture in air to water.

Convection: The movement of heat that either occurs naturally due to temperature differences or is forced by a fan or pump.

Dew point: The temperature at which air becomes oversaturated with moisture and the moisture condenses.

Diffusion: The movement, often contrary to gravity, of molecules of gas in all directions, causing them to intermingle.

Perm: The measure of water vapor permeability. It equals the number of grains squared of water vapor passing through a 1-square-foot piece of material per hour, per inch of mercury difference in vapor pressure.

Permeability: The measure of a material's capacity to allow the passage of liquids or gases.

Permeable: Porous; having small openings that permit liquids or gases to seep through.

Permeance: The ratio of water vapor flow to the vapor pressure difference between two surfaces.

R-value: The resistance to conductive heat flow through a material or gas.

Vapor barrier: A material used to retard the flow of vapor and moisture into walls and prevent condensation within them. The vapor barrier must be located on the warm side of the wall.

Water stop: Thin sheets of rubber, plastic, or other material inserted in a construction joint to obstruct the seepage of water through the joint.

Water vapor: Water in a vapor (gas) form, especially when below the boiling point and diffused in the atmosphere.

Appendix

RECOMMENDED REGIONAL R-VALUES

Prior to 2006, the International Code Council (ICC) recommended insulation R-values based on the region's average low temperature (see *Figure A-1* and *Table A-1*). Today, insulation values are based on the region's average temperature and humidity levels. Interestingly, the United States Department of Energy makes more stringent recommendations based on climate zones. It is important for you to remember that the ICC and other agencies, as well as the federal government may recommend insulation requirements for an area, but the local building code is the final authority.

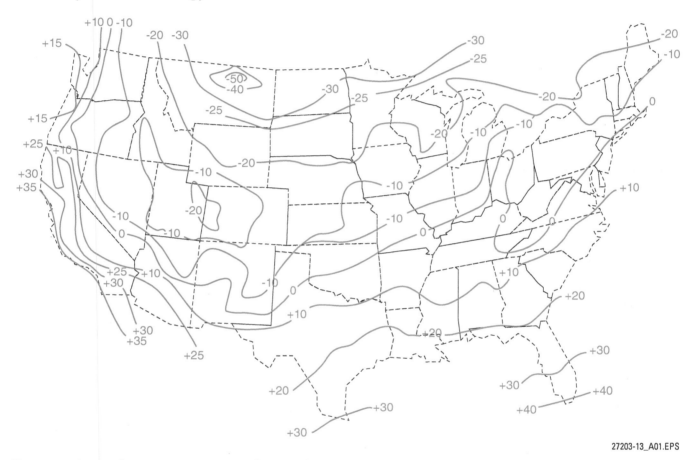

Figure A-1 Average low temperatures across the United States.

Table A-1 Recommended R-Values of Insulation

Average Low Temperature	Floors	Walls	Ceilings
+10°F to +40°F	0	11	19
0°F to +10°F	19	17	30
0°F and below	19	17	38

Additional Resources

This module presents thorough resources for task training. The following resource material is suggested for further study.

International Energy Conservation Code®. 2006. International Code Council.
US Department of Energy website. **www.eere.energy.gov**.

Figure Credits

Courtesy of Certain Teed Corporation, CO01, SA01, Figures 6, 7, 10

Excerpted from the 2006 International Energy Conservation Code, Copyright 2009 Washington DC: International Code Council. Reproduced with permission. All rights reserved. **www.ICCSAFE.org**, Figure 4

Copyright 2008 Certain Teed Corporation, used with permission, Figure 9

Owens Corning, Figure 15

DuPont, Figures 31 (top), 31 (bottom)

DuPont™ Building Innovations™, Figure 32, SA05

Section Review Answer Key

Answer	Section Reference	Objective
Section One		
1. b	1.2.0	1b
2. a	1.1.0	1a
Section Two		
1. c	2.1.0	2a
2. b	2.2.0	2b
3. a	2.3.0	2c
4. d	2.4.0	2d
5. b	2.5.0	2e
6. c	2.6.0	2f
7. c	2.5.0	2e
Section Three		
1. b	3.1.0	3a
2. a	3.2.0	3b
3. b	3.3.0	3c
Section Four		
1. d	4.1.0	4a
2. a	4.2.1	4b
Section 5.0.0		
1. a	5.0.0	5

This page is intentionally left blank.

NCCER CURRICULA — USER UPDATE

NCCER makes every effort to keep its textbooks up-to-date and free of technical errors. We appreciate your help in this process. If you find an error, a typographical mistake, or an inaccuracy in NCCER's curricula, please fill out this form (or a photocopy), or complete the online form at **www.nccer.org/olf**. Be sure to include the exact module ID number, page number, a detailed description, and your recommended correction. Your input will be brought to the attention of the Authoring Team. Thank you for your assistance.

Instructors – If you have an idea for improving this textbook, or have found that additional materials were necessary to teach this module effectively, please let us know so that we may present your suggestions to the Authoring Team.

NCCER Product Development and Revision
13614 Progress Blvd., Alachua, FL 32615

Email: curriculum@nccer.org
Online: www.nccer.org/olf

❏ Trainee Guide ❏ Lesson Plans ❏ Exam ❏ PowerPoints Other _____

Craft / Level: _____ Copyright Date: _____

Module ID Number / Title: _____

Section Number(s): _____

Description: _____

_____

_____

_____

_____

Recommended Correction: _____

_____

_____

_____

_____

Your Name: _____

Address: _____

_____

Email: _____ Phone: _____

This page is intentionally left blank.

Roofing Applications

Overview

As you travel, take note of various types of residential and commercial structures. Note the many kinds of roof construction, and how many different types of roofing materials are used. Part of your work as a carpenter will involve preparing roof decks to receive the finish roofing material, and you may even install roofing material. The roof is the most vulnerable part of a building. If it is not properly installed, it will leak. In some situations, it could even collapse. Safety is always a major consideration when working on a roof.

Module 27202

Trainees with successful module completions may be eligible for credentialing through NCCER's National Registry. To learn more, go to **www.nccer.org** or contact us at **1.888.622.3720**. Our website has information on the latest product releases and training, as well as online versions of our *Cornerstone* magazine and Pearson's product catalog.

Your feedback is welcome. You may email your comments to **curriculum@nccer.org**, send general comments and inquiries to **info@nccer.org**, or fill in the User Update form at the back of this module.

This information is general in nature and intended for training purposes only. Actual performance of activities described in this manual requires compliance with all applicable operating, service, maintenance, and safety procedures under the direction of qualified personnel. References in this manual to patented or proprietary devices do not constitute a recommendation of their use.

Copyright © 2013 by NCCER, Alachua, FL 32615, and published by Pearson, New York, NY 10013. All rights reserved. Printed in the United States of America. This publication is protected by Copyright, and permission should be obtained from NCCER prior to any prohibited reproduction, storage in a retrieval system, or transmission in any form or by any means, electronic, mechanical, photocopying, recording, or likewise. To obtain permission(s) to use material from this work, please submit a written request to NCCER Product Development, 13614 Progress Blvd., Alachua, FL 32615.

27202 V5

From *Carpentry, Trainee Guide*. NCCER.
Copyright © 2013 by NCCER. Published by Pearson. All rights reserved.

27202
ROOFING APPLICATIONS

Objectives

When you have completed this module, you will be able to do the following:

1. Explain the safety requirements for roofing projects.
   a. Identify potential hazards when working on roofs.
   b. Discuss the fall protection equipment required when working on roofs.
   c. Identify proper personal protective equipment (PPE) and hazard control devices used when working on roofs.
2. Identify the tools and fasteners used in roofing.
   a. Identify the hand tools used when working on roofing projects.
   b. Identify the power tools used when working on roofing projects.
   c. Identify fasteners used on roofing projects.
3. Identify the different roofing systems and their associated materials.
   a. Identify composition shingles and their applications.
   b. Identify roll-roofing applications.
   c. Identify wood shakes and shingles and their applications.
   d. Identify tile/slate roofing materials and their applications.
   e. Identify metal roofing and its applications.
   f. Identify built-up roofing and its applications.
   g. Identify single-ply roofing and its applications.
   h. Explain the purpose of underlayment and waterproof membrane.
   i. Discuss the purpose of drip edge, flashing, and roof ventilation.
4. Describe the installation techniques for common roofing systems.
   a. Describe how to properly prepare a roof deck.
   b. Explain how to install composition shingles.
   c. Explain how to install metal roofing.
   d. Describe how to install roll roofing.
   e. Discuss roof projections, flashing, and ventilation.
5. Describe the estimating procedure for roofing projects.

Performance Tasks

Under the supervision of your instructor, you should be able to do the following:

1. Demonstrate how to install composition shingles on a specified roof and valley.
2. Demonstrate the method to properly cut and install the ridge cap using composition singles.
3. Lay out, cut, and install a cricket or saddle.
4. Demonstrate the techniques for installing other selected types of roofing materials.

Trade Terms

Asphalt roofing cement	Pitch	Side lap	Valley flashing
Base flashing	Ridge	Slope	Vent-stack flashing
Bundle	Roof sheathing	Square	Wall flashing
Cap flashing	Saddle	Top lap	
Exposure	Scrim	Underlayment	
Overhang	Selvage	Valley	

Code Note

Codes vary among jurisdictions. Because of the variations in code, consult the applicable code whenever regulations are in question. Referring to an incorrect set of codes can cause as much trouble as failing to reference codes altogether. Obtain, review, and familiarize yourself with your local adopted code.

Industry-Recognized Credentials

If you're training through an NCCER-accredited sponsor, you may be eligible for credentials from NCCER's Registry. The ID number for this module is 27202. Note that this module may have been used in other NCCER curricula and may apply to other level completions. Contact NCCER's Registry at 888.622.3720 or go to **www.nccer.org** for more information.

Contents

Topics to be presented in this module include:

1.0.0 Roofing Safety 5
  1.1.0 Roofing Hazards 5
  1.2.0 Fall Protection Equipment 6
    1.2.1 Body Harnesses 6
    1.2.2 Lanyards 7
    1.2.3 Deceleration Devices 7
    1.2.4 Lifelines 8
    1.2.5 Anchoring Devices and Equipment Connectors 8
    1.2.6 Wearing a Full-Body Harness 8
    1.2.7 Selecting an Anchorage Point and Tying Off 9
    1.2.8 Rescue after a Fall 9
    1.2.9 Inspecting and Testing Fall Protection Systems and Equipment ... 10
  1.3.0 PPE and Hazard Control 10
    1.3.1 Scaffolding and Staging 10
    1.3.2 Ladders 11
    1.3.3 Material Movement 13
    1.3.4 Roofing Brackets 13
2.0.0 Tools and Fasteners 16
  2.1.0 Hand Tools 16
  2.2.0 Power Tools 17
    2.2.1 Power Nailers 19
  2.3.0 Fasteners 20
    2.3.1 Roofing Nails 20
    2.3.2 Cold Asphalt Roofing Cement 20
3.0.0 Roofing Systems and Materials 22
  3.1.0 Composition Shingles 22
  3.2.0 Roll Roofing 24
  3.3.0 Wood Shakes and Shingles 24
  3.4.0 Tile/Slate Roofing 26
    3.4.1 Tile Roofing 26
    3.4.2 Slate Roofing 27
  3.5.0 Metal Roofing 27
  3.6.0 Built-Up Roofing 28
    3.6.1 Modified Bitumen Membrane Roofing Systems 29
  3.7.0 Single-Ply Roofing 30
  3.8.0 Underlayment and Waterproof Membrane 31
  3.9.0 Drip Edge, Flashing, and Roof Ventilation 32
4.0.0 Roof Installation 36
  4.1.0 Preparing the Roof Deck 36
    4.1.1 Protection against Ice Dams 37
  4.2.0 Installing Composition Shingles 38
    4.2.1 Gable Roofs 40
    4.2.2 Gable Roofs—Long Runs 41
    4.2.3 Gable Roofs—Short Runs 43

Contents (continued)

- 4.2.4 Hip Roofs 44
- 4.2.5 Valleys 45
- 4.2.6 Open Valley 45
- 4.2.7 Closed-Woven Valley 46
- 4.2.8 Closed-Cut Valley 47
- 4.3.0 Installing Metal Roofing 48
- 4.3.1 Corrugated Metal Roofing 48
- 4.3.2 Simulated Standing-Seam Metal Roofing 49
- 4.3.3 Snug-Rib System 52
- 4.4.0 Installing Roll Roofing 53
- 4.4.1 Single-Coverage Roll-Roofing Installation 53
- 4.4.2 Exposed Nail Method 53
- 4.4.3 Concealed Nail Method 54
- 4.4.4 Double-Coverage Roll-Roofing Installation 57
- 4.5.0 Roof Projections, Flashing, and Ventilation 58
- 4.5.1 Soil Stacks 58
- 4.5.2 Vertical Wall Flashing 59
- 4.5.3 Dormer Roof Valley 61
- 4.5.4 Chimneys 61
- 4.5.5 Hip or Ridge Row (Cap Row) 66
- 4.5.6 Installing Box Vents 69
- 4.5.7 Installing Ridge Vents 69
- 5.0.0 Estimating Roofing Materials 73
- Appendix Gutters and Downspouts 79

Figures and Tables

Figure 1   Full-body harness 7
Figure 2   Typical shock-absorbing lanyard 7
Figure 3   Deceleration devices 8
Figure 4   Position D-ring properly 9
Figure 5   Ladder jack and aluminum stage 11
Figure 6   Pump jacks 12
Figure 7   Safe ladder placement 13
Figure 8   Roofing brackets 14
Figure 9   Roofing hand tools 17
Figure 10   Roofing equipment 18
Figure 11   Typical composition shingle nails 20
Figure 12   Typical wood shake or shingle nails 21
Figure 13   Typical composition shingles 22
Figure 14   Typical architectural shingle 23
Figure 15   Typical shingle or roll-roofing applications for various
            roof slopes/pitches 24
Figure 16   Typical roll roofing 24
Figure 17   Typical wood shake and shingle 24
Figure 18   Panelized shakes or shingles and wood-fiber hardboard panel ... 25
Figure 19   Typical tile roofing styles 26
Figure 20   Slate roofing 27
Figure 21   Common metal roofing styles 28
Figure 22   Example of architectural metal fascia/roofing system 29
Figure 23   Conventional built-up roofing membrane 29
Figure 24   Flame heating equipment for BUR and torch-down roofing 30
Figure 25   Single-ply membrane installation 30
Figure 26   Typical methods of anchoring single-ply membrane to a roof 31
Figure 27   Various types of drip edges 32
Figure 28   W-metal valley flashing 33
Figure 29   Residential roof vents 33
Figure 30   Typical ridge vents 34
Figure 31   Typical roof installation 37
Figure 32   Drip edge and waterproof membrane placement 37
Figure 33   Underlayment or waterproof membrane placement
            over roof deck 38
Figure 34   Ice dam 38
Figure 35   Types of ice edging 39
Figure 36   Typical composition shingle characteristics 41
Figure 37   Roofing terminology used in instructions 42
Figure 38   Nailing points 42
Figure 39   Shingle layout—6" pattern 43
Figure 40   Correct and incorrect nailing 43
Figure 41   Shingle layout—4" pattern 45
Figure 42   Ribbon courses 45
Figure 43   Hip and ridge layout 46

Figures and Tables (continued)

Figure 44 Open valley flashing (steep pitch) 46
Figure 45 Closed-woven valley 47
Figure 46 Closed-cut valley 48
Figure 47 Corrugated roofing 49
Figure 48 Corrugated roof layout 50
Figure 49 Eave trim 51
Figure 50 Placing, securing, and sealing panels 51
Figure 51 Nailing a T-clip to a roof deck 51
Figure 52 Rake edge 52
Figure 53 Channel strips and valley flashing 52
Figure 54 Ridge flashing 52
Figure 55 Typical single-coverage roll-roofing installation 54
Figure 56 First course of exposed-nail roll roofing 54
Figure 57 Second and subsequent courses of exposed-nail roll roofing 55
Figure 58 Roll roofing in a valley 56
Figure 59 Roll-roofing starter strips 56
Figure 60 First course of concealed-nail roll roofing 56
Figure 61 Third and subsequent courses of concealed-nail roll roofing 56
Figure 62 Covering hip or ridge joints 57
Figure 63 Coating the starter strip 58
Figure 64 Cementing a vertical seam 58
Figure 65 Vent-stack flashing 59
Figure 66 Layout around stack 59
Figure 67 Placement of flashing 59
Figure 68 Covering flashing 60
Figure 69 Wall (step) flashing 60
Figure 70 Continuous flashing 61
Figure 71 Bending continuous flashing 61
Figure 72 Covering flashing 61
Figure 73 Dormer flashing 62
Figure 74 Dormer valley flashing 62
Figure 75 Dormer valley coverings 63
Figure 76 Simple chimney cricket 63
Figure 77 Cricket frame 63
Figure 78 Front base flashing 64
Figure 79 Step-flashing method 64
Figure 80 Cricket flashing 65
Figure 81 Cap flashing methods at sides 65
Figure 82 Flashing cap and lap 66
Figure 83 Counter (cap) flashing installation 66
Figure 84 Architectural cap shingle 66
Figure 85 Cutting cap shingles 67
Figure 86 Applying the last course of ridge shingles 67
Figure 87 Installing a ridge cap 67
Figure 88 Hip and ridge end cap 69

Figure 89 Typical box vent 69
Figure 90 Example of slot cutout placement 69
Figure 91 Example of a continuous roof line 70
Figure 92 Determining total slot width 70
Figure 93 Snapping a chalkline for slot width 70
Figure 94 Cutting slots 70
Figure 95 Exposed-end shingle caps 71
Figure 96 Positioning vent material 71
Figure 97 Typical rigid vent section 71
Figure 98 Applying cap shingles over a vent 72
Figure 99 Roof example (including overhangs) 73
Figure A-1 Debris guard 79
Figure A-2 Typical metal K-style sizes 80
Figure A-3 Typical metal K-style drainage system and accessories 81
Figure A-4 Placement of gutter 82
Figure A-5 Typical vinyl K-style drainage system 82
Figure A-6 Typical vinyl C-style drainage system 83

Table 1 Sizes, Weights, and Coverage of Asphalt-Saturated Felt 32
Table 2 Typical Weights, Characteristics, and Recommended Exposures for Roll Roofing 53

This page is intentionally left blank.

Section One

1.0.0 Roofing Safety

Objective

Explain the safety requirements for roofing projects.

a. Identify potential hazards when working on roofs.
b. Discuss the fall protection equipment required when working on roofs.
c. Identify proper personal protective equipment (PPE) and hazard control devices used when working on roofs.

Trade Terms

Bundle: A package containing a specified number of shingles or shakes. The number is related to square-foot coverage and varies with the product.

Pitch: The ratio of the rise to the span, indicated as a fraction. For example, a roof with a 6' rise and a 24' span will have a ¼ pitch.

Roof sheathing: Usually 4 × 8 sheets of plywood, but can also be 1 × 8 or 1 × 12 roof boards, or other new products approved by local building codes. Also referred to as decking.

Slope: The ratio of rise to run. The rise in inches is indicated for every foot of run.

Underlayment: Asphalt-saturated felt protection for sheathing; 15-lb roofer's felt is commonly used. The roll size is 3' × 144', or a little over four squares.

Roofing materials are used to protect a structure and its contents from the elements. Besides rain protection, some materials are especially suitable for use in areas where fire, high wind, or extreme heat problems exist, or in areas where cold weather, snow, and ice are problems. Materials can contribute to the attractiveness of the structure with the careful selection of texture, color, and pattern. However, the design of the structure as well as local building codes may limit the choice of materials because of the pitch of the roof or because of other considerations at a particular location. In every case, the project specifications must be checked to determine the type of roofing materials to be used.

1.1.0 Roofing Hazards

Roofing projects have many inherent hazards. Roofing materials are installed at elevated locations. The roof load capacity must be known so that the weight of roofing materials does not exceed the anticipated load. When possible, place loads directly over rafters or trusses.

Overhead power lines must be identified and carefully considered when working on roofing projects, not only for the workers, but also for equipment operators moving materials to various parts of the roof. Per OSHA (Occupational Safety and Health Administration), a safe distance of at least 10 feet must be maintained between a power line and a piece of equipment for line voltages of 50kV or less. (For voltages above 50kV, the distance must be increased by 0.4 feet for each 1kV of voltage.) Contact the power company to see if the power can be shut down on the lines. If this is not possible, ask if they can place insulation over the lines during the time you will be working in the location. If overhead power lines are present, ensure they are included in the job hazard analysis (JHA). Use nonconductive ladders when working around power lines.

Tools and materials must be properly secured when working at elevations, to protect workers and others below. In addition, hazards caused by tools and materials must be considered before beginning a roofing project. Always complete a JHA before beginning a roofing project to identify the potential hazards and recommend action that can be taken to eliminate or minimize each hazard. In addition to working at elevations, the following other potential hazards may be encountered on a roofing project:

- Injuries that may occur when cutting roofing materials
- Respiratory hazards when applying certain types of roofing materials such as asphalt-based roofing materials
- Flying debris as a result of cutting or handling roofing materials
- Tools falling from a roof, ladder, or other elevated surface onto workers below

Roofing materials are commonly lifted manually, resulting in twisting or lifting injuries. Many major contractors have instituted stretching programs to minimize these types of injuries. The programs involve stretching activities, similar to those of an athlete's, prior to and/or during the workday.

Skylights or roof windows are commonly installed in both residential and commercial structures. Although they cover the open space below, skylights and roof windows are not structural elements and should be treated with care when installing roofing materials. Do not sit or stand on skylights and roof windows, and do not place tools or supplies on them.

1.2.0 Fall Protection Equipment

Roofers spend a major part of their time working on sloped roofs. Most construction injuries and deaths are caused by falls. Falls from high places can cause serious injury or death when the wrong type of fall protection equipment is used, or when the right equipment is used improperly. A fall protection plan must be prepared for any project where workers will be more than 6' off the ground or on an elevated working surface.

There are three common types of fall protection equipment: guardrails, personal fall arrest systems, and safety nets. This section will focus on personal fall arrest systems. These devices and their use are governed by *OSHA Safety and Health Standards for the Construction Industry, Part 1926, Subpart M*. The rules covering guardrails on scaffolds are contained in Subpart L. Basically, OSHA requires that all workers use guardrail systems, safety-net systems, or personal fall arrest systems to protect themselves from falling more than 6' and hitting the ground or a lower working surface.

Another method of fall protection is the use of hole covers. Holes are commonly placed in floors for chutes or other purposes. Hinged floor-opening covers of standard strength and construction, equipped with standard railings leaving one exposed side, can be used to cover the holes. When the hole is not in use, the cover must be closed and the exposed side must be properly guarded by railings.

When describing personal fall arrest systems, the following terms must be understood:

- *Free-fall distance* – The vertical distance a worker moves after a fall before a deceleration device is activated.
- *Deceleration device* – A device such as a shock-absorbing lanyard, rope grab, or self-retracting lifeline that brings a falling person to a stop without injury.
- *Deceleration distance* – The distance it takes before a person comes to a stop. The required deceleration distance for a fall arrest system is a maximum of 3'-6".
- *Arresting force* – The force needed to stop a person from falling. The greater the free-fall distance, the more force is needed to stop, or arrest, the fall.

The following sections discuss equipment used in personal fall arrest systems:

- Body harnesses
- Lanyards
- Deceleration devices
- Lifelines
- Anchoring devices and equipment connectors

1.2.1 Body Harnesses

Full-body harnesses (*Figure 1*) with a fixed back D-ring are used in personal fall arrest systems. They are made of straps that are designed to be worn securely around the user's body. This allows the arresting force to be distributed via the harness straps throughout the body, including the shoulders, legs, torso, and buttocks. This distribution decreases the chance of injury. When a fall occurs, the D-ring moves to the nape of the neck, keeping the worker in an upright position and helping to distribute the arresting force. This keeps the worker in a relatively comfortable position while awaiting rescue.

Selecting the right full-body harness depends on a combination of job requirements and personal preference. Harness manufacturers normally provide selection guidelines in their product literature. Other types of full-body harnesses can be equipped with front chest D-rings, side D-rings, or shoulder D-rings. Harnesses with front chest D-rings are typically used in ladder climbing and personal positioning systems. Those with side D-rings are also used in personal positioning systems. Personal positioning systems are systems that allow workers to hold themselves in place, keeping their hands free to accomplish a task. Per OSHA regulations, a personal positioning system should not allow a worker to free-fall more than 2', and the anchoring device, also call an anchor or anchorage, to which it is attached should be able to support at least twice the impact load of a worker's fall or 5,000 pounds, whichever is greater. Harnesses equipped with shoulder D-rings are typically used with a spreader bar or rope yoke for entry into and retrieval from confined spaces.

Note that in the past, body belts were frequently used instead of a full-body harness as part of a fall arrest system. As of January 1, 1998, OSHA banned them from such use. This is because body belts concentrate all of the arresting force in the abdominal area. Also, after a fall, the worker hangs in an uncomfortable and potentially dangerous position while awaiting rescue.

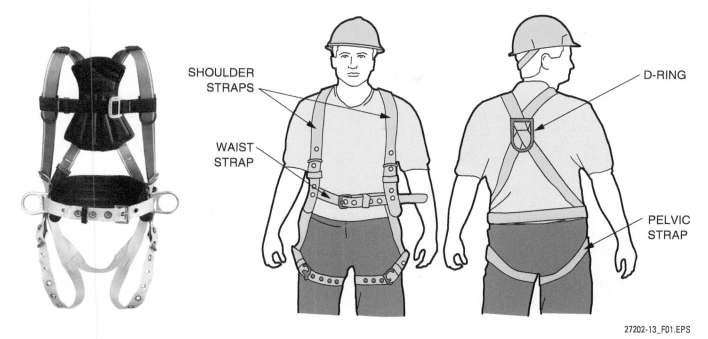

Figure 1 Full-body harness.

1.2.2 Lanyards

Lanyards are short, flexible lines with connectors on each end. They are used to connect a body harness to a lifeline, deceleration device, or anchoring device. There are many kinds of lanyards made for different uses and climbing situations. All must have a minimum breaking strength of 5,000 pounds. They come in both fixed and adjustable lengths, and are made out of steel, rope, or nylon webbing. Some have a shock absorber (*Figure 2*), which absorbs up to 80 percent of the arresting force when a fall is being stopped. When using a lanyard, always follow the manufacturer's recommendations.

WARNING! When activated during the fall-arresting process, a shock-absorbing lanyard stretches as it acts to reduce the fall-arresting force. This potential increase in length must always be taken into consideration when determining the total free-fall distance from an anchorage point.

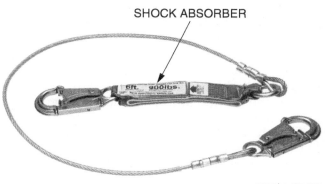

Figure 2 Typical shock-absorbing lanyard.

1.2.3 Deceleration Devices

Deceleration devices limit the arresting force that a worker is subjected to when the fall is stopped suddenly. Rope grabs with shock-absorbing lanyards and self-retracting lifelines are two common deceleration devices. A rope grab (*Figure 3*) connects to a shock-absorbing lanyard and attaches to a lifeline. In the event of a fall, the rope grab is

Fall Protection Equipment Precautions

Do not mix or match equipment from different manufacturers. All substitutions must be approved by your supervisor. All damaged or defective parts must be taken out of service immediately and tagged as unusable or destroyed. If the equipment is subjected to impact from a fall, remove it from service until it can be inspected by a qualified person.

ROPE GRAB   **RETRACTABLE LIFELINE**

Figure 3 Deceleration devices.

pulled down by the attached lanyard, causing it to grip the lifeline and lock in place. Some rope grabs have a mechanism that allows the worker to unlock the device and slowly descend the lifeline to the ground or surface below.

Self-retractable lifelines (*Figure 3*) allow unrestricted movement and fall protection while climbing and descending ladders or when working on multiple levels. Typically, they have a 25' to 100' galvanized steel cable that automatically takes up the slack in the attached lanyard, keeping the lanyard out of the worker's way. In the event of a fall, a centrifugal braking mechanism engages to limit the worker's fall. Per OSHA requirements, self-retracting lifelines and lanyards that limit the free-fall distance to 2' or less must be able to support a minimum tensile load of 3,000 pounds. Those that do not limit the free-fall distance to 2' or less must be able to hold a tensile load of at least 5,000 pounds.

1.2.4 Lifelines

Lifelines are ropes or flexible steel cables that are attached to an anchorage point. They provide a means for tying off personal fall protection equipment.

Vertical lifelines are suspended vertically from a fixed anchorage point at the upper end to which a fall arrest device such as a rope grab is attached. Vertical lifelines must have a minimum breaking strength of 5,000 pounds. Each worker must use his or her own line. This is because if one worker falls, the movement of the lifeline during the fall arrest may also cause the other workers to fall. Vertical lifelines must be terminated in a way that will keep the worker from moving past its end, or they must extend to the ground or the next lower working level.

Horizontal lifelines are connected horizontally between two fixed anchorage points to which a fall arrest device is attached. Horizontal lifelines must be designed, installed, and used under the supervision of a qualified and competent person. The required strength of a horizontal line and its anchors increases substantially for each worker attached to it.

1.2.5 Anchoring Devices and Equipment Connectors

Anchoring devices, commonly called tie-off points (not to be confused with anchorage points), support the entire weight of the fall arrest system. The anchoring device must be capable of supporting 5,000 pounds for each worker attached. Eye bolts, overhead beams, and integral parts of building structures are all types of anchorage points for anchoring devices.

The D-rings, buckles, and snap hooks that fasten and/or connect the parts of a personal fall arrest system are called connectors. OSHA regulations specify how they are to be made, and require D-rings and snap hooks to have a minimum tensile strength of 5,000 pounds. All such components should be designed for use with the attached hardware. As of January 1, 1998, only locking-type snap hooks are permitted for use in personal fall arrest systems.

1.2.6 Wearing a Full-Body Harness

Before using fall protection equipment on the job, your employer must provide you with training in the basics of fall protection and the proper use of the equipment. In addition, a written, job-specific fall protection plan must be available for the project. All equipment supplied by your employer must meet OSHA standards for strength. Before each use, always read the instructions and warnings on any fall protection equipment. Inspect the equipment using the following guidelines:

- Examine harnesses and lanyards for mildew, wear, damage, and deterioration.

- Make sure no straps are cut, broken, torn, or scraped.
- Check for damage due to fire, chemicals, or corrosives.
- Check that hardware is free of cracks, sharp edges, or burrs.
- Check that snap hooks close and lock tightly, and that buckles work properly.
- Check ropes for wear, broken fibers, pulled stitches, and discoloration.
- Make sure lifeline anchoring devices and mountings are not loose or damaged.

The general procedure for using a full-body harness is as follows:

Step 1 Hold the harness by the back D-ring, then shake the harness, allowing all the straps to fall into place.

Step 2 Unbuckle and release the waist and/or leg straps.

Step 3 Slip the straps over your shoulders so that the D-ring is located between your shoulder blades (*Figure 4*).

Step 4 Fasten the waist strap. It should be snug, but not binding.

Step 5 Pull the straps between each leg and buckle the straps.

Step 6 After all the straps have been buckled, tighten all friction buckles so that the harness fits snugly but allows a full range of movement.

Step 7 Pull the chest strap around to the shoulder straps and fasten it in the mid-chest area. Tighten it enough to pull the shoulder straps taut.

1.2.7 Selecting an Anchorage Point and Tying Off

Once the full-body harness has been put on, the next step is to connect it either directly or indirectly to a secure anchorage point by the use of a lanyard or lifeline. This is called tying off. Tying off is always done before you get into a position from which you can fall. Follow the manufacturer's instructions on the best tie-off methods for your equipment. When tying off, ensure that your anchorage point has the following characteristics:

- Directly above you
- Easily accessible
- Damage-free and capable of supporting 5,000 pounds per worker
- Never on the same point as a workbasket tie-off

Be sure to check the manufacturer's equipment labels and allow for any equipment stretch and deceleration distance.

When tying off, consider the following:

- Tie-offs that use knots are weaker than other methods of attachment. Knots can reduce the lifeline or lanyard strength by 50 percent or more. A stronger lifeline or lanyard should be used to compensate for this effect.
- To protect equipment from cuts, do not tie off around rough or sharp surfaces. Tying off around H-beams or I-beams can weaken the line because of the cutting action of the beam's edge. This can be prevented by using a webbing-type lanyard or wire-core lifeline.
- Never tie off in a way that would allow you to fall more than 6'.
- A shorter fall can reduce your chances of falling into obstacles, being injured by the arresting force, and damaging your equipment. To limit your fall, a shorter lanyard can be used between the lifeline and your harness. Also, the amount of slack in your lanyard can be reduced by raising your tie-off point on the lifeline. The tie-off point to the lifeline or anchoring device must always be higher than the connection to your harness.

1.2.8 Rescue after a Fall

Every elevated job site should have a written rescue and retrieval plan in case it is necessary to rescue a fallen worker. Planning is especially im-

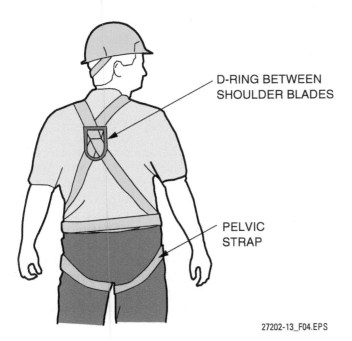

Figure 4 Position D-ring properly.

portant in remote areas that are not readily accessible to a telephone. Before there is a risk of a fall, make sure that you know what your employer's rescue plan calls for you to do. Find out what rescue equipment is available and where it is located. Learn how to use the equipment for self-rescue and the rescue of others.

If a fall occurs, any employee hanging from the fall arrest system must be rescued safely and quickly. Your employer should have previously determined the method of rescue for fall victims, which may include equipment that lets the victim perform the rescue without assistance, a system of rescue by co-workers, or a way to alert a trained rescue squad. If a fall rescue depends on calling for outside help, such as the fire department or rescue squad, all the needed phone numbers must be posted in plain view at the work site. In the event a co-worker falls, follow your employer's rescue plan. Call any special rescue service needed. Communicate with the victim and monitor him or her constantly during the rescue. When in doubt, call 9-1-1.

1.2.9 Inspecting and Testing Fall Protection Systems and Equipment

The inspection of fall arrest equipment should be performed regularly to make sure it complies with OSHA requirements. Guidelines for inspecting personal fall arrest equipment and systems are given in *OSHA Safety and Health Standards for the Construction Industry, Part 1926, Appendices C* and *Appendices D to Subpart M*. A good practice is to tag or label all items of fall protection equipment with the date when the equipment was last inspected and the date it is due for the next inspection.

Safety nets should be drop-tested at the job site after the initial installation, whenever relocated, after a repair, and at least every six months if left in one place. The drop test consists of dropping a 400-pound bag of sand into the net from at least 42" above the highest walking/working surface at which workers are exposed to fall hazards.

1.3.0 PPE and Hazard Control

The following guidelines must be observed to ensure your safety and the safety of others:

- Wear boots or shoes with rubber or crepe soles that are in good condition.
- Always wear fall protection devices, even on shallow-pitch roofs.
- Rain, frost, and snow are all dangerous because they make a roof slippery. If possible, wait until the roof is dry; otherwise, wear special roofing footwear with skid-resistant cleats in addition to fall protection.
- Brush or sweep the roof periodically to remove any accumulated dirt or debris.
- Install any required underlayment as soon as possible. Underlayment usually reduces the danger of slipping. On sloped roofs, do not step on underlayment until it is properly fastened.
- On pitched roofs, install necessary roof brackets as soon as possible. They can be removed and repositioned as shingle-type roofing is installed.
- Remove any unused tools, cords, and other loose items from the roof. They can be a serious hazard.
- Check and comply with any federal, local, and state code requirements when working on roofs.
- Be alert to any other potential hazards such as live power lines.
- Use common sense. Taking chances can lead to injury or death.

When working outdoors or in high-heat conditions for extended periods of time, take precautions to avoid heat exhaustion and exposure to the sun's ultraviolet rays. Preventive measures include the following:

- Wear a hard hat.
- Wear light clothing that is made of natural fibers.
- If possible, wear tinted glasses or goggles.
- Use a sun protection factor (SPF) 30 or higher sunblock on exposed skin.
- Drink adequate amounts of water to prevent dehydration, especially in arid parts of the country.

1.3.1 Scaffolding and Staging

Scaffolding and staging have been the causes of many minor and serious accidents due to faulty or incomplete construction or inexperience on the part of the designer or craftsperson constructing them. Therefore, to avoid hazards caused by faulty or incompetent construction, all scaffolding and staging should be designed and constructed by competent, certified persons. Scaffolding and staging must be inspected on a daily basis by a certified, competent person. The inspector must tag the scaffolding/staging for safety every morning or at every change of shift.

Even though you have no part in the design or construction of scaffolding, for safety's sake you should be familiar with safety rules and regulations that govern its construction. If you are going to use it, you should know how it is built. The following safety factors should be thoroughly understood and adhered to by everyone on the job site:

- Any type of scaffold used should have a minimum safety factor ratio of four to one; that is, it should be constructed so that it will carry at least four times the load for which it is intended. Roofing material weight adds up quickly when placed in one location.
- All staging or platform planks must have end bearings on scaffold edges with adequate support throughout their lengths to ensure the minimum safety factor ratio of four to one.
- Scaffolding timbers, if used, must be carefully selected and maximum nailing used for added strength.
- Because scaffolds are built for work that cannot be done safely from the ground, makeshift scaffolds using unstable objects for support such as boxes, barrels, or piles of bricks are prohibited.

When the scaffolding is placed on a solid, firm base and erected correctly, the roofer should be able to work with confidence.

Any scaffolding assembled for use should be tagged. Three tag colors are used:

- *Green* – A green tag identifies a scaffold that is safe for use. It meets all OSHA standards.
- *Yellow* – A yellow tag means the scaffolding does not meet all applicable standards. An example is a scaffold where a railing cannot be installed because of equipment interference. A yellow-tagged scaffold may be used; however, a safety harness and lanyard are mandatory. Other precautions may also apply.
- *Red* – A red tag means a scaffold is being erected or taken down. You should never use a red-tagged scaffold.

Other more common types of scaffolding include ladder jacks (*Figure 5*) and pump jacks (*Figure 6*). Pump jacks and, to a lesser extent, ladder jacks are useful for applying the starter strip and lower courses of roofing. Ladder jacks can usually support a 2'-wide adjustable-length platform up to 10' or 18' long. Ladder jacks, which must not be used for heights over 20', can be attached to either side of a ladder and must be separated by not more than 8' intervals along the length of the platform. Pump jacks using aluminum posts for heights under 50' are movable platform supports that are raised or lowered vertically. They are operated by a foot lever and can raise a person plus a rated load.

Figure 5 Ladder jack and aluminum stage.

1.3.2 Ladders

Ladders are useful and necessary pieces of equipment for roofing application and do not cause accidents if properly used and maintained. Ladder accidents are caused by:

- Improper use of ladders
- Ladders not secured properly
- Structural failure of ladders
- Improper handling of objects while on a ladder
- Lack of training on proper ladder use

All workers who will be using a ladder must be trained on the proper use of the ladder prior to its use.

Figure 7 shows a ladder erected correctly in relation to the roof eaves. If the ladder is to be left standing for a long period of time, it should be securely fastened at both the top and bottom.

The normal purpose of a ladder used in a roofing application is solely to gain access to the roof itself. For ladder safety, follow these precautions:

- Always use an OSHA-rated ladder sized appropriately for the job to be undertaken.
- Always inspect a ladder for defects or damage before use.
- Fiberglass ladders should always be used to reduce the possibility of accidental electrocution resulting from contact with power lines. Avoid using aluminum ladders. Aluminum ladders, if used, should never be raised or placed in situations where they can fall or accidentally come into contact with power lines.
- For longer ladders, two people should carry, position, and erect the ladder.

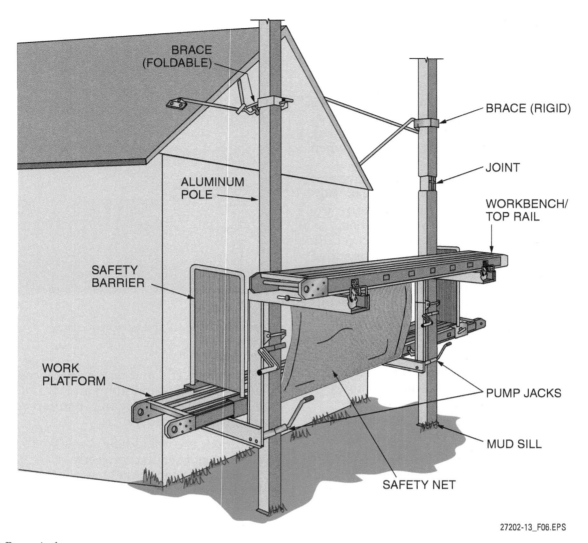

Figure 6 Pump jacks.

- Always face a ladder and grasp the side rails or rungs with both hands when going up or down.
- Take one step at a time. Always maintain three points of contact with the ladder.
- Remember that an ordinary straight ladder is built to support only one person at a time.
- Before using a ladder, be sure there is no oil, grease, or sand on the soles of your shoes. Due to the tread composition, some shoe types easily attract foreign objects that can cause you to slip on a ladder.
- Never carry tools or materials up or down a ladder. A rope or other device should be used to raise or lower everything so that you can always have both hands free when climbing the ladder.
- Make sure that the base of the ladder is level and has adequate support. Shim the legs or use levelers, if necessary.
- Make sure the ladder is at the correct angle. Always tie off the ladder prior to use (see *Figure 7*). The ladder needs to be secured at its top to a rigid support.
- Never overreach.
- Fixed ladders must be provided with cages, wells, ladder safety devices, or self-retracting lifelines where the length of climb is less than 24' but the top of the ladder is at a distance greater than 24' above lower levels.

> **NOTE**
> A protective board should be placed along the outside edges of an elevated workbench to prevent tools, equipment, and materials from falling to a lower level and possibly injuring someone.

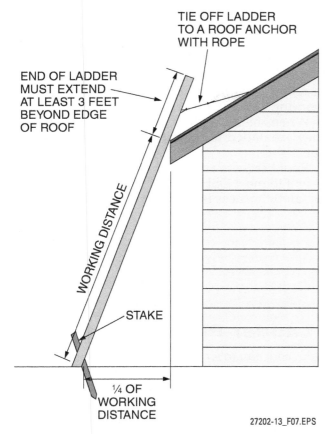

Figure 7 Safe ladder placement.

1.3.3 Material Movement

Many mechanical devices are used to move materials on a job site. These devices include conveyor belts, power ladder conveyors (attached to a ladder), forklifts, and truck-mounted hydraulic lifts. Make sure all safety devices are in place before starting.

Exercise extreme caution to ensure that the lift does not make contact with the roof surface or the person unloading the bundles. Immediately distribute the bundles around the roof area. Do not pile them in any one spot. Roof structures have been designed to carry a specific dead load. Placing unnecessary strain on the roof structure by concentrating the shingles in one place on the roof can jeopardize the safety of the workers. The end result might prove to be disastrous, with the collapse of the roof itself. An immediate dispersal of the bundles will prevent any problems from occurring and also make the installation more efficient because you only want to move and place the load once.

1.3.4 Roofing Brackets

Roofing brackets provide firm footing and material storage points on steep-slope roofs. If any type of roofing bracket is to be used, the decision will be based on the slope of the roof. Most roofers feel comfortable on a roof with a 4 in 12 or a 5 in 12 slope. When the slope increases to 6 in 12, more strain is placed on the feet and the body. Therefore, the roofer has to be very conscious of the height off the ground and be careful with each and every movement. *Figure 8* shows two types of roofing brackets that can be used with a 10' to 14', defect-free, 2"-thick plank.

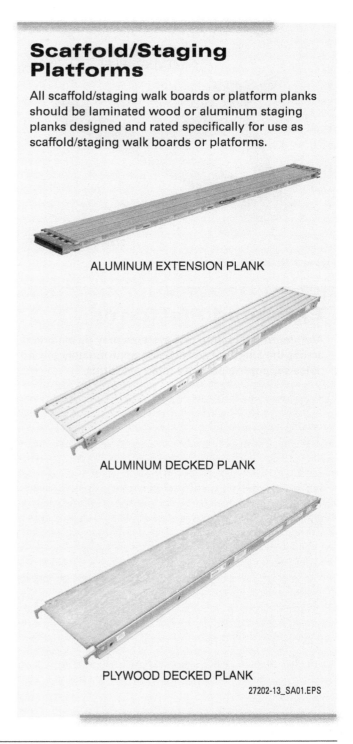

Scaffold/Staging Platforms

All scaffold/staging walk boards or platform planks should be laminated wood or aluminum staging planks designed and rated specifically for use as scaffold/staging walk boards or platforms.

Figure 8 Roofing brackets.

Both types of brackets can be nailed firmly to the roof, but the adjustable bracket, which is installed to correspond to the slope of the roof, makes standing and moving around more comfortable. Never get overconfident due to bracket usage. Be aware of the height at which you are working and be cautious.

When installing roof brackets, make sure that they are nailed to the rafters, not just to the roof sheathing.

Roof brackets and toeboards alone are not sufficient to meet OSHA fall standards. A proper rail or safety harness is required above 6 feet.

Ladder Placement

To check for proper ladder placement, stand straight with your toes touching the base of the ladder and with your arms extended straight out toward the ladder. If you can just touch a rung of the ladder, it is probably positioned properly for climbing.

Extended Platforms

Aluminum-pole pump jack systems may be extended across the length of a wall using appropriately placed poles supporting properly rated platforms.

Ladder Conveyors

For residential or light commercial work, a ladder conveyor system greatly reduces manual lifting of roofing materials to a roof. Note the safety railings and danger signs.

1.0.0 Section Review

1. When doing roofing work, the minimum distance between any piece of equipment and a live power line of 50kV or less is at least _____.
   a. 3 feet
   b. 10 feet
   c. 4 yards
   d. 20 feet

2. A short, flexible line used in a fall protection system to connect a body harness to an anchorage point is called a _____.
   a. lifeline
   b. bungee cord
   c. safety line
   d. lanyard

3. Scaffolding with a yellow tag may be used if applicable precautions are observed.
   a. True
   b. False

Section Two

2.0.0 Tools and Fasteners

Objective

Identify the tools and fasteners used in roofing.
  a. Identify the hand tools used when working on roofing projects.
  b. Identify the power tools used when working on roofing projects.
  c. Identify fasteners used on roofing projects.

Trade Terms

Asphalt roofing cement: An adhesive that is used to seal down the free tabs of strip shingles. This plastic asphalt cement is mainly used in open valley construction and other flashing areas where necessary for protection against the weather.

Saddle: An auxiliary roof deck that is built above the chimney to divert water to either side. It is a structure with a ridge sloping in two directions that is placed between the back side of a chimney and the roof sloping toward it. Also referred to as a cricket.

Square: The amount of shingles needed to cover 100 square feet of roof surface. For example, square means 10' square, or 10' × 10'.

Valley: The internal part of the angle formed by the meeting of two roofs.

A variety of hand and power tools are used to install roofing. Some of these tools are unique to the roofing, while others are commonly used for other carpentry tasks. Depending on the type of roof being installed, different fasteners may be required.

2.1.0 Hand Tools

Many of the tools used for roofing are common to other trades. Some of these common tools are:

- Backsaw
- Crowbar
- Handsaw
- Carpenter's level
- Nail apron
- Sliding T-bevel
- Keyhole saw
- Pop riveter
- Chalkline
- Tape measure
- Angle square
- Caulking gun
- Tin snips
- Prybar
- Scribing compass
- Utility knife
- Framing square
- Claw hammer
- Flat spade or spud bar (for roofing material removal)

Other tools that are specific to the installation of certain types of roofing are also used. Some of these tools are shown in *Figure 9*.

The roofing hammer, also referred to as a shingle hatchet, is used primarily for wood shingle and shake installation. The hatchet end is used to split shingles or shakes, and the top edge is marked or equipped with a sliding gauge to set a dimension for the amount of weather exposure for the shingle or shake.

A composition shingle knife is used to trim or cut all types of composition shingles, including architectural shingles, during installation. It is also used to cut underlayment, cap shingles, roll roofing, and membrane roofing.

Slate-roofing installation usually requires three specialized tools: a slater's hammer, a nail ripper, and a slate cutter. The slater's hammer is equipped with a sharp edge for cutting slate and a point for poking nail holes through the slate. The nail ripper has sharp-edged barbs on one end that are used to shear off nails under a piece of slate. To shear nails, the ripper is struck on the face of the anvil with a hammer. The slate cutter aids in the trimming of slate and the punching of nail holes.

Various types of tile cutters and nibblers are used in the installation of tile roofs for splitting or trimming tiles (*Figure 10*). A nail ripper, like the one used for slate roofing, can be used to shear off nails under a tile. A wet saw with a diamond wheel can be used on large projects for flat or shaped tile cutting.

Portable brakes are used for the custom bending of flashing material for any type of roof installation. Some roofers use heavy rollers to flatten underlayment to eliminate buckling under the finish roof and for the application of cold-cement, fully-adhered roll roofs, built-up roofing (BUR), or single-ply membrane roofing. These rollers are sold as vinyl flooring rollers in weights ranging from 75 to 150 pounds.

Hand grinders with diamond wheels are used to cut slots in masonry for flashing installation.

In addition to tools, other equipment such as scaffolding, material-handling equipment, ladders, and ladder jacks or pump jacks may be re-

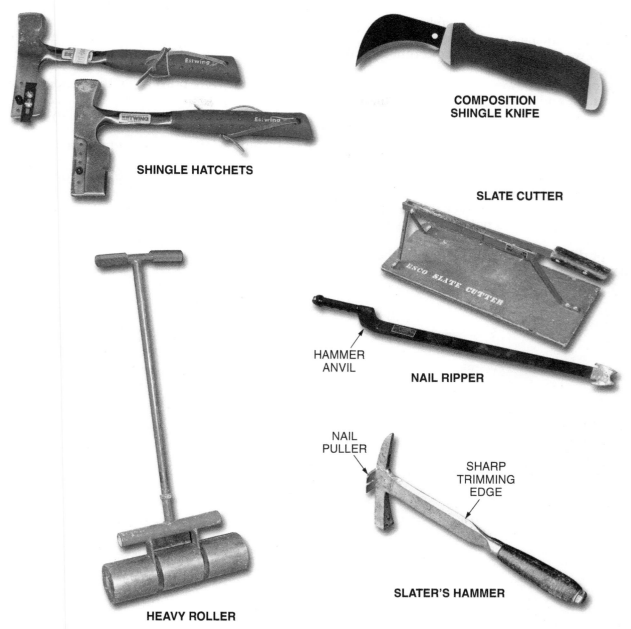

Figure 9 Roofing hand tools.

quired. All roofing installation jobs will require some type of fall protection system.

Worker safety is important on a construction site. Every work site must have a fall protection plan for working on roofs or at certain heights off the ground.

2.2.0 Power Tools

Just like the hand tools, many power tools used in roofing are also used in other trades, including:

- Power circular saw
- Power saber saw
- Power drill and drill bit sets (regular and masonry)
- Pneumatic nailers

Rules for the safe use of all power tools include the following:

- Keep all tools in good condition with regular maintenance.
- Do not attempt to operate any power tool before being trained by the instructor or a competent person on that particular tool.
- Use only equipment that is approved to meet Occupational Safety and Health Administration (OSHA) standards.

- Examine each tool for damage before use and do not use damaged tools.
- Always wear eye protection and other appropriate personal protective equipment (PPE) when operating power tools.
- Wear face and hearing protection when required.
- Wear proper respiratory equipment when necessary.
- Wear the appropriate clothing for the job being done. Always wear tight-fitting clothing that cannot become caught in the moving tool. Roll up or button long sleeves, tuck in shirttails, and tie back long hair. Do not wear any jewelry or watches.
- Do not distract others or let anyone be a distraction while operating a power tool.
- Do not engage in horseplay.
- Do not throw objects or point tools at others.
- Consider the safety of others, as well as yourself.
- Do not leave a power tool unattended while it is running.
- Do not carry, raise, or lower a tool by its electrical cord.
- Assume a safe and comfortable position before using a power tool.
- Do not remove ground plugs from electrical equipment or extension cords.
- Be sure that a power tool is properly grounded and connected to a ground fault circuit interrupter (GFCI) circuit before using it.

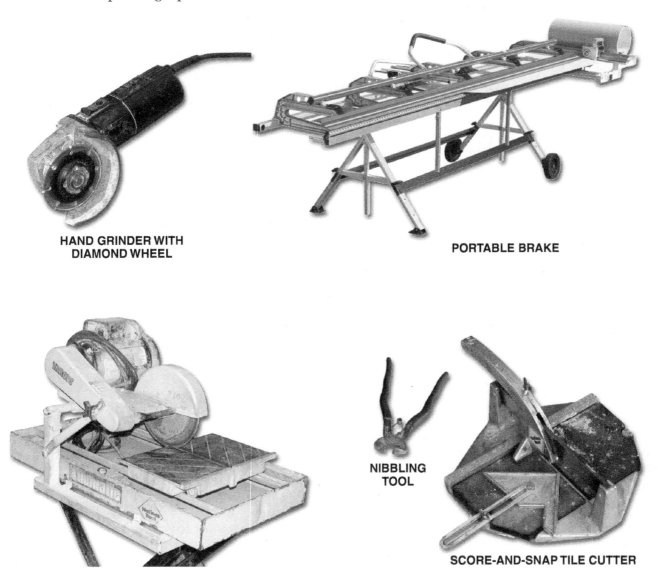

Figure 10 Roofing equipment.

- Be sure that portable or stationary power tools are unplugged at the power source or disabled before performing maintenance or changing accessories.
- Do not use a dull or broken tool or accessory.
- Use a power tool only for its intended purpose.
- Keep your feet, fingers, and hair away from the blade and/or other moving parts of a power tool.
- Do not use a power tool with guards or safety devices removed or disabled.
- Do not operate a power tool if your hands or feet are wet.
- Keep the work area clean at all times.
- Become familiar with the correct operation and adjustments of a power tool before attempting to use it. Always follow the manufacturer's instructions as it pertains to its intended use.
- Keep a firm grip on the power tool at all times.
- Use electric extension cords of sufficient rating to energize the particular power tool being used.
- Do not use worn or frayed extension cords.
- If a tool or extension cord is defective, it must be brought to the attention of the supervisor so it can be tagged and immediately removed from service.
- Extension cords should not be hung by nails or wire, or fastened with staples.
- Report unsafe conditions to your instructor or supervisor.

2.2.1 Power Nailers

There are many types of power nailers. In the application of asphalt or fiberglass shingles, the use of one of these tools can cut the labor time in half. Portable units are used by carpenters on the construction site. Portable units are electric or pneumatic (powered by air or carbon dioxide gas).

Some of the listed safety practices for power nailers are as follows:

- Always wear safety glasses. They must be an OSHA-approved and ANSI-designated type, and are usually recommended by the manufacturer of the unit being used.
- Tool operator training is required before using a power nailer.
- Because operating principles vary, study the manufacturer's operating manual.
- Be certain to use the type of fastener required by the manufacturer.
- If pneumatic, make sure that the pressure can be adjusted at the nailer.
- Treat the machine as you would a gun. Do not point it at yourself or others.
- Always keep the unit tight against the surface to drive the fastener correctly.
- When not in use, disconnect the unit from the power source to prevent accidental release of fasteners.
- Keep air lines untangled on the roof to prevent tripping.

Power Nailers

Make sure that the nailer is equipped with a flush-mount attachment or that the impact pressure of the tool can be regulated at the tool to prevent overdriving the nails and cutting through the roofing material.

PNEUMATIC ROOFING NAILER WITH PLASTIC WASHER ATTACHMENT FOR UNDERLAYMENT APPLICATION

TYPICAL PNEUMATIC ROOFING NAILER

27202-13_SA04.EPS

2.3.0 Fasteners

Roofing materials are typically fastened to the underlayment using roofing nails or roofing cement. Always refer to the roofing manufacturer to determine the appropriate size and type of fasteners to be used or the proper roofing cement to be used.

2.3.1 Roofing Nails

Common roofing materials must be fastened with nails of the proper length and made of a material that is compatible with or the same as the drip edges and flashing. *Figures 11* and *12* show the most common nails used for composition shingles and wood shakes or shingles. These nails are usually available in galvanized steel; some other types are aluminum, stainless steel, or copper. Normally, copper slater's nails are used for slate roofs, and stainless steel nails are used for tile roofs. About 1 pound of nails per square (100 square feet) is required to fasten composition shingles. For wood shakes and shingles, 2 to 4 pounds of nails per square will be required. For other types of nailed roofing, follow the manufacturer's recommendations. Check your local code for nail penetration requirements.

For composition roofs or underlayment, nails can be installed using pneumatic-powered or electric-powered nailing guns.

2.3.2 Cold Asphalt Roofing Cement

Cold asphalt roofing cements, consisting of modified asphalt and/or coal tar products, are used in the installation of composition shingles, underlayment, roll roofing, and cold asphalt BUR. They are available in liquid nonfibered form or in plastic-fibered form. The nonfibered forms are usually used in the lapped installation of underlayment and roll roofing, and are spread over large areas with a spreader or mop. The plastic-fibered cement is used for spot repairs or cementing nail heads, shingle tabs, valley overlaps, and saddle materials, and as an exposed sealer for gaps or flashing. Asphalt cements are generally available in 1-gallon and 5-gallon pails.

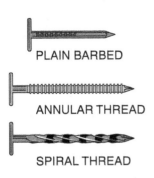

NAILING APPLICATION	⅜" PLYWOOD OR WAFERBOARD	1" SHEATHING
STRIP OR SINGLE (NEW CONSTRUCTION)	⅞"	1¼"
OVER OLD ASPHALT LAYER	1"	1½"
REROOFING OVER WOOD SHINGLES	–	1¾"

Figure 11 Typical composition shingle nails.

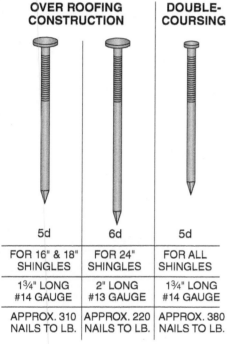

Figure 12 Typical wood shake or shingle nails.

2.0.0 Section Review

1. Which of the tools listed below is *not* used when roofing with slate?
    a. slater's hammer
    b. keyhole saw
    c. slate cutter
    d. nail ripper

2. Using a power nailer to install asphalt or fiberglass shingles can reduce labor time by ____.
    a. 10 percent
    b. 25 percent
    c. 35 percent
    d. 50 percent

3. Tile roofs are usually installed using nails made from ____.
    a. copper
    b. galvanized steel
    c. stainless steel
    d. aluminum

4. Asphalt roofing cement must be heated before application.
    a. True
    b. False

Section Three

3.0.0 Roofing Systems and Materials

Objective

Identify the different roofing systems and their associated materials.

a. Identify composition shingles and their applications.
b. Identify roll-roofing applications.
c. Identify wood shakes and shingles and their applications.
d. Identify tile/slate roofing materials and their applications.
e. Identify metal roofing and its applications.
f. Identify built-up roofing and its applications.
g. Identify single-ply roofing and its applications.
h. Explain the purpose of underlayment and waterproof membrane.
i. Discuss the purpose of drip edge, flashing, and roof ventilation.

Trade Terms

Exposure: The distance (in inches) between the exposed edges of overlapping shingles.

Ridge: The horizontal line formed by the two rafters of a sloping roof that have been nailed together. The ridge is the highest point at the top of the roof where the roof slopes meet.

Scrim: A loosely knit fabric.

Selvage: The section of a composition roofing roll or shingle that is not covered with an aggregate.

Valley flashing: Watertight protection at a roof intersection. Various metals and asphalt products are used; however, materials vary based on local building codes.

Many types of roofing materials as well as commercial roofing systems are available today. This section will cover the most common materials used on residential and small commercial structures. Commercial roofing systems used on larger or more expensive structures will be described in more detail in a later level.

3.1.0 Composition Shingles

Composition shingles (*Figure 13*) are the most common roofing material in North America. They are available in a wide variety of colors, textures, types, and weights (thicknesses). Three-tab shingles are made of a fiber or fiber-mat material, coated or impregnated with asphalt, and then coated with various mineral granules to provide color, fire resistance, and ultraviolet protection.

In the past, composition shingles were made using asbestos fiber or organic fiber and were commonly referred to as asphalt shingles. The manufacture of asbestos-fiber shingles has been prohibited because the asbestos poses a cancer risk and environmental disposal hazard.

Organic fiber shingles, which have a life span of 15 to 20 years, have been largely replaced by asphalt-coated, fiberglass mat shingles, simply called fiberglass shingles, which have a life span of 20 to 25 years.

More expensive architectural shingles with life spans of 25 to 40 years or more are also available. Architectural shingles are constructed of multiple layers of fiberglass that are laminated together when manufactured or job-applied in layers to create a heavy shadow effect (*Figure 14*). Manufacturers may also provide their shingles in fungus- and/or algae-resistant versions for use in damp locations where unsightly fungus growth or black streaks caused by algae tend to be a problem. The fungus and algae resistance is provided

ARCHITECTURAL SHINGLE

THREE-TAB SHINGLE

27202-13_F13.EPS

Figure 13 Typical composition shingles.

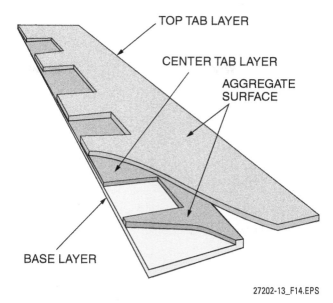

Figure 14 Typical architectural shingle.

by copper granules incorporated with the mineral aggregates that are bonded to the surface of the shingle.

Composition shingles generally suit every climate in North America and can normally be applied to any roof with a slope of 4 in 12 (*Figure 15*) up to 21 in 12, provided they have factory-applied, seal-down adhesive strips. By applying double-lap underlayment, they can be used on roofs with a slope as low as 2 in 12. However, unless appearance is a problem, roll roofing or membrane roofing is usually recommended on slopes lower than 4 in 12.

Standard three-tab composition shingles are supplied in squares (100 square feet) with three bundles per square. They are also labeled by their weight per square (210-lb, 230-lb, 240-lb, or 250-lb shingles). The heavier the shingle, the longer its life.

Special Architectural Shingles

These shingles are very expensive and are used primarily in elegant residential applications. Various styles of special architectural shingles that provide different pattern effects on a roof are available.

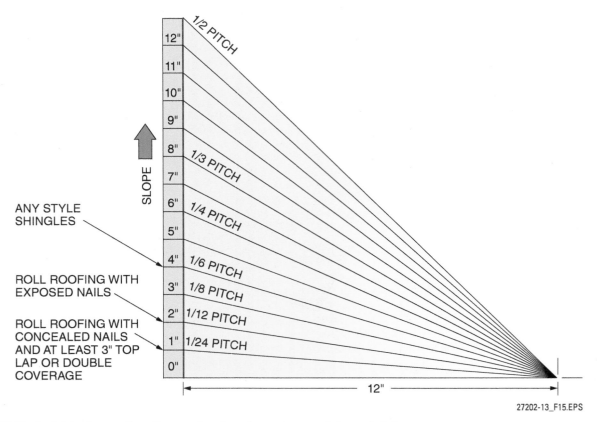

Figure 15 Typical shingle or roll-roofing applications for various roof slopes/pitches.

3.2.0 Roll Roofing

Roll roofing (*Figure 16*) is available in 50-lb to 90-lb weights with the same materials and colors as composition shingles. While it is inexpensive and quick to apply, the life of rolled roofing is typically only 5 to 12 years. It is recommended for use on shallow-slope roofs with slopes of less than 4 in 12 where appearance is not a problem. It can be used on slopes of 2 in 12 with the nails revealed, or on slopes as low as 1 in 12 with the nails concealed. Roll roofing can also be used as **valley flashing** to match the color of the roof shingles. It is supplied in three types. One type is a smooth-surface roofing not covered with any granules. It is used on roofs that are subsequently covered with a hot or cold asphalt and separately applied aggregate. The other two types are for finish roofs. One is completely covered with granules and the other type, called **selvage** or double-coverage roll roofing, is only half covered with granules and is designed to be applied with a cemented-down half lap, with concealed fasteners used on very low slopes. The rolls are normally 36" wide. However, granule-covered, half-width, starter-strip rolls are also available.

3.3.0 Wood Shakes and Shingles

Wood shingles and shakes (*Figure 17*) are among the oldest materials used in shingling, and both, particularly shakes, provide a pleasing and desirable visual effect. In fact, many of the architectural fiberglass shingles attempt to reproduce the rustic

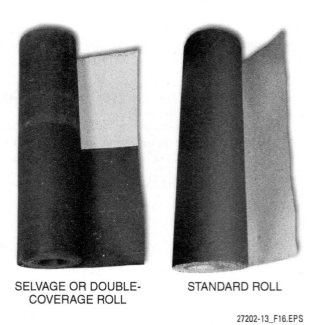

Figure 16 Typical roll roofing.

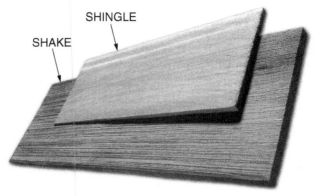

Figure 17 Typical wood shake and shingle.

Machine-sawn and grooved shakes that simulate hand-split shingles are also available.

Wood shingles and shakes are also available as 4' to 8' prebonded panels that are in two-ply and three-ply sections with **exposures** of 5½" to 9" (*Figure 18*). These panelized shingles and shakes, with course-separating underlayment pre-installed, are more expensive than traditional shakes and shingles. However, they are claimed to be up to two times faster to install than composition shingles and up to four times faster to install than traditional wood shakes and shingles. A simulated wood shingle made of wood-fiber composition hardboard is also available in panel form. The panels are 12" × 48" and are applied lengthwise across the roof. They are embossed

visual effect of wood shakes. Most wood shingles and shakes are made using western red cedar, cypress, or redwood; however, red cedar shingles and shakes are the most popular. Typically, they have twice the insulation value of composition shingles, are lighter in weight than most other roofing materials, and are very resistant to hail damage. They can also withstand the freeze-thaw conditions of variable climates. Given periodic coatings of wood preservative, wood shingles and shakes should last for 50 years or more.

The drawbacks of using wood shingles and shakes are that they are expensive, slow to install, a fire hazard, and subject to insect damage and rot. Building codes may prohibit the use of wood shingles in certain areas for various reasons; however, fire-retardant material that is pressure-injected during manufacture, as indicated by the industry designation Certi-Guard™, is supposed to conform to all state and local building codes for use in fire hazard regions.

Wood shingles are machine cut and smoothed on both sides. They are uniform in thickness and length but vary in width. Hand-split wood shakes, which are more expensive, are either hand split on one side and machine smoothed on the other side, or hand split on both sides. The hand splitting produces a rough, rustic surface. Hand-split shakes vary in thickness and width, but are uniform in length. Because of the varying thickness, as well as the rough surface of at least one face of the shakes, an underlayment is used between each course of the shakes to prevent wind-driven rain from being forced back under the shakes.

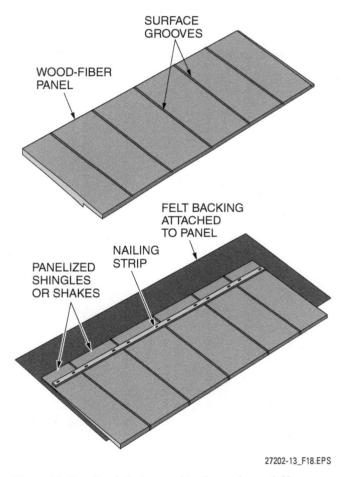

Figure 18 Panelized shakes or shingles and wood-fiber hardboard panel.

Metric Shingles

Most shingles manufactured today are available in foot-pound (English) and metric dimensions. The standard foot-pound, three-tab shingle or laminated architectural shingle is 36" long. The metric versions are 39⅜" long.

with deep shadow lines and random-cut grooves that mimic the look of shakes. The panels overlap with a shiplap joint between courses and between panels in the same course. After exposure, they weather to a gray that is similar to cedar shingles. Simulated shingle and shake metal roofing panels that are 4' long and completely fireproof are also available. These metal panels are very much like metal siding. They are applied horizontally and have an interlocking joint at the top and bottom edges.

3.4.0 Tile/Slate Roofing

Roofing materials, such as tile and slate, are expensive and heavier than composition shingles. Slate is rarely used today due to its expense. Although tile and slate roofs are not common roofing materials, it is important to be able to recognize the materials and understand the proper methods of installation.

3.4.1 Tile Roofing

Like slate, glazed and unglazed clay and ceramic tile roofing products (*Figure 19*) are fireproof and rot proof, and last from 50 to 100 years. Tile roofing is expensive and heavy (7 to 10 pounds per square foot), and requires appropriate roof framing to support the tile weight and any other anticipated loads. It is used in the South and to a great extent in the Southwest areas of the country because it is fireproof and impervious to damage caused by intense sunlight. It is available in Spanish, Mission (barrel or S-style), and other styles known by various names such as French, English, Roma, and Villa. Other styles include flat tiles that may resemble slate or wood shakes. Matching hip and ridge tiles, rake/barge tiles, and hip starter tiles are also available. Every tile for a particular style furnished by a manufacturer is a uniform size in width, thickness, and length.

Tiles are fastened with noncorrosive nails (copper, galvanized, or stainless steel). Copper nails have the advantage of being soft enough to allow expansion and contraction without causing the tiles to crack if they are fastened too tight. They

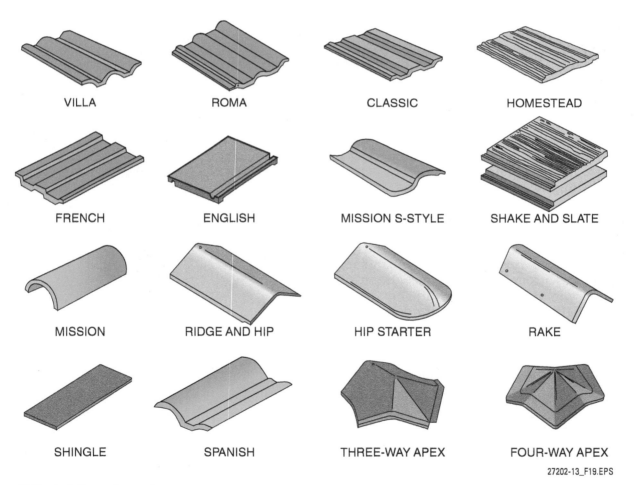

Figure 19 Typical tile roofing styles.

also allow easier replacement of any broken tiles. Because of its durability and resistance to any alkaline corrosion, copper is generally used as flashing.

3.4.2 Slate Roofing

Today, real slate roofing (*Figure 20*) is probably the most expensive roofing option in terms of roof framing materials, roofing materials, and increased installation time. Even though it has a very long service life of 60 to 100 years or more, it is usually not financially practical to use slate unless it is required or desired strictly for architectural purposes. No other material can match the high-quality look of slate. Unfortunately, besides being expensive, slate is heavy—about 7 to 10 pounds per square foot. As a result, the roof framing must be engineered to be substantially stronger to support the slate load, as well as any anticipated snow and ice loads.

Slate is completely fireproof and rot proof. It is available in a wide variety of grades, thicknesses, and colors ranging from gray or black to shades of green, purple, and red. The colors are qualified as unfading or, if subject to some change, weathering. The industry often uses the old federal grading system of A (best) through C to indicate the quality of slate; however, architectural specifications normally use the American Society for Testing and Materials (ASTM) International testing numbers to specify the desired slate quality. New York and Vermont slate types are very durable and have a uniform color and a straight, smooth grain running lengthwise. Slate types with streaks of color usually contain intrusions of sand or other impurities and are not as durable. Premium-quality slate in various colors can be obtained from quarries from Maine to Georgia. Copper slater's nails and flashing are normally used with slate roofs.

Synthetic slate is made of fiber mat (usually fiberglass) that has been impregnated and coated with cement. It looks like real slate, but is lighter in weight. Even though it is lighter than slate, it still requires strong roof framing for support. Synthetic slate is fireproof and can last 40 years or more. Synthetic slate is still a relatively expensive material and is typically used on structures where historically appropriate materials are required. The synthetic slate material is difficult to cut and must be carefully fastened to avoid cracking the shingles.

3.5.0 Metal Roofing

Metal roofing is available in a great variety of materials and styles. These materials can be purchased with a baked enamel, ceramic, or plastic coating. They can also be purchased without a coating so that a separate roof coating can be applied after installation. The following are some common metal roofing materials:

- Aluminum (plain or coated)
- Galvanized steel (plain or coated)
- Terne metal (heat-treated, copper-bearing steel hot-dipped in metal consisting of 80 percent lead and 20 percent tin)
- Aluminum- or zinc-coated steel
- Stainless steel

Besides the common residential panel roofing styles (*Figure 21*), many new engineered/preformed, architectural metal fascia/roofing

Figure 20 Slate roofing.

Steep Slope Applications

On steep-slope roofs greater than 21 in 12, such as a mansard or simulated mansard roof, the built-in sealant strips on shingles do not generally adhere to the underside of the overlaying shingles. Because of this, each shingle must be sealed down with spots of quick-setting asphalt cement at installation. On standard three-tab shingles, two spots of cement are used under each tab. On strip shingles, three or four spots are used under the strip. On laminated architectural shingles, extra nails and four spots of cement are usually used for the shingle.

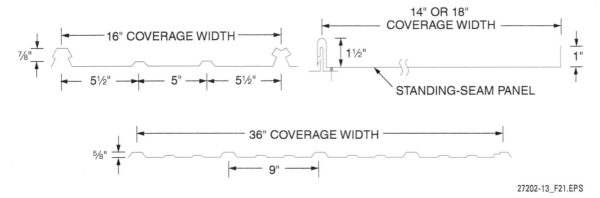

Figure 21 Common metal roofing styles.

systems (*Figure 22*) are available for commercial or residential use in a wide selection of colors and in styles that include shingles, panels, and tiles.

3.6.0 Built-Up Roofing

Conventional built-up roofing (BUR) membrane (*Figure 23*) has been used for over 100 years on very low-slope roofs of residential and commercial structures. While it is still a viable form of roofing, it is gradually being replaced by premanufactured membrane roofing systems.

Built-up roofing, which is field fabricated, consists of three to five layers of heavy, asphalt-coated polyester or fiberglass felt embedded in alternate layers of hot-applied or cold-applied bitumens (coal tar-based or asphalt based) that are the waterproofing material. Most commercial applications use hot asphalt as the waterproofing material. The top surface layer of bitumen is

---

Synthetic Tiles, Shakes, and Shingles

One alternative to clay and ceramic tiles is synthetic tile made of fiber-reinforced concrete or shredded tire material. These tiles are very durable and lighter than clay or ceramic tiles. In some cases, they may be light enough to install on roofs intended to support composition shingles. Synthetic tiles are available in many colors and styles. Like clay and ceramic tiles, synthetic tiles are fireproof and rot proof. They last almost as long as clay and ceramic tiles. Other alternatives to clay, ceramic, or concrete tiles are interlocking baked-enamel steel and vinyl-coated aluminum tiles and shingles. These types of tiles and shingles have a life expectancy exceeded only by real slate or tile. They are available in styles similar to shakes, slate, and tile roofing and are light as well as less expensive.

SLATE

SHAKE

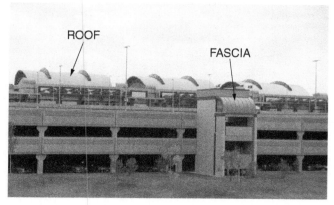

Figure 22 Example of architectural metal fascia/roofing system.

sometimes left smooth but is most often covered with either a mineral-coated cap sheet or embedded with a loose mineral surface consisting of small aggregates such as washed gravel, pea rock, or crushed stone.

Hot asphalt applications require special heating equipment along with some method of transporting the hot asphalt to the roof unless the heating equipment is lifted and positioned onto the roof. Installing this type of roof is labor-intensive and requires experienced roofers. Because the membrane is created in the field, its quality is subject to many variables, including the weather, application techniques, and the experience of the roofers. Most BUR is placed over one or more insulation boards that are bonded or fastened to the roof substrate.

The life span of a correctly applied, built-up roof is about 10 to 20 years, depending on the number of layers. Generally, specific damage to this type of roof can be easily repaired. However, the normal, gradual deterioration of the roof, which may result in deep splits over much of the surface or delamination of the layers, will generally require that the roof be completely removed and a new roof applied.

3.6.1 Modified Bitumen Membrane Roofing Systems

Modified bitumen roofing systems can be classified as either styrene-butadiene-styrene (SBS) or atactic polypropylene (APP) modified bitumen products. The SBS products are usually a composite of polyester or glass fiber and modified asphalts coated with an elastomeric (rubberlike) blend of asphalt and SBS rubber. The APP prod-

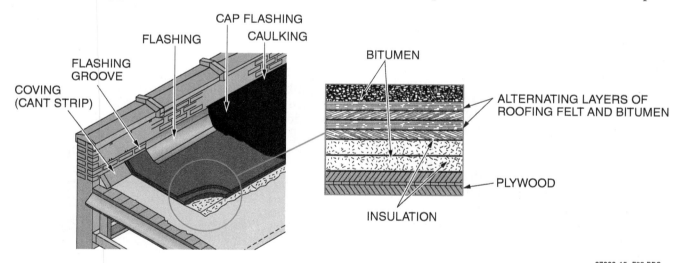

Figure 23 Conventional built-up roofing membrane.

The Hazards of Working on Slate Roofs

Extreme care must be used when walking on slate roof surfaces. Fall protection and soft-soled shoes must be used. A slightly damp or wet slate roof is very slippery and should never be walked on. Numerous roof brackets and boards (roof jacks) should be used to keep roofers and roofing material on the roof. An eave debris safety net and a roped-off zone must be maintained away from the eaves. Falling slate can cause severe injury or death to persons under the eaves. Because of the weight, do not stack large quantities of slate on any one roof jack. Spread smaller quantities of slate out over numerous roof jacks.

ucts are coated with an elastomeric blend of asphalt and atactic polypropylene. Both products are the weatherproofing medium, and are used in a hybrid built-up roof as a cap sheet over one or more base/felt plies that have been secured with hot asphalt or cold adhesive to a deck board covering the insulation or directly to the insulation.

Depending on the manufacturer, SBS products are either supplied with a pre-applied adhesive and the product is rolled for adhesion, or the product is secured with hot asphalt or a cold adhesive applied separately. APP products, called torch-down roofing, are usually secured to the layers below by heat welding. Using flame heating equipment, the back of the product roll is heated as the product is unrolled and pressed down. The back coating of the product is heated to the point where the bitumen coating acts as an adhesive, bonding the product to the layer below and the overlapped edges of the adjacent sheet. Flame heating equipment (*Figure 24*) can be used for BUR and torch-down roofing systems. Both types of systems require a mineral covering or protective coating to prevent ultraviolet destruction of the modified bitumen materials. In some cases, the product is available with various colors of ceramic roofing granules pre-applied to the exposed surface.

3.7.0 Single-Ply Roofing

There are two general categories of premanufactured membrane roofing systems: modified bitumen systems or single-ply systems. The single-ply membrane systems are wholly synthetic roofing materials that exhibit elastomeric properties to various degrees. There are a number of types within these two general systems. Today, there are many manufacturers of both systems, and it is important to note that each manufacturer requires specific, compatible materials and accessories for their versions of each type within each system. These materials and accessories include flashing, fasteners, drain and vent boots, inside/outside corners, cant strips (coving), cleaners, solvents, adhesives, caulking, sealers, and tapes. To obtain the maximum performance from the system and for warranty purposes, compatible materials and accessories must be used and manufacturer-specified application procedures must be strictly followed. Normally, these systems are installed over insulation boards by roofers experienced with a specific manufacturer's product.

Single-ply roofing systems can be classified as either thermoplastic (plastic polymer) or thermoset (rubber polymer) systems. Within these classifications, a number of types exist. One of the most

Figure 24 Flame heating equipment for BUR and torch-down roofing.

common thermoset polymer membranes is an ethylene-propylene-diene monomer (EPDM) product. This polymer, usually reinforced with polyester scrim, retains its flexibility over a wide range of temperatures and is very resistant to ozone and ultraviolet-ray damage. The membrane is usually supplied in large sheets or rolls and is spliced together with compatible adhesives or tapes.

Thermoplastic single-ply membranes have become very popular for commercial and industrial roofing. Two of the most common types of thermoplastic membranes available are polyvinyl chloride (PVC) and thermoplastic polyolefin (TPO). PVC membrane is usually reinforced with a polyester scrim and can be joined using solvent or hot-air welded seams that are extremely durable (*Figure 25*). PVC membrane is lightweight and aesthetically pleasing. It is very resistant to ozone and ul-

Figure 25 Single-ply membrane installation.

traviolet-ray damage. It is also puncture and tear resistant. TPO membrane combines the advantages of the solvent or hot-air seam-welding capability of PVC with the greater weatherability and flexibility benefits of the more traditional EPDM membranes. TPO membranes may also be reinforced with a polyester scrim. Both PVC and TPO membranes are supplied in sheets or wide rolls.

Single-ply roofing systems are clean and economical to install because they do not use hot-asphalt installation techniques common to BUR and modified bitumen membrane roofing systems. Most single-ply membranes are underlaid with insulation boards or protective mats along with a fireproof slip sheet, if necessary.

Single-ply membranes are usually anchored to a roof structure in one of four ways:

- *Loose laid/ballasted* – The perimeter is anchored with adhesives or mechanical fasteners, and the entire surface is weighted down with a round stone ballast or walking pavers (thin concrete blocks). This method is used only on very low-slope roofs capable of supporting the ballast load. It is fast and economical. See *Figure 26*.
- *Partially adhered* – The entire area of the membrane is spot-adhered to the roof with mechanical fasteners and/or adhesives. This method produces a membrane with a dimpled, wind-resistant surface.
- *Mechanically adhered* – The entire area of the membrane is spot-adhered to the roof with mechanical fasteners. This method produces a dimpled, wind-resistant installation and allows easy removal of the membrane during future roof replacement.
- *Fully adhered* – The entire area of the membrane is completely cemented down with an adhesive. This method produces a smooth, windproof surface.

3.8.0 Underlayment and Waterproof Membrane

Underlayment is available as 60-lb rolls of nonperforated 15-lb, 30-lb, and 60-lb asphalt-saturated felt (*Table 1*). The felt used under roofing materials must allow the passage of water vapor. This prevents the accumulation of moisture or frost between the underlayment and the roof deck. The correct weight of felt to be used is usually specified by the manufacturer of the finish roofing material. Roofs with a slope of more than 4 in 12 normally use 15-lb felt. Wood shakes usually require the use of 30-lb felt to separate the courses.

In areas of the country where water backup under the finish roof is a problem due to wind-driven rain, ice, and snow buildup, a waterproof membrane is available from a number of roofing material manufacturers under such names as Storm-Guard™, Water-Guard™, and Dri-Deck™. The membrane is usually made of a modified asphalt-impregnated fiberglass mat, coated on the bottom side with an elastic-polymer sealer that is also an adhesive. The membrane must be applied directly to the roof deck, not the underlayment. One of the top side edges is also covered with an adhesive so that the next course of overlapping membrane will adhere to the previous course. Any nails or staples driven through the membrane are sealed by the membrane, thus preventing water leakage. Because the membrane is impermeable, any moisture from inside the structure condensing under the membrane will eventually damage any wood directly under the membrane. As a result, some manufacturers suggest that the membrane only be applied along the bottom edges of the eaves and up the roof to at least 24" above the outside wall, along the rake edges, up any valleys, around skylights, and on any saddles or other problem areas. The rest of

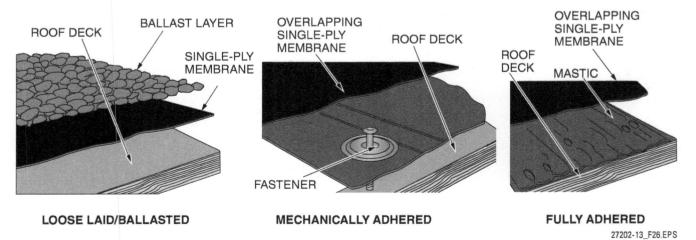

Figure 26 Typical methods of anchoring single-ply membrane to a roof.

Table 1 Sizes, Weights, and Coverage of Asphalt-Saturated Felt

Approximate Weight per Roll	Approximate Weight per Square	Squares per Roll	Roll Length	Roll Width	Side or End Laps	Top Lap	Exposure
60 lb	15 lb	4	144'	36"	4" to 6"	2"	34"
60 lb	30 lb	2	72'	36"	4" to 6"	2"	34"
60 lb	60 lb	1	36'	36"	4" to 6"	2"	34"

the roof is covered with conventional underlayment. However, other manufacturers indicate that the membrane can be applied over the entire roof deck, if adequate ventilation exists under the deck and a vapor barrier is installed on any inside ceiling under the deck.

Installing waterproof membrane is a two-person job. After a manageable strip is unrolled and cut off, one person must hold the material in place while another peels away a protective sheet covering the bottom adhesive. The membrane adhesive remains tacky during installation and the membrane can be lifted off the roof deck and repositioned, if necessary. However, after a short time it will set up and cannot be removed without damaging the membrane.

3.9.0 Drip Edge, Flashing, and Roof Ventilation

Figure 27 shows various types of drip edges that are available. Some are used in reroofing applications only. Drip edges and any other flashing must be made of materials that are compatible with the roofing and any items that are being flashed. They must last as long as the finish roof material. Today, aluminum, galvanized steel, copper, vinyl, and stainless steel are the most common materials used for drip edges and flashing. Galvanized steel, copper, or stainless steel must be used with any cement-based roofing material or against masonry materials, due to the corrosive nature of cement. Copper is normally used with slate roofing because of its long life and resistance to corrosion.

For best results, any galvanized drip edge or flashing should be coated with an appropriate primer before installation. All drip edges and flashing must be fastened with nails or staples made of a compatible material to prevent electrolytic corrosion between the fasteners and the flashing. The roofing material manufacturer's recommendations for flashing must be followed.

Figure 28 shows a W-metal (so named because its end profile looks like a letter *W*) or standing-seam valley flashing. This type of valley flash-

Wraparound end cap flashing covers the edges of old roofing layers.

This type of drip edge is designed to contain pea gravel on a built-up roof and is commonly called a gravel stop.

Often called style D or dripcap, this flashing adds a lip to the roof edge that overlaps the gutter or rake edge of the roof.

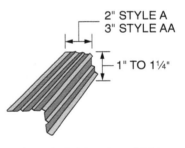
A canted strip edge of this variety carries the water away from the fascia. It is used on old roofs to hide old shingles or on new roofs. It provides a clean edge for new shingles. Style AA is used so that nails penetrate into wood when previous roof used a style A edging.

For roofing trimmed flush with the fascia, this type of end cap covers the edges of layers and keeps water and ice from backing up under the old shingles.

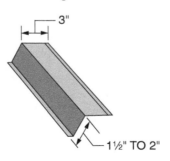
Angled gutter apron used at eaves to divert water into gutters to protect fascia.

Figure 27 Various types of drip edges.

ing is available in 8' to 10' lengths in widths of 20" to 24". If desired, it can be field-fabricated using flat roll flashing and a metal brake. In any case, it must be wide enough so that the finish roofing overlaps the metal by more than 6".

The W-metal valley flashing is preferred if open valleys will be used. Open valleys are defined as valleys where the valley flashing material will be visible after the finish roofing is applied. Open valleys are more difficult to install but accommodate higher rainfall rates than closed valleys. They can also be used with any type of finish roofing. Closed valleys can be used only on composition shingle roofs. The ridge in the middle of the W-metal valley (about ¾" to 1" high) prevents water rushing down the slope of one roof from washing under the shingles of the intersecting roof. On short valleys or relatively low-slope roofs, ordinary flat-roll flashing can be used in the valleys.

Proper attic ventilation is necessary to allow heat and moisture to escape so that damage to the roofing and roof deck does not occur. In the winter, ventilation keeps the roof deck cold and reduces the buildup of ice on the eaves. This helps prevent water penetration through the roof and subsequent water damage to the structure. It also carries away moisture so that it does not condense on the roof deck, which can cause rotting. In the summer, excess heat can cause overheating of composition shingles, resulting in early failure of the roof.

Residential attics are generally ventilated by convection vents in the form of gable vents or roof-mounted ridge or box vents. Sometimes, electric-powered or wind-powered turbine vents or fans are used (*Figure 29*). The air used for ventilation is provided by soffit vents at the eaves of the roof. The amount of soffit ventilation (in square feet or inches) must be equal to or greater than the amount of roof ventilation.

In residential structures, the proper amount of convection-type ventilation for an unheated attic space is usually defined as 1 sq ft of ventilation for every 300 sq ft of attic area with 50 percent in the roof for exhaust and 50 percent in the eaves for intake. The attic area is calculated using the exterior foundation dimensions of the structure. For example, the amount of ventilation for a residence with an exterior foundation measurement of 40' × 60' would be calculated as follows:

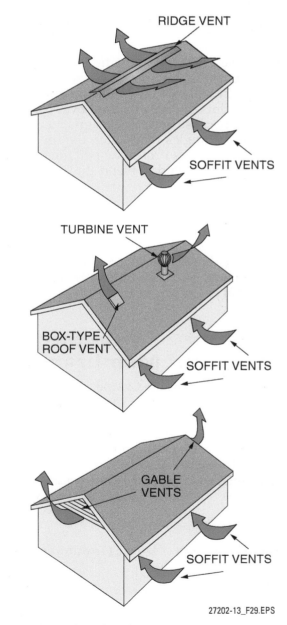

Figure 29 Residential roof vents.

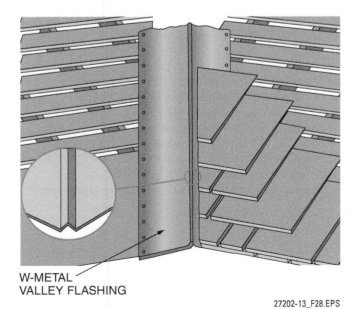

Figure 28 W-metal valley flashing.

40' × 60' = 2,400 sq ft (attic area)

2,400 sq ft ÷ 300 sq ft =
8 sq ft (total ventilation required)

8 sq ft × 144 sq in per sq ft = 1,152 sq in

1,152 sq in ÷ 2 = 576 sq in

for the ridge and 576 sq in for the soffits

Based on the calculated ventilation requirement, appropriately sized roof and soffit convection ventilation devices would be selected and installed.

Ridge vents, box vents, and turbine vents are available in a variety of styles and sizes for residential use. Different types of ridge vents are illustrated in *Figure 30*. The metal or plastic ridge vent is available in several patterns and colors. It is sold in 10' lengths and can be installed over most roofing materials.

The flexible, plastic composition vent is available in 4' lengths. An inert, coarse fiber vent is available in rolls. Flexible plastic and rolled coarse-fiber vents can only be used over composition shingles because both are designed to be covered with cap shingles that match the roof. Most residential customers prefer the shingle-covered vents because they tend to blend into the overall roof.

The manufacturer's specifications for a ventilation device must be consulted to determine the amount of free-air ventilation (in square feet or inches) that the device will provide. Ridge ventilators are probably the most efficient and are usually rated in square inches of free-air ventilation per linear foot of the product.

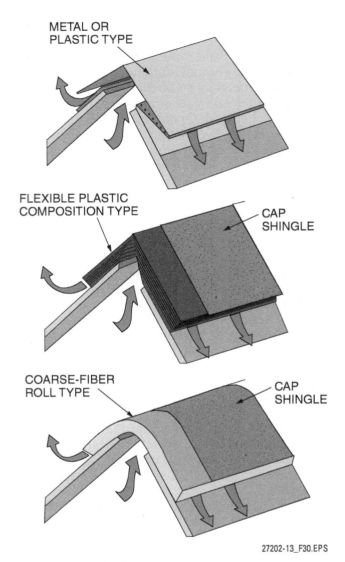

Figure 30 Typical ridge vents.

Waterproof Membranes

Waterproof membranes used on roof edges and valleys are self-healing. This means that if the membrane is intentionally or accidentally penetrated by a screw or nail, a modified asphalt coating will flow to seal the penetration when the roof and membrane are heated by the sun. Because they have an adhesive that secures them to the roof deck, waterproof membranes are virtually impossible to remove once the adhesive is heat-set by the sun. This makes reroofing of a structure very expensive if the membrane must be removed. This is because the roof deck must be replaced if it is wood or a wood product. Newer versions of membranes are available with a granular surface that allows the overlying shingles to be removed without damaging the membrane.

Ventilation Requirements

Always check local codes for the proper ventilation requirements of a structure. In some areas, the amount of air exchange required may dictate fan-assisted ventilation if the capacity of free-air ventilation devices is not adequate.

Open-Flame Heat Welding

Today, most seam-sealing methods and equipment recommended by manufacturers of single-ply membrane roofing systems make use of hot-air welding methods that are quite safe. However, some contractors still use open-flame heating equipment for seam welding. Open-flame seam sealing can be hazardous and may cause damage to the membrane. For these reasons, open-flame seam sealing of polymer membrane roofing should be avoided.

3.0.0 Section Review

1. Architectural shingles have a lifespan of _____.
    a. 10 to 25 years
    b. 25 to 40 years
    c. 40 to 70 years
    d. 70 to 100 years

2. With nails concealed, roll roofing can be used on slopes as low as _____.
    a. 1 in 12
    b. 2 in 12
    c. 3 in 12
    d. 4 in 12

3. If coated periodically with a preservative, wood shingles and shakes should last _____.
    a. at least 10 years
    b. 25 to 30 years
    c. 50 years or more
    d. approximately 100 years

4. For tile and slate roofing installations, use nails and flashing made of _____.
    a. stainless steel
    b. copper
    c. galvanized steel
    d. aluminum

5. A metal roofing material with a coating that is 80 percent lead and 20 percent tin is called _____.
    a. composite-coated sheeting
    b. pot metal
    c. galvanized steel
    d. terne metal

6. Depending upon the number of layers, a correctly applied built-up roof should not need replacement for _____.
    a. 5 to 10 years
    b. 10 to 20 years
    c. 15 to 25 years
    d. 20 to 30 years

7. Thermoplastic single-ply membranes are a popular material used for roofing single-family residences.
    a. True
    b. False

8. Rolls of asphalt-saturated felt used for roofing underlayment weigh _____.
    a. 15 pounds
    b. 30 pounds
    c. 60 pounds
    d. 75 pounds

9. Metal drip edge and flashing must be installed with fasteners made of a compatible material to avoid _____.
    a. metal fatigue
    b. electrolytic corrosion
    c. distortion
    d. rusting

Section Four

4.0.0 Roof Installation

Objective

Describe the installation techniques for common roofing systems.
 a. Describe how to properly prepare a roof deck.
 b. Explain how to install composition shingles.
 c. Explain how to install metal roofing.
 d. Describe how to install roll roofing.
 e. Discuss roof projections, flashing, and ventilation.

Performance Tasks

Demonstrate how to install composition shingles on a specified roof and valley.

Demonstrate the method to properly cut and install the ridge cap using composition singles.

Lay out, cut, and install a cricket or saddle. Demonstrate the techniques for installing other selected types of roofing materials.

Trade Terms

Base flashing: The protective sealing material placed next to areas vulnerable to leaks, such as chimneys.

Cap flashing: The protective sealing material that overlaps the base and is embedded in the mortar joints of vulnerable areas of a roof, such as a chimney.

Side lap: The distance between adjacent shingles that overlap, measured in inches.

Top lap: The distance, measured in inches, between the lower edge of an overlapping shingle and the upper edge of the lapping shingle.

Vent-stack flashing: Flanges that are used to tightly seal pipe projections through the roof. They are usually prefabricated.

Wall flashing: A form of metal shingle that can be shaped into a protective seal interlacing where the roof line joins an exterior wall. Also referred to as step flashing.

Roofing projects must be properly planned prior to installation. The manufacturer's recommendations should be referenced to ensure the proper fasteners are being used for the type of roofing material being applied. Prior to installing the finished roofing material, the roof deck must be properly prepared.

4.1.0 Preparing the Roof Deck

A typical roof installation is shown in *Figure 31*. Before the finish roofing is applied, the roof deck must be flashed with a drip edge along the eaves and any valleys must be flashed. Then, an underlayment and/or a waterproofing membrane is usually installed and capped at the rake edges with metal drip edge flashing. On bare wood roof decks, the underlayment/membrane must be applied on dry wood as soon as possible. If the wood is moist due to rain or morning dew, allow it to dry before applying the underlayment/membrane. If the roof deck is damp, the membrane may not adhere to the roof deck, or the underlayment will buckle and cause the final roof to appear wavy. The underlayment/membrane prevents the finish roof materials from having direct contact with any damaging resinous or corrosive areas of the roof deck, and helps resist or eliminate any water penetration into the roof deck. *Figures 32* and *33* show the recommended underlayment/waterproof membrane placement and drip edge installation.

Normally, the drip edge is installed along the length of the eaves first, followed by any valley flashing. The drip edge should be held against the fascia and nailed to the roof deck every 8" to 10". When installing valley flashing, it should overhang the valley at the upper and lower ends. The flashing is nailed every 6" to 8" on both sides, ½" from the edges.

After the flashing is secured, both ends are carefully trimmed flush with the roof deck. After the eave drip edge is in place, the exposed nail heads are covered with asphalt. Starting at the bottom of the roof, the underlayment and/or a waterproof membrane is rolled out and flattened with a roof roller before being tacked to the roof. In valleys, the waterproof membrane should extend over the flashing nails. The membrane will adhere and seal to the valley flashing; however, the underlayment should be trimmed to cover the flashing nails and should be cemented to the valley. In some cases on lower-sloped roofs, the underlayment is half lapped and cemented with asphalt to provide more of a water barrier. After the underlayment/waterproof membrane is in place, the rake edges of the roof are capped with a drip edge that is nailed every 8" to 10" to the roof deck. The bottom end of the rake drip edge overlaps the eave drip edge, and the fascia flange is cut to interlock behind the fascia flange of the

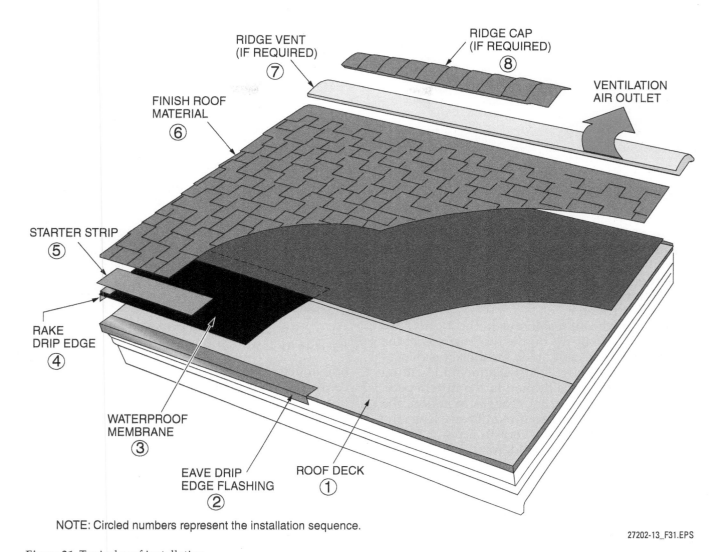

Figure 31 Typical roof installation.

eave drip edge. The nail heads should be covered with asphalt cement.

After the roof preparation is complete, the finish roof materials can be lifted and distributed equally over the roof deck.

4.1.1 Protection against Ice Dams

In areas subject to heavy snow, the snow will accumulate on the roof. Heat rising through the roof from inside the structure will melt the snow and cause ice to build up on the edge of the roof and in the rain gutters, creating an ice dam (*Figure 34*). As snow continues to melt, the water will be trapped by the ice dam and will be forced under the shingles. Eventually, it will find its way into the building. Information regarding metal and vinyl gutters and downspouts is included in the *Appendix*.

This ice-dam problem can be eliminated by a combination of attic insulation, roof venting, and

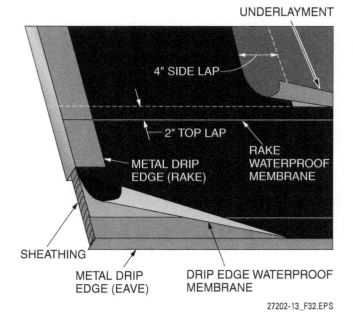

Figure 32 Drip edge and waterproof membrane placement.

Module 27202   Roofing Applications   33

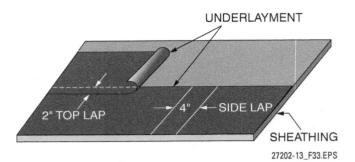

Figure 33 Underlayment or waterproof membrane placement over roof deck.

the use of a waterproof shingle underlayment. This underlayment comes in 36"-wide rolls. The material has a sticky side and is designed to stick to the roof deck, forming a tight seal against water penetration. It will seal around any nails that are driven through it.

Many sloped-roof residences in the North have ice-damming problems on the roof, usually at the eaves. This is especially true for those residences with finished attics where insulation and venting under the roof deck is limited to the rafter space. In most cases, when a new roof is installed, the application of a waterproof membrane from the eaves up the slope of the roof to a point that is at least 24" beyond the inside wall will prevent water backed up behind any ice dams from penetrating into the structure. However, on a problem residence with an existing roof that lacks a waterproof membrane underlayment, it may be desirable to install an ice edge at the eaves. On most roofing materials, ice dams cannot be easily removed. In addition, ice dams can damage some roofing materials, including slate.

An ice edge is an exposed metal sheeting mounted from the eaves up the slope of the roof from 18" to 36". It provides a shield against water penetration and allows any ice dams that form to be quickly shed from the roof during any brief thaws, thus reducing the chances of a large ice-dam buildup. The width of the ice edging used is determined by how steep the roof is and how thick the ice usually becomes, which determines

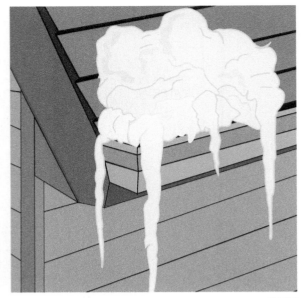

Figure 34 Ice dam.

how much water backs up on the roof. Normally, continuous sheets of plain or tinted/painted aluminum flashing or special standing-seam aluminum panels are used as the ice edge (*Figure 35*); however, copper flashing can be used where corrosion is a problem. The standing-seam panels are more expensive and are not subject to buckling due to temperature extremes. Their surface appearance is more uniform and does not appear wavy.

There are disadvantages to the use of ice edges. One is that they are generally not considered attractive unless colored to match the roof. The possible exception is plain copper used with slate roofing. Another disadvantage is that for ice edges to be effective in shedding ice, eave troughs cannot be used on the building. This can lead to several severe hazards such as injury or death to people or damage to property or foliage, including damage to the siding of the residence caused by falling ice. In addition, water draining from the roof can collect and penetrate under the foundation and/or into a basement.

Protruding Nails and Debris

Before any roofing material is applied to a roof deck, walk the nail pattern on the bare deck sheathing and check that all nails are driven flush with the surface. Also, make sure that all debris including small pebbles has been removed. After all flashing, underlayment, and drip edges are installed, again walk the nail patterns to make sure all nails are driven flush. Check that all debris has been removed. Any protruding nails or debris under a composition shingle may eventually penetrate the shingle and cause leaks.

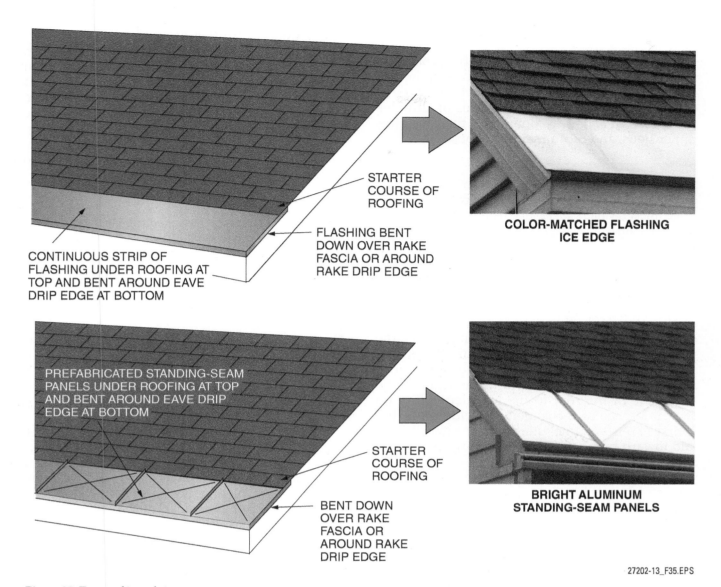

Figure 35 Types of ice edging.

Residential Secondary Roof Systems

An alternative for roofs that cannot be insulated properly from inside the residence is an insulated, secondary roof system installed over the original roof deck. These systems have spacing between a rigid insulation layer and a second roof deck to allow free airflow over the insulation and under the second roof deck. This greatly reduces the melting of snow and the resulting ice dam on the second roof deck. These secondary roof systems are available from several manufacturers. They can also be used in hot climates to reduce the heat load in a residence with similar insulation problems. The airflow of these systems can also reduce the heat load on the roofing materials used for the second deck.

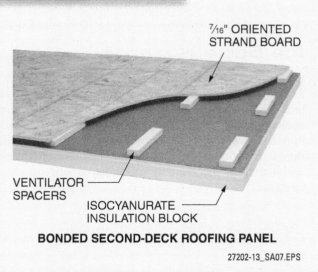

4.2.0 Installing Composition Shingles

At one time, the three-tab, square-butt composition fiberglass shingle was the most common. The architectural shingle is the most common type of composition shingle. Various types of shingles are shown in *Figure 36*, along with their weights, dimensions, and recommended exposures.

> **NOTE**
> The manufacturer provides a set of instructions with each bundle of shingles. These instructions must be followed. Failure to do so may void the manufacturer's warranty.

The following general instructions pertain to a standard three-tab fiberglass shingle.

The instructions for the installation of all types of shingles, including wood, composition, and slate, use a standard terminology to describe the placing of the shingles. This terminology is explained in *Figure 37*. Roof shingles can be placed from the left side of the roof to the right or the right side of the roof to the left, depending on the preference of the roofer. On wide roofs, the courses are sometimes started in the middle and laid toward both ends. In this module, the left-to-right convention is used.

All strip shingles are started with a double first row, which may be made up of a starter row and a row of shingles or a double course of shingles in which two joints have been offset. A common practice is to place a starter row of shingles with the tabs cut off. The type of starter course used will depend upon the type of shingle being used, the availability of materials, and local building codes.

Unopened bundles of shingles are usually placed at various points on the roof for the roofer.

Under normal circumstances, the shingle can be fastened with four aluminum or galvanized roofing nails positioned at a nailing line from the bottom of the shingle, one at each end and the other two above and adjacent to each cutout (*Figure 38*). Depending upon the area of the country in which you live, this cutout may be referred to as a notch or gusset.

> **CAUTION**
> Failure to scatter unopened shingle bundles can result in a broken rafter or collapsed roof, due to the weight of the bundles.

> **NOTE**
> If you are going to stand on a scaffold, it should be erected in such a manner that you will be working approximately waist high with the eave line. This places you in a safe, comfortable position to install the critical double course of shingles along the eaves.

In areas with very high winds, the number of nails above the cutout can be increased to two, positioned about 3" apart and forming a triangle with the top of the cutout as a low point. In high-wind areas, staples are not normally used.

When laying a shingle, butt the shingle to the previous shingle in the course and align the shingle. Then fasten the butted end to the roof. Keep the shingle aligned and fasten across the shingle to the other end. Alignment can be done by using the guide built into a roofing hammer or by chalking lines.

Ice Edge Hazards

Before ice edges are considered for use on the eaves of a building, the hazard of falling ice causing injury, death, or property damage under each eave must be carefully evaluated. With ice edging, a long, heavy ice dam (1,000 lb or more) along an entire roof eave can be released from the eave without warning. In addition, attempts to remove icicles may cause an ice dam along an entire eave to be released.

27202-13_SA08.EPS

	PRODUCT*	CONFIGURATION	APPROXIMATE SHIPPING WEIGHT	PER SQUARE		SIZE		EXPOSURE, INCHES
				SHINGLES	BUNDLES	WIDTH, INCHES	LENGTH, INCHES (NOMINAL)	
ARCHITECTURAL	WOOD-APPEARANCE STRIP SHINGLE; MORE THAN ONE THICKNESS PER STRIP 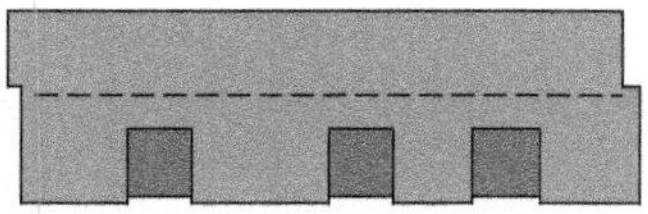 — PRELAMINATED OR JOB-APPLIED LAYERS	VARIOUS EDGE SURFACE TEXTURE AND APPLICATION TREATMENTS	285 lb TO 390 lb	67 TO 90	4 OR 5	11½ TO 15	36 OR 40	4 TO 6
ARCHITECTURAL	WOOD-APPEARANCE STRIP SHINGLE; SINGLE THICKNESS PER STRIP 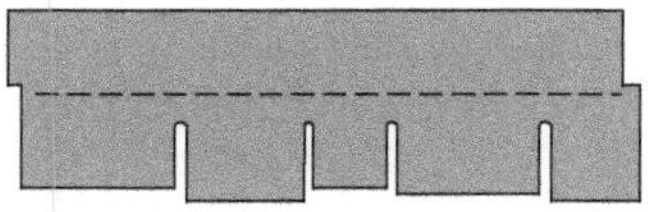	VARIOUS EDGE SURFACE TEXTURE AND APPLICATION TREATMENTS	VARIOUS, 250 lb TO 350 lb	78 TO 90	3 OR 4	12 OR 12¼	36 OR 40	4 TO 5⅛
STANDARD	3-TAB SELF-SEALING STRIP SHINGLE 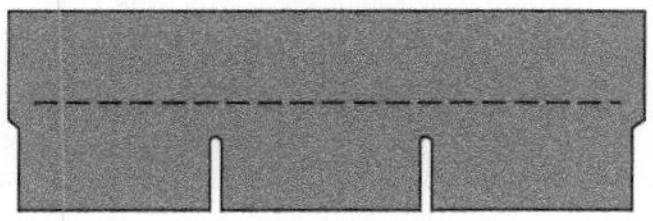	CONVENTIONAL 3-TAB	205 lb TO 240 lb	78 OR 80	3	12 OR 12¼	36	5 OR 5⅛
STANDARD	2-TAB (OR 4-TAB) VERSION 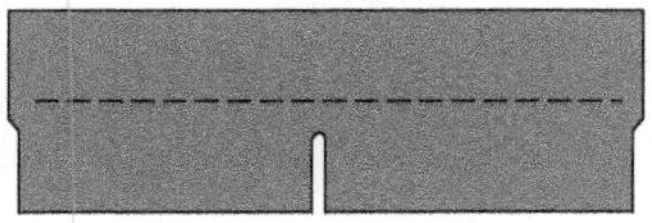	2- OR 4-TAB	VARIOUS, 215 lb TO 325 lb	78 OR 80	3 OR 4	12 OR 12¼	36	5 OR 5⅛
STANDARD	SELF-SEALING STRIP SHINGLE 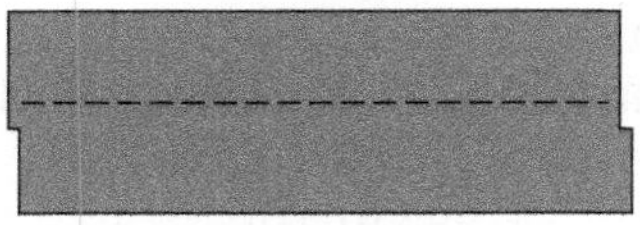 — NO CUTOUT	VARIOUS EDGE AND TEXTURE TREATMENTS	VARIOUS, 215 lb TO 290 lb	78 TO 81	3 OR 4	12 OR 12¼	36 OR 40	5

*Other types available from some manufacturers in certain areas of the country. Consult your regional asphalt roofing manufacturers' association.

Figure 36 Typical composition shingle characteristics.

4.2.1 Gable Roofs

This section describes the installation of long and short runs of standard shingles on gable roofs. It does not include instructions for metric or architectural shingles.

4.2.2 Gable Roofs—Long Runs

On large roofs, start applying the shingles at the center of the long run. The following procedure explains how to lay out and mark the roof. By beginning in the center, there is less chance of

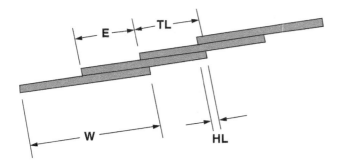

- **W = Width:** The total width of strip shingles or the length of an individual shingle.
- **E = Exposure:** The distance between the exposed edges of overlapping shingles.
- **TL = Top Lap:** The distance that a shingle overlaps the shingle in the course below.
- **HL = Head Lap:** The distance from the lower edge of an overlapping shingle to the upper edge of the shingle in the second course below.

Figure 37 Roofing terminology used in instructions.

misalignment as you proceed in both directions. Shingle manufacturers suggest that you run shingles horizontally first rather than stacking them one row above another. The color of the shingles will blend better that way.

To install a long run on a gable roof, proceed as follows:

Step 1 Measure along the length of the roof and find the halfway point. Do this along the ridge and along the eaves. Mark both of these places. Snap a chalkline vertically up the roof at these two marks.

Step 2 Measure 6" to the right and left of this center line at the ridge and at the eaves. Snap a line vertically for these marks as well. This will provide three lines to start the rows on.

Step 3 Starting at the eaves at both ends of the roof, measure 6" up from the eave drip edge and place a mark. Then go up to 11" and make a mark there. Proceed up the tape and mark every 5" interval. Once both sides are done, snap lines horizontally across the roof to align the shingles.

Step 4 For the starter shingles, cut the tabs off a tabbed shingle. Place the remaining part of the shingle so that the tar strip is nearest the edge of the roof. With standard shingles, you should have a starter row that is 7" wide by 36" long. If the starter strip is not done this way, then the first full row of shingles will not be sealed

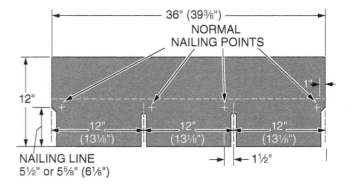
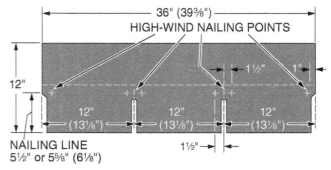
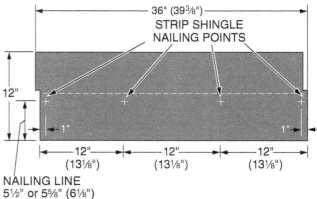

NOTE: Numbers in parentheses represent metric size; shingle dimensions in English units.

Figure 38 Nailing points.

down. The starter strip can be started on one of the vertical lines in the center of the roof. Make sure that the starter shingle stays on the first line snapped horizontally across the roof. If drip edges have been installed, position the starter strip with a ½" overhang on the eave and rake drip edges; otherwise, position the strip with a ¾" overhang. Place and fasten the starter strip both ways from center.

Step 5 Take a full shingle and place it directly over the starter strip and 6" to the right or left of the vertical line that was used for the starter strip. This will ensure that the cutouts will be spaced 6" away from the joints of the starter strip. This is considered the first full row of shingles.

Step 6 The second row is in the center, 6" to the right or left of the starting place where the first row was started. Proceed right and left from that point.

> **NOTE**
> Starting 6" over helps to minimize waste. Most of the time, the scrap you have left over on the right end will work on the left and vice versa.

4.2.3 Gable Roofs—Short Runs

To install a short run on a gable roof, proceed as follows:

Step 1 Cut the tabs off of the number of strip shingles it will take to go across the roof. Mark a spot 6" up on both eaves. Snap a line across the roof at this location. This line will keep the starter strip running straight across the roof. With the first left starter strip shortened by 6", place the starter strip shingles so that the tar strip is nearest the eave. If drip edges have been installed, position the starter strip with a ½" overhang on the eave and rake drip edge; otherwise, position the strip with a ¾" overhang. Nail the starter close to the top in four locations. After the starter strip is laid as far to the right as possible, return to the rake on your left and start to double up this first course by placing a full shingle directly over the top of the first upside-down shingle. See *Figure 39*. Continue this process with full shingles as you move to the right, and end before reaching the last starter strip shingle. You will observe that all cutouts and joints are covered with the full 12" tab. Make sure to nail the shingles correctly, as shown in *Figure 40*.

Step 2 Start the second course with a full strip minus 6". This layout is called a 6", half-tab pattern, or a 6-up/6-off layout, which means that half a tab is deliberately cut off and produces vertically aligned cutouts (refer to *Figure 39*). Overhang the cut edge at the rake by the same ¼" margin as the first course. Nail this shingle in place and proceed to the right with full shingles. The gusset openings can be

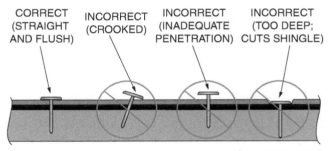

Figure 40 Correct and incorrect nailing.

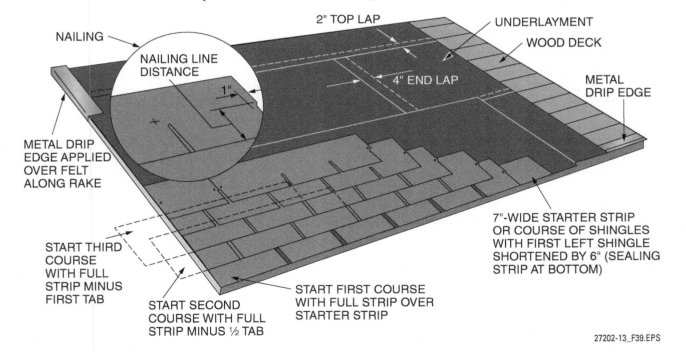

Figure 39 Shingle layout—6" pattern.

used as a checking procedure to obtain the proper 5" exposure. End the second course before reaching the end of the first course. This is called stair stepping and allows the maximum number of shingles to be placed before moving ladders or scaffolding.

Step 3 Start the third course with a full strip minus a full tab (12"). Follow the same procedure with the same rake overhang. Again, nail the shingles in place moving to the right and gauging your exposure by the 5" gusset.

Step 4 Returning now to the fourth course, start this row with half a strip (for example, with 18" removed). This will show a 6" tab. Nail this in place with the same overhang margin on the rake. Proceed with full shingles, nailing to the right and ending short of the previous course while constantly checking the alignment.

Step 5 Start the fifth course. This row starts with a full tab only (12"). Use the same overhang margin on the rake and full shingles as you move to the right. End before reaching the last shingle on the previous course.

Step 6 The sixth course starts with a 6" tab, the same overhang margin at the rake, and then full shingles. It continues to the right as the nailing proceeds. Again, end before reaching the last shingle on the previous course.

Step 7 As the process is repeated, the seventh course starts with a full shingle. Each successive course of shingles is shortened by an additional 6". This continues as previously described until the twelfth course.

Depending upon the individual or working team, two or more courses may be carried or nailed at a time as the shingling proceeds across the roof. When two people are working together, they usually work out their own system for speedy, accurate installation. The procedure described above uses the left rake, when facing the roof, as a starting point. Keep in mind that the entire application can be reversed by starting from the right rake (this applies to gable roof construction). On small roofs, strip shingles may be laid starting at either end with a successful result since the roof measurement is usually symmetrical.

To obtain different variations of roof patterns using tabbed shingles, only a change of starting measurement is required. *Figure 41* shows one possibility using a course with a full shingle (36") followed by a second course using a reduced-size shingle (32"). The third course would be reduced again to a shorter measurement (28"). Repeat these three measurements starting with the fourth course and continuing up the rake. This is called a 4" pattern and produces a diagonal cutout pattern.

Ribbon courses (*Figure 42*) are a way to add interest to a standard 6" pattern. After six courses have been applied, cut a 4"-wide strip lengthwise off the upper section of a full course of shingles. Fasten the strips as the seventh course is correctly aligned with the cutouts of the sixth course. Then reverse the 8"-wide leftover pieces of shingle and align them directly over the 4" strip. Fasten the 8" pieces to the roof deck at the top of the tabs. Cover both with a full-width seventh course of shingles. This creates a three-ply edge known as the ribbon. Repeat the pattern every seventh course.

Due to its simplicity, the first pattern mentioned (half tab = 6") is the most commonly used in the field. The full-strip asphalt shingle eliminates all pattern problems and alignment concerns on the vertical plane because it contains no gussets or cutouts.

4.2.4 Hip Roofs

When you encounter a hip roof, the basic nailing procedures remain the same, but the shingle layout starting point has to be at the center of the roof, as described for long gable roofs.

To begin, the starter strip is applied as previously described for long gable roofs. Return to the vertical line and use a full shingle for the doubling of this starter course. Offset this shingle 6" on either side of this vertical line. This will automatically cover the seam and gussets underneath,

---

Correct Nailing

Improper setting of nails and crooked nails can prevent the shingles from tabbing correctly and allow the wind to lift or tear the tabs of the shingles. Practice your fastening procedures so you drive the nail straight. The head should be flush with the surface of the shingle. Since your goal is to make the roof watertight, no pinholes or breaks are acceptable. If an accident should happen, a dab of asphalt cement spread with a putty knife will remedy the problem.

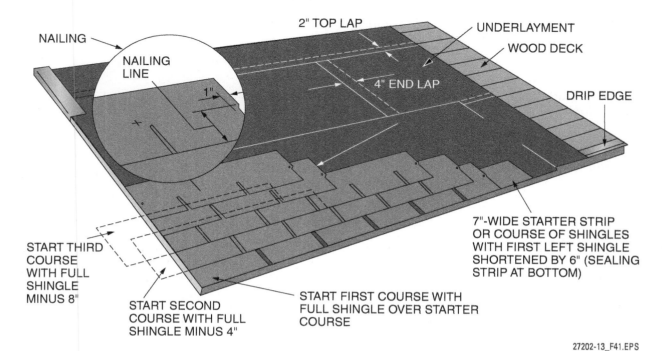

Figure 41 Shingle layout—4" pattern.

and seal the roof against water penetration. You have the option of continuing this dual-shingle starting course in either direction until it terminates at the hip rafter. See *Figure 43*.

At this point, the shingles should be cut to match the angle of the hip rafter and covered with a hip cap (the same as a ridge cap), completing the installation of the shingles. The hip cap is centered on the hip rafter and usually consists of a 12" tab showing a 5" exposure.

Exposed nails in the last caps should be covered with roof sealant. The ridge cap should cover the hip cap to prevent leaks.

4.2.5 Valleys

If the building you are constructing is not a perfect rectangle, you may encounter an L or T shape, which calls for another variation of shingling procedures. Where two sloping roofs meet, this intersecting valley has to be able to carry a high concentration of water drainage. Shingling becomes very critical, and the application must be done with extreme care.

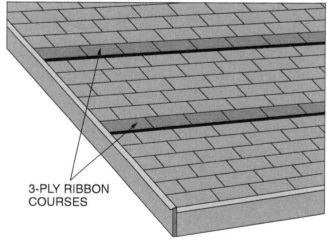

Figure 42 Ribbon courses.

4.2.6 Open Valley

With the valley flashing installed as previously described, snap two chalklines the full length of the valley. They should be 6" apart at the ridge or uppermost point. This means they should mea-

Starter Strips

Precut starter strips are available from many manufacturers. In many areas of the country, they are available in two sizes: 5" wide for roofing over existing shingles and 7" wide for new and tear-off installations.

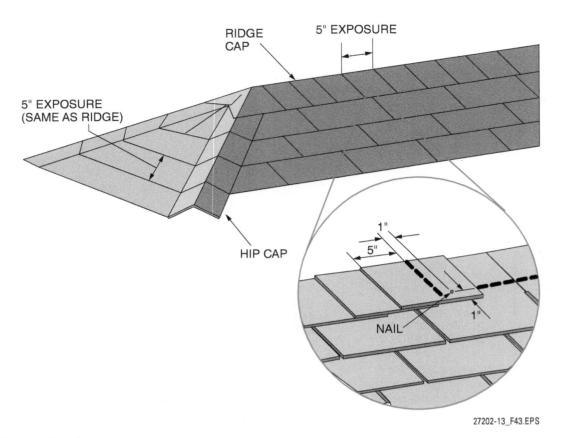

Figure 43 Hip and ridge layout.

sure 3" apart when measured from the center of the valley. The marks diverge at the rate of ⅛" per foot as they approach the eaves. For example, a valley 8' in length will be 7" wide at the eaves; one 16' long will be 8" wide at the eaves. The enlarged spacing provides adequate flow as the amount of water increases, passing down the valley. See *Figure 44*.

The chalkline you have snapped serves as a guide in trimming and cutting the last shingle to fit in the valley. This ensures a clean, sharp edge and a uniform appearance in the valley. The upper corner of each end shingle is clipped slightly on a 45-degree angle. This keeps water from getting in between the courses. The roofing material is cemented to the valley lining and to itself where an overlap occurs. Use plastic asphalt cement and spread a 6" to 8" bed. Do not overdo it, and clean up all excess cement so no tar shows.

4.2.7 Closed-Woven Valley

Some roofers applying asphalt shingles prefer to use a closed-woven valley design, sometimes called a full-weave or laced valley. It is faster to install, and some feel it gives a tighter bond. Others believe closed valleys are inferior to open valleys because they do not shed high volumes of water

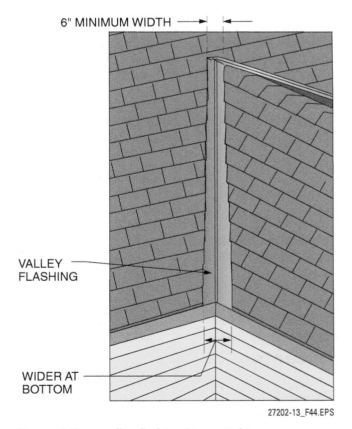

Figure 44 Open valley flashing (steep pitch).

very well. Composition shingles are the only type that can be used for this pattern. See *Figure 45*.

It is essential that a shingle be of sufficient width to cross the lowest point of the valley and continue upward on each roof surface a minimum of 12". Because of the skill required for this process, it is suggested that the two converging roofs be completed to a point 4' to 5' from the center of the valley. Then the weaving process can be accomplished carefully.

To create the 12" extension, it may be necessary to cut some of the preceding shingles in the course back to two tabs. To ensure a watertight valley, either a strip of 36"-wide, waterproof membrane, or 50-lb or heavier roll roofing over the standard 15-lb felt is placed in the valley, as shown in *Figure 45*.

For the weaving process, the first course is placed and fastened in the normal manner. Note that no fasteners are located closer than 6" to the valley center line. An extra fastener is placed at the high point at the end of the strip where it extends the extra 12". The first course on the opposite side is then laid across the valley over the previously applied shingles. Succeeding courses alternate, first along one roof area and then the other, as shown in *Figure 44*. Extreme care must be taken to maintain the proper exposure and alignment. As the shingles are woven over each other, they must be pressed tightly into the valley to provide a smooth surface where the roof surfaces join.

4.2.8 Closed-Cut Valley

In a closed-cut valley, sometimes called a half-weave or half-laced valley, the underlayment and valley flashing materials are the same as for the woven application.

To create a closed-cut valley, proceed as follows:

Step 1 Lay the first double-course of shingles along the eaves of one roof area up to and over the valley. See *Figure 46*. Extend it up along the adjoining roof section. The distance of this extension should be at least 12" or one full tab. Follow the same procedure when applying the next course of shingles. Make sure that the shingles are pressed tightly into the valley.

Step 2 This procedure is followed up the entire length of one side of the valley. If there is a high and a low slope, the first application should always be done on the low-slope side.

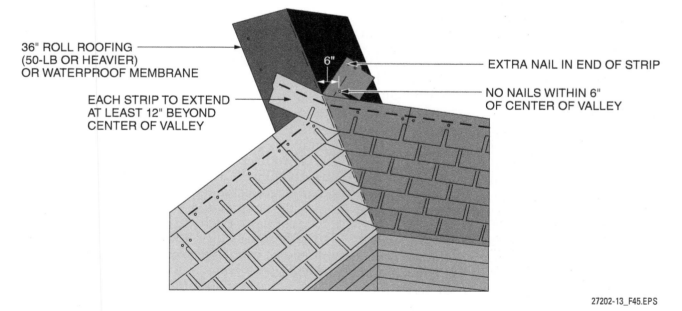

Figure 45 Closed-woven valley.

Other Shingle Alignment Methods

Several other alignment methods can be used for exposure. One popular method begins by snapping a chalkline along the top edge of the shingle of the first course, or 12" above and parallel with the eave line. Snap several other chalklines parallel with this first line. If you make the lines 10" apart, they can be used to check every other course by aligning the top of the shingle.

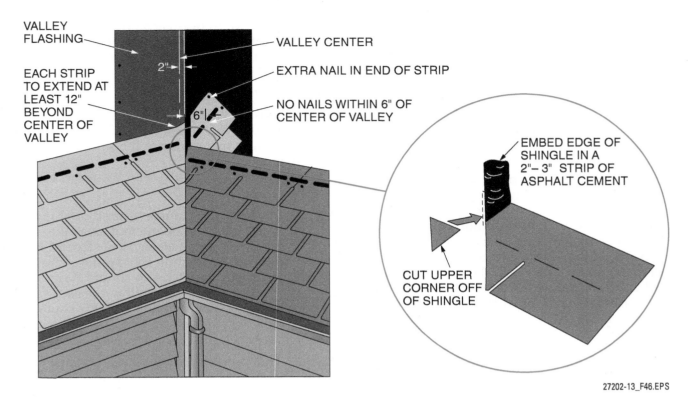

Figure 46 Closed-cut valley.

Step 3 When this roof surface is complete, you are ready to proceed with the intersecting roof surface, which will overlap the preceding application. Measure over 2" from the center line of the valley in the direction of the intersecting roof. Carefully snap a chalkline from top to bottom. This will be your guideline for the trim cut on the shingles.

Step 4 Now apply the first course of shingles on the intersecting roof. Use extreme care to match your chalkline angle exactly. Also, trim off the upper corner of the shingle to prevent water from running back along the top edge. Embed the end of the shingle in a 2" to 3" strip of plastic asphalt cement, being careful to allow no tar to show on the original shingle opposite. Succeeding courses are applied and completed as shown in *Figure 45*, making the valley watertight.

4.3.0 Installing Metal Roofing

This section covers various types of metal roofing systems.

4.3.1 Corrugated Metal Roofing

Another type of roofing material is corrugated metal roofing, or galvanized metal roofing. Only galvanized sheets that are heavily coated with zinc are recommended for permanent construction.

Galvanized sheets may be laid on slopes as low as a shallow 3" rise to the foot (⅛ pitch). If more than one sheet is required to reach the top of the roof, the ends should overlap by at least 8". When the roof has a pitch of ¼ or more, 4" end laps are usually satisfactory. To make a tight roof, sheets should be overlapped by 1½ corrugations at either side (*Figure 47A*).

When using roofing that is 27½" wide with 2½" corrugations and a corrugation lap of 1½,

High-Wind Areas

In high-wind areas, the fifth and sixth courses of the 6" pattern are usually eliminated because of the possibility of the small starter tabs being torn off the roof. The fifth course tab (12" long) should be saved for use as a ridge cap. The sixth course tab (6" long) would be discarded. The shingles would also be secured using double fastening at each cutout.

each sheet covers a net width of 24" on the roof. If 26-gauge galvanized sheets are used, supports may be 24" apart. If 28-gauge galvanized sheets are used, supports should not be more than 12" apart. The heavier gauge has no particular advantage except its added strength, because the zinc coating is what gives this type of roofing its durability.

When 27½" roofing is not available, sheets of 26" width may be used. When laying the narrower sheets, every other one should be turned upside down so that each alternate sheet overlaps the two intermediate sheets, as shown in *Figure 47A*.

For best results, galvanized sheets should be fastened with neoprene-headed nails, galvanized nails and neoprene washers, or screws with neoprene washers. Fasteners are used only in the tops of the corrugations to prevent leakage. To avoid unnecessary corrosion, use the fasteners specified by the roofing manufacturer.

Corrugated metal roofing panels (*Figure 48*) are used on garages, storage buildings, and farm buildings. They are available in widths up to 4'-0" and lengths up to 24'-0". Normally, these panels are used on roofs with slopes of 4 in 12 or steeper. They can be used on 2 in 12 roofs if a single panel reaches from the ridge to the eave.

The panels are fastened to purlins. A purlin is a structural member running perpendicular to the rafters. See *Figure 48*. Usually, 2 × 4 wood stock is used. The spacing should follow the directions specified by the manufacturer. Filler strips are sold with the panels. They are set at the eave and the ridge. Normally, the panel is cut so it overhangs 2" to 3" at the eave. The installation of metal roofs should be done in accordance with the manufacturer's specifications.

4.3.2 Simulated Standing-Seam Metal Roofing

The character of each standing-seam roof system dictates the amount and type of planning. Each system has different components and slightly different requirements for tools and equipment. In some cases, where seaming machines are required, supervisors may have to decide not only when to lease the machines, but how many machines to have on the project. A backup machine is always a good idea, especially if the system requires seaming shortly after panel installation.

Power-tool requirements also differ from system to system. Most systems use screw guns to install self-drilling and self-tapping screws, but some systems require impact wrenches or bulb rivet guns for fastening. It is important to have a sufficient number of power tools on the job to allow the work to move smoothly and efficiently.

It is also important that the assemblers carry oversize fasteners. Standing-seam roof systems minimize the number of through-the-panel fasteners by up to 90 percent. Therefore, it is crucial that the fasteners be installed correctly. If a fastener is stripped during installation, it should be

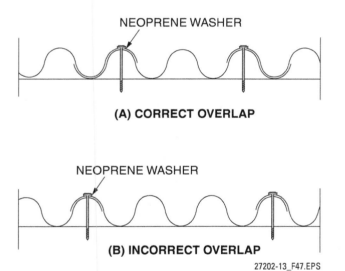

Figure 47 Corrugated roofing.

Unequally Pitched Roof Intersections

If the roof valley is formed by an intersection of two unequally pitched roofs, the woven valley will creep up one side, making it nearly impossible to maintain the correct overlap of shingles. The open valley or the closed-cut valley should be used with unequally pitched roof intersections.

Closed-Cut Valley

If a valley is formed by two different roof pitches, the two sides will climb at different rates. The closed-cut valley will give a much neater appearance in this situation.

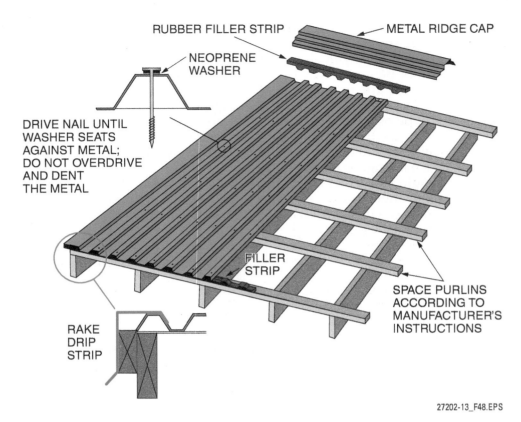

Figure 48 Corrugated roof layout.

removed immediately and replaced by a fastener of the next larger size.

The direction in which the sheeting takes place has to be considered. Some systems have strict requirements; others are more flexible. The sheeting direction can be found on the construction drawings.

Perhaps the biggest demand each standing-seam roof system makes is to be installed by competent assemblers. Improper installation techniques cause the majority of standing-seam roof failures. This underscores the need for special training in the particular system being used.

Before roofing can begin, the structure must be plumb and level. The purlins must also be straight. Z-purlins, in particular, have a tendency to roll. If there are no purlin braces, use wood blocking. Most manufacturers suggest that wood blocks be driven tightly between purlins to ensure proper spacing, and recommend either 2 × 6 or 2 × 8 lumber, depending upon the purlin depth. Place at least one row of blocks in the center of the bay. The erection drawings contain the proper purlin spacing.

Purlins may also be straightened by adjusting the sag rods, if the structure has them. Many sag rods are cut to set a specific width automatically.

Until at least one run of panels has been installed, there is no safe place to work from unless a work platform is constructed. A work platform should be made by stacking two panels on top of each other and placing walk boards in the center. The panels should be attached to the structure with locking C-clamp pliers or some other means to prevent the panels from moving. This platform can be used to store the insulation required for the first run, as well as the necessary tools and components.

The specific sequence of erection of standing-seam roof systems is determined by the manufacturer of the given system. What is recommended for some systems may not be recommended for others. This fact emphasizes the importance of knowing and understanding the particular system before installing it.

Usually, a blanket of insulation is stretched across the structural members prior to sheeting. The blanket width depends on the panel. Keep the stapling edge ahead of the panel, but no more than a foot ahead of it.

As previously mentioned, installing metal roofing requires special tools and experience. The following is an overview of the installation procedure for one type of metal roofing that is designed to be laid over a closed, fully-sheathed roof deck. It is applied in panels that are 12" to 16½" wide and that are precut to run from the eaves to the ridge of the structure. The joints between panels are weatherproofed by means of a C-clip with a neoprene seal.

Cover the roof with a 30-lb felt or waterproof membrane. The membrane or felt must overlap the edges of the eaves and rakes.

The eave trim (*Figure 49*) is then screwed in place before panels are applied. The panels are placed one at a time and secured to the roof deck with a T-clip on one side of the panel.

The next panel is then inserted under the T-clip that is holding the first panel and secured on the opposite side with another T-clip (*Figures 50 and 51*). At the joint of any two panels, the joint is weather-sealed with a C-clip running the full length of the joints, as shown in *Figures 50* and *51*. The T-clips are used every 12" in high-wind areas and every 18" elsewhere. After panels are placed under each side of the T-clips, the T-clip wings are bent down, and the C-clip is forced down on the seam. The neoprene flaps inside the C-clip seal the seam against water penetration, and the bent-down wings of the T-clip keep the C-clip from being dislodged from the seam.

After the first panel is placed at one of the rakes, a rake edge is applied, as shown in *Figure 52*. If necessary at the opposite rake, the panel is trimmed off and Z-strips are sealed and secured to the panel before the rake edge is installed.

At valleys, the panels are trimmed to the angle of the valley. Channel strips running parallel in the valley are sealed and screwed to the valley and hold the edges of the panels (*Figure 53*).

At the ridge, Z-strips are fastened and sealed to the panels between the seams. They are also sealed to the seams. The cover flashing is then sealed and secured to the Z-strips (*Figure 54*).

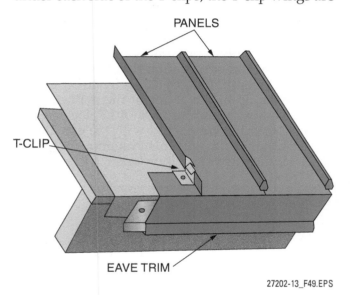

Figure 49 Eave trim.

Figure 51 Nailing a T-clip to a roof deck.

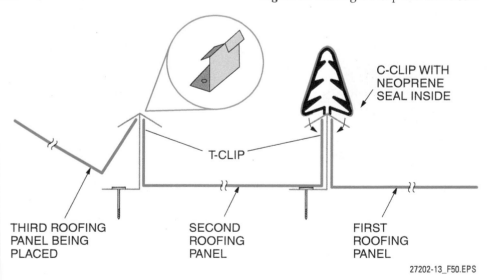

Figure 50 Placing, securing, and sealing panels.

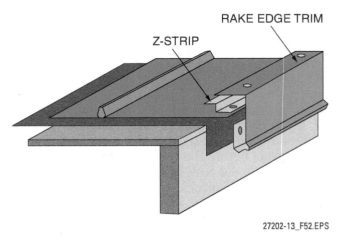

Figure 52 Rake edge.

4.3.3 Snug-Rib System

Another type of roofing sheet is called the snug-rib system. It utilizes a concealed fastener, which is leak resistant and eliminates through fasteners. This combination of a V-beam industrial sheet and the snug seam joint makes for a greater beam strength and a deeper corrugation than other roofing profiles. Because of the greater strength of this type of roofing panel, the purlin spacing can be increased.

The snug-rib joint is a highly efficient watertight joining system of a simple nature. The joint is created by engaging the hooked edges of two panels into a Y-shaped extruded spline, previously measured and anchored to the purlins with a self-templating clip. A neoprene gasket is then rolled into the extrusion between the panel edges, where it holds the panel edges securely in place and creates a watertight seal.

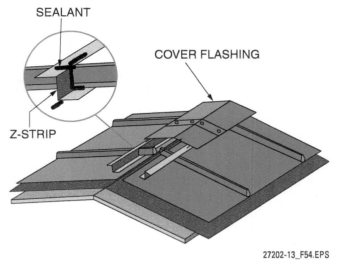

Figure 54 Ridge flashing.

There is a 19½" covering width. The material has a V-shaped corrugation that has a 4⅞" pitch and a 1¾" depth. Lengths vary from 77" to 163", depending upon the gauge of the material. The end laps must be a minimum of 12", located over roof purlins, and staggered with the end laps in adjacent panels. Install this system according to the manufacturer's specifications.

> **NOTE:** No primary fasteners penetrate the weatherproofing membrane.

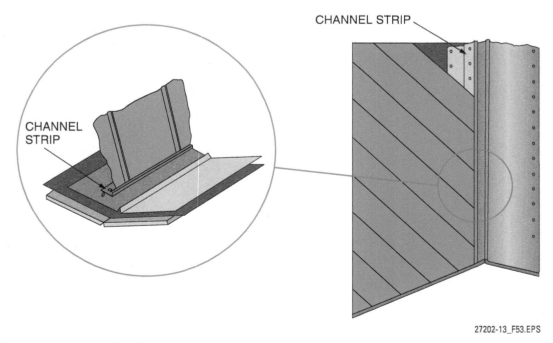

Figure 53 Channel strips and valley flashing.

4.4.0 Installing Roll Roofing

Nearly flat roofs can be roofed with hot-asphalt BUR, a single-ply membrane system, or roll roofing. Roll roofing can be installed on underlayment by itself, as part of a cold asphalt built-up roof, or on a waterproof membrane. The weights, characteristics, side lap, top lap, and recommended exposures for double-coverage, single-coverage, and uncoated roll roofing are shown in *Table 2*.

All flat roofs must have some pitch, either to an edge or roof drains, so that water does not collect. Composition shingles are not used on roofs with a slope of less than 2 in 12, wood shingles are not used with a slope of less than 3 in 12, and shakes are not used with a slope of less than 4 in 12. Flat roofs should have a minimum slope of ¼" per foot. When not installed as part of a cold-asphalt built-up roof, roll roofing can be installed as single-coverage roofing with exposed or concealed nails, or as double-coverage roofing with concealed nails. Single-coverage roofing with exposed nails is generally used on slopes of 2 in 12 or more. Single-coverage roofing with concealed nails is used on slopes of 1 in 12 or more. Double-coverage roofing with concealed nails is used on slopes of more than ¼ in 12.

Apply a drip edge and waterproof membrane, or as a minimum, apply a 15-lb underlayment to the roof deck. Make sure all debris is removed from the roof deck and all nails are flush before applying the waterproof membrane/underlayment, and that it is clean before applying roll roofing. Even a very small pebble or protruding nail will eventually poke a hole through single-layer roofing. Flashing of roof projections is accomplished in the same way as described for composition shingles in the section *Roof Projections, Flashing, and Ventilation*.

4.4.1 Single-Coverage Roll-Roofing Installation

This section describes the installation of single-coverage roll roofing using the exposed and concealed nail methods.

4.4.2 Exposed Nail Method

Single-coverage roll roofing with exposed nails is generally applied horizontally over underlayment, as shown in *Figure 55*. To install roll roofing using the exposed-nail method, proceed as follows:

Step 1 Protect each valley with 18"-wide metal flashing.

Step 2 For horizontal application (*Figure 56*), snap a chalkline 35½" above the eaves. Apply a 2" band of roof cement to the eaves and rake edges.

Step 3 Using the chalkline as a reference, run the first course so that it overhangs the eaves by ½" and the rake edges by 1". Cement and overlap any vertical seams by 6". Nail all seams and the bottom and rake edges of the first course every 3" with galvanized or aluminum roofing nails. Use a utility knife to trim the roofing edges to the drip edges.

Step 4 Snap another chalkline 3" down from the top edge of the first course and apply a 2" band of roof cement within the band and up the rake edges. Lay the second course to the chalkline and nail every 3" along all seams, the bottom edge, and the rake edges, as shown in *Figure 57*.

Step 5 The third and subsequent courses are applied in the same manner. Trim the edges at all rakes and eaves.

Step 6 At all valleys, hips, and ridges, apply the roofing to the point that it overlaps. Then trim the roofing to the center of the valley, hip, or ridge (*Figure 58*). Apply a 6" band of roofing cement to each side of the valley flashing, and press the edges of the roofing into the cement. Do not nail the edges of the roofing in the valley, and do not nail horizontal seams within 6" of the center of the valley.

Table 2 Typical Weights, Characteristics, and Recommended Exposures for Roll Roofing

Product	Approximate Shipping Weight		Squares per Package	Length	Width	Side or End Laps	Top Lap	Exposure
	Per Roll	Per Squares						
Mineral-surface roll, double coverage	75 lb to 90 lb	75 lb to 90 lb	1	36' / 38'	36" / 38"	6"	2" / 4"	34" / 32"
Mineral-surface roll, single coverage	55 lb to 70 lb	55 lb to 70 lb	1/2	36'	36"	6"	19"	17"
Uncoated roll	50 lb to 65 lb	50 lb to 65 lb	1	36'	36"	6"	2"	34"

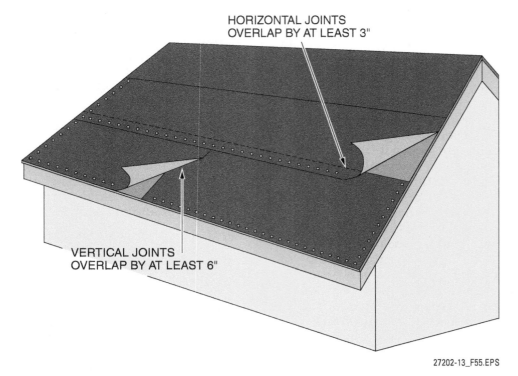

Figure 55 Typical single-coverage roll-roofing installation.

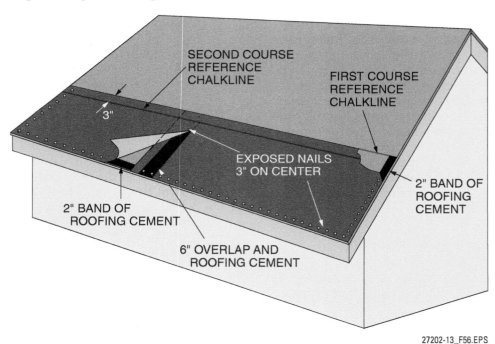

Figure 56 First course of exposed-nail roll roofing.

Step 7 At hips or ridges, snap a chalkline 6" on both sides. Apply a 2" band of roofing cement just above the 6" chalklines. Cut 12"-wide strips of roofing and nail down the strips through the cement on both sides of the hip or ridge. As necessary, overlap the strips by 6" and cement.

4.4.3 Concealed Nail Method

To install roll roofing using the concealed-nail method, use the following procedure.

Step 1 Cut and install 9"-wide roofing material starter strips along the eaves and rakes (*Figure 59*). Nail the strips to the roof deck on both edges with galvanized or alumi-

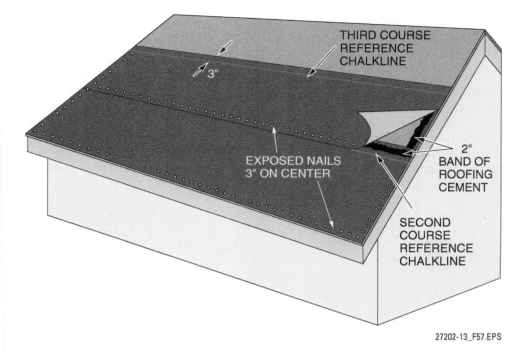

Figure 57 Second and subsequent courses of exposed-nail roll roofing.

Corrugated Roofing Light Panels

Corrugated translucent fiberglass light panels can be used to allow daylight into the interior of a structure. They can also be used as the entire roof, if desired. These panels are the same size and are installed in the same manner as corrugated metal roofing.

num nails spaced ¾" from the edges and 4" apart. These strips provide a surface for adherence of roofing cement. Protect each valley with 18"-wide metal flashing.

Step 2 Snap a chalkline 35½" above the eaves. Using the chalkline as a reference, run the first course so that it overhangs the eaves by ½" and the rake edges by 1". Nail along the top edge, ¾" from the edge, every 4". Then apply a 6" band of roof cement to the eaves and rake edges (*Figure 60*). Press the roofing down into the cement with a roof roller. Cement and overlap any vertical seams by 6". Use a utility knife to trim the roofing edges to the drip edges.

Step 3 Snap another chalkline 6" down from the top edge of the first course. Lay the second course to the chalkline and nail along the top edge, ¾" from the edge, every 4". Apply a 6" band of roof cement under the bottom of the second course and up the rake edges. Press the roofing down into the cement with a roof roller.

Step 4 The third and subsequent courses are applied in the same manner (*Figure 61*). Trim the edges at all rakes and eaves.

Step 5 At all valleys, hips, and ridges, apply the roofing to the point that it overlaps. Then trim the roofing to the center of the valley,

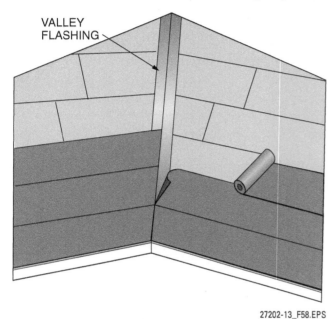

Figure 58 Roll roofing in a valley.

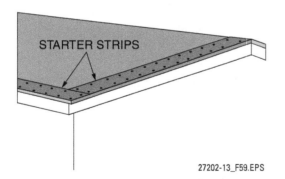

Figure 59 Roll-roofing starter strips.

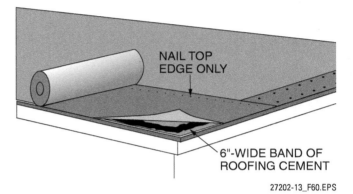

Figure 60 First course of concealed-nail roll roofing.

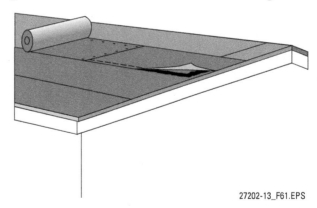

Figure 61 Third and subsequent courses of concealed-nail roll roofing.

Removing the Curl from Roll Roofing

Before installing roll roofing in cool to cold ambient temperatures, cut the roofing into 12' to 18' sections and stack it for a sufficient length of time to remove any curl. The length of time required will depend on the ambient temperature.

hip, or ridge. Apply a 6" band of roofing cement to each side of the valley flashing and press the edges of the roofing into the cement. Do not nail horizontal seams within 6" of the center of the valley. At hips or ridges, nail the last course at the top of the hip or ridge. Then, snap a chalkline 6" on both sides.

Step 6 Apply a 6" band of roofing cement just above the 6" chalklines (*Figure 62*). Using 12"-wide strips cut from the roofing material, press the strips down into the cement on both sides of the hip or ridge with a roof roller. As necessary, overlap the strips by 6" and cement.

4.4.4 Double-Coverage Roll-Roofing Installation

Double-coverage roll roofing is available for both hot and cold asphalt application. Make sure that only the cold asphalt version is used for the following installation. Double-coverage roll roofing has a 19" overlap (called the selvage) and a 17" mineral-coated exposure.

To apply double-coverage roll roofing, proceed as follows:

Step 1 Cut the 17" mineral-coated exposure from enough strips of the roofing to extend across all eaves. The mineral-coated pieces will be used as starter strips. Save the 19" selvage strips for use when the ridge caps are installed.

Figure 62 Covering hip or ridge joints.

Step 2 Snap a chalkline 18½" from the eaves and use it to position the top of the starter strip so that it overlaps the eaves by ½" and the rakes by 1". Nail down the starter strip at both edges with nails spaced 3" at the bottom edge and 12" at the top edge. Nail the center at 12" intervals.

Step 3 Starting from one end, position one strip of the first course over the starter strip and nail it into place using two rows of nails spaced 4½" and 13" from the top of the strip in the selvage area. Space the nails in the rows about 12" apart.

Simulated Standing-Seam Roofing Systems

A number of different methods of joining metal roofing panels are employed for these types of systems. Older systems used galvanized steel or copper standing seams that were soldered together to form a weather seal. Other systems require crimping machines or use snap-type seals.

MACHINE-CRIMPED PANELS

SNAP-TYPE SEAL

Module 27202   Roofing Applications   53

Step 4 Roll back the strip and thickly coat the starter strip underneath with nonfibered liquid roofing cement (*Figure 63*). Roll the strip back onto the cement and press it down using a roof roller.

Step 5 Overlap the next strip by 6" and repeat the nailing and cementing procedure. Vertical seams are cemented, as shown in *Figure 64*, and are not nailed except in the selvage area.

Step 6 Continue applying strips and courses until the roof deck is covered. Trim the edges at all rakes and eaves.

Step 7 At all valleys, hips, and ridges, apply the roofing to the point that it overlaps. Then trim the roofing to the center of the valley, hip, or ridge.

Step 8 Apply a 6" band of roofing cement to each side of the valley and press the edges of the roofing into the cement. Do not nail horizontally to within 6" of the center of the valley.

Step 9 At hips or ridges, nail the last course at the top of the hip or ridge. Snap a chalkline 6" on both sides.

Step 10 Cut 12"-wide pieces from a roll of double-coverage roofing; include both the selvage and the mineral-coated exposure. These pieces are treated like the shingle tabs used to make a ridge cap, but are applied in double-coverage just like the roof.

Step 11 To begin, cut the selvage from one shingle. Then, starting at one end of the ridge or the bottom of a hip, place the selvage piece over the ridge and nail it down, spacing the nails 1" from the edges and at 4" intervals. Coat the selvage piece with cement.

Step 12 Next, place a full shingle over the cement with the mineral side up and press it into the cement. Nail the selvage of this shingle like the starting piece of selvage. Coat the selvage of that shingle with cement and apply another shingle.

Step 13 Repeat the process until the hip or ridge is completed. The junction of two hips and a ridge can be end-capped in the same way as for a shingle roof.

4.5.0 Roof Projections, Flashing, and Ventilation

The following sections describe various types of roof projections and flashing.

4.5.1 Soil Stacks

Another roofing task is waterproofing around soil or vent stacks. Most building roofs have pipes or vents emerging from them. Most are circular. They call for special flashing methods. Asphalt products combined with metals may be used for this purpose. A soil pipe made of cast iron, cop-

Figure 63 Coating the starter strip.

Figure 64 Cementing a vertical seam.

per, or other approved materials is used as a vent for plumbing. Various types of **vent-stack flashing** are available for this purpose. See *Figure 65*.

To apply vent-stack flashing, proceed as follows:

Step 1 First, apply the roofing up to where the stack projects. Use extreme care when cutting and fitting the shingles around the stack. See *Figure 66*.

Step 2 Slip the flange over the stack and place it down into a bed of asphalt cement that has been carefully spread to the same size as the flange. The flange, sometimes called a boot or collar, is usually made of metal, but can also be made of plastic or rubber. See *Figure 67*. Prefabricated boots that slip over the stack are available in different pitches.

Step 3 Mold the flange boot to the soil stack to ensure a snug fit. Use the manufacturer's recommended sealant to close up any opening. When the next course of shingles is laid, it covers the upper portion of the flange. Prior to this course, a bed of cement can be spread on the top of the flange. The end result seals and waterproofs the vented stack opening. See *Figure 68*.

Step 4 After the installation is completed, install the remainder of the shingles as previously described.

4.5.2 Vertical Wall Flashing

In the process of roof construction, there are times when the roof abuts a vertical wall horizontally or at an angle. This is a very critical spot to make watertight. Extreme care must be taken to follow correct procedures. An example of a sloped or angled abutment is shown in *Figure 69*. The initial step is to let the underlayment turn up on the vertical wall a minimum of 3" to 4". As an alternative, a strip of waterproof membrane may be applied to the roof deck and turned up on the wall.

This turn-up bend has to be done very carefully to maintain the seal and overlap of material without creasing or tearing the felt or membrane. Regular shingling procedures are used as each course is brought close to the vertical wall. **Wall flashing** (step flashing) is used when the rake of the roof abuts the vertical wall.

Figure 65 Vent-stack flashing.

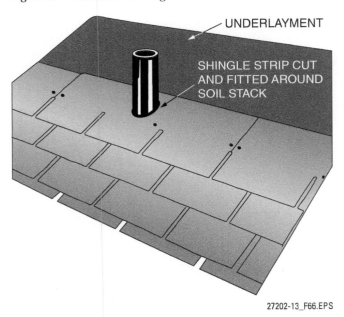

Figure 66 Layout around stack.

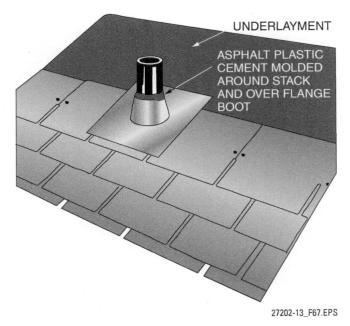

Figure 67 Placement of flashing.

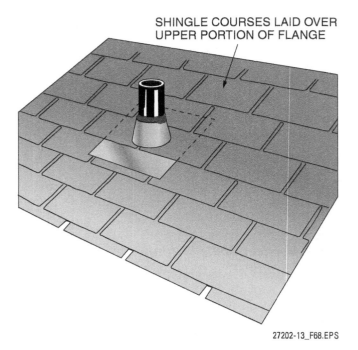

Figure 68 Covering flashing.

Metal step-flashing shingles are applied over the end of each course of shingles and covered by the next succeeding course. The flashing shingles are usually rectangular. They are approximately 6" to 7" long and from 5" to 6" wide. When used with shingles laid 5" to the weather, they are bent so half of the flashing piece is over the roof deck with the remaining half turned up on the wall. The 7" length enables one to completely seal under the 5" exposure with asphalt cement and provide a 2" overlap up the entire length of the rake.

A careful study of *Figure 69* shows that each flashing shingle is placed just up the roof from the exposed edge of the shingle that overlaps it. It is secured to the deck sheathing with one nail in the top corner.

When the finished siding or clapboards are brought down over this flashing, they serve as a cap flashing, sometimes referred to as counter flashing. Usually, a 1" reveal margin is used and the ends of the boards are fully painted or stained to exclude dampness and prevent rot. With proper application of flashing and shingles, the joint between a sloping roof and a vertical wall should be watertight.

On a horizontal abutment (*Figure 70*), continuous flashing must be applied horizontally across the entire top of the abutting roof and against the vertical wall under the siding.

Continuous flashing can be formed with a metal brake or by hand, as shown in *Figure 71*. The flashing should be at least 9" wide and bent to match the angle of the joint to be flashed. Position the bend so that there will be at least 4" of flashing on the roof and 5" on the wall.

Before applying the flashing, adjust the last two courses of shingles so that the last course, which will be trimmed to butt against the wall, is at least 8" wide. After this abutting course is installed, place roofing cement on top of the last course of shingles. Place the flashing against the wall (slipping it under any siding, if necessary) and press it into the cement. Do not nail the flashing to the wall or roof deck. If desired, apply several beads of roofing cement to the top of the flashing. Press the tabs cut from shingles into the cement to cover and hide the flashing (*Figure 72*). Position the tabs

NAILING STEP FLASHING TO ROOF DECK

INSTALLED STEP FLASHING

Figure 69 Wall (step) flashing.

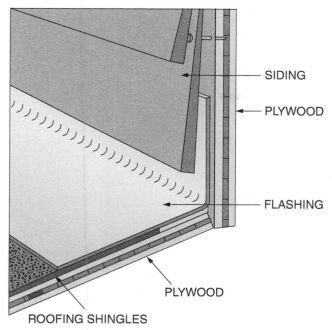

Figure 70 Continuous flashing.

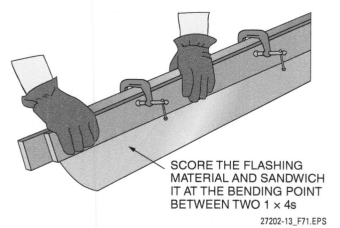

Figure 71 Bending continuous flashing.

the same distance apart as the cutouts on the shingles and stagger them to match the pattern on the roof deck.

4.5.3 Dormer Roof Valley

The installation of a dormer roof valley will require you to combine some of the procedures previously covered. *Figure 73* shows an open valley for a gable dormer roof. Note that the shingles

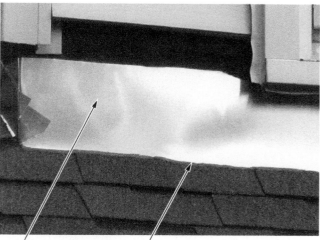

Figure 72 Covering flashing.

have been laid on the main roof up to the lower end of the valley.

Extreme care must be used during the installation of the last course against the vertical wall to ensure a tight, dry fit. *Figure 74* displays the standard valley procedures for dormer flashing and shows how the valley material overlaps the course of shingles to the exposure line for a watertight seal.

Regular valley nailing procedures are used until work proceeds past the dormer ridge and resumes a full in-line course of shingles, as determined by a reference chalkline. See *Figure 75*.

4.5.4 Chimneys

Chimneys are subject to varying loads and certain opposing structural movements due to winds, temperature changes, and settling. Therefore, roofing materials and base flashing should not be attached or cemented to both the chimney and roof deck. The process of shingling around chimneys must be approached with extreme care. Due to the size of the opening in the roof and of the chimney itself, additional work must be done on the roof deck around chimneys prior to shingling. A cricket, also called a saddle, must be made. See *Figure 76*.

Nailing Step Flashing

Step flashing should only be nailed to the roof deck, never to the wall sheathing. This will allow settling or shifting of the structure without tearing the flashing and roofing away from the roof deck.

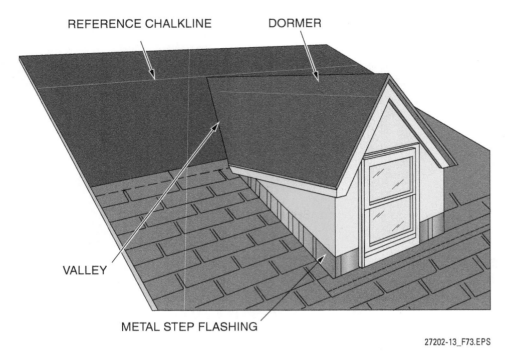

Figure 73 Dormer flashing.

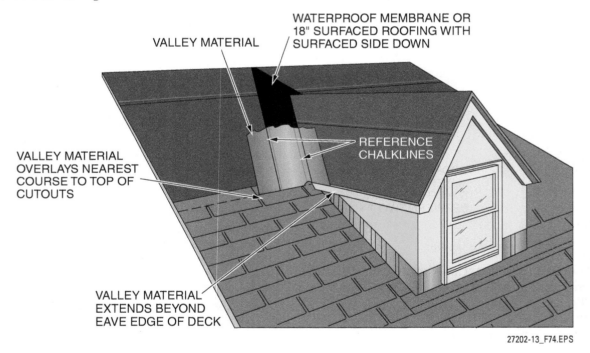

Figure 74 Dormer valley flashing.

Gable Dormer Roof Valleys

Besides an open valley, a closed-cut valley can be used on a gable dormer that has a different roof pitch than the main roof. A woven valley may be used if the gable and main roof pitches are equal.

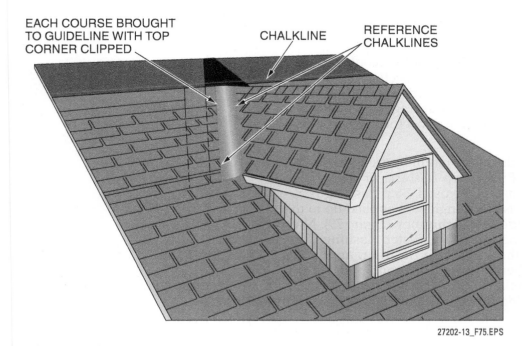

Figure 75 Dormer valley coverings.

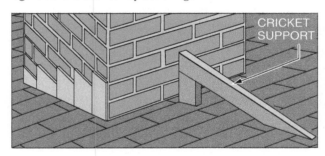
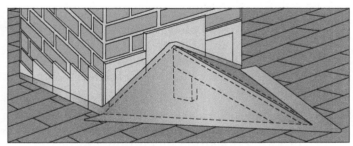

Figure 76 Simple chimney cricket.

A cricket placed behind the chimney keeps rainwater or melting snow/ice from building up in back of the chimney. It steers flowing water around the chimney. The cricket is usually supported by a horizontal ridge piece and a vertical piece at the back of the chimney, as shown in *Figure 76*. The height of the cricket is typically half the width of the chimney, although these requirements will vary. Check local codes. The ridge, which is level, extends back to the roof slope. On wide chimneys, it may be necessary to frame the cricket, as shown in *Figure 77*. Either type of cricket may be covered with two triangular pieces of ¾" exterior plywood cut to fit from the ridge to the edge of the chimney and the roof slope. Heavy-gauge metal can also be used to form the cricket. The covering is then nailed to the support and roof deck.

Flashing at the point where the chimney comes through the roof requires something that will allow movement without damage to the water seal.

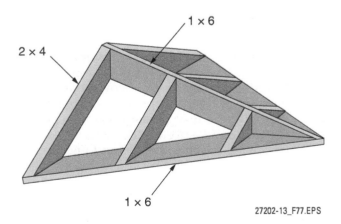

Figure 77 Cricket frame.

It is necessary to use base flashing. The counter or cap flashing is secured to the masonry. Metal is used for base flashing and cap flashing.

To apply the chimney flashing, proceed as follows:

Step 1 Apply shingles over the roofing felt up to the front face of the chimney (*Figure 78*).

Step 2 The base flashing for the front cut is applied first. The lower section is laid over the shingles in a bed of plastic asphalt cement. Bend the triangular ends of the upper section around the corners of the chimney.

Step 3 The sides of the chimney are base-flashed next. Either step flashing or continuous flashing can be used. Step flashing is applied as the shingles are applied up to the top side of the chimney (*Figure 79*). Note that the first piece of step flashing overlaps the front base flashing, and is cut and bent around the front of the chimney. Like the front base flashing, the step flashing is fastened only to the roof deck with a nail or, if desired, roof cement. Bend and cement the flashing to the underlayment along the slope at the sides of the chimney, with the lower end overlapping the front base flashing. Bend the triangular end pieces around the chimney. Apply shingles up the roof to the top side of the chimney. Cement the shingles to the side base flashing to form a waterproof joint.

Step 4 It is now necessary to go to the top side of the chimney and complete the waterproofing operation by cutting and fitting base flashings over the cricket, known as cricket flashing, as shown in *Figure 80*.

Step 5 Using a pattern, cut the cricket base flashing.

Step 6 Bend the base flashing to cover the entire cricket, as shown at C in *Figure 80*. Extend the flashing laterally to cover part of the side base flashing previously installed. Set it tightly using plastic asphalt cement. Use care to spread the cement in the proper location to be covered and not extend out onto the finished shingles. Neatness is a must. Bend the ends around the chimney.

Step 7 Cut a rectangular piece of flashing, as shown at D in *Figure 80*, and make a V-cutout on one side to conform to the rear angle of the cricket. Center it over that part of the cricket flashing extending up to the deck. Set it tightly using plastic asphalt cement. This piece provides added protection where the ridge of the cricket meets the deck.

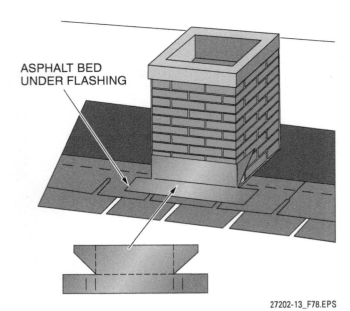

Figure 78 Front base flashing.

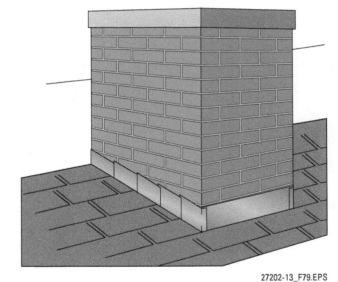

Figure 79 Step-flashing method.

Step 8 Cut a second small rectangular piece of flashing. Cut a V on one side to conform to the pitch of the cricket, as shown at E in *Figure 80*. Place it over the cricket ridge and against the flashing that extends up the chimney. Embed it in plastic asphalt cement to the cricket flashing. Nail the edges of the flashing. In most cases, similarly colored pieces of roll roofing are cut to overlap the entire cricket and extend onto the roof deck. The roll roofing is cemented to the cricket flashing and sealed with cement at the chimney edge.

Step 9 For completion, cap flashing (also called counter flashing) is installed. It is usually made of sheet copper 16 ounces or heavier. It can also be made of 24-gauge galvanized steel. If steel is used, it should be painted on both sides. *Figures 81* and *82* show metal cap flashing on the face of the chimney and on the sides. Cap flashing is secured to the brickwork, as shown in *Figure 83*.

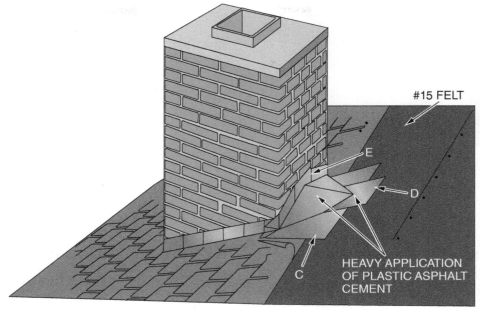

Figure 80 Cricket flashing.

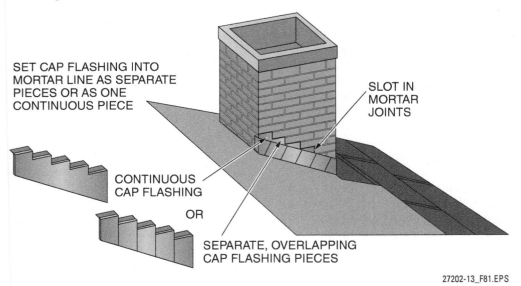

Figure 81 Cap flashing methods at sides.

Combustible-Material Spacing Requirements for a Cricket

Some building codes require that the wood framing and sheathing for a cricket must be spaced up to 1" from the chimney masonry. Always check your local codes for spacing requirements pertaining to combustible materials near chimneys.

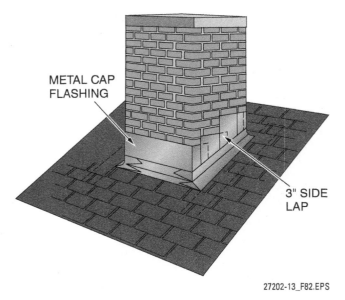

Figure 82 Flashing cap and lap.

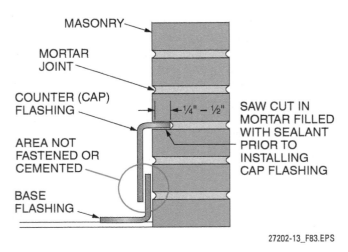

Figure 83 Counter (cap) flashing installation.

Step 10 *Figure 83* shows a good method of securing the cap flashing. Cut a slot in the mortar joint to a depth of ¼" to ½". Insert a 90-degree bent edge of the flashing into the cleared slot between the bricks using an elastomeric sealant or mortar in the slot to secure the flashing to the masonry. When installed, the cap should lie snugly against the masonry. The front unit of the cap flashing should be one continuous piece. On the sides and the rear, the sections are similar in size. They are cut to conform to the locations of mortar joints and the pitch of the roof. If the sides are lapped, they must lap each other by at least 3". The slots are refilled with the brick mortar mix or elastomeric sealant and conform to the original brickwork. Patient installation of the flashing will provide the watertight seal necessary for a dry roof. Do not cement the counter (cap) flashing to the base flashing.

Step 11 Once the chimney flashing and shingling have been accomplished and the shingles next to the cricket are cemented under the edges to make a waterproof joint, the regular shingling process resumes on the next full course above the cricket. Another method of finishing the cricket is to extend the horizontal composition shingles of the roof deck up the pitch of the cricket. Then use step flashing and cement shingles parallel with the cricket ridge to form a half-weave valley at the edges of the cricket. This second method requires cementing cap shingles over the ridge of the cricket. The application of the shingles continues until the roof ridge is reached.

4.5.5 Hip or Ridge Row (Cap Row)

Special shingles are required to complete the hip or ridge rows. In some cases, ridge caps and hip caps are premanufactured.

Architectural shingles have a matching cap-row shingle (*Figure 84*) that must be used. Cap-row shingles cannot be cut from architectural shingles. Most of the time if the roof was shingled with standard three-tab shingles, cap rows are cut from the shingles and the 12 × 12 tab is used (*Figure 85*). The tab can be reduced to 9 × 12, but nothing less. Since the hip or ridge is a potential spot for water leaks, precautions must be taken. If ridge venting will be used, do not apply the ridge caps.

To install a ridge or hip row, proceed as follows:

Step 1 Butt and nail shingles as they come up on either side of a hip or ridge. On a ridge, lay the last course and trim the shingles, as shown in *Figure 86*. On a hip, trim the shingles at an angle on the hip line.

Figure 84 Architectural cap shingle.

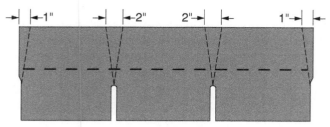

Figure 85 Cutting cap shingles.

Step 2 After the cap shingles are cut, bend them lengthwise in the center line. In cold weather, warm the shingles before bending to prevent cracks. Begin at the bottom of any hips. Cut the first tab to conform to the dual angle at the eaves.

Step 3 Lap the units to provide a 5" exposure of the granular surface. See *Figure 87*. Secure with one nail on each side, 6" back from the exposed end and 1" from the edge. As each succeeding tab is nailed going up the hip, the nail penetrates and secures two tabs. This tight bond prevents the wind from getting underneath and lifting the tab.

Nailing of the ridge row is similar to that described for the overall hip. Nailing takes place from both ends of the ridge. A final cap piece joins the ridge together in the center, and the ex-

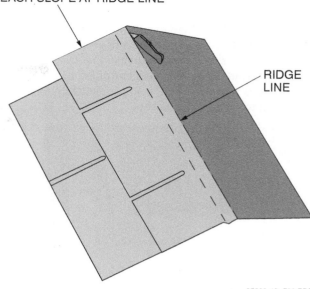

Figure 86 Applying the last course of ridge shingles.

posed nails are covered with roof cement. An exception to this may occur in a very windy area. In that case, the ridge cap would be started at the point on the roof opposite the wind direction. As each ridge shingle is placed, it automatically allows the wind to pass over it, and there is no possibility of shingles blowing off. The junction of the roof ridge and any hip ridges can be capped with a special molded cap or by a fabricated end cap, as shown in *Figure 88*. Bed the final ridge cap or

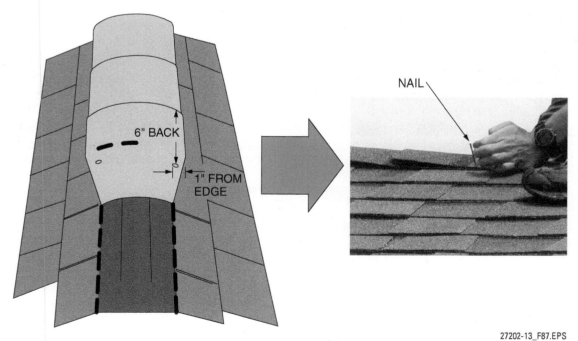

Figure 87 Installing a ridge cap.

Preformed Cap Flashing

Commercial preformed cap flashing may be installed on vertical brick, concrete, or block surfaces including chimneys. This flashing is made of an aluminum-coated steel. As shown in the sequenced photographs, a slot is cut in all sides of a chimney using a ¼"-thick diamond-impregnated steel wheel mounted in a small, high-speed electric grinder. The flashing is trimmed to shape and the V-edge of the flashing is pressed into the groove. The flashing is formed to shape and sealed with an elastomeric sealant. When set, the sealant and flashing may be painted to blend with the roof.

1. PREFORMED CAP FLASHING

2. CUTTING ¼" GROOVE

3. COMPLETED GROOVES

4. TRIMMING FRONT FLASHING IN GROOVE TO SIZE

5. SEATING V-EDGE OF FRONT FLASHING IN GROOVE

6. FITTING SIDE FLASHING

7. SEATING V-EDGE OF SIDE FLASHING IN GROOVE

8. SIDE FLASHING INSTALLED AND FORMED TO FRONT FLASHING

9. SEALANT APPLIED TO GROOVE AND FLASHING

10. FLASHING PAINTED TO MATCH ROOFING

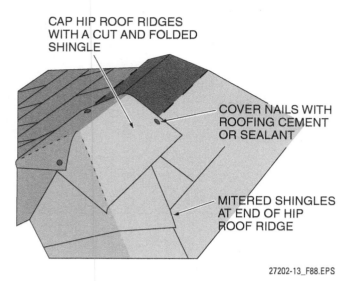

Figure 88 Hip and ridge end cap.

hip/ridge caps in asphalt and secure with nails, as shown in the figure. Cover the nail heads with roof cement or sealant.

4.5.6 Installing Box Vents

Box vents (*Figure 89*) are easily installed on most roofing materials. On new roofs, the proper size hole is cut in the roof and the hole is surrounded with at least a 24"-wide waterproof membrane. When the roofing courses reach the area of the ventilator, the ventilator is installed with the lower edge of its flashing overlapping the course below it. The flashing is fastened to the roof at the top and sides. Then, roofing courses are applied over and cemented to the flashing at the top and sides of the ventilator.

On existing roofs, the hole is cut, and the upper flashing of the ventilator is slid under the courses above the hole and fastened to the roof. The side and bottom flashing is not fastened; however, the side flashing of the ventilator is completely cemented down to the roofing under the sides. The bottom flashing is usually not cemented.

4.5.7 Installing Ridge Vents

Use the following procedure to install ridge vents:

Step 1 Determine where the roof vent slots will be cut on the ridges and any hips (*Figure 90*). Slots cut along a ridge should start and stop approximately 12" from the gable (terminal) ends, any vertical walls, any higher intersecting roof, any roof projections at the ridge, any valleys, and any hip joints. Slots cut in hips should end 24" above the eaves and should be in 24" sections separated by 12" to maintain maximum roof strength. For appearance, the roof ridges and hips should be covered completely with the roof vent material to maintain a continuous roof line, as shown in *Figure 91*.

> **WARNING!** Always wear proper personal protection equipment (PPE), including fall protection equipment, when working on roofs.

Step 2 Next, refer to the manufacturer's specifications and note the width of the slot required for the vent being used. Determine the roof construction (ridgeboard or no ridgeboard). If a ridgeboard is used

Figure 89 Typical box vent.

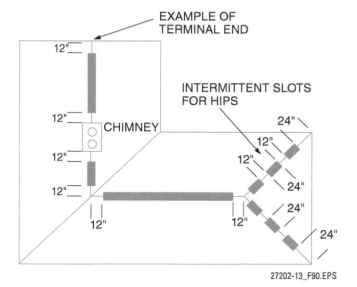

Figure 90 Example of slot cutout placement.

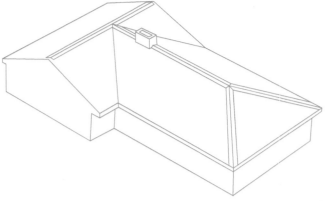

Figure 91 Example of a continuous roof line.

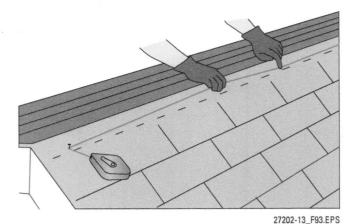

Figure 93 Snapping a chalkline for slot width.

(*Figure 92*), add 1½" to the required slot width and divide the result by two to find the total slot width to be cut on each side of the peak. If no ridgeboard is used, only divide the required slot width by two to find the slot width to be cut on each side of the peak.

Step 3 At the peak, measure down the roof on both sides for half of the required slot width as determined above and snap a chalkline along any ridges or hips, at both sides of the peak or hip (*Figure 93*). Mark the start and stop point for each slot, as previously specified.

Step 4 Using a knife, cut away the shingles along the chalklines between each start and stop point.

Step 5 Using a power saw set to the thickness of the shingles plus the roof sheathing, carefully cut into the sheathing and along the chalklines to remove the sheathing without damaging the rafters or, if present, the ridgeboard (*Figure 94*).

Step 6 Place a shingle ridge cap at each gable end, before and after each roof projection, at each vertical wall, at each higher intersecting roof joint, and at each eave end for any hips (*Figure 95*). This prevents any water intrusion at or under the exposed ends of the vent from penetrating into or through the sheathing.

Step 7 If packaged in rolls, unroll the vent material and temporarily tack it in place over the ridges and hips, or secure the flexible plastic composition sections to the roof over the ridges and hips. Make sure that the vent covers the entire length of all ridges and hips (*Figure 96*). If rigid vent sections (*Figure 97*) are being used, make sure that vents are secured through the prepunched holes.

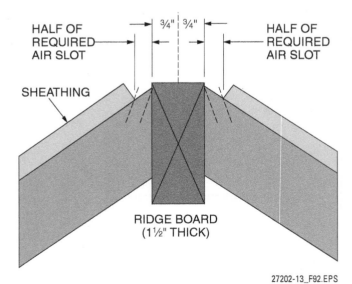

Figure 92 Determining total slot width.

Figure 94 Cutting slots.

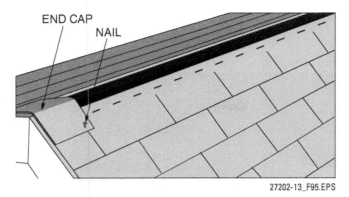

Figure 95 Exposed-end shingle caps.

ROLLED VENT MATERIAL   RIGID VENT MATERIAL

Figure 96 Positioning vent material.

Step 8 Cut and taper the tabs from a number of three-tab standard or architectural cap shingles for use as cap shingles.

Step 9 Starting from the exposed ends of a ridge or hip and using sufficiently long nails, fasten the cap shingles to the roof through the vent material nailing line (*Figure 98*). The cap shingles are applied and mated over the ridges and hips in the same manner as described in the section *Hip or Ridge Row (Cap Row)*. See *Figure 98*. However, make sure that the shingles are nailed snugly without compressing the vent material. It is advisable to place a bead of roofing cement under the overlapping sections of the shingles to help secure them.

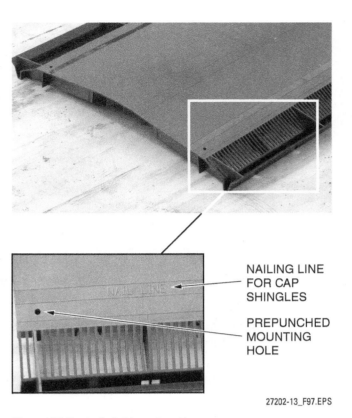

Figure 97 Typical rigid vent section.

Plastic or Metal Ridge Vent

If using a rigid plastic or metal vent not intended to be capped with shingles, fasten the vent to the roof through the flanges using a noncorrosive fastener with a neoprene seal washer; then insert an end cap in the exposed ends of the vents, if required.

Precautions When Cutting a Ridge Vent

Use goggles and make sure your footing is solid along the ridges when cutting with a power saw. Make sure the power cord is laid out so that there is no chance of getting entangled with it during the cutting operation. Make sure that the saw depth is set so that the ridgeboard or rafters are not cut.

Alternate End Treatments of a Ridge Cap

Some contractors cut the ridge vent further in from the end of the roof so that the ridge cap is not flush with the end of the roof. This hides the end view of the cap from the ground for a more attractive appearance.

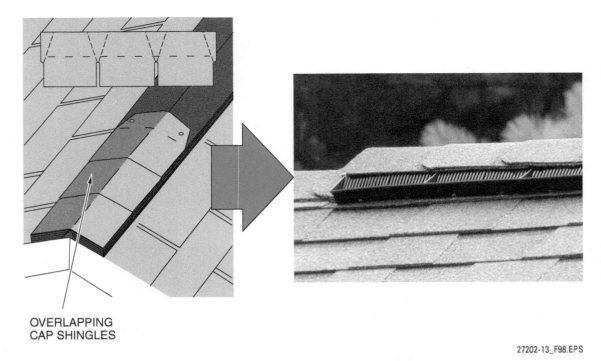

Figure 98 Applying cap shingles over a vent.

4.0.0 Section Review

1. Drip edge should be nailed at an interval of _____.

   a. 4" to 6"
   b. 6" to 9"
   c. 8" to 10"
   d. 8" to 12"

2. A row of shingles with the tabs cut off is often the first row laid at the eave edge of the roof. These shingles are known as a _____.

   a. starter row
   b. initial row
   c. base row
   d. underlay row

3. To form a tight roof, adjacent corrugated metal roofing panels should be overlapped by _____.

   a. 1 corrugation
   b. 1½ corrugations
   c. 2 corrugations
   d. 2½ corrugations

4. Double-coverage roll roofing has a 17" mineral-coated exposure, and an overlap (selvage) of _____.

   a. 15"
   b. 17"
   c. 19"
   d. 21"

5. Step flashing should be nailed to the roof deck, but not to the wall sheathing.

   a. True
   b. False

Section Five

5.0.0 Estimating Roofing Materials

Objective

Describe the estimating procedure for roofing projects.

Trade Terms

Overhang: The part that extends beyond the building line. The amount of overhang is always given as a projection from the building line on a horizontal plane.

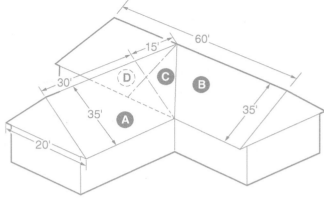

Figure 99 Roof example (including overhangs).

Regardless of the type of roofing to be installed, the amount of material required must first be estimated. After the amount of material has been determined and obtained, the roof deck must be prepared before the finish roofing material is raised to the roof deck and installed.

If the building plans for the structure are available, the roof dimensions can be determined directly from the plans. If the plans are not available, the length and width of each section of the roof can be measured. Then, the area of each section of the roof is calculated and a percentage is added for waste. The result is converted into the number of squares (100 sq ft) of material required.

Step 1 Measure the length and width of each triangular and rectangular roof section of the structure, including any overhangs (*Figure 99*).

Step 2 Calculate the area for one-half of each roof section and add the areas together. Then, multiply by 2 to obtain the total roof area and subtract any triangular areas covered by roof intersections.

To calculate the total area of the roof shown in *Figure 99*, proceed as follows:

30' × 35' = 1,050.0 sq ft (Area A)

60' × 35' = + 2,100.0 sq ft (Area B)

15' × 35' × ½ = + 262.5 sq ft (Area C)

= 3,412.5 sq ft (½ roof area)

3,412.5 × 2 = 6,825.0 sq ft

35' × 20' × ½ = 350.0 sq ft (Area D*)

= 6,475.0 sq ft (roof area)

Step 3 Add an average of 10 percent for hips, valleys, and waste. For a complicated roof with a number of valleys or hips, more than 10 percent may be required. For a plain, straight gable roof, less is usually required. In addition, for wood or slate roofs, an additional 100 square feet is usually required for each 100 lineal feet of hips and valleys. For our example, a relatively simple roof with standard three-tab composition shingles, a 10 percent waste factor is included.

Total material = (10% × 6,475) + 6,475

= 648 + 6,475

= 7,123 sq ft

Step 4 Convert the total square feet of material required to squares by dividing by 100 sq ft (if using wood roofing, round up or down to the nearest bundle or square). For standard three-tab composition shingles, the rounding would be to the nearest ⅓ or ⅔ of a square (one or two bundles) or whole square.

Squares = 7,123 ÷ 100 = 71.23 = 71⅓

Step 5 The number of rolls of underlayment required is determined from the same total material requirement of 7,123 square feet. Starter strips, eave flashing, valley flashing, and ridge shingles must be added to complete the estimate. All of these are determined with linear measurements.

5.0.0 Section Review

1. When estimating the material needed for a building with a hip roof, increase the calculated quantity by _____.
   a. 5 percent
   b. 7.5 percent
   c. 10 percent
   d. 15 percent

SUMMARY

Many types of roofing materials and commercial roofing systems are available. They are used to protect a structure and its contents from the elements, and vary depending on geographic location. This module covered the most common materials used on residential and small commercial structures, including composition shingles, roll roofing, wood shingles/shakes, slate, tile, metal, and membrane roofing.

While carpenters may not usually be required to install roofing, they can be involved in preparing the roof surface before the installation of roofing materials. Therefore, carpenters should be familiar with the basic preparation and installation of common roofing materials.

Review Questions

1. To help minimize injuries due to manually lifting materials, some contractors have developed _____.
   a. yoga classes
   b. fitness groups
   c. stretching programs
   d. calisthenics routines

2. A vertical lifeline used in a fall protection system must have a minimum tensile strength of _____.
   a. 2,500 lb
   b. 5,000 lb
   c. 7,500 lb
   d. 10,000 lb

3. For safe use, a ladder must extend above the edge of a roof by a distance of _____.
   a. 1'
   b. 2'
   c. 3'
   d. 4'

4. A scaffold that meets OSHA requirements and is safe to use is marked with a(n) _____.
   a. blue ribbon
   b. orange tape
   c. checkered flag
   d. green tag

5. The standard roofing hammer is also called a _____.
   a. shingle hatchet
   b. slater's hammer
   c. claw hammer
   d. cricket hammer

6. A slater's hammer can be used to _____.
   a. remove nails
   b. bend metal flashing
   c. punch nail holes
   d. cut tile

7. Organic-fiber composition shingles have been largely replaced by shingles made with _____.
   a. asbestos
   b. wood
   c. fiberglass
   d. synthetic resin

8. Fiberglass shingles have a typical life span of _____.
   a. 5 to 12 years
   b. 15 to 20 years
   c. 20 to 25 years
   d. 25 to 40 years

9. The most common type of composition shingle applied to residential structures is the _____.
   a. two-tab
   b. three-tab
   c. architectural
   d. strip

10. Roll roofing is available in weights of _____.
    a. 20 to 40 lb
    b. 30 to 70 lb
    c. 50 to 90 lb
    d. 70 to 120 lb

11. The most popular wood shingles and shakes are made from _____.
    a. yellow pine
    b. red cedar
    c. cypress
    d. redwood

12. Per square foot, both tile roofing and slate roofing weigh about _____.
    a. 3 to 7 lb
    b. 5 to 9 lb
    c. 7 to 10 lb
    d. 9 to 12 lb

13. The abbreviation BUR stands for _____.
    a. builder's union representative
    b. bundle-unitized roofing
    c. built-up roofing
    d. building unit resources

14. Premanufactured membrane roofing systems are grouped into _____.
    a. two categories
    b. three categories
    c. four categories
    d. five categories

15. One of the most common thermoset single-ply membranes is known as _____.
    a. APP
    b. SBS
    c. PVC
    d. EPDM

16. Two common thermoplastic single-ply membranes are _____.
    a. APP and BUR
    b. SBS and PVC
    c. PVC and TPO
    d. EPDM and TPO

17. The attic ventilation area is normally divided between exhaust and inlet vents in the proportion of _____.
    a. 50 percent exhaust, 50 percent inlet
    b. 60 percent exhaust, 40 percent inlet
    c. 70 percent exhaust, 30 percent inlet
    d. 80 percent exhaust, 20 percent inlet

18. The 6" pattern of laying three-tab shingles means that 6" is _____.
    a. added to each course
    b. subtracted from each course
    c. subtracted from every other course
    d. subtracted from the second course, 12" from the third, 18" from the fourth, and so on

19. An open valley is wider at the bottom than at the top _____.
    a. to provide a more visually pleasing joint
    b. for easier cleaning and maintenance
    c. to accommodate the higher volume of water that will be present near the bottom
    d. to allow for variations in the width of shingles

20. There are two types of closed valleys, cut and _____.
    a. stepped
    b. woven
    c. staggered
    d. interleaved

21. If a fastener is stripped during the installation of a standing-seam metal roof, you should _____.
    a. remove the stripped fastener and replace it with a fastener of the next larger size
    b. remove the stripped fastener, seal the hole with roofing cement, and drill a hole 2" from the original for a new fastener
    c. leave the stripped fastener in place and put another fastener alongside it
    d. remove the stripped fastener and replace it with a fastener of the same size

22. Double-coverage roll roofing with concealed nails is used on slopes of more than _____.
    a. ¼ in 12
    b. 1 in 12
    c. 2 in 12
    d. 3 in 12

23. Step flashing is used on _____.
    a. valleys
    b. slopes against walls
    c. hips
    d. vent pipes

24. The additional structure required to properly apply roofing around a chimney is a _____.
    a. cricket
    b. dormer
    c. fascia
    d. valley

25. A square of three-tab composition shingles consists of _____.
    a. one bundle
    b. two bundles
    c. three bundles
    d. four bundles

Trade Terms Quiz

Fill in the blank with the correct term that you learned from your study of this module.

1. A _____ is a package consisting of enough shingles or shakes to cover a specified square footage.
2. The highest point at the top of a roof, where two slopes meet, is the _____.
3. _____ is the distance between the exposed edges of shingles that are overlapped.
4. The protective sealing material applied to the intersection of a chimney and roof or similar areas subject to leaks is called _____.
5. On a slanted surface, such as a roof, the ratio of rise to run is the _____.
6. A structure with a ridge sloping in two directions, a _____ (also called a cricket) diverts water to either side of a chimney.
7. Prefabricated _____ is used to seal pipe projections through the roof surface.
8. The distance between overlapped adjacent shingles, in inches, is called _____.
9. A _____ is the internal angle formed by the meeting of two roof slopes.
10. _____ is a plastic adhesive used to seal flashing and the free tabs of strip shingles.
11. A loosely knit fabric is referred to as _____.
12. The quantity of shingles required to cover an area of 100 square feet is called a _____.
13. _____ is embedded in mortar joints and overlaps the base of leak-prone structures such as a chimney.
14. Protective sheeting, often made from metal, that is placed at roof intersections to make them watertight is called _____.
15. Expressed as a fraction, _____ is the ratio of rise to span.
16. _____ is asphalt-saturated felt, in roll form, used to protect roof sheathing.
17. Metal sheet material, shaped to form a seal between the roof surface and an exterior wall, is described as _____.
18. The area of a composition shingle or roll roofing that is not coated with aggregate is referred to as _____.
19. _____ is the distance between the lower edge of an overlapping shingle and the top edge of the underlying shingle.
20. Also referred to as decking, _____ usually is a plywood sheet material, but may be in the form of boards.
21. An _____ is any part of a structure that extends beyond the building line.

Trade Terms

Asphalt roofing cement
Base flashing
Bundle
Cap flashing
Exposure
Overhang
Pitch
Ridge
Roof sheathing
Saddle
Scrim
Selvage
Side lap
Slope
Square
Top lap
Underlayment
Valley
Valley flashing
Vent-stack flashing
Wall flashing

Trade Terms Introduced in This Module

Asphalt roofing cement: An adhesive that is used to seal down the free tabs of strip shingles. This plastic asphalt cement is mainly used in open valley construction and other flashing areas where necessary for protection against the weather.

Base flashing: The protective sealing material placed next to areas vulnerable to leaks, such as chimneys.

Bundle: A package containing a specified number of shingles or shakes. The number is related to square-foot coverage and varies with the product.

Cap flashing: The protective sealing material that overlaps the base and is embedded in the mortar joints of vulnerable areas of a roof, such as a chimney.

Exposure: The distance (in inches) between the exposed edges of overlapping shingles.

Overhang: The part of a structure that extends beyond the building line. The amount of overhang is always given as a projection from the building line on a horizontal plane.

Pitch: The ratio of the rise to the span, indicated as a fraction. For example, a roof with a 6' rise and a 24' span will have a ¼ pitch.

Ridge: The horizontal line formed by the two rafters of a sloping roof that have been nailed together. The ridge is the highest point at the top of the roof where the roof slopes meet.

Roof sheathing: Usually 4 × 8 sheets of plywood, but can also be 1 × 8 or 1 × 12 roof boards, or other new products approved by local building codes. Also referred to as decking.

Saddle: An auxiliary roof deck that is built above the chimney to divert water to either side. It is a structure with a ridge sloping in two directions that is placed between the back side of a chimney and the roof sloping toward it. Also referred to as a cricket.

Scrim: A loosely knit fabric.

Selvage: The section of a composition roofing roll or shingle that is not covered with an aggregate.

Side lap: The distance between adjacent shingles that overlap, measured in inches.

Slope: The ratio of rise to run. The rise in inches is indicated for every foot of run.

Square: The amount of shingles needed to cover 100 square feet of roof surface. For example, square means 10' square, or 10' × 10'.

Top lap: The distance, measured in inches, between the lower edge of an overlapping shingle and the upper edge of the lapping shingle.

Underlayment: Asphalt-saturated felt protection for sheathing; 15-lb roofer's felt is commonly used. The roll size is 3' × 144', or a little over four squares.

Valley: The internal part of the angle formed by the meeting of two roofs.

Valley flashing: Watertight protection at a roof intersection. Various metals and asphalt products are used; however, materials vary based on local building codes.

Vent-stack flashing: Flanges that are used to tightly seal pipe projections through the roof. They are usually prefabricated.

Wall flashing: A form of metal shingle that can be shaped into a protective seal interlacing where the roof line joins an exterior wall. Also referred to as step flashing.

Appendix

Gutters and Downspouts

Metal and Vinyl Gutters and Downspouts

Occasionally, you may be called upon to install gutters and downspouts after constructing the cornice. Therefore, you should be familiar with available products and have a general knowledge of how they are installed. Always check the local codes. Some localities specify the size and capacity of gutters applied to commercial and/or residential structures.

Gutters are constructed of vinyl or metal, usually aluminum or galvanized steel, but sometimes copper, stainless steel, or baked-on enamel steel. Vinyl and aluminum gutters are prefinished and ready to install. Galvanized metal gutters are usually unfinished and must be painted after they are in place.

Many gutters installed on residential and light commercial buildings are seamless and are installed by companies specializing in the field. The final product is measured, formed, fabricated, and installed in place in the field by the same company.

Some gutters are fabricated on site with a debris guard that requires no cleaning and sheds roof debris, such as twigs and leaves, while directing water into the gutter (*Figure A-1*). Another similar product is available for application to existing gutters. Both types depend on the principle of liquid cohesion and adhesion. Water running off a roof will follow the sharp curve and drop into the gutter because of adhesion to the metal and the surface tension caused by cohesion of the water molecules. Debris, on the other hand, passes by the sharp curve and is shed over the edge of the gutter.

Prefabricated Metal Gutters and Downspouts

Gutters are manufactured in several shapes. The most common shape is the ogee, or K-style, as shown in *Figure A-2*. Another is an older C-shaped (half-round) style.

Figure A-1 Debris guard.

Downspouts also come in a variety of shapes to match the type of gutter. The most common is the rectangular corrugated type. They are made in standard 8' to 10' lengths. In some cases, longer lengths are available to reduce the number of joints required.

Accessories include elbows of various angles and straps for attachment. One end of each piece is smaller than the other, and each piece is installed with the smaller end down.

There are also many accessories used in the installation of a complete roof drainage system. A typical metal K-style drainage system, along with some of its accessories, is shown in *Figure A-3*.

The size of gutters and downspouts will depend on the intensity of rainfall and the amount of roof area to be drained. Furthermore, the slope of the gutter and the number of outlets will affect the gutter size. The required sizes will be specified in the construction drawings. The drawings will also indicate where the downspouts are to be located.

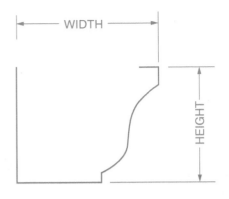

Stock Metal Gutter Section Ogee or Style "K"

Stock Sizes:
2¾" H × 4" W   5¼" H × 7" W
3¾" H × 5" W   6" H × 8" W
4¾" H × 6" W

Figure A-2 Typical metal K-style sizes.

The relationship between the gutter and the edge of the finish roofing material is important. A gutter should be centered under the edge of the roof to catch the water off the roof. Gutters should always be placed below the slope of the roof to catch water that runs off, but not to catch snow and ice that slide off, if in areas with regular snowfall. It is necessary that the gutter slope slightly toward the outlet, usually 1" or 2" in 40'. It is a good idea to check on local codes to determine the correct slope.

The exact height varies with the slope of the roof and is measured from a continuation of the slope to the outer lip of the gutter. These dimensions apply to the high point of the gutter, as shown in *Figure A-4*.

When shedding of ice is desired, gutters are sometimes not used on roofs with wide metal ice edging installed on the roof slope over the eaves.

Gutters should be supported on both sides of all outlets, at all ends and joints, and at intermediate points. The maximum spacing of supports is 36" in climates free of snow and ice, while 18" spacing is recommended in areas that have snow.

The tools required to install a typical drainage system are a soldering iron or caulking gun, hacksaw or power saw with a metal-cutting blade, metal shears, hammer, steel tape or folding rule, carpenter's level, putty knife, chalkline, screwdriver, rivet gun, and ladders.

Prefabricated Vinyl Gutters and Downspouts

Vinyl gutters and downspouts are manufactured in various colors with smooth or patterned finishes in essentially the same styles as metal gutters. Downspouts are manufactured in a variety of shapes. *Figures A-5* and *A-6* show two of the most common styles, the K-style and the C-style, along with their downspouts, fittings, and accessories.

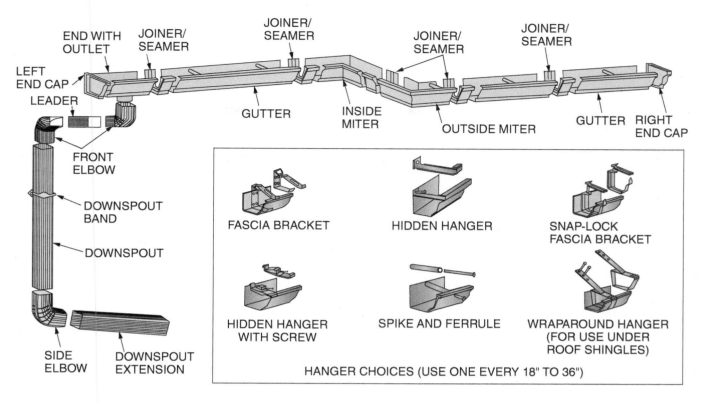

Figure A-3 Typical metal K-style drainage system and accessories.

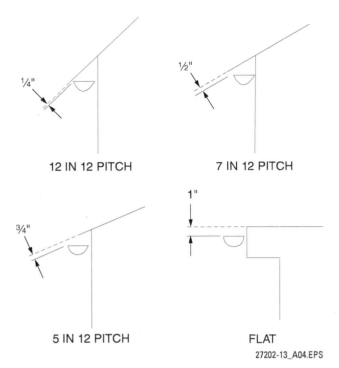

Figure A-4 Placement of gutter.

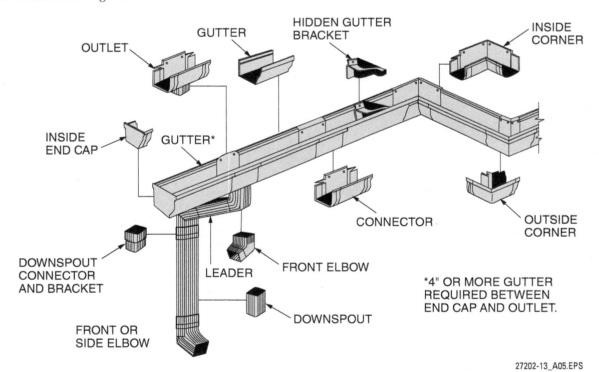

Figure A-5 Typical vinyl K-style drainage system.

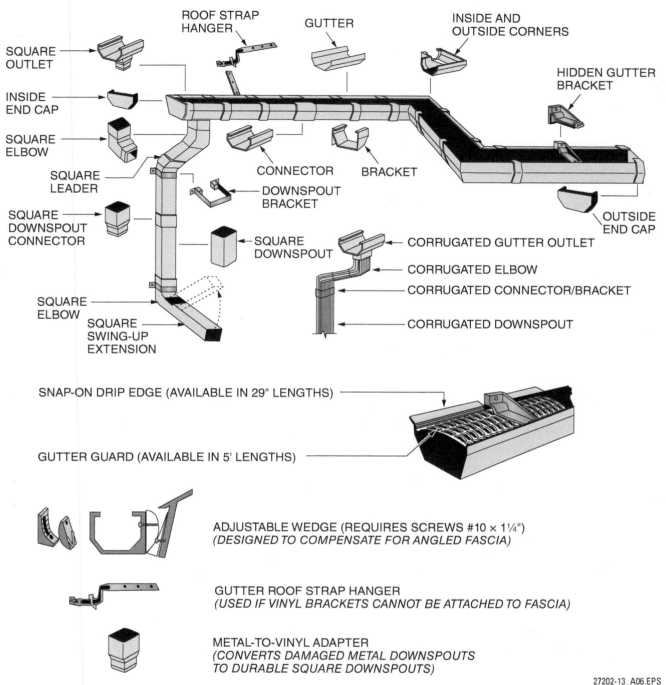

Figure A-6 Typical vinyl C-style drainage system.

Additional Resources

This module presents thorough resources for task training. The following resource material is suggested for further study.

Asphalt Manufacturers Association website. **www.asphaltroofing.org**
National Roofing Contractors Association website. **www.nrca.net**
Roof Coating Manufacturers Association website. **www.roofcoatings.org**
OSHA Safety and Health Standards for the Construction Industry, Part 1926, Subpart M. **www.osha.gov**
OSHA Safety and Health Standards for the Construction Industry, Part 1926, Appendices C and D to Subpart M. **www.osha.gov**

Figure Credits

Courtesy of DaVinci Roofscapes, CO01, SA06
Courtesy of Werner Co., Figure 1 (photo), Figure 2, SA01, Figure 5
Alum-A-Pole Corporation, SA02
Reimann & Georger Corporation, SA03
The Stanley Works, Figure 9
Courtesy of CertainTeed Corporation, Figures 13-14, SA05
Copyright 2006 CertainTeed Corporation used with permission, Figure 14

Follansbee Steel, Figure 22
Johns Manville, Figure 25
Cornell Corporation, SA07
Courtesy of ATAS International, Inc., SA10
Englert Inc., A01
Amerimax Home Products, Inc., A03
Bemis Manufacturing Company, A05-06

Section Review Answer Key

Answer	Section Reference	Objective
Section One		
1. b	1.1.0	1a
2. d	1.2.2	1b
3. a	1.3.2	1c
Section Two		
1. b	2.1.0	2a
2. d	2.2.1	2b
3. c	2.3.1	2c
4. b	2.3.2	2c
Section Three		
1. b	3.1.0	3a
2. a	3.2.0	3b
3. c	3.3.0	3c
4. b	3.4.1	3d
5. d	3.5.0	3e
6. b	3.6.0	3f
7. b	3.7.0	3g
8. c	3.8.0	3h
9. b	3.9.0	3i
Section Four		
1. c	4.1.0	4a
2. a	4.2.0	4b
3. b	4.3.1	4c
4. c	4.4.4	4d
5. a	4.5.2	4e
Section Five		
1. c	5.0.0	5

This page is intentionally left blank.

NCCER CURRICULA — USER UPDATE

NCCER makes every effort to keep its textbooks up-to-date and free of technical errors. We appreciate your help in this process. If you find an error, a typographical mistake, or an inaccuracy in NCCER's curricula, please fill out this form (or a photocopy), or complete the online form at **www.nccer.org/olf**. Be sure to include the exact module ID number, page number, a detailed description, and your recommended correction. Your input will be brought to the attention of the Authoring Team. Thank you for your assistance.

Instructors – If you have an idea for improving this textbook, or have found that additional materials were necessary to teach this module effectively, please let us know so that we may present your suggestions to the Authoring Team.

NCCER Product Development and Revision
13614 Progress Blvd., Alachua, FL 32615

Email: curriculum@nccer.org
Online: www.nccer.org/olf

❏ Trainee Guide ❏ Lesson Plans ❏ Exam ❏ PowerPoints Other _____

Craft / Level: _____ Copyright Date: _____

Module ID Number / Title: _____

Section Number(s): _____

Description: _____

_____

_____

_____

_____

Recommended Correction: _____

_____

_____

_____

_____

Your Name: _____

Address: _____

_____

Email: _____ Phone: _____

This page is intentionally left blank.

Doors and Door Hardware

Overview

The installation of interior and exterior doors and their hardware is an important part of a carpenter's work. Although many doors are prehung in a frame, they must be installed level and plumb, which is a skill that takes considerable practice. Not all doors are prehung, however. Sometimes the carpenter has to install the hinges and hang the door in a door opening. This is a specialized skill that requires careful measuring and skillful use of tools.

Module 27208

Trainees with successful module completions may be eligible for credentialing through NCCER's National Registry. To learn more, go to **www.nccer.org** or contact us at **1.888.622.3720**. Our website has information on the latest product releases and training, as well as online versions of our *Cornerstone* magazine and Pearson's product catalog.

Your feedback is welcome. You may email your comments to **curriculum@nccer.org**, send general comments and inquiries to **info@nccer.org**, or fill in the User Update form at the back of this module.

This information is general in nature and intended for training purposes only. Actual performance of activities described in this manual requires compliance with all applicable operating, service, maintenance, and safety procedures under the direction of qualified personnel. References in this manual to patented or proprietary devices do not constitute a recommendation of their use.

Copyright © 2013 by NCCER, Alachua, FL 32615, and published by Pearson, New York, NY 10013. All rights reserved. Printed in the United States of America. This publication is protected by Copyright, and permission should be obtained from NCCER prior to any prohibited reproduction, storage in a retrieval system, or transmission in any form or by any means, electronic, mechanical, photocopying, recording, or likewise. To obtain permission(s) to use material from this work, please submit a written request to NCCER Product Development, 13614 Progress Blvd., Alachua, FL 32615.

27208 V5

From *Carpentry, Trainee Guide*. NCCER.
Copyright © 2013 by NCCER. Published by Pearson. All rights reserved.

27208
Doors and Door Hardware

Objectives

When you have completed this module, you will be able to do the following:

1. Describe the safety hazards related to working with doors.
2. Identify the different types and composition of residential and commercial doors.
   a. Identify the different types and composition of residential doors.
   b. Identify the different types and composition of commercial doors.
3. Identify the various types of door jambs and frames.
   a. Describe the uses and benefits of wood door jambs and frames.
   b. Describe the uses and benefits of metal door jambs and frames.
4. Identify the different types of door hardware.
   a. Identify the different types of door hardware used in residential applications.
   b. Identify the different types of door hardware used in commercial applications.
5. Describe the various installation techniques for doors and hardware.
   a. Describe the various installation techniques for residential doors and hardware.
   b. Describe the various installation techniques for commercial doors and hardware.
6. List and identify specific items included on a typical door schedule.
   a. Describe the hardware finish classifications.
   b. Describe the information included in a typical door schedule.

Performance Tasks

Under the supervision of your instructor, you should be able to do the following:

1. Demonstrate the proper installation of a hollow metal frame and door using the proper safety precautions.
2. Install a prehung door unit or door hanging system using the proper safety precautions.
3. Lay out and cut hinges in an instructor-selected project.
4. Install a door closer using the proper safety precautions.

Trade Terms

Access	Finishing sawhorse	Mortise	Sound attenuation
Astragal	Flush bolt	Mortise lockset	Sound transmission class (STC)
Butt	Flush door	Panic hardware	Stile
Casing	Hanging stile	Plumb	Strike plate
Catches	Hardware	Prefinished	Sweep
Coordinator	Head	Prehung door	Template
Cylindrical lockset	Hinge	Rabbet	Threshold
Deadbolt	Jamb	Rail	Transom
Door closer	Kerf	Rough opening	Weather stripping
Door frame	Knob lockset	Sanitary stop	
Door stop	Latch bolt	Shim	
Dust-proof strike	Lockset	Sill	
Finish hardware	Molding	Smoke gasket	

Code Note

Codes vary among jurisdictions. Because of the variations in code, consult the applicable code whenever regulations are in question. Referring to an incorrect set of codes can cause as much trouble as failing to reference codes altogether. Obtain, review, and familiarize yourself with your local adopted code.

Industry-Recognized Credentials

If you're training through an NCCER-accredited sponsor, you may be eligible for credentials from NCCER's Registry. The ID number for this module is 27208. Note that this module may have been used in other NCCER curricula and may apply to other level completions. Contact NCCER's Registry at 888.622.3720 or go to **www.nccer.org** for more information.

Contents

Topics to be presented in this module include:

1.0.0 Door Installation Safety 1
2.0.0 Types of Residential and Commercial Doors 3
   2.1.0 Residential Doors 6
      2.1.1 Bypass Doors 7
      2.1.2 Bifold Doors 8
      2.1.3 Pocket Doors 8
      2.1.4 Wood Folding Doors 10
      2.1.5 Metal Doors 12
   2.2.0 Commercial Exterior Doors 13
      2.2.1 Fire Doors and Fire Ratings 15
3.0.0 Door Jambs and Frames 17
   3.1.0 Wood Door Jambs and Frames 17
      3.1.1 Double Wood Door Frames 21
      3.1.2 Door-Stop Strips 21
   3.2.0 Metal Door Frames 22
      3.2.1 Installing Welded Metal Door Frames in Wood-Framed Construction 22
      3.2.2 Sound Attenuation and Grouting for Metal Frames 25
      3.2.3 Installing Welded Metal Door Frames in Steel-Framed Construction 27
      3.2.4 Installing Unassembled Metal Door Frames in Drywall Construction 30
      3.2.5 Installing Welded Metal Door Frames in Existing Masonry Construction 34
      3.2.6 Installing Welded Metal Door Frames in New Masonry Construction 36
4.0.0 Door Hardware 39
   4.1.0 Residential Door Hardware 39
      4.1.1 Door Hinges 39
      4.1.2 Locksets 40
   4.2.0 Commercial Hardware 41
      4.2.1 Hinges 41
      4.2.2 Interior Locksets 41
      4.2.3 External Door Stops, Door Holders, and Door Closers 43
      4.2.4 Security Hardware 44
      4.2.5 Weather Stripping 47
      4.2.6 Thresholds 47
      4.2.7 Touch-Bar or Crossbar Hardware 47
      4.2.8 Flush Bolts 48
      4.2.9 Dust-Proof Strike 50
      4.2.10 Door Coordinator 50
      4.2.11 Smoke Gasket 50

Contents (continued)

5.0.0 Installing Doors and Door Hardware 53
   5.1.0 Residential Doors 53
      5.1.1 Door Jack 53
      5.1.2 Manufactured Prehung Door-Unit Installation 53
   5.2.0 Commercial Doors 55
      5.2.1 Handling, Job Finishing, and Installation Instructions 56
      5.2.2 Fitting Doors in Prepared Openings 56
      5.2.3 Door-Hinge Installation 57
      5.2.4 Butt-Hinge Template for Doors and Jambs 59
      5.2.5 Selecting and Preparing a Router 60
      5.2.6 Installing Hinges and Hanging the Door 61
      5.2.7 Lockset Installation 61
      5.2.8 Installation of Mortise Locksets 66
      5.2.9 Installation of Door Closers 67
   6.0.0 Door Schedules 69
   6.1.0 Hardware Finish 69
   6.2.0 Door Schedules 69

Figures and Tables

Figure 1   Typical door components 4
Figure 2   Typical hollow-core and solid-core doors 5
Figure 3   Panel door 6
Figure 4   Door swing when facing the outside of a door 7
Figure 5   Hinge-location method of determining door swing 8
Figure 6   Door lights 9
Figure 7   Snap-in glazing strip 9
Figure 8   Plan view and elevation of bypass doors 9
Figure 9   Head section through bypass doors 9
Figure 10 Plan view and elevation of bifold doors 10
Figure 11 Plan view and elevation of a pocket door 10
Figure 12 Door guide strip 10
Figure 13 Elevation of a common wood folding door 10
Figure 14 Track molding and head stop of a wood folding door 11
Figure 15 Wood-folding-door hangers 11
Figure 16 Wood-folding-door spring hinging 11
Figure 17 Crossover switch track 11
Figure 18 Curved switch track 11
Figure 19 Curved track 12
Figure 20 Metal doors 12
Figure 21 Standard and engineered door sections 13
Figure 22 Metal door construction 14
Figure 23 Revolving door with ornate facade 15
Figure 24 Thermally efficient door on a loading dock 15
Figure 25 Labeled fire-door application chart 16

Figure 26 Fixed- and adjustable-width door jambs 18
Figure 27 Top left corner of interior door jamb 19
Figure 28 Finding the width and length of the head and side jambs 19
Figure 29 Placing a spreader board 22
Figure 30 Plan section at the side jamb 22
Figure 31 Door jambs shimmed in the rough opening 22
Figure 32 Elevation of door showing hardware locations 23
Figure 33 Elevation of door jambs showing shim locations 23
Figure 34 Plan view of additional jamb nailing 25
Figure 35 View of a door jamb and door 25
Figure 36 Typical fixed-width, single-piece jamb and casing metal frames 26
Figure 37 Typical fixed-width or adjustable-width metal frame with snap-on casing (1 of 2) 27
Figure 37 Typical fixed-width or adjustable-width metal frame with snap-on casing (2 of 2) 28
Figure 38 Wood stud wall anchors 29
Figure 39 Install anchors and closer reinforcement sleeve 29
Figure 40 Replace metal spreader bar with a wood spreader 29
Figure 41 Fasten a floor anchor 30
Figure 42 Install the spreader 30
Figure 43 Bend the jamb anchors around the wood studs 30
Figure 44 Anchors 31
Figure 45 Place the anchor 31
Figure 46 Assembled metal door frame moved into position 31
Figure 47 Replace the metal spreader bar 32
Figure 48 Fasten the floor anchor 32
Figure 49 Wood spreader installed 32
Figure 50 Attach the steel studs to anchors 32
Figure 51 Door opening height and width 32
Figure 52 Attach the sill anchors 33
Figure 53 Alternate base anchor 33
Figure 54 Slip the hinge jamb into position 33
Figure 55 Slip the head jamb into position 33
Figure 56 Slip the strike jamb into position 33
Figure 57 Shimming angle 34
Figure 58 Adjust the grip-lock anchor 34
Figure 59 Fasten the sill anchor 34
Figure 60 Wood spreader 34
Figure 61 Head jamb fastened to the side jambs 35
Figure 62 Masonry wall anchors 35
Figure 63 Placing anchors and closer reinforcement sleeve into position ... 35
Figure 64 Rough-opening dimensions 35
Figure 65 Wood spreader bar and middle spreader bars 36
Figure 66 Drilling into masonry 36
Figure 67 Using a sleeve anchor 36
Figure 68 Masonry anchors 37

Figures and Tables (continued)

Figure 69 Wood spreader bar installation 37
Figure 70 Floor anchor installation 37
Figure 71 Bracing 37
Figure 72 Loose-pin hinge (butt) 23
Figure 73 Plan-view section of a door jamb and door 40
Figure 74 Basic knob-type locksets and knobs 41
Figure 75 Ball-bearing hospital hinge with a rounded tip 41
Figure 76 Various lever-handle locksets with activation devices 42
Figure 77 Cutaway view of a typical mortise lockset 42
Figure 78 Door stops and holders 43
Figure 79 Regular arm installation on the pull side only (door mounted) 44
Figure 80 Regular or parallel arm installation on the push side only
          (jamb mounted) 44
Figure 81 Concealed closer with soffit plate 44
Figure 82 Concealed closer 45
Figure 83 Arm and track fully concealed 46
Figure 84 Typical electrical strikes 45
Figure 85 Electric bolt locks 46
Figure 86 Electric locksets 46
Figure 87 Electromagnetic locks 47
Figure 88 Touch-sensitive bars and handles, and armored cables 47
Figure 89 Delayed-exit alert locks 48
Figure 90 Rubber fixed-bottom sweep 48
Figure 91 Interlocking threshold 49
Figure 92 Vinyl bulb 49
Figure 93 Automatic door bottom 49
Figure 94 Aluminum threshold 49
Figure 95 Surface-mounted, double-latch touch-bar device 49
Figure 96 Surface-mounted touch-bar or crossbar device 50
Figure 97 Dust-proof strikes 50
Figure 98 Door coordinator 50
Figure 99 Diagram of an astragal set for a pair of doors 52
Figure 100 Door jack 53
Figure 101 Typical unassembled prehung door unit... 54
Figure 102 Typical assembled split-jamb, prehung door unit 54
Figure 103 Unpack and separate split jambs 54
Figure 104 Position and plumb the split-jamb half with the door 55
Figure 105 Shimming the door half of a split-jamb frame 55
Figure 106 Fastening the jamb to the trimmer stud and removing
           the spacers 56
Figure 107 Installing the second half of a split-jamb frame 56
Figure 108 Door-sawing template shown in position 57
Figure 109 Section of door showing door bevel 58
Figure 110 Plan view of door showing swing 58
Figure 111 Hinge backset 59

Figure 112	Butt marker	59
Figure 113	Scoring the wood in the mortise	60
Figure 114	Removing the excess wood from the mortise	60
Figure 115	Template kit and tools for door hanging: router, planer, and butt-hinge templates	60
Figure 116	Routing the door for hinges	60
Figure 117	Self-centering screw-hole punch	61
Figure 118	Adjusting the hinge-side jamb clearance using thin cardboard shim strips	62
Figure 119	Entrance lockset, passage latch, and privacy lockset	62
Figure 120	Exploded view of a typical tubular lockset	63
Figure 121	Wedges under the lockset edge of a door	64
Figure 122	Use of an installation template	64
Figure 123	Center-line location of the strike	64
Figure 124	Disassembled lockset	65
Figure 125	Heavy-duty cylindrical locksets	66
Figure 126	Exploded view of a heavy-duty cylindrical lockset	66
Figure 127	Typical knurled knob	66
Figure 128	Example of a closer mounting template	68
Figure 129	Adjusting closer tension	68
Figure 130	Typical door schedule	70

Table 1	Hinge Height	40
Table 2	Hinge Width	40
Table 3	Common Hardware Symbols and Finishes	69

This page is intentionally left blank.

SECTION ONE

1.0.0 DOOR INSTALLATION SAFETY

Objective

Describe the safety hazards related to working with doors.

At first glance, it appears that there would be few, if any, safety hazards involved in door installation, but this is not the case. Some of the potential safety hazards include back injuries resulting from lifting heavy doors or overreaching, eye injuries resulting from improper tool use or misuse of personal protective equipment (PPE), and injuries due to falls from ladders.

A good carpenter is always aware of the safety rules in every area of the construction field. The interior of a building nearing completion sometimes gives the worker a false sense of security. However, the element of danger is always present.

- *Keep work area clean* – Cluttered areas and benches invite injuries.
- *Avoid dangerous equipment* – Do not expose power tools to rain. Do not use power tools in damp or wet locations. Keep the area well lit. Avoid chemical or corrosive environments. Do not use tools in the presence of flammable liquids or gases.
- *Guard against electric shock* – Avoid body contact with grounded surfaces such as pipes, radiators, ranges, or refrigerator enclosures.
- *Keep visitors away* – Do not let visitors come in contact with tools or extension cords.
- *Store idle tools* – When not in use, tools should be stored.
- *Secure the work* – Use clamps or a vise to hold the work. It is safer than using your hand, and it frees both hands for operating the tool.
- *Use the right tools* – Do not force a smaller tool or attachment to do the job of a heavier tool. Do not use a tool for any purpose for which it was not intended.
- *Dress properly* – Do not wear loose clothing or jewelry. Loose clothing, drawstrings, and jewelry can be caught in moving parts. Rubber gloves and nonskid footwear are recommended. Wear a protective covering to protect long hair.
- *Use safety goggles* – Wear safety glasses or goggles while operating power tools. Wear a face or dust mask if the operation creates dust. All persons in the area where power tools are being operated should also wear safety glasses and face or dust masks.
- *Do not abuse extension cords or power tool cords* – Never carry any tool by the cord or yank it to disconnect it from the receptacle. Keep the cord away from heat, oil, and sharp edges. Have damaged or worn power cords and strain relievers replaced immediately.
- *Do not overreach* – Keep proper footing and balance at all times.
- *Maintain tools with care* – Keep tools sharp and clean for safer performance. Follow instructions for lubricating and changing accessories. Inspect tool cords periodically, and if they are damaged, have them repaired by an authorized service facility. Have all worn, broken, or lost parts replaced immediately. Keep handles dry, clean, and free from oil and grease.
- *Disconnect (unplug) tools* – Always disconnect tools when not in use, before servicing, and when changing accessories such as blades, bits, and cutters.
- *Remove keys and adjusting wrenches* – Check to make certain that keys and adjusting wrenches are removed from a tool before turning it on.
- *Avoid unintentional starting* – Do not carry a plugged-in tool with a finger on the switch. Be sure the switch is off when plugging in a tool.
- *Outdoor extension cords* – When a tool is used outdoors, use only extension cords marked as suitable for use with outdoor appliances and store them indoors when not in use.
- *Check damaged parts* – Before using any tool, inspect it to be sure that it will operate properly and perform its intended function. Check for alignment of moving parts, binding of moving parts, breakage of parts or mountings, and any other conditions that may affect its operation. A guard or other part that is damaged should be properly repaired or replaced by an authorized service center unless otherwise indicated in the instruction manual. Inspect tool cords periodically, and if they are damaged, have them repaired by an authorized service facility. Have all worn, broken, or lost parts replaced immediately. Have defective switches replaced by an authorized service center.

- *Stay alert* – Watch what you are doing and use common sense. Do not operate a tool when you are tired or while under the influence of medication, alcohol, or drugs.
- *Inspect extension cords* – Inspect extension cords periodically and replace them if damaged.

WARNING! Remember that an extension cord with cracked or cut outer insulation or a loose or missing ground pin can cause serious injury or death. Be careful. Always unplug power tools when changing accessories or making adjustments. Plan ahead. Always put safety first.

1.0.0 Section Review

1. If the ground pin of a tool's electrical plug is loose, it should be removed so you can continue to use the tool.
    a. True
    b. False

Section Two

2.0.0 Types of Residential and Commercial Doors

Objective

Identify the different types and composition of residential and commercial doors.

a. Identify the different types and composition of residential doors.
b. Identify the different types and composition of commercial doors.

Trade Terms

Butt: Any kind of hinge, except a strap or T-hinge.

Casing: The exposed finish material around the edge of a door or window opening.

Door frame: The surrounding case of a door into which a door closes. It consists of two upright pieces called jambs and a horizontal top piece called the head.

Door stop: The strip against which a door closes on the inside face of the frame or jamb. It can also be a hardware device used to hold the door open to any desired position or a hardware device placed against the baseboard to keep the door from marring the wall.

Flush bolt: A sliding bolt mechanism that is mortised into a door at the top and bottom edge. It is used to hold an inactive door in a fixed position on a pair of double doors.

Flush door: A door of any size that has a totally flat surface.

Hanging stile: The door stile to which the hinges (butts) are fastened.

Hardware: Components, such as hinges, locksets, and closers, used to attach a door to its frame or operate the door.

Hinge: The hardware fastened to the edge of a door that allows the frame to pivot around a steel pin, permitting the operation of the door.

Jamb: One of the vertical members on either side of a door or window opening.

Lockset: The entire lock unit, including locks, strike plate, and trim pieces.

Molding: Long strips of material used for finishing and decorative trim.

Mortise: A measured portion of wood removed to receive a piece of hardware such as a lock or butt.

Mortise lockset: A rectangular metal box that houses a lock. It usually has a latch and deadbolt as part of the unit. Options consist of a locking cylinder and/or thumb turn by which it can be secured. It is used on residential entry doors and commercial doors.

Plumb: A true vertical position.

Prefinished: Material such as molding, doors, cabinets, and paneling that has been stained, varnished, or painted at the factory.

Prehung door: A door that is delivered to the job site from the mill already hung in the frame or jamb. In some instances, the trim may be applied on one side.

Rail: A horizontal member of a door or window sash.

Rough opening: Any unfinished door or window opening in a building.

Sound transmission class (STC): The rating by which the sound attenuation of a door is determined. The higher or greater the number, the better the sound reduction.

Stile: The vertical edge of a door.

Strike plate: A metal plate screwed to the jamb of a door so that when the door is closed, the latch bolt strikes against it. The bolt is then retracted and slides along the metal plate. When the door is fully closed, the latch bolt slides into a hole in the plate to hold the door securely in place. Also called a strike.

Except for some welded metal door frames, interior doors are usually installed after the walls have received their finish covering. Like exterior doors, interior door components consist of the door, jamb, door stops, and casing (*Figure 1*), along with the door hardware such as the hinges and locksets. The jamb is the frame in which the door hangs. The stops halt the door swing when the door is being closed, and position the door so that the lockset bolt engages the strike plate properly after the door is closed. The casing provides the trim for the door frame.

There are many types of interior doors made of various materials, or combinations of materials, ranging from wood and metal to synthetic materials. However, most doors can be categorized as flush or panel types.

Flush doors consist of a frame covered with a smooth skin made of wood, metal, or a synthetic material. Some are equipped with door lights (windows). The window material can be glass or plastic. The two main types of flush doors are hollow-core doors and solid-core doors.

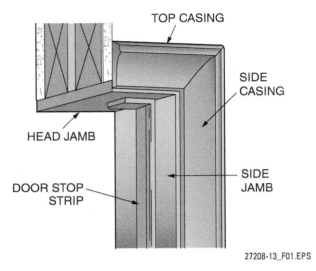

Figure 1 Typical door components.

- *Hollow-core doors* – These doors are usually less expensive than solid-core doors and are used more often. *Figure 2* (top left) shows a typical hollow-core door.
- *Solid-core doors* – These doors are heavier than hollow-core doors. They provide better sound insulation and have less tendency to warp. The core material is usually particleboard or a staggered-block core, also known as a staved-wood core, as shown in *Figure 2*. Special solid-core doors include fire doors that use a mineral core, radiation-blocking doors that use a lead lining, and sound transmission class (STC) doors that use special sound-reducing materials and gaskets.

Wood panel doors are also known as stile-and-rail doors because stiles (vertical members) and rails (cross members) are fastened together by dowels or mortise-and-tenon joints and glued to make up the frame holding the panels (*Figure 3*). The panels fit into the grooved edges of the frame and are usually not glued in place to allow for expansion and contraction of the door elements. These doors are available in many designs with raised or flat panels and can be glazed with glass or plastic.

Many doors, especially residential doors, are available today as assembled or unassembled prehung door units. The hinges are premortised and installed on the jamb and door, the lockset holes and associated mortises have been precut, and the door stops have been installed.

The direction in which a door opens is called the hand or swing of the door. Prior to hanging a door, the carpenter must have some knowledge about the hand or swing. The doors that are delivered to the job site must be checked and compared to the door schedule so each door's location

Simulated Panel Doors

In many cases, a flush door can be made to resemble a panel door by applying molding to the surface of a hollow-core or solid-core door to simulate the appearance of stiles and rails. Some metal or synthetic hollow-core doors have a simulated panel effect molded or stamped into the surface material.

MOLDED PANEL DOOR

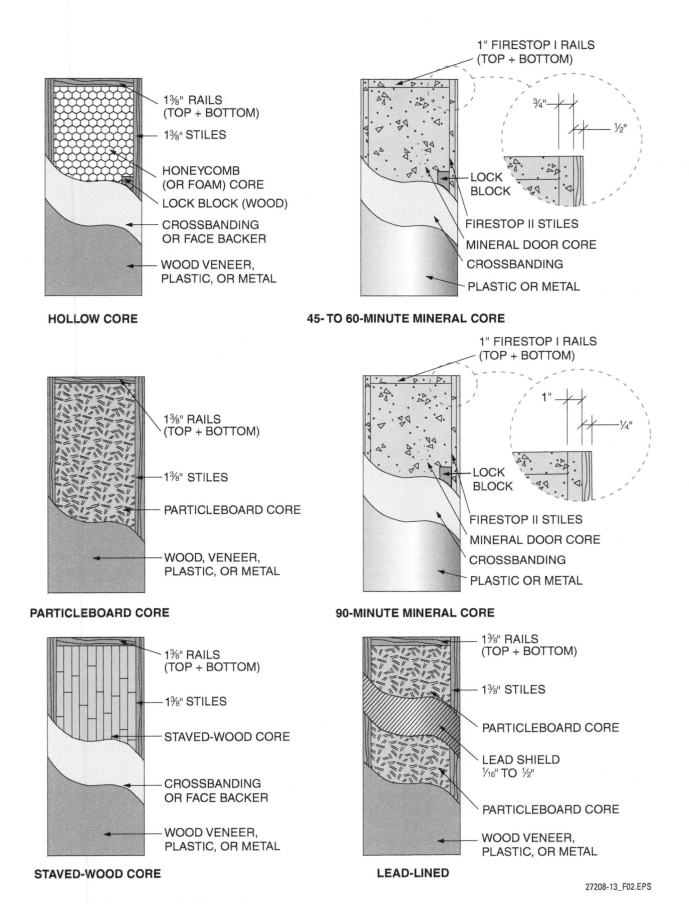

Figure 2 Typical hollow-core and solid-core doors.

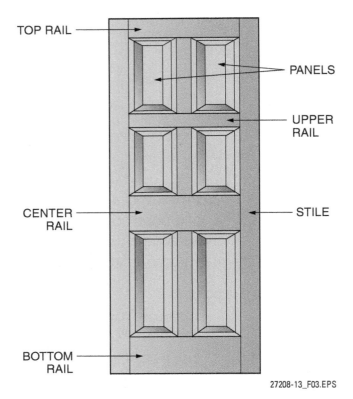

Figure 3 Panel door.

may be identified on the floor plan of the drawings and to ensure the hand or swing is correct according to the plans. A good carpenter should have some knowledge of how the swing is determined. It is very important when ordering assembled or unassembled prehung door units, and it is also important in some instances when ordering hardware to fit a door. Knowing the direction of a particular door swing is critical when giving or receiving instructions on hanging it in the opening. The incorrect determination of the door swing is one of the most common mistakes in construction due to the fact that there is no universal standard for specifying door swing. In some cases, the hand can be defined as either the handle location or the hinge location.

One method of identification dictates that a door swing be described in four ways, as shown in the plan-view drawing in *Figure 4*. The door swing is the direction in which a door swings and is determined when facing the door from the outside (public side). This method is used by many contractors and door manufacturers, and is also used by most lockset manufacturers to specify locksets.

In this method, a door can be either left hand or right hand, and either of these types can be a reverse (one that opens out of the room instead of into the room). Reverse doors are prepared for hardware differently than standard doors.

In another method, the hand or swing is determined by mentally placing oneself in the floor plan of the building at the doorway in question so that your back is against the hinge locations. The side of the jamb on which the hinges (butts) are located determines the swing of the door, as shown in a plan view in *Figure 5A*. A term used by many carpenters on the job to describe this method is *butt-to-butt*. Some door manufacturers use other methods and each manufacturer must be checked to determine the specific method used. As a general rule, door swing is based on which side the handle or hinge is on when the door is pulled closed. See *Figure 5B*.

To mark the swing of the door for a particular opening, mark the jamb on the side that will have the hinges. Mark the location of the hinges on the hanging stile of the door that corresponds with that opening.

Door lights, also referred to as vision lights, are simply windows within a door unit. See *Figure 6* for examples of door lights. Door lights are constructed of different materials, but the primary product used is glass. When a door light is installed in a door, this is referred to as glazing. Glass is not a good sound-retarding material, and double glazing may actually increase the noise level. To reduce the noise level, manufacturers can install sound-absorptive material around the perimeter in the space between the two panes. They can also reduce sound by using laminated glass of different thicknesses, isolating glass from the frame with flexible gaskets, or using resilient isolation material between frame sections. Non-parallel panes reduce annoying glare and reflections. Refer to *Figure 7* for an example of snap-in glazing in a steel door panel.

2.1.0 Residential Doors

Common types of residential doors include bypass doors, bifold doors, pocket doors, wood folding doors, and metal doors.

Delivery Checks

Every interior door must be checked for straightness and damage upon delivery so that any returns or claims against the manufacturer or dealer may be justified. All parties concerned must be notified immediately upon delivery of any defective or damaged doors.

LEFT-HAND SWING:
HINGES AT LEFT.
DOOR OPENS INWARD.
HANDED LOCK = LH

LEFT-HAND REVERSE:
HINGES AT LEFT.
DOOR OPENS OUTWARD.
HANDED LOCK = LHR

RIGHT-HAND SWING:
HINGES AT RIGHT.
DOOR OPENS INWARD.
HANDED LOCK = RH

RIGHT-HAND REVERSE:
HINGES AT RIGHT.
DOOR OPENS OUTWARD.
HANDED LOCK = RHR

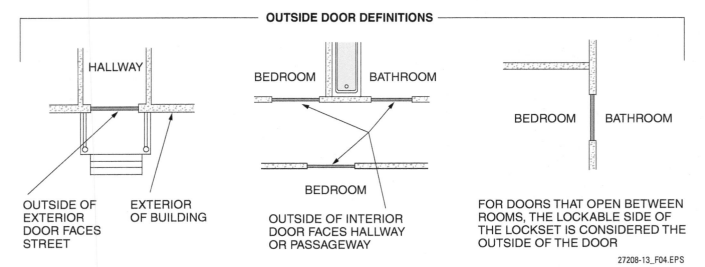

Figure 4 Door swing when facing the outside of a door.

2.1.1 Bypass Doors

Bypass doors, as shown in *Figure 8*, are usually hung in pairs, but there may be more doors in an opening. These doors are suspended from an overhead track and move on rollers, as shown in *Figure 9*.

The method of hanging bypass doors varies from one hardware manufacturer to another. The basic method is to screw the head track into position. The rollers, which are adjustable, are screwed to the top edge of the door. Look at the installation instructions to see if the doors must be cut off on the bottom edge to shorten them. If this is the case, it is usually a result of the excessive space taken up by the hardware at the top of the door. Preplanning when framing should prevent the necessity of cutting the doors. Hang the doors in position by placing the rollers in the track. Hang the door closest to the rear first. Mark the location of the doors on the floor and screw the nylon guide strip to the floor. The door pulls are mounted 38" above the floor and on the stiles nearest the jambs. The pulls must be flush with the face of the door.

It is a good practice to mount the doors in such a manner so that when a person enters the room, the door nearest that person is mounted in the front of the opening. The person sees that door first, and it covers the edge of the door mounted behind it.

Prehung Door-Unit Swing

When ordering assembled or unassembled prehung door units from the manufacturer, make sure there is an understanding of which method is to be used to determine the swing. This will facilitate delivery of the correct door unit to the job site. Most manufacturers use the convention that when facing the hinge barrel (visible) side of the prehung door unit and the hinge is on the left, the door is a right hand. If the hinge is on the right, it is a left hand.

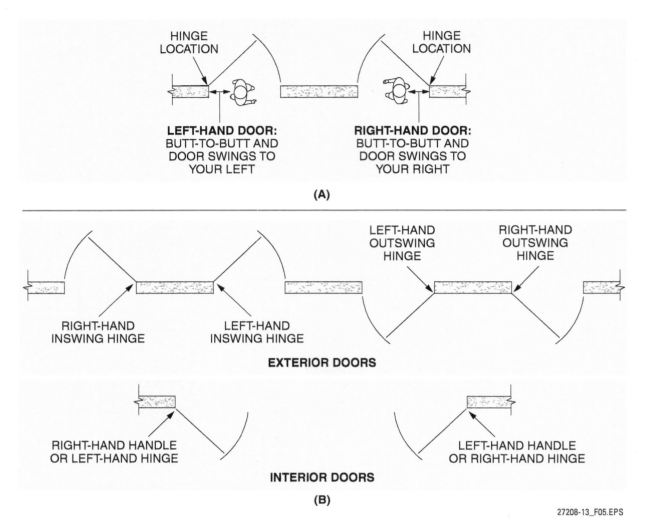

Figure 5 Hinge-location method of determining door swing.

2.1.2 Bifold Doors

Bifold doors (*Figure 10*) allow full access to a closet and, at the same time, permit furniture to be placed very close to the side trim of the closet opening. Usually, this door unit is used with four doors or two pairs, with each pair operating independently of the other. This four-door unit is used for a 6'-wide opening. For openings of less than 4', only two doors, or one pair, are used.

The door unit is supplied with wood, metal, plastic, or mirrored doors. Hardware such as hinges, pulls, pivot pins, and guide pins are installed on the doors at the factory.

Hanging the unit in a prepared opening is simple. A slide guide track is screwed to the bottom of the head jamb according to the manufacturer's instructions. An adjustable jamb bracket is screwed in place at the base of the side jambs against the floor on each side.

The pivot pins and guide pins are spring loaded, so it is a simple matter to mount the doors on the hardware in the opening. A pivot pin and bracket secure the door next to the jamb. The first door and second door are hinged together. If the doors are not plumb or do not meet properly in the center, an adjustment may have to be made at the jamb bracket on the floor. The second door slides along a track on a roller guide. This track is secured to the head jamb.

2.1.3 Pocket Doors

A pocket-door unit is delivered to the job site partially knocked down and is rapidly and easily assembled. Pocket-door units are designed to fit in a framed opening of 3½" studs. The hardware consists of a track and rollers that are partially concealed in the head pocket of the unit. When the unit is positioned in the rough opening, any 1⅜"-thick interior door of the correct size may be hung on the track. The portion of the unit called the pocket or the right-hand side of the unit, as shown in *Figure 11*, may be covered with practi-

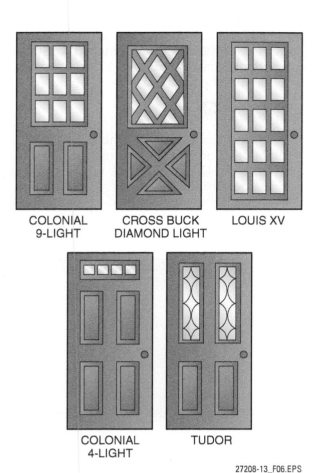

Figure 6 Door lights.

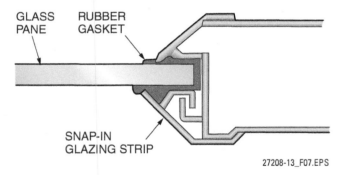

Figure 7 Snap-in glazing strip.

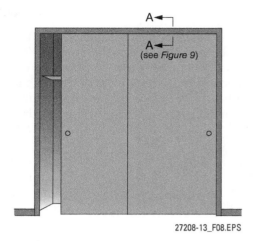

Figure 8 Plan view and elevation of bypass doors.

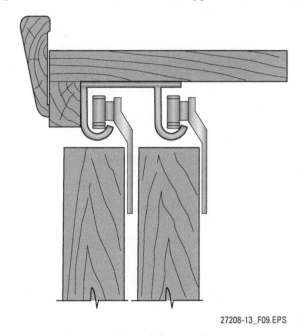

Figure 9 Head section through bypass doors.

cally any interior wall finish. The wall finish is applied to the horizontal members of the unit, creating the pocket for the door to slide into.

After the door is positioned in the opening and the rollers are adjusted so that it hangs plumb with the jambs and square with the head, a guide strip is fastened to the bottom edge, as shown in *Figure 12*.

Stops are then nailed on either side of the pocket opening. The thickness of the stops decreases the width of the opening, preventing the door from

Fastening Finish Wall Material to a Pocket Door Unit

When fastening the wall finish to the horizontal members of a pocket door unit, make sure the nails, staples, or screws are not so long that they extend into the pocket area where they might scratch the door or even prevent the movement of the door.

being pushed out of the opening. The spacing of the stops also acts as a guide for the guide strip on the door bottom. The purpose of the guide strip is to position the door so that the surfaces of the door are not scratched when being slid in and out of the pocket. When the door is completely inside the pocket, the edge of the door facing the opening is flush with the face of the stops.

A special piece of hardware called an edge pull is installed on the edge of the door so that it may be pulled out of the pocket far enough to use the conventional flush pulls mounted on the face of the door. There are many variations of the edge pull, but the simplest is a small metal plate with a finger hole in it that is mortised to the edge of the door, 38' above the floor.

2.1.4 Wood Folding Doors

Wood folding doors are assembled of vertical, prefinished, wood door panels operating similar to an accordion (*Figure 13*). The door can be installed to open from either side.

The width of the panels, called the stack width, is 4¼" so it will fit within a typical door frame.

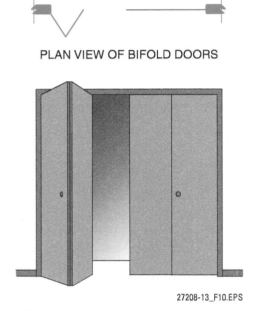

Figure 10 Plan view and elevation of bifold doors.

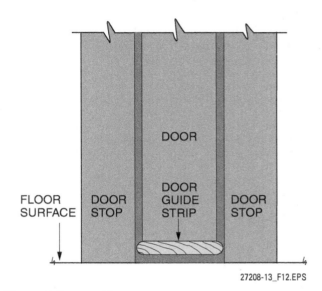

Figure 12 Door guide strip.

PLAN VIEW OF POCKET DOOR UNIT

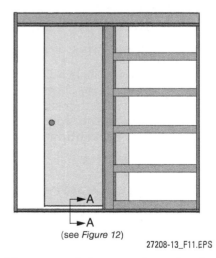

Figure 11 Plan view and elevation of a pocket door.

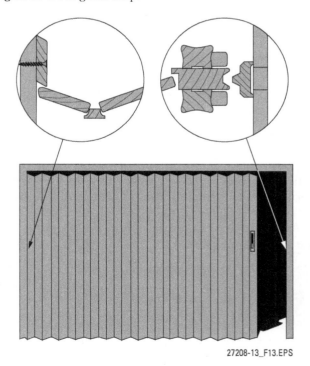

Figure 13 Elevation of a common wood folding door.

A beveled wood molding at the head conceals a surface-mounted head track. The track and molding are drilled for mounting screws supplied with the door. A spring-action catch at the top of the end post (*Figure 14*) engages the nylon head stop for tight stacking of the panels.

The hangers, with free-riding nylon rollers, are attached to alternate panels for quiet, easy operation. The end post hanger has two sets of rollers to keep it aligned with the track (*Figure 15*). The positioning of the rollers allows operation even in openings that are slightly out of plumb.

The panels have special steel alloy springs that are run horizontally through the panels and connect to vertical wood moldings, as shown in *Figure 16*. The spring connectors can be detached to disassemble the door or to replace damaged panels.

Wood folding doors are available in opening widths from 1'-9" to 30'-2¼".

Wood folding doors can be mounted directly to the ceiling to divide large rooms into smaller rooms as in an office or meeting room area. To do this, the track is simply screwed to supporting members in the ceiling and the beveled wood molding is applied to both sides. A crossover switch track permits the door tracks to cross at right angles, as shown in *Figure 17*.

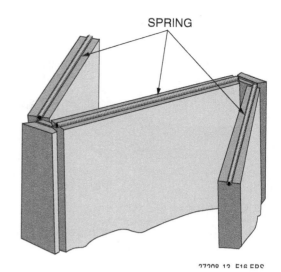

Figure 16 Wood-folding-door spring hinging.

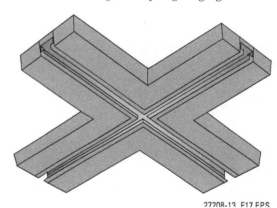

Figure 17 Crossover switch track.

A curved switch track (*Figure 18*) is available in 90-degree left or right arcs, and seven radii from 1'-6" to 10'. This type of switch permits the doors to divide a room at several designated points. It may also be used to store stacked doors parallel to a wall.

A simple curved track (*Figure 19*) permits doors to move around structural members and to create curved wall effects. The track is available in standard radii from 1'-6" to 10' and 90-degree left or right arcs.

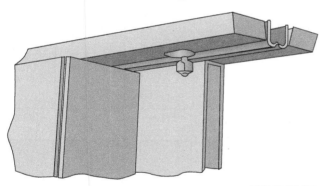

Figure 14 Track molding and head stop of a wood folding door.

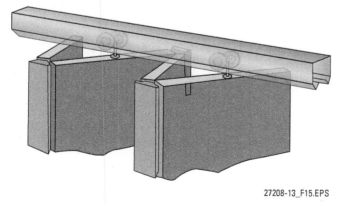

Figure 15 Wood-folding-door hangers.

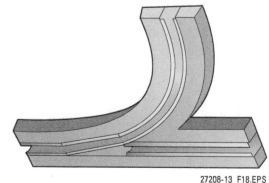

Figure 18 Curved switch track.

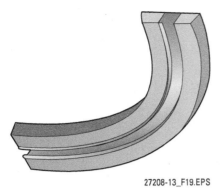

Figure 19 Curved track.

Wood folding doors are simple to hang and all hardware is supplied by the manufacturer.

2.1.5 Metal Doors

Metal doors are available in practically any variation and are used mostly in residential exterior and commercial construction. They are identified in architectural specifications by a defined coding system, as shown in *Figure 20*.

Metal door sizes range in thickness from 1⅜" to 1¾", in width from 2' to 4', and in height from 6'-8" to 8'. If the specifications call for a size other than stock, a door can be made to meet those requirements.

Metal doors are available with butts, cylindrical locksets, mortise locksets, unit (mono) locksets, panic devices, flush bolts, and other installed hardware. Door lights and louvers are available. Doors and frames are available in stainless steel to be used in corrosive atmospheres as in chemical plants, sewage treatment plants, or swimming pool areas. Stainless steel doors and frames, because of their sanitary qualities, are also used in meat-processing plants, hospital operating theaters, and laboratories. Metal doors and frames are manufactured to meet the requirements of practically any architectural opening that can be imagined.

The internal portion of the door, or core, is also designed to meet most specification requirements, as shown in *Figure 21*. For example, a metal door with a lead core should be used in radiation areas.

To determine the rough opening, check the manufacturer's instructions for the number of inches to add to the door width and the door height.

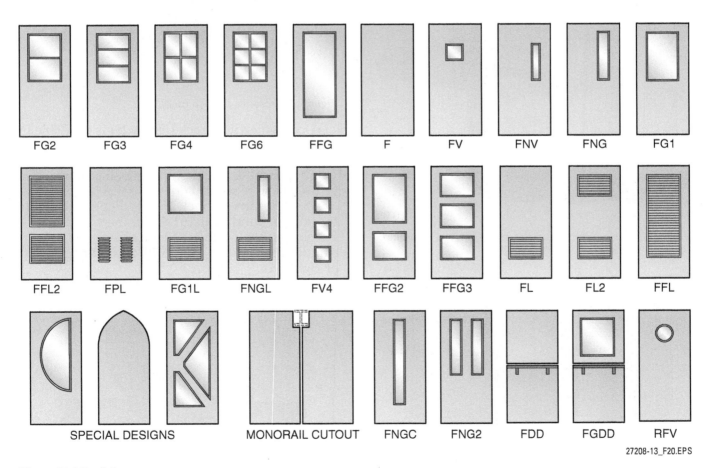

Figure 20 Metal doors.

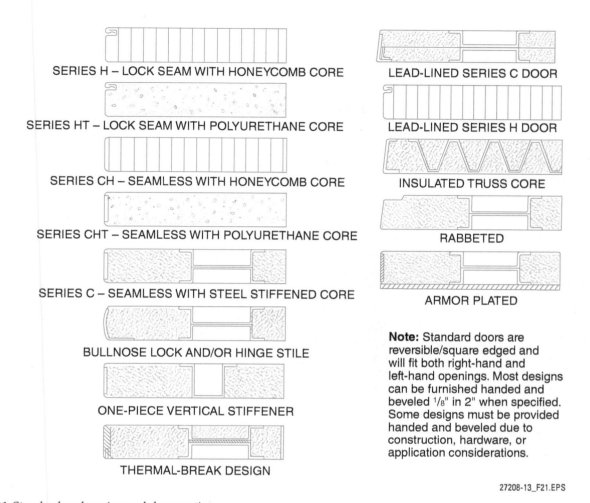

Figure 21 Standard and engineered door sections.

Steel doors have prepared hinge mortises and lock holes. Some hollow metal doors are reinforced with a honeycomb pattern core. Others only have steel reinforcement sections for the locks and hinges or other hardware. The holes are drilled and tapped to receive machine screws. *Figure 22* shows an example of metal door construction.

2.2.0 Commercial Exterior Doors

Commercial exterior doors are designed to provide a secure and convenient means of entering and leaving a building. A large variety of exterior doors is available. The type and style selected for an exterior door is based on its purpose. Doors used exclusively by employees are usually no-frills doors, while those used on loading docks are strictly utilitarian. The doors used by the public are selected to project a particular image and to make their use convenient. A trendy retail store will use a very different style than a supermarket, and an upscale hotel in New York City will have a different-style door than a budget motel located near an interstate highway.

Figure 23 shows a revolving door on the Wrigley Building in Chicago. This beautiful entrance is perfect for a building that is headquarters for major corporations, as well as investment management companies, advertising agencies, marketing firms, a bank, and foreign consulates.

Figure 24 shows a thermally efficient door on a loading dock. This type of door is used when temperature control of the interior of the building is critical.

As you can imagine, the installation of some commercial doors is very complicated. Because of their complexity, commercial exterior doors are often installed by teams that have been specially trained by the door manufacturer. In these cases, a carpenter's job is to prepare the installation site according to the building plans and specifications. If you are required to install a commercial exterior door, it is very important that you obtain the door vendor's installation instructions and then carefully follow the instructions.

A. EDGE SEAM IS EXPOSED ON BOTH EDGES

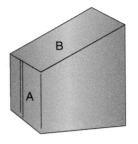

B. FLUSH TOP REINFORCED WITH 16-GAUGE CHANNEL

G. 18-GAUGE DOOR CLOSER REINFORCEMENT BOX LAMINATED TO INSIDE OF DOOR SKIN

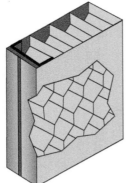

E. ¾" CELL HONEYCOMB CORE

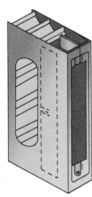

F. CONTINUOUS 14-GAUGE MORTISE LOCK REINFORCEMENT WITH PROVISIONS FOR GOVERNMENT SERIES 86 MORTISE LOCKSET AND *ANSI A115.1* LOCK FRONT (1¼" × 8")

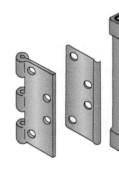

C. CONTINUOUS 11-GAUGE STEEL INTEGRAL HINGE REINFORCEMENT WITH PROVISION FOR 3½" × 3½" FULL MORTISE TEMPLATE-TYPE HINGES

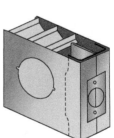

F. CONTINUOUS 14-GAUGE STEEL INTEGRAL CYLINDRICAL LOCK REINFORCEMENT WITH PROVISIONS FOR GOVERNMENT SERIES 160 OR 161 CYLINDRICAL LOCKSETS (2¾" BACKSET) AND *ANSI A115.2* LOCK FRONT (1⅛" × 2¼")

AVAILABLE IN 1⅜" OR 1¾" THICKNESSES. EASILY MODIFIED FOR LIGHTS, LOUVERS, AND OTHER OPTIONS

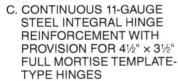

C. CONTINUOUS 11-GAUGE STEEL INTEGRAL HINGE REINFORCEMENT WITH PROVISION FOR 4½" × 3½" FULL MORTISE TEMPLATE-TYPE HINGES

C. STANDARD DOORS ARE NON-HANDED USING HINGE FILLERS

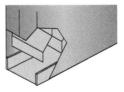

D. BOTTOM REINFORCED WITH 16-GAUGE CHANNEL

Figure 22 Metal door construction.

Figure 23 Revolving door with ornate facade.

Figure 24 Thermally efficient door on a loading dock.

CAUTION: When you are installing a door, you should perform the work that you are trained and qualified to perform. Other work, such as electrical installations, must be completed by a qualified worker.

You can help to prevent problems during door installation by ensuring that wall construction is completed to specification. The wall angle must be plumb and the door opening must be square and level. Although there are a number of fixes the installation team can use to compensate for tolerance variations, it is much better to take the time to ensure the work is correct, rather than rely on someone else to fix errors.

Regardless of whether you will be installing the doors or not, one important task that may fall to you is to accept delivery of the door. It is important to carefully inspect the door for damage before accepting it. Note any damage to the door packaging on the delivery invoice. Do not accept delivery of any severely damaged doors, but report the damage to a supervisor. Don't remove any packaging unless the product is suspected to be damaged. The packaging material will help to protect the door during storage.

Once the door has been accepted, store it in a secure location. The door should be stored in the unopened shipping container to protect it. Store the door in an area that permits good air circulation. Doors should never be stored in an area where they are exposed to the elements.

2.2.1 Fire Doors and Fire Ratings

Fire doors are manufactured to meet specific conditions, and because of this, the openings are referred to as labeled openings. The letters A, B, C, D, and E are codes used by labeling agencies and refer to the opening locations. Refer to the door schedule for the project. The classification is used to state the opening for which the fire door is considered suitable. The rating time associated with each classification indicates the duration of the fire test the door is capable of enduring. Each location classification requires a specific door rating depending on the fire hazard involved. An example of these classifications and ratings is shown in *Figure 25*.

- *Label A* – Refers to doors located in walls between separate buildings, fire walls, and division walls or curtain walls leading to highly flammable contents such as a paint or chemical storage area.
- *Label B* – Refers to doors located in openings into vertical shafts, such as fire stairs, elevator shafts, incinerator chutes, and walls separating garages from living quarters.
- *Label C* – Refers to doors located in room partition openings and openings in corridors.
- *Label D* – Refers to doors located in openings in exterior walls subject to extreme fire exposure.
- *Label E* – Refers to doors located in openings in exterior walls subject to moderate fire exposure.

NOTE: A labeled door may not be modified and the label must not be painted.

Doors are also available for sound-insulating purposes. X-ray doors, security doors, armor-plated doors, and pressure-resistant doors are also available from most manufacturers.

LOCATION	CLASS	RATING	MAXIMUM GLASS AREA	REGENT	MEDALLION	FUEGO	VERSADOOR
Openings in walls separating buildings or dividing a building into fire areas	A	3 HR.	NONE	NTR	NTR	250° MAXIMUM TEMP. RISE	NA
Openings in enclosures of vertical communication: elevators, stairwells, and in 2-hour rated partitions providing horizontal fire separation	B	1½ HR.	100 SQ. IN. LITE PER DOOR	NTR	NTR	250° MAXIMUM TEMP. RISE	NTR (100 SQ. IN. PER DOOR)
Corridor and room partitions	C	¾ HR.	1296 SQ. IN. PER LITE (EXCEPT VERSADOOR)	NTR	NTR	NTR	NTR (100 SQ. IN. PER DOOR)
Exterior walls subject to severe fire exposure from outside	D	1½ HR.	NONE	NTR	NTR	250° MAXIMUM TEMP. RISE	NTR (100 SQ. IN. PER DOOR)
Exterior walls subject to moderate fire exposure from outside	E	¾ HR.	1296 SQ. IN. PER LITE (EXCEPT VERSADOOR)	NTR	NTR	NTR	NTR (100 SQ. IN. PER DOOR)
Openings where smoke control is a primary consideration ... in partitions between a habitable room and a corridor when the wall is constructed to have a fire resistance rating of more than one hour ... or across corridors where a smoke partition is required		½ HR. ⅓ HR.	1296 SQ. IN. PER LITE (EXCEPT VERSADOOR)	NTR	NTR	NTR	NTR (100 SQ. IN. PER DOOR)

NTR = NO TEMPERATURE RISE

Figure 25 Labeled fire-door application chart.

2.0.0 Section Review

1. In comparison to hollow-core doors, solid-core doors _____.
    a. provide less sound insulation
    b. are lighter
    c. are less prone to warping
    d. are less expensive

2. The type of labeled opening indicated by fire class C is one that will _____.
    a. open to a vertical shaft
    b. be located in a wall between separate buildings
    c. be located in room partitions or corridor openings
    d. be located in an exterior wall subject to extreme fire exposure

Section Three

3.0.0 Door Jambs and Frames

Objective

Identify the various types of door jambs and frames.
a. Describe the uses and benefits of wood door jambs and frames.
b. Describe the uses and benefits of metal door jambs and frames.

Trade Terms

Door closer: A mechanical device used to check a door and prevent it from slamming when it is being closed. It also ensures the closing of the door.

Kerf: A slot or cut made with a saw.

Rabbet: A rectangular groove cut in the edge of a board.

Sanitary stop: Door stop with a 45-degree angle cut at the bottom on the vertical jambs.

Shim: A thin, tapered piece of wood such as a wooden shingle used to fill in gaps and to level or plumb structural components.

Sill: The lowest member at the bottom of a door or window opening. It also refers to the lowest member of wood framing supporting the framing of a building.

Sound attenuation: The reduction of sound as it passes through a material.

Weather stripping: Strips of metal or plastic used to keep out air or moisture that would otherwise enter through the spaces between the outer edges of doors and windows and the finish frames.

Door jambs and frames provide the wall opening for doors. Door jambs are attached to the trimmer studs, which provide the structural framing for the door.

3.1.0 Wood Door Jambs and Frames

The majority of door jambs delivered to the job site from the millwork supply house are ¾" thick. In most cases, when a paint finish is required, the jambs will be finger jointed. When supplied for a natural finish, the jambs will usually be of the same wood as the rest of the interior trim. The width of the door jambs must be the same as the thickness of the wall or partition in which they are to be placed. In some cases, adjustable as well as fixed-width jambs are available. The types of jambs that are most often used on prehung door units are shown in *Figure 26*.

When the jambs are delivered to the job site, they may be assembled or unassembled (knocked down). If they are knocked down, place the head jamb into the dado joint on the side jambs, and fasten it to the side jambs with 8d box or casing nails. Saw **kerfs** or grooves help prevent warping of jambs. See *Figure 27*.

The opening into which the jamb is to be placed must be checked for height, width, and squareness. For wood jambs, the rough opening should normally be 1½" higher and 2" wider than the door. Make sure that the face of the wall is plumb. If the jambs are slightly higher than the rough opening, the lugs can be trimmed off the top so the jamb will fit. If they are trimmed off flush with the head jamb, shimming of the head jamb at the corners during installation is recommended to prevent separation of the head and side jambs. Place the jambs in the rough opening and check the head jamb to see if it is level. If it is not level, cut off the bottom of the highest side jamb by the amount required to make the head jamb level. Then, adjust the jamb height as follows:

- Lift the jamb/casing up off the subfloor at a height equal to the approximate thickness of the finished floor. For floors to be covered with carpeting, it is recommended that the jamb and casing be located approximately ½" off the subfloor.
- Do not place the wood jamb directly on a concrete floor.

The door opening is built to the exact size specified, and the door is cut in the field, if necessary. As shown in *Figure 28*, the inside width of the jambs should be the width of the door specified for the opening plus 3⁄16" for clearance. For example, for a 2'-6" door, the jamb-to-jamb width would be 2'-6 3⁄16" (2'-6" + 3⁄16"). Check the specifications for required clearances. The inside height of the jambs, from the bottom of the finished floor to the bottom of the head jamb, should be the door height plus 3⁄8" to ½" plus the finished floor thickness. Be certain to verify the height and width of the jambs.

Step 1 Cut a strip of wood the same width as the jambs and the same length as the distance between the jambs at the head. This piece is used to keep the jambs the correct distance apart during installation (*Figure 29*).

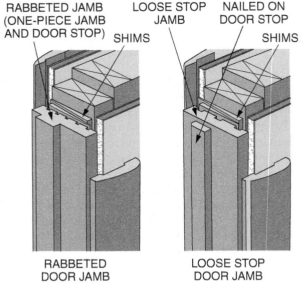

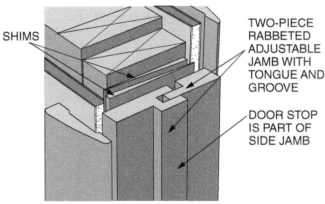

Figure 26 Fixed- and adjustable-width door jambs.

Step 2 Measure in from the edge of the jambs by a distance equal to the thickness of the door plus one half the width of the stops. Do this on the side of the jambs where the door will be located. Draw a pencil line from the bottom of both side jambs to the head jamb. This will be the nailing location through the jambs over which the door stop will be placed. The door stop will conceal both the line and the nail heads (*Figure 30*).

Step 3 Using a long level as a straightedge, check again to make sure the faces of the wall on both sides of the opening are plumb. If the wall is not plumb, the door will strike the floor and not open all the way, or the door will swing closed of its own accord, depending on which way the wall is out of plumb. If the wall is not plumb, do not attempt to install the door jambs. If it is a stud wall, it can sometimes be made plumb by bumping the bottom plate with a sledgehammer. Make sure a piece of blocking is used to prevent marring the wall when attempting this procedure.

Step 4 If the wall faces are plumb, proceed to plumb and level the jambs in the opening. Using shims as wedges, place two on each side between the jamb and the trimmer stud at the top and bottom. Make sure the edges of the jamb are flush with the finished wall (*Figure 31*).

Step 5 Check the jamb side where the hinges will be placed, using a long level as a straightedge. Make sure the butt jamb is plumb. Adjust the top and bottom shims as required. Fasten the butt jamb with 8d or larger casing nails at the top and bottom, using the nailing line to locate the nails. Don't drive the nails flush. Check the lock-strike side jamb for plumb and nail it in the same manner. Place three pairs of shims behind the butt jamb; one pair goes behind the location of each butt. If only two butts (one pair) are being used, three pairs of shims will still be used. Place shims at the recommended butt locations shown in *Figure 32*. Use 8d or larger casing nails, ½" on each side of the shims along the nailing line to secure the shims. Don't drive the nails flush.

Step 6 Use three pairs of shims on the lock side, with one pair at the lock location and two pairs halfway between the lock location and the top and bottom, as shown in *Figure 33*. Temporarily fit the door to the opening, making sure the jambs are plumb and square and the correct space exists around the door. Remove the door and complete the nailing of the jambs (*Figure 33*).

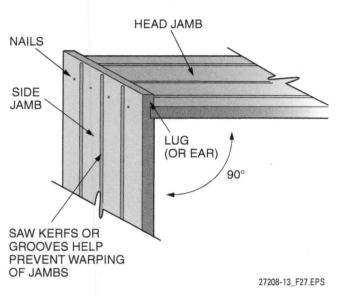

Figure 27 Top left corner of interior door jamb.

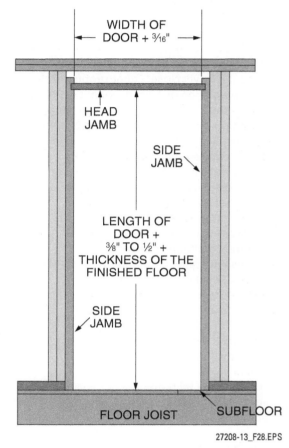

Figure 28 Finding the width and length of the head and side jambs.

Checking and Correcting the Plumb and Square of Wood Door Framing

Before installing any doors, it is extremely important to check the plumb of all wood door king studs and trimmers in two planes. One plane is parallel to a wall and the other is perpendicular to a wall. If the king studs and trimmers are plumb, the door header must be checked for squareness. If the framing for a door is out of plumb or the king studs and trimmers are badly bowed or crowned, the door cannot be hung properly. Even if it is hung, the door would have to be trimmed at an angle on the top and bottom so that it will close. Other problems are that the door will swing open or closed when it is unlatched or it may contact the floor somewhere in its swing radius.

The best time to check door framing is before the finish walls are installed so that the doorways can be easily reframed if required. Because most doors are installed after the finished walls have been completed, correcting a major framing problem at that point will require removal and replacement of some of the finish wall material in order to reframe the door. However, most minor out-of-plumb conditions can be corrected as illustrated, with minor repair to the finish wall material. In addition, a minor crown in the king studs and trimmers can be corrected by cutting both the trimmer and king stud and closing the saw kerf with screws. Several cuts spaced along the length of both studs may be required to achieve an almost-straight condition. This type of correction can be accomplished with the finish wall material installed; however, the wall material will require some minor repair.

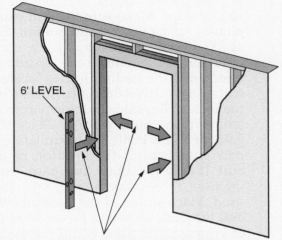

CHECK PLUMB IN TWO PLANES ON BOTH SIDES OF DOOR FRAMING WITH A LONG LEVEL

CHECKING PLUMB AND SQUARE OF WOOD DOOR FRAMING

27208-13_SA02.EPS

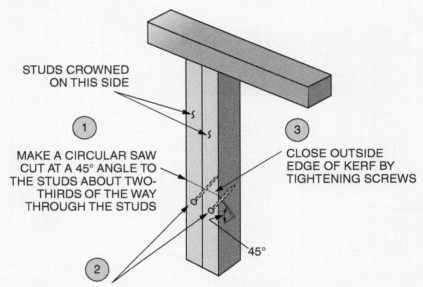

CORRECTING A CROWN IN THE KING STUD AND TRIMMER

27208-13_SA04.EPS

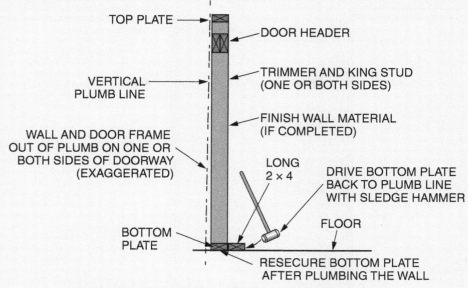
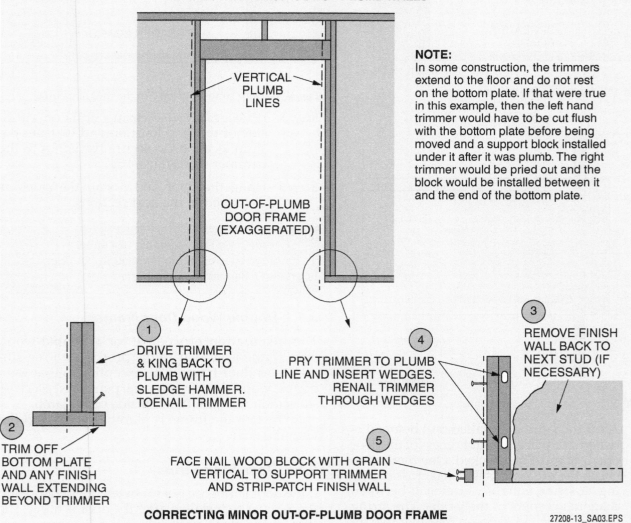

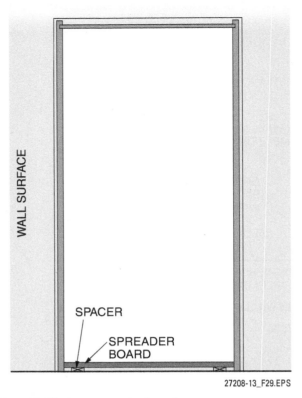

Figure 29 Placing a spreader board.

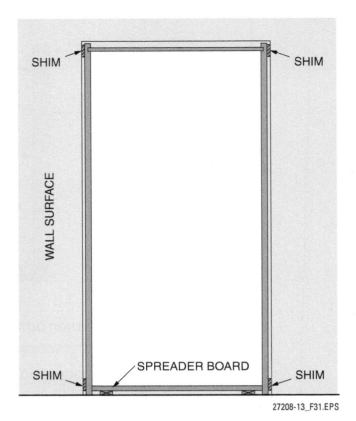

Figure 31 Door jambs shimmed in the rough opening.

Step 8 If necessary, use casing nails on either side of the stop location and ½" from the shim locations to secure the jambs to the trimmer studs (*Figure 34*).

Step 9 Hang the door and install the stops as described in the next sections.

> **NOTE:** Many premachined doors use equal hinge spacing at top and bottom for door interchangeability.

3.1.1 Double Wood Door Frames

The installation procedure for a double wood door frame is similar to the procedure for a single wood door frame. The exception is that for a double wood door frame, a carpenter uses a cross story (double-X) method outside of the jamb.

Be certain the frame is plumb, square, and flat with the wall.

3.1.2 Door-Stop Strips

If the door jambs are not **rabbeted**, loose door-stop strips must be installed after the door is hung. The stop on the hinge-jamb side of the door must have a $\frac{1}{16}$" clearance between it and the door, as shown in *Figure 35*. The stop on the lock-jamb side of the

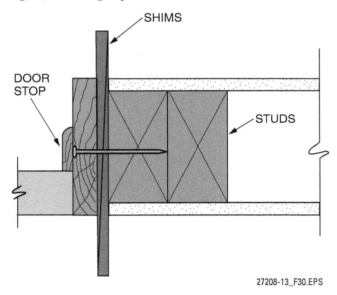

Figure 30 Plan section at the side jamb.

Step 7 After the jamb is installed, but before it is nailed into place, the squareness of the door opening must be checked. One method to check the squareness is called cross stringing. A string is tightly stretched between diagonal corners of the jambs. The string is then stretched between the other two diagonal corners. If the distances are equal, the jamb is square.

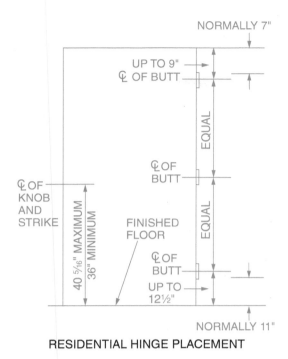
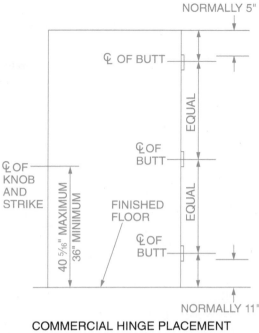

Figure 32 Elevation of door showing hardware locations.

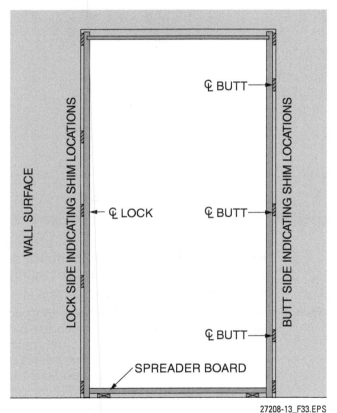

Figure 33 Elevation of door jambs showing shim locations.

door must fit flush, so the face of the door is flush with the face edge of the jamb. The stop on the head jamb of the door should line up with both the side-jamb stops. Note that by cutting the bottom of the stops on the vertical jambs at a 45-degree angle, a sanitary stop can be created. To fasten the door stops to the jambs, use 4d finish nails.

3.2.0 Metal Door Frames

Metal door frames are used extensively in commercial and institutional applications. *Figures 36* and *37* show some of the styles and types of typical metal door frames.

The two basic types of metal door frames are a pre-assembled, welded unit and an unassembled or knockdown (KD) unit. Either type is available in fixed wall widths or adjustable wall widths. The fixed widths typically range from 2½" to 7½" in ⅛" increments. Adjustable units are available to accommodate wall thicknesses ranging from 3¾" to 9¼". Some frames have manufactured and finished snap-on casings made of wood, metal, or synthetic materials. Others can be obtained with sanitary door stops.

Any type of wood door or a matched metal door can be attached to the frames, and the frames can be fastened to metal structural studs, wood studs, or masonry. This section will cover the typical installation of one type of assembled and unassembled metal frame in wood or masonry construction.

Clinching a Hinge-Side Trimmer

For heavy doors such as wide solid-wood entrance doors or French doors, some carpenters clinch the hinge-side trimmer to the king stud at the top and bottom to prevent pulling or twisting the trimmer loose from the king stud. Clinching was and still is accomplished using nails on both sides of the trimmer; however, metal straps are easier to install. If the finish wall material has been installed, a small piece at each clinch point will have to be removed and repaired after the clinching is completed.

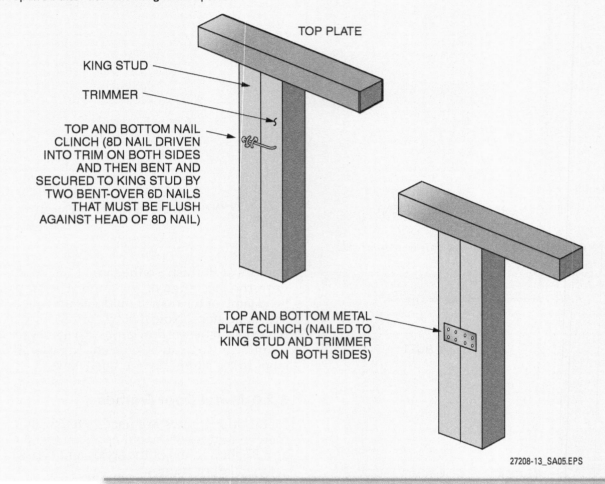

3.2.1 Installing Welded Metal Door Frames in Wood-Framed Construction

Installing a welded metal door frame in wood stud construction is not a difficult task if the following steps are completed. Usually, the metal frame will have wood stud anchors welded to it. Some wood stud anchors will be furnished loose, as shown in *Figure 38*. In some cases, reinforced closer sleeves will be required if the reinforcements have not been welded into the frame.

The loose anchors are installed inside this particular metal frame by forcing or expanding the anchors between the two casing flanges of the frame.

Step 1 Place the anchor into position at an angle and twist it into a horizontal position, as shown in *Figure 39*. If required, place a closer sleeve inside the head jamb frame at the proper location. It should fit tightly to prevent movement.

Step 2 The metal frame may have been shipped using a temporary metal spreader bar that must be removed. Replace the metal spreader bar with a wood spreader bar at the sill. Cut notches in the ends of the wood spreader for the stops (*Figure 40*). Make sure that the spacing of the jambs at the sill is the same as at the top of the frame.

Step 3 Check the head jamb for level and shim the bottom of the side jambs as required, subject to the manufacturer's recommen-

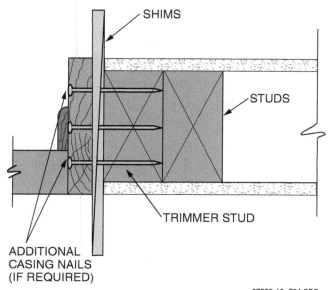

Figure 34 Plan view of additional jamb nailing.

dations. Check the location of the metal frame again, and using an appropriate fastener, attach the floor anchors (*Figure 41*).

Step 4 The bottom of the door frame must be attached to the floor or the framed opening per the manufacturer's instructions or local code requirements. Add a wood spreader at the center (*Figure 42*).

Step 5 Using an accurate level, plumb the jambs and make sure all studs are square and true with the metal frame. Bend the jamb anchors around the wood studs, nail them in place, and recheck for plumb and squareness (*Figure 43*). Install a wood door or matched metal door.

3.2.2 Sound Attenuation and Grouting for Metal Frames

Check the jamb details, specifications, and fire codes to see if grouting is required. Grout is in-

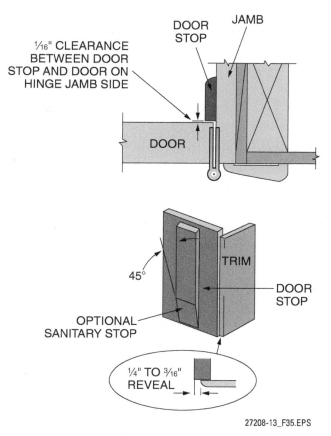

Figure 35 View of a door jamb and door.

stalled in frames to provide sound attenuation and/or fireproofing. Grout may be either a sand and cement mixture or perlited gypsum, per the specifications. Some frames may be grouted prior to installation; other frames may have to be grouted in place. Any attachment clips that are necessary must be installed in the door frame prior to grouting. All butt plates must have duct tape placed over them on the inside, and Styrofoam™ must be placed inside at the approximate location of the door closer so screw holes can be drilled and tapped in that location. If possible and necessary, install screws for weather stripping so you can back them out and install the weather stripping later.

The Importance of a Plumb and Level Metal Door Frame

Care must be taken to ensure that a metal frame is absolutely plumb and square. An improper frame installation will affect the performance of the door and make fitting the door difficult or impossible. A little extra time spent installing the frame will make door installation much easier.

Plumbing a metal door frame can be easily accomplished with a magnetic plumb bob device. The magnetic bed of the device is attached to the frame. A short arm extending from the base supports the string from the plumb bob. The plumb bob is allowed to hang from the arm and measurements are taken from the frame to the string at the top and between the frame and tip of the plumb bob at the bottom. If the two measurements are the same, the frame is plumb.

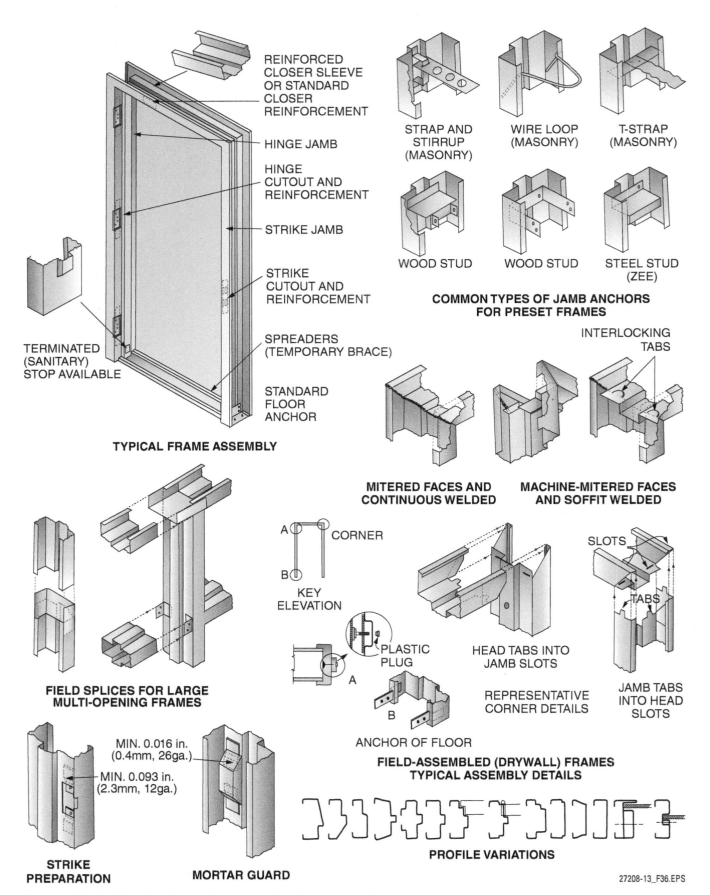

Figure 36 Typical fixed-width, single-piece jamb and casing metal frames.

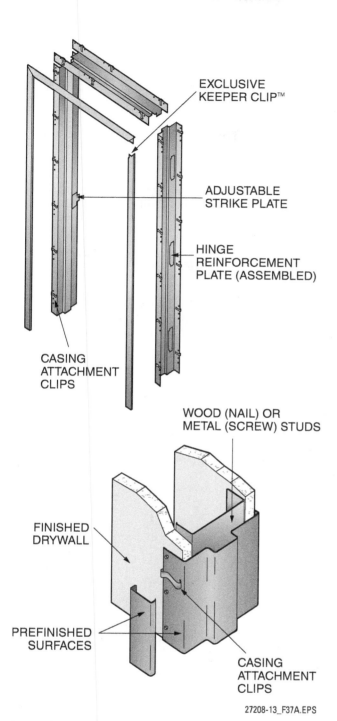

Figure 37 Typical fixed-width or adjustable-width metal frame with snap-on casing (1 of 2).

3.2.3 Installing Welded Metal Door Frames in Steel-Framed Construction

Installing a welded metal door frame in steel-framed construction is done in much the same manner as with wood studs. Steel stud anchors are furnished with the metal door frame. Some anchors are welded to the frame and some are loose, as shown in *Figure 44*.

Step 1 The loose anchors are installed inside the metal frame by forcing the anchor between the two casing flanges of the frame. Place the anchor into position at an angle and twist it into a horizontal position, as shown in *Figure 45*. If required, place a closer sleeve inside the head jamb frame at the proper location. It should fit tightly to prevent movement.

Step 2 Move the assembled metal door frame into position within the correct location of the metal stud partition, as shown in *Figure 46*. The metal frame may have been shipped using a temporary metal spreader bar. This metal spreader bar must be removed. Replace the metal spreader bar with a wood spreader bar at the sill. Cut out notches for the stops in the wood spreader (*Figure 47*). Ensure that the spacing of the jambs at the sill is the same as at the top of the frame.

Step 3 Check the head jamb for level and shim as required, subject to the frame manufacturer's recommendations. Check the location of the metal frame again and fasten the floor anchor, as shown in *Figure 48* and in accordance with local code requirements.

Step 4 Add a wood spreader and vertical support at the center of the frame (*Figure 49*).

Step 5 Using an accurate level, plumb the jambs and make sure all the studs are square and true with the metal frame. Magnetic levels are available for metal door frames.

Metal Door Frame Support

Be sure to use at least three spreaders spaced equally within a metal frame when grouting a frame that has been installed. Make sure that the spreaders are accurately cut to maintain adequate door clearance.

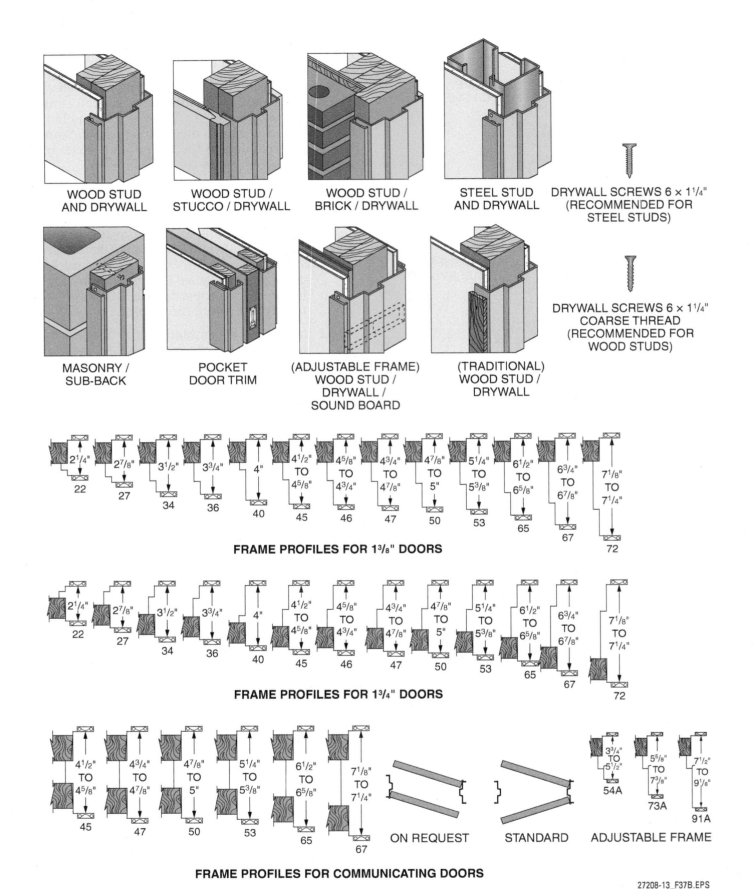

Figure 37 Typical fixed-width or adjustable-width metal frame with snap-on casing (2 of 2).

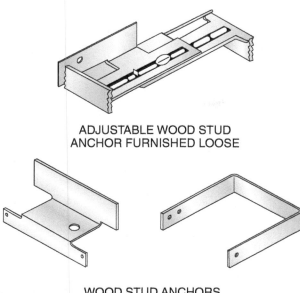

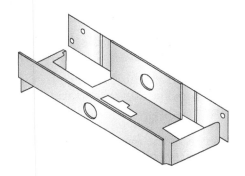

Figure 38 Wood stud wall anchors.

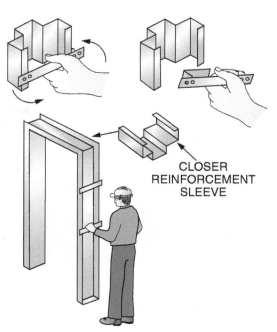

Figure 39 Install anchors and closer reinforcement sleeve.

Figure 40 Replace metal spreader bar with a wood spreader.

Checking Squareness of a Metal Frame

Check for accuracy before and during the process. One way to check squareness is by measuring diagonally from top to bottom at the corners. Equal measurements indicate that the frame is square.

Figure 41 Fasten a floor anchor.

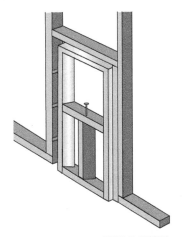

Figure 42 Install the spreader.

Care should be taken to ensure that the frame is absolutely plumb and square. An improper frame installation will affect the performance of the door in the opening. Attach the steel studs to the anchors, as shown in *Figure 50*, in which an open stud is shown being wired to an anchor and a closed stud is shown being screwed to an anchor. This will complete the frame installation. A wood door or a matched metal door is installed after finishing work is complete.

Figure 43 Bend the jamb anchors around the wood studs.

> **NOTE**
> Knockdown frames are usually installed after drywall is installed.

3.2.4 Installing Unassembled Metal Door Frames in Drywall Construction

An unassembled metal door frame is shipped to the job site knocked down and must be installed in the opening one piece at a time, so do not assemble the frame. The wood stud opening must be 1" higher than the metal door and 2" wider, giving it a 1" clearance on the three sides, as shown in *Figure 51*.

Step 1 Attach the sill anchors (*Figure 52*) to the bottom of each jamb. Some frame designs are furnished with a screw hole at the base of each jamb in lieu of the standard base anchor, as shown in *Figure 53*. The screws are usually provided only when the frame is factory finished so they will match the finish.

Step 2 Remove the hinge jamb from the knockdown frame parts and slip the jamb into position over the wall, as shown in *Figure 54*. Hold the top in place, then push the bottom in so that it moves toward and over the wall. Check the plans to be sure that the hinge jamb is on the correct side. If a closer will be required and the head

Using a Door Buck

A door buck is a cross braced open frame that is the same size as the finish door. Use a door buck in the frame opening often to check that the door fits during the metal frame installation.

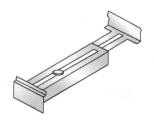

ADJUSTABLE STEEL STUD
ANCHOR FURNISHED LOOSE

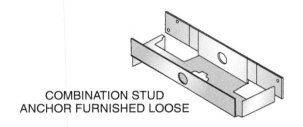

COMBINATION STUD
ANCHOR FURNISHED LOOSE

SNAP OFF FOR 2½" (63.5mm) OR
3½" (88.9mm) STEEL STUD

Z CLIP CLOSED STUD ANCHOR
WELDED TO FRAME

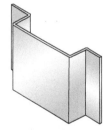

HAT CLIP CLOSED STUD
ANCHOR WELDED TO FRAME

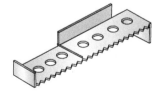

OPEN STUD ANCHOR
WELDED TO FRAME

Figure 44 Anchors.

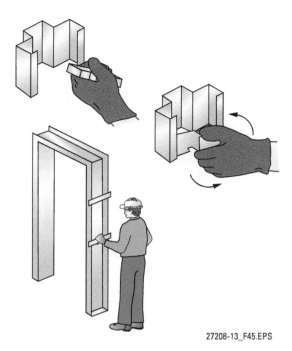

Figure 45 Place the anchor.

Figure 46 Assembled metal door frame moved into position.

jamb frame does not have reinforcement, install a closer sleeve at the proper location. Position the head jamb frame over the wall. Align the head tabs with the jamb slots, then slide the head jamb toward the hinge jamb and engage the tabs in the slots (*Figure 55*).

Step 3 The strike jamb is then slipped over the wall, as shown in *Figure 56*. Push the top of the remaining jamb over the wall and mate the jamb slots and head jamb tabs. Push the bottom of this jamb in so it moves toward and over the wall. Level the head jamb.

Figure 47 Replace the metal spreader bar.

Figure 48 Fasten the floor anchor.

Figure 49 Wood spreader installed.

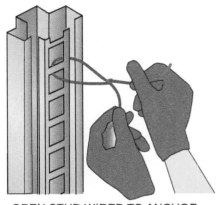

OPEN STUD WIRED TO ANCHOR

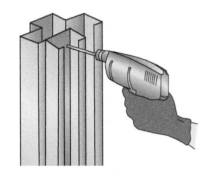

CLOSED STUD SCREWED TO ANCHOR

Figure 50 Attach the steel studs to anchors.

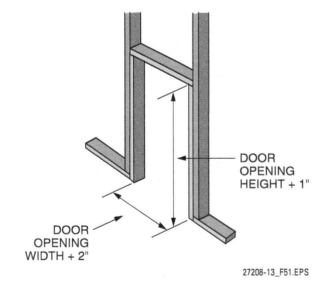

Figure 51 Door opening height and width.

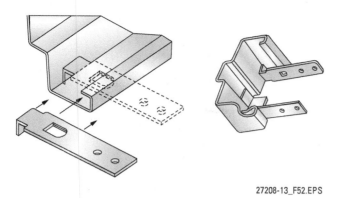

Figure 52 Attach the sill anchors.

Figure 54 Slip the hinge jamb into position.

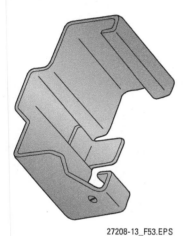

Figure 53 Alternate base anchor.

Figure 55 Slip the head jamb into position.

Step 4 At the base of each side jamb is a shimming angle to be used for vertical shimming (*Figure 57*). If necessary, level the head using these shimming angles.

Step 5 Adjust the grip-lock anchors in both jambs by turning the anchor screw counterclockwise until the anchors hit the studs (*Figure 58*). Keep turning the anchors until the metal door frame is rigid and secure.

Step 6 Plumb and square the jambs with a level. When the hinge jamb is plumb, fasten the sill anchor at the hinge jamb with nails or screws, depending on the manufacturer's instructions, as shown in *Figure 59*.

Step 7 Cut a temporary wood spreader the same width as at the top of the frame and install it between the jambs at the base (see *Figure 60*). Adjust the strike jamb to fit firmly against the spreader. Fasten the sill anchor at the strike jamb with nails or screws, depending on the manufac-

Figure 56 Slip the strike jamb into position.

Figure 57 Shimming angle.

Figure 59 Fasten the sill anchor.

Figure 58 Adjust the grip-lock anchor.

Figure 60 Wood spreader.

turer's instructions. On some Underwriters Laboratory-classified frames, the head may have to be fastened to the side jambs with two screws or 8d casing nails placed through a hole at the top corners of the head, as shown in *Figure 61*.

The metal door frame installation is now complete. Remove the temporary spreader and install a wood door or a matched metal door.

3.2.5 Installing Welded Metal Door Frames in Existing Masonry Construction

The welded metal door frame will be furnished with an appropriate quantity of one type of existing masonry wall anchor, furnished loose or welded to the frame, as shown in *Figure 62*.

Step 1 The loose anchors are installed in the metal frame by forcing the anchor between the two casing flanges of the frame. Place the anchor into position at an angle and twist it into a horizontal position, making sure the holes through the anchor are aligned with the hole in the frame stop (*Figure 63*). If required, insert a closer reinforcement sleeve inside the head jamb frame at the proper location.

> **NOTE**
> As shown in *Figure 64*, the recommended rough-opening height in an existing masonry wall should be the frame height plus ¼". The rough-opening width should be the frame width plus ½", giving the frame a ¼" clearance on the top and each side.

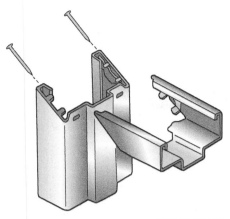

Figure 61 Head jamb fastened to the side jambs.

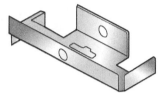

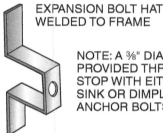

Figure 62 Masonry wall anchors.

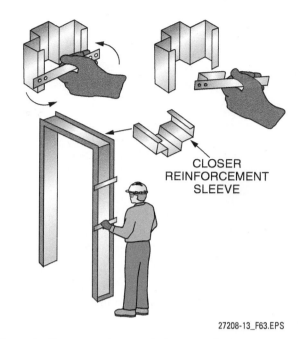

Figure 63 Placing anchors and closer reinforcement sleeve into position.

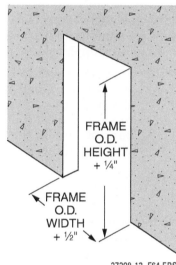

Figure 64 Rough-opening dimensions.

Step 2 Place the metal door frame in the existing rough masonry opening. The frame may have a spreader bar welded across the sill for shipping purposes. Remove this spreader bar and discard it. Place a wood spreader bar at the sill to match the width of the top of the frame. Two middle spreader bars should be installed only if the frame must be grouted in place. See *Figure 65*.

Step 3 Plumb the jambs with a level. Level the head and square the corners of the metal frame. Ensure that the frame is absolutely plumb and square. Improper frame installation will affect the performance of the door opening and closing.

Step 4 If the frame is to be installed with expansion shields and machine bolts, the location of the shields must be marked at this time and the frame removed from the wall. Using the marked locations, drill the depth and diameter hole recommended by the shield manufacturer. Install the shields, reinstall the metal frame, and fasten it into place. If expansion shields are not to be used, place a ⅜" drill bit through the ⅜" existing holes in the metal frame, and drill no less than 1⅜" into the masonry (*Figure 66*).

Step 5 Using the sleeve anchor shown in *Figure 67*, fasten the metal door frame in the opening. The installation should be complete. Remove the spreaders and install a wood door or a matched metal door.

Figure 65 Wood spreader bar and middle spreader bars.

3.2.6 Installing Welded Metal Door Frames in New Masonry Construction

The welded metal door frame will be furnished with an appropriate quantity of one type of masonry wall anchor, furnished loose or welded to the frame, as shown in *Figure 68*. If necessary, install a closer reinforcement sleeve in the head jamb frame.

Step 1 Move the welded metal door frame to the correct location. If the frame has a spreader bar welded across the sill, remove and discard it. Cut a wood spreader bar notched on the ends and place it into position at the sill (*Figure 69*).

Step 2 Check the head jamb for level and shim; adjust as required, subject to the manufacturer's recommendations. Check the location of the metal frame again and fasten the floor anchors, as shown in *Figure 70* and in accordance with code requirements.

Step 3 Place and fasten blocking on the floor. Use telescoping braces to support the frame, as shown in *Figure 71*. Using a level, plumb the jambs and check the level of the head jamb. Square the corners and add two wood spreaders in the center of the frame.

Step 4 Care should be taken to ensure that the frame is absolutely plumb, level, and square. Check it often during placement of the masonry. An improper frame installation will affect the performance of the door opening.

Figure 66 Drilling into masonry.

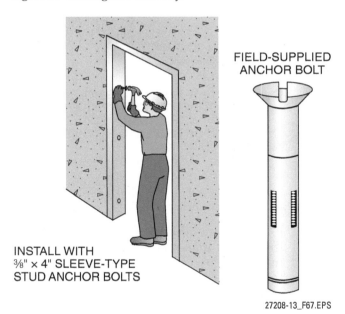

Figure 67 Using a sleeve anchor.

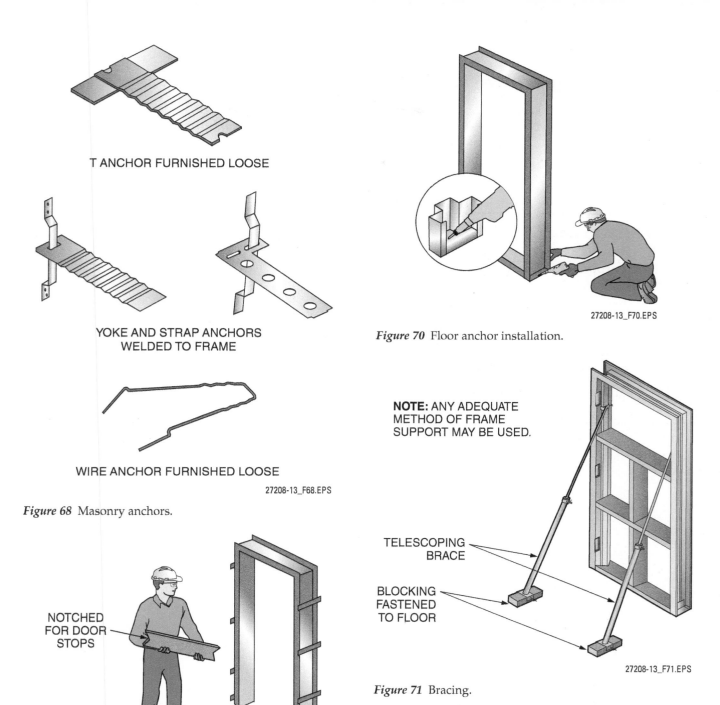

Figure 68 Masonry anchors.

Figure 70 Floor anchor installation.

Figure 69 Wood spreader bar installation.

Figure 71 Bracing.

Door Bucks for New Concrete or Masonry Construction

For concrete and masonry walls, a wood or metal door buck is used during wall construction to provide the correct-size opening for a metal door frame. Once the walls are completed, the door buck is removed and the metal door frame is installed.

Module 27208   Doors and Door Hardware   37

Completing Metal Door Frame Installation in New Masonry

Erection of a block wall around the metal door frame should complete the installation. Prepare the frame for grouting, as described earlier, then install a wood door or a matched metal door.

3.0.0 Section Review

1. The purpose of the kerfs on the back side of wood jambs is to _____.
   a. allow alignment of the head jamb with the side jambs
   b. space the nails used to fasten the head jamb to the side jambs
   c. mark the cutoff lines to be used when reducing the width of a jamb
   d. help prevent warping of the jambs

2. When grouting a metal door frame that has been installed, be sure to use at least three equally spaced _____.
   a. braces
   b. spreaders
   c. shims
   d. bucks

Section Four

4.0.0 Door Hardware

Objective

Identify the different types of door hardware.
  a. Identify the different types of door hardware used in residential applications.
  b. Identify the different types of door hardware used in commercial applications.

Trade Terms

Access: A passageway or a corridor between rooms.

Astragal: A piece of molding attached to the edge of an inactive door on a pair of double doors. It serves as a stop for the active door.

Catches: Spring bolts used to secure a door when shut.

Coordinator: A device for use with exit features to hold the active door open until the inactive door is closed.

Deadbolt: A square-head bolt in a door lock that requires a key to move it in either direction.

Dust-proof strike: A metal piece used to receive the latch bolt when a door is closed. The piece has a spring-action cover to keep particles out of the strike when the door is open.

Knob lockset: A lock for a door, with the locking cylinder located in the center of the knob.

Latch bolt: A spring-loaded bolt in a lock, with a beveled head that is retracted when hitting the strike.

Panic hardware: Hardware that provides an emergency escape exit.

Smoke gasket: A rubber strip that goes all the way around the door to keep smoke from penetrating an area in case of fire.

Sweep: A type of weather stripping. A sweep is a felt or rubber flap mounted in a metal channel to seal door bottoms to prevent air infiltration.

Threshold: A piece of wood, metal, or stone that is set between the door jamb and the bottom of a door opening.

Transom: A panel above a door that lets light and/or air into a room or is used to fill the space above a door when the ceiling heights on both sides of the door opening allow.

Door hardware is available in a wide variety of styles, sizes, and finishes. While similar basic door hardware is used for both residential and commercial construction, commercial door hardware is manufactured for greater strength and durability. Commercial door hardware also includes devices that allow ease of entering and leaving a building.

4.1.0 Residential Door Hardware

Basic residential door hardware includes door hinges and locksets. Door hinges provide a pivot point for a door within the frame, while locksets provide a means for securing a door.

4.1.1 Door Hinges

Door hinges are sometimes referred to as butts or butt hinges. The flat portions of the door hinge that contain the countersunk holes for the screws are called the leaves (*Figure 72*). The round portion in the center of the hinge that joins the leaves is called the knuckle. The knuckle is usually divided into five sections, two on one leaf and three on another. A pin running through these five knuckle sections joins the two leaves, creating a five-knuckle butt hinge.

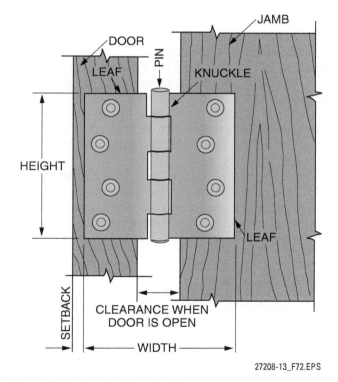

Figure 72 Loose-pin hinge (butt).

Common hinges used in residential construction include loose- and fixed-pin models. Loose-pin hinges, commonly used on interior doors, have a pin that can easily be removed, making the door installation and removal much easier. For fixed-pin hinges, the pin is nonremovable, providing more security than loose-pin hinges. Fixed-pin hinges are also called fast-joint hinges or NRP (nonremovable pin) hinges.

Prior to installing a door hinge, its size must be determined. The basic dimensions of a hinge are taken with the hinge in the open position. The two main dimensions are the width and the height, as shown in *Figure 72*. The size of a door hinge is always stated with the height first and the width last. The height of the hinge is determined by the width and thickness of the door, as shown in *Table 1*.

The width of the hinge is determined by the door thickness and the clearance required for the trim (*Figure 73*). This clearance will allow the door to open parallel to the wall and therefore not mar the door trim.

To determine the width of the hinge, use the door thickness and clearance required, as shown in *Table 2*.

4.1.2 Locksets

Locksets include the working components of the lock and the trim pieces. Trim pieces include the doorknob or lever, strike plate, and cylinder. Some types of locksets are reversible, while others are designed to be installed in one direction.

The entrance lock (*Figure 74*) is used for entrance doors where locking is required. Turning either the outside or inside knob operates the latch bolt. Both knobs can be locked or unlocked by rotating the turn button on the inside knob or by using the key in the outside knob.

The patio lock is used for entrance doors with limited entry. Turning either the outside or inside knob operates the latch bolt. Both knobs can be locked or unlocked by rotating the turn button on the inside knob.

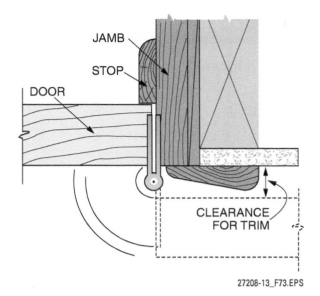

Figure 73 Plan-view section of a door jamb and door.

The passage latch is used for doors that do not require locking, such as a closet door. Turning either knob operates the latch bolt at all times.

The privacy lock is used primarily for bedrooms and bathrooms. Turning either knob operates the latch bolt, and rotating the turn button on the inside knob locks the outside knob. The inside knob is always active. An emergency release knob is used as an outside knob in case of accidental locking. Some of these locksets are furnished with different inside and outside finishes.

Dummy trim is a nonoperational knob used for decorative trim, as shown in *Figure 74*. It may be used on a door that operates but is held in a closed position with magnetic catches instead of a latch bolt. It may also be used on the nonoperating half of a pair of doors to balance the hardware.

Table 1 Hinge Height

Door Thickness	Door Width	Hinge Height
1⅜"	To 36"	3½"
1¾"	To 36"	4"
	36" to 41"	4½"
2", 2¼", 2½"	42" to 48"	4½" Extra Heavy
	To 42"	5"
	Over 42"	6"

Table 2 Hinge Width

Door Thickness	Clearance Required	Hinge Width
1⅜"	1¼"	3½"
	1¾"	4"
1¾"	1"	4"
	1½"	4½"
	2"	5"
	3"	6"
2"	1"	4½"
	1½"	5"
	2½"	6"
2¼"	1"	5"
	2"	6"
2½"	¾"	5"
	1¾"	6"

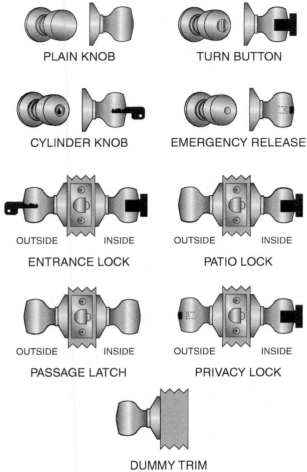

Figure 74 Basic knob-type locksets and knobs.

4.2.0 Commercial Hardware

A wide variety of commercial door hardware is available to fit the needs of commercial structures such as hospitals, schools, and shopping malls. In addition to the basic door hardware, such as hinges and locksets, specialized door hardware such as touch bars or door coordinators may be required.

4.2.1 Hinges

For commercial and institutional doors, three hinges per door are typically specified to accommodate the additional size and weight of the doors. Extra-heavy, high-frequency hinges should be used on doors that have high traffic patterns. This type of door is usually specified for commercial buildings such as an entrance to a department store or mall. Examples of institutional doors that receive high-frequency usage are school entrance or school lavatory doors.

On some jobs, such as a hospital, doors may be quite wide and heavy, and will require a ball-bearing hinge. A hinge of this type is permanently lubricated and has a fast joint. The tip of the hinge is usually rounded to avoid snagging clothing, etc., as shown in *Figure 75*.

4.2.2 Interior Locksets

A wide variety of locksets and other door hardware is available. Besides the basic knob locksets shown in *Figure 74*, there are lever-handle locksets and deadbolts with and without activation devices

Understanding knob functions will help in understanding lockset functions. The first knob function is the plain knob, as shown in *Figure 74*. It can be rotated clockwise or counterclockwise, and the operation of the latch bolt is its only function.

The second knob function is the turn button. The knob function is the same as that of the plain knob except that by using the turn button, the opposite knob or both knobs can be locked, depending on the lockset function.

The third knob function is the cylinder knob. The knob function is the same as the plain knob except that by using the correct key, the unit can be locked.

The fourth knob function is the emergency release. This knob function is the same as the plain knob except when an accidental locking occurs (such as a child turning the turn button on the room side of the lock), the latch bolt can be released by inserting a special key or pin in the hole of the outside knob.

Figure 75 Ball-bearing hospital hinge with a rounded tip.

operated by push buttons, a magnetic pass card, or a remote (*Figure 76*). Many of the activation device locksets are used in commercial and institutional buildings such as hotels and hospitals.

Mortise locksets are very expensive, extremely well made, and available in many styles and finishes. They are usually only used in high-profile construction, such as churches and some offices, and it takes an experienced finish carpenter to install them. A cut-away of a mortise lockset is shown in *Figure 77*.

The keying of door locks is necessary in large office buildings, hotels, or motels. This is done so one key will open a number of doors in a given location. It eliminates the need for a person to carry more keys than needed.

The following terminology should explain most keying methods used in construction today. Most of the same principles can be applied to magnetic pass card locks.

There are several types of master key systems, as outlined below:

- *Simple master key systems* – Each lock has its own key, which will not operate any other lock in the system, but all locks in the system can be operated by the master key.
- *Grand master systems* – Each lock has its own individual key, as in the simple master key system. The locks are divided into two or more groups, with each group being operated by a master key, and all locks in the system can be operated by the grand master key.
- *Great grand-master systems* – Each lock has its own individual key, as in the simple master key system. The locks are divided into additional

Figure 77 Cut-away view of a typical mortise lockset.

subgroups as needed, including a master key for each subgroup and a grand master key for each group. All locks in the system can then be operated by a great grand-master key.

Conventional cylinders for all locksets can be provided with construction keying for use by architect and contractor personnel during building construction. A special breakout key is used to permanently void the construction key. This eliminates further use of the construction key.

Common hotel keying subgroups are as follows:

- The guest's key operates only the lock or locks of one room or suite.
- The maid's master key operates one group of locks, generally the guest-room entrances and linen closets on one floor served by one maid.
- The housekeeper's master key operates a number of maids' groups, generally the entire guest-room portion.
- The emergency/shutout key operates all guest-room locks even when they are locked from the inside. It is also a shutout key, locking a guest room so it cannot be opened by any other keys in the system except the individual display key.
- The display key locks a guest-room door against all other keys except the emergency shutout key. This key is used for guest rooms when the rooms are used as sample rooms or when extra security or privacy is required.
- The grand master key operates all locks in the hotel. The grand master key does not operate as an emergency/shutout key.

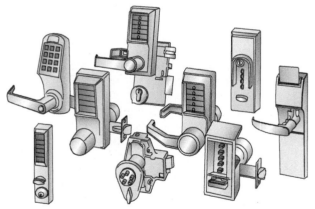

Figure 76 Various lever-handle locksets with activation devices.

Commercial/Institutional Mortise Deadbolts

Heavy-duty armored-front auxiliary mortise deadbolts are available for use on metal or wood doors. Armored-front mortise locksets with deadbolts are also available.

4.2.3 External Door Stops, Door Holders, and Door Closers

An external door stop is a rubber-tipped device fastened near the bottom of the door, the wall base, or the floor that prevents the door from striking the wall when the door is fully opened (*Figure 78*). One type of door holder has a plunger device, which is released by foot pressure. The spring holds the rubber tip to the floor. Another type of door holder is screwed into the door. This type wedges the door open when the lever is dropped to the floor.

Many building codes require the use of magnetic door holders such as the one shown on the right in *Figure 78* to control hallway access doors in office buildings and apartment houses. These devices contain an electromagnet that is wired into the building fire alarm system. Under normal conditions, the magnets hold the door open. If there is a fire alarm, however, the electrical power to the magnet is automatically turned off, and any open doors will close to inhibit the passage of fire and smoke.

Prior to installing a door closer, the hand or swing of the door must be determined. Closer hardware is available for regular arm installation on the pull side of the door only (*Figure 79*). It is also available for regular or parallel arm installation on the push side (transom bar or top jamb) of the door only (*Figure 80*). Head-frame mounting provides leverage and power to control exceptionally wide doors or doors subject to pressure problems.

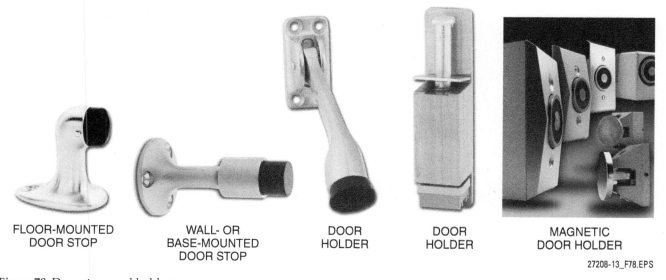

Figure 78 Door stops and holders.

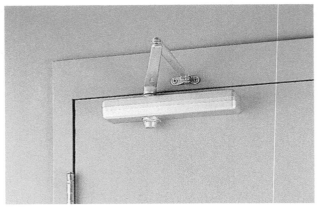

Figure 79 Regular arm installation on the pull side only (door mounted).

Concealed door closers may be used on interior doors where appearance is important. *Figure 81* shows a narrow concealed closer for use in an overhead transom. As shown in *Figure 82*, this concealed closer is mounted inside the head frame, and the track for the arm is in the top of the door. *Figure 83* shows an assembly in which the closer is mounted inside the top of the door and the track is mounted in the head frame.

Manufacturers offer a wide variety of door closers, some of which are nonhanded to permit installation on doors of either hand. Some of the closers can be mounted in different ways. Closers are available in a variety of finishes to complement other door hardware.

New closers offer spring-power adjustment, enabling fine tuning of the closer's power. Many manufacturers also offer closers that can be adjusted to independently regulate closing and latch speed.

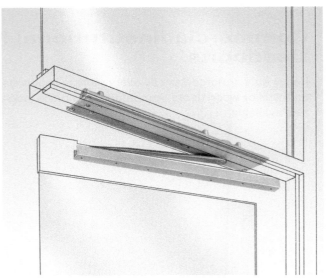

Figure 81 Concealed closer with soffit plate.

4.2.4 Security Hardware

The following sections cover various locking devices and accessories used for entry doors and gates. The electrically operated devices covered are all 12VDC (volts direct current) or 24VDC units.

Figure 84 shows electric strikes, along with a typical local power module. Electric strikes are very common and provide a remote release of a locked door without requiring the retraction of a latch bolt. They are available as fail-safe or fail-secure. Fail-secure means that on loss of power, the strike remains locked. Fail-safe means that on loss of power, the strike opens. With the possible exception of prisons or mental institutions, most local jurisdictions require a locking device on an exterior exit door to fail-safe upon a fire alarm, sprinkler alarm, or loss

TOP JAMB CLOSER

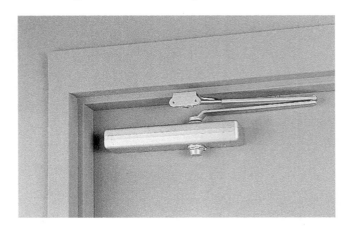

PARALLEL ARM CLOSER

Figure 80 Regular or parallel arm installation on the push side only (jamb mounted).

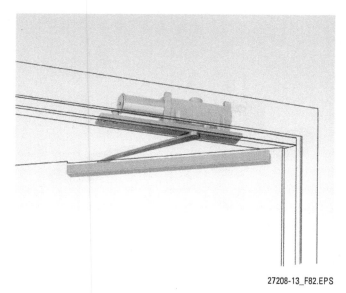

Figure 82 Concealed closer.

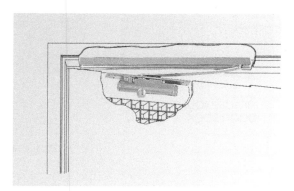

Figure 83 Arm and track fully concealed.

Figure 84 Typical electrical strikes.

of alternating current (AC) power. For interior fire doors or stairway doors, to prevent the spread of fire and smoke, most codes prohibit the use of fail-safe strikes. In these applications, electric locks or latches must be used to allow the doors to remain latched when they are electrically unlocked.

Most electric strikes can be mounted on either side of the door frame for a right-hand or left-hand opening door. With these devices, a portion of the outer edge of the door frame must be removed to accommodate the back box that allows the strike lip to swing open. This can weaken the door frame, allowing easier penetration.

Electric bolt locks (*Figure 85*) are an alternative to electric strikes or magnetic locks because the bolting device that locks a door or gate is mounted on the top and/or sides of the door frame. The door itself has no latch. Multiple electric bolt locks can be used on a door to provide very strong security. The locks are available as fail-secure or fail-safe units. In most cases, neither can be used on exterior exit doors because of code restrictions.

Some of these devices are designed to fit in narrow door frames and do not require removal of a portion of the door-frame edge. Others are designed for surface mount or for use as sliding or swinging gate locks.

The primary use for electric locksets (*Figure 86*), also called electric latches, is in stairway fire doors on each floor of a building. Building codes generally require that stairway fire doors not be locked on the stair side unless they may be remotely unlocked without unlatching. Electric locksets provide the required locking, unlocking, and latching features. While providing controlled access and remote unlocking capability, the doors stay latched even when unlocked, maintaining fire door integrity. Because these locksets are mounted in the door, power transfer hinges like those shown in *Figure 86* are required. The hinges are available with 2-, 4-, and 10-wire conductors that are protected on the inner face of the hinge.

Electromagnetic locks (*Figure 87*) are fail-safe and can be used on interior doors and exterior exit doors. However, they cannot be used on interior or stairway fire doors. They have no moving parts and are not subject to wear. Electromagnetic locks are available as direct-hold and shear-hold (concealed) styles. The direct-hold styles are graded for use by ANSI (American National Standards Institute) as follows:

- *Grade 1* – 1,650 pounds direct holding force, medium security
- *Grade 2* – 1,200 pounds direct holding force, light security
- *Grade 3* – 650 pounds direct holding force, door holding only

EXTRA HEAVY-DUTY COMMERCIAL/INDUSTRIAL GRADE ELECTRIC BOLT LOCK

SPACESAVER CONCEALED NARROW ELECTRIC BOLT LOCK

CONCEALED DIRECT THROW DESIGN ELECTRIC BOLT LOCK

GATE LOCK

SURFACE MOUNT DOOR LOCK

Figure 85 Electric bolt locks.

FRAME ACTUATOR CONTROLLED MORTISE LOCKSET — ELECTRIC MORTISE LOCKSET

ELECTRIC CYLINDRICAL LOCKSET

ELECTRIC POWER TRANSFER HINGE

Figure 86 Electric locksets.

There are some electromagnetic locks with 2,000 pounds or more of direct holding force. These locks will stay joined even when the door they secure is destroyed. Shear types have holding forces of 2,700 pounds, but they are rated as only grade 1 because of the 90-degree pulling angle. Most electromagnetic locks have integral door position switches to indicate that the door is locked and secure. The shear locks have relock delay timers activated by the position switch so that the door is at rest before the lock reactivates.

In place of exit switches or readers, touch-sensitive bars and handles (*Figure 88*) and switch bars are commonly used to turn off electromagnetic locks. The touch-sensitive bars and handles are capacitive touch-sensitive switches and have no moving parts. The switch bars have mechanical switches. Some touch-sensitive bars have electronic timers that delay de-energizing an emergency exit for a set amount of time. Armored cables are used to connect the bars or handles to the hinge side of the door frame, where the wir-

Figure 87 Electromagnetic locks.

ing is routed to a controller or to the electromagnetic lock. These touch bars/handles are part of a group of exit devices that are sometimes called request-to-exit (RTE or REX) devices.

Delayed-exit alert locks (*Figure 89*), sometimes called RTE or REX locks, legally delay an exit through exterior exit doors in secure facilities. (By law, exit doors in other public facilities may not be locked or delayed.) When an exit is attempted, an alarm is sounded and a signal is sent to guards for closed-circuit television (CCTV) or physical monitoring purposes. After 15 seconds, the exit door is unlocked, permitting an exit. A signal from a fire alarm system can also release the lock, allowing unrestricted exits during an emergency.

4.2.5 Weather Stripping

Weather stripping is the application of materials in spaces between the outer edges of doors and windows and the finished frames to keep out air or moisture that would otherwise enter.

There are a variety of methods for achieving this purpose on doors. *Figure 90* shows a rubber fixed-bottom sweep, which is one way to prevent air infiltration.

Other forms of weather stripping are an interlocking threshold, as shown in *Figure 91*; a vinyl bulb that compresses when the door is closed, as shown in *Figure 92*; and an automatic door bottom, as shown in *Figure 93*. Common forms of weather stripping might also include a spring metal V-strip, a wood-backed foam-rubber strip, or rolled vinyl.

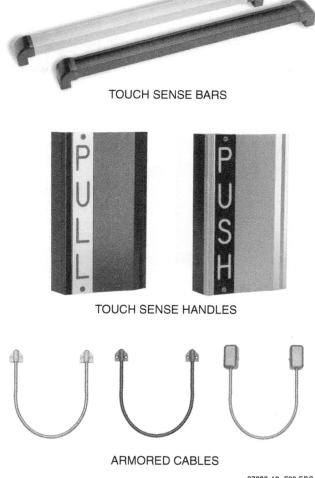

Figure 88 Touch-sensitive bars and handles, and armored cables.

4.2.6 Thresholds

A threshold is a piece of wood, metal, or stone that is set between the door jamb and the bottom of a door opening. An example of an aluminum threshold is shown in *Figure 94*, along with stop-strip weather seals. The threshold must be set in a bead of caulk to keep air from getting under the door.

4.2.7 Touch-Bar or Crossbar Hardware

Touch-bar or crossbar hardware, sometimes called panic hardware, provides secure locking of single or double doors in one direction (from the opposite side of the touch bar), but allows easy passage in the other direction (from the touch-bar side). Several types are shown in *Figures 95* and

Figure 89 Delayed-exit alert locks.

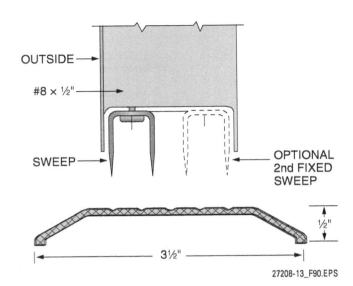

Figure 90 Rubber fixed-bottom sweep.

96. Touch-bar and crossbar hardware is available with or without an opposite side latch and release lock. Touch bars are primarily used on emergency or normal exit doors where local codes prohibit the locking of the doors on the touch-bar side at any time. They are also used on internal doors of buildings to control access to an area and yet allow rapid egress from the area. Touch-bar or crossbar hardware is usually installed in commercial and institutional buildings such as stores, hospitals, schools, and government buildings.

4.2.8 Flush Bolts

A flush bolt is a sliding bolt mechanism that is mortised into the door at the top and/or bottom edge. It is used to hold an inactive door in a fixed position on a pair of double doors. For automatic and semi-automatic flush bolts on metal, composite, and wood doors, a door coordinator is required. An automatic flush bolt retracts without manual actuation. A semi-automatic flush bolt engages a door latch when an inactive door closes without any use of a triggering mechanism. The bolt remains extended until it is retracted by manual release of the bolt-actuating lever.

Electric Plunger Strike

This type of electric strike is activated in the same manner as electric lip strikes. However, instead of a lip, the device uses a motor-driven plunger to depress the latch bolt and unlock the door. When de-energized, the plunger retracts. When the door closes, the beveled latch bolt rides over the strike plate and falls into the latch pocket. The advantage of this device is that it does not require removing part of the outer frame edge to accommodate the strike.

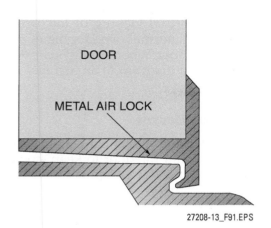

Figure 91 Interlocking threshold.

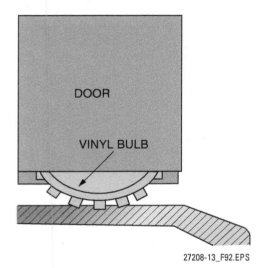

Figure 92 Vinyl bulb.

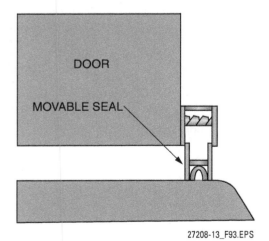

Figure 93 Automatic door bottom.

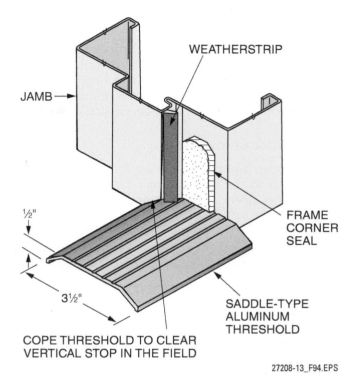

Figure 94 Aluminum threshold.

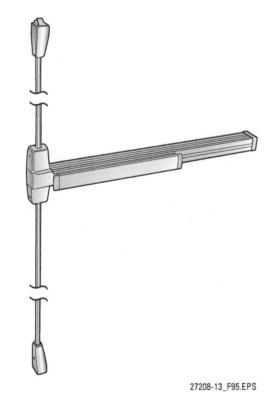

Figure 95 Surface-mounted, double-latch touch-bar device.

Module 27208   Doors and Door Hardware   49

4.2.9 Dust-Proof Strike

A **dust-proof strike** is a metal piece used to receive the latch bolt when the door is closed. The piece has a spring-action cover to keep particles out of the strike when the door is open. *Figure 97* shows two examples of dust-proof strikes.

4.2.10 Door Coordinator

A door coordinator is a device for use on double doors to hold the active door open until the inactive door is closed. This trip action then gives the active door the function of a single-acting door, allowing the normal locking or latching of the lock bolt and the normal overlapping of an **astragal** or rabbeted door, when used (*Figure 98*). Refer to *Figure 99* for a diagram of an astragal set for a pair of doors.

4.2.11 Smoke Gasket

A **smoke gasket** is a rubber cord or strip that goes all the way around the door to keep smoke from penetrating an area in case of fire.

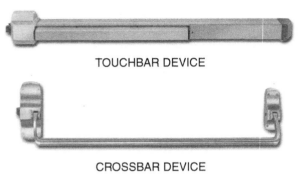

Figure 96 Surface-mounted touch-bar or crossbar device.

Figure 97 Dust-proof strikes.

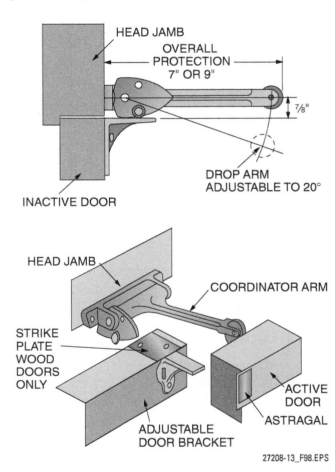

Figure 98 Door coordinator.

Alarmed Emergency-Exit Touch Bars

Alarmed emergency-exit touch bars, equipped with a large deadbolt, provide secure, alarmed, code-compliant protection for emergency-only exits. They prevent the door from being used for unauthorized facility entrance or exit. The touch bar can be armed/disarmed and opened from the inside or outside by a key using a standard rim cylinder.

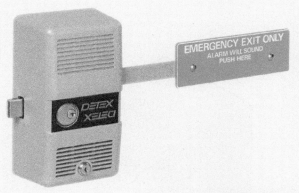

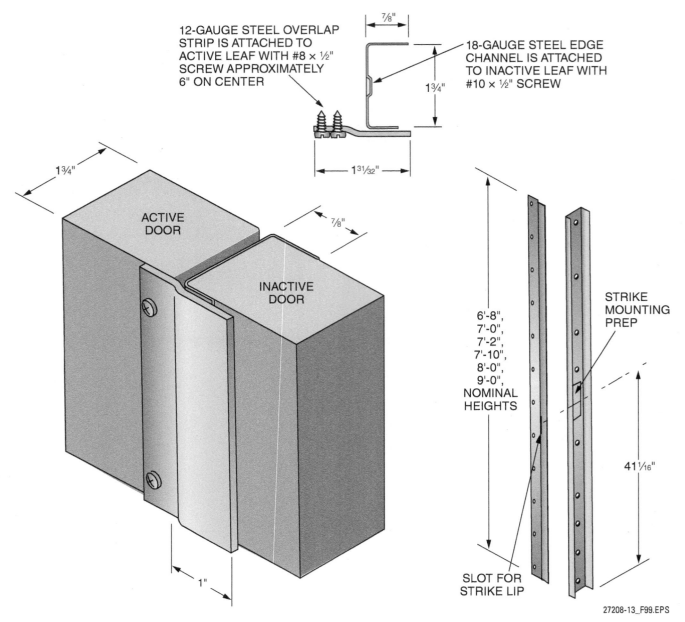

Figure 99 Diagram of an astragal set for a pair of doors.

4.0.0 Section Review

1. A latch bolt that can be released by inserting a pin in a hole in the outside knob is found in a(n) _____.

   a. emergency release knob
   b. turn button knob
   c. passage latch
   d. cylinder knob

2. Relative to an electric door lock, fail-safe means _____.

   a. backup power is available
   b. it will not fail in emergency
   c. the strike opens in the event of a power loss
   d. the strike stays locked if power is lost

Section Five

5.0.0 Installing Doors and Door Hardware

Objective

Describe the various installation techniques for doors and hardware.
a. Describe the various installation techniques for residential doors and hardware.
b. Describe the various installation techniques for commercial doors and hardware.

Performance Tasks 1 through 4

Demonstrate the proper installation of a hollow metal frame and door using the proper safety precautions.

Install a prehung door unit or door hanging system using the proper safety precautions.

Lay out and cut hinges in an instructor-selected project.

Install a door closer using the proper safety precautions.

Trade Terms

Cylindrical lockset: A lock in which the keyhole and tumbler mechanism are in a cylinder that is separate from the lock case and can be removed to change the keying of the lock.

Finishing sawhorse: A pair of trestles used to support lumber, molding, doors, stair parts, and other materials while they are being fitted and shaped for installation.

Template: A thin piece of material such as plastic or heavy paper with a shape cut out of it, or the whole of it cut to a shape on its perimeter. A template is used to transfer that shape to another object by tracing it with a pencil or scribe.

While residential and commercial doors serve the same purpose—allowing persons to enter and exit a building—the installation of residential and commercial doors can be drastically different. Due to the number of inhabitants, building codes have more stringent regulations for commercial buildings.

5.1.0 Residential Doors

Most residential doors currently being installed in new-home construction and remodeling work are prehung doors, providing ease of installation for the carpenter or homeowner. Residential doors are typically lighter weight and do not require some of the complex security hardware used in commercial construction.

5.1.1 Door Jack

A door jack is a piece of equipment that holds a door securely on edge while it is being prepared for installation. Some tool manufacturers make a metal, spring-loaded door jack; however, a simpler door jack (*Figure 100*) can be easily fabricated. The uprights that receive the door are lined with scrap carpeting to protect the door finish and provide a snug fit for the door.

5.1.2 Manufactured Prehung Door-Unit Installation

Manufactured prehung door units may be supplied unassembled, as shown in *Figure 101*, or assembled, as shown in *Figure 102*. They are available in fixed jamb widths or with adjustable width split jambs.

Unassembled prehung door units can be assembled and installed in the same manner as other assembled prehung units, or the jambs can be installed separately and the door hung after the jambs have been secured. In either case, the hinges and lockset mortises and holes are cut by the manufacturer. Most prehung door units do not have lugs at the top of the side jambs. Consequently, the head jamb is not dadoed into the side

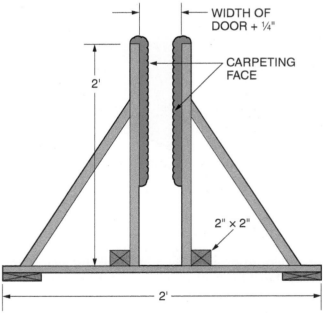

Figure 100 Door jack.

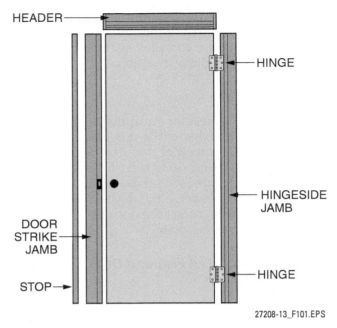

Figure 101 Typical unassembled prehung door unit.

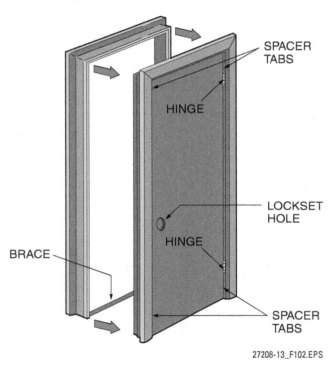

Figure 102 Typical assembled split-jamb, prehung door unit.

jambs, and as a precaution, the head jamb should be shimmed at both top corners when the door is installed, to prevent separation of the head jamb from the side jambs. Before installing an assembled unit, make sure that the lockset side of the door is not nailed in place through the back side of the strike-plate jamb. Also, make sure that any shipping braces or tie straps holding the frame together at the bottom of the door will not interfere with the installation and can be removed or cut away when the door is installed and secured in the rough opening.

The following installation guidelines are for a typical split-jamb, milled-stop, prehung door. However, always refer to the door manufacturer's instructions for specific information pertaining to any door being installed. Note that the rough opening required for prehung units is typically specified as 2" wider and 1" higher than the door size.

Step 1 Unpack the split-jamb door unit and separate the two halves. One of the halves will have the door attached to the jamb by the hinges, and the other will consist of only the assembled top and side jambs (*Figure 103*). Both halves will normally have the trim casing installed.

Step 2 Determine the clearance of the door above the floor, and if necessary, trim the bottom of the jambs and casings to obtain the correct spacing.

Step 3 Position the split-jamb half containing the door into the rough opening, and plumb the jamb by placing a level against the side of the casing on the hinge side of the jamb (*Figure 104*). Check that both jambs rest on the floor. In some cases, the jambs will not rest on the floor. It depends on the type of finished floor. If they do not and a metal door is being hung, insert shims under the jamb and casing that is off the floor. If a wood door is being hung,

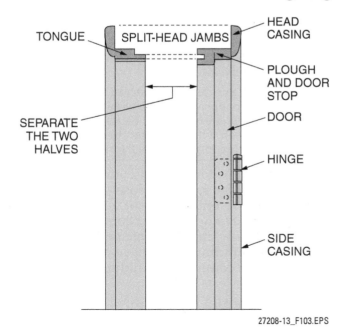

Figure 103 Unpack and separate split jambs.

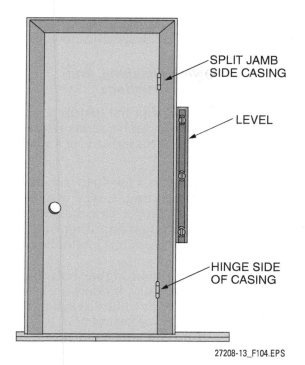

Figure 104 Position and plumb the split-jamb half with the door.

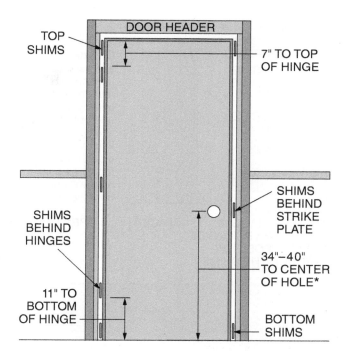

Figure 105 Shimming the door half of a split-jamb frame.

remove the door unit and trim the bottom of the appropriate jamb and casing so that both jambs will touch the floor (except on concrete). Then, replace the door unit in the opening and replumb the unit. Temporarily nail the casing to the trimmer stud at the top and bottom on the hinge side to hold the door in place. Then, temporarily nail the strike-side jamb in place after verifying that the head clearance is correct. Do not drive the nails flush.

Step 4 Move to the other side of the wall, and place shims at the top and bottom of both side jambs to hold the unit in place. Shims should span the face of the trimmer studs. Check that the jambs are plumb and re-nail casings and/or adjust the shims as necessary. Also, place shims behind each hinge location and the strike-plate location (*Figure 105*).

Step 5 Remove the spacer tabs and open the door (*Figure 106*). After checking the plumb of the hinge jamb, nail through the jamb behind the edge of the door (not the stop) to the trimmer stud. Nail above and below each hinge and at the top and bottom of the jamb. Close the door to check for proper and even spacing at the header and strike-plate jamb. If necessary, adjust the header and the strike-side jamb. Then, nail the strike-side jamb above and below the strike-plate location and at the top and bottom of the jamb. If no lugs exist on the side jambs or they were cut off, shim the top corners of the header jamb.

Step 6 Remove any tie straps or bracing from the installed half of the split-jamb frame and any bracing from the other half of the split-jamb frame.

Step 7 Slide the tongue of the second half of the split-jamb frame into the plough of the installed half of the frame (*Figure 107*). Nail the casing to the trimmer studs. Then, nail the jamb to the trimmer studs at approximately the same locations as the door half of the jamb. Do not nail through the door-stop strip.

Step 8 Complete the nailing of the casing on both sides of the door, then install the appropriate lockset and other hardware as specified.

5.2.0 Commercial Doors

When doors are delivered to the job site, they should be checked with the plans of the building to make sure they are the correct thickness,

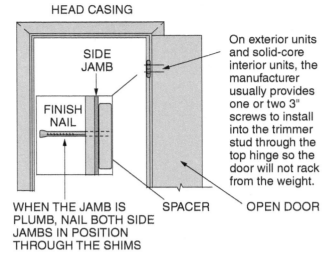

Figure 106 Fastening the jamb to the trimmer stud and removing the spacers.

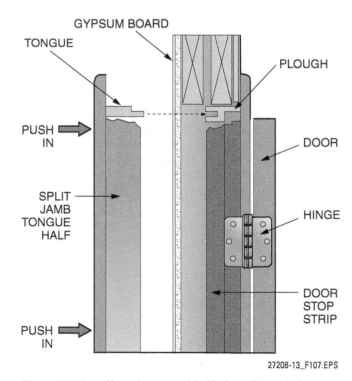

Figure 107 Installing the second half of a split-jamb frame.

width, and height. Make sure the style and type of wood correspond to the specifications.

5.2.1 Handling, Job Finishing, and Installation Instructions

The following is typical of the handling, job finishing, and installation instructions required to comply with one door manufacturer's warranty policy:

- Store the door flat on a level surface in a dry, well-ventilated building. Cover it to keep it clean, but allow for proper air circulation.
- Wear clean gloves when handling doors, and do not drag doors across one another or across other surfaces.
- Deliver doors to the building site only after any plaster or concrete is dry. If doors are stored at the job site for more than one week, the top and bottom edges should be sealed.
- Doors should not be subjected to abnormal heat, dryness, humidity, or sudden changes thereof. They should be conditioned to the average prevailing humidity of the locality before hanging.
- The utility or structural strength of the door must not be impaired when fitting the door, during the application of hardware, or when cutting and altering the door for lights, louvers, panels, or any other special details.
- Use three hinges per door on doors 7'-0" in height or less, and four hinges per door on doors over 7'-0" in height. Hinges should be set flush with the edge surfaces. Be sure that the hinges are set in a straight line to prevent distortion. Allow approximately ³⁄₁₆" clearance for swelling of the door or frame during future damp weather periods.
- Immediately after fitting, apply the appropriate weather stripping and/or threshold, and before hanging any interior or exterior door on the job, the top and bottom edges should receive two coats of paint, varnish, or sealer to prevent undue absorption of moisture.

5.2.2 Fitting Doors in Prepared Openings

After making sure the door is the correct one for the particular opening, measure the opening width and height, and compare it to the size of the door to ensure the size matches. Having determined the hinge side of the door, fit it to the hinge jamb of the opening. The butt stile is usually left square and will not interfere with the operation of the door. Some carpenters, however, will slightly bevel the butt stile even though it is not necessary. The use of a power plane to fit the door in the opening will

Handling Doors

Be very careful when handling and installing doors. The finish can be easily damaged. Make sure the door is oriented correctly before attempting to install it. Some smooth-finish doors have a definite top and bottom and should not be installed upside down.

save time and energy. If a power plane is not available, it will have to be done with a hand plane. The fore plane is usually recommended. The fore plane is 18" in length, and its size is somewhere between the larger jointer plane and the smaller jack plane.

The industry standard is 1/8" for the clearance between the door and the jamb. The clearance at the bottom of the door is determined by the floor finish and, if required, clearance for air circulation. For example, if the floor is a hardwood floor, the clearance over it should be 3/8". If the floor finish is a carpet, the clearance should be the carpet clearance plus 3/8". If there is a threshold under the door, refer to the clearance recommended by the hardware manufacturer. This information is usually included in the hardware manufacturer's installation instructions. Whatever clearance is used between the floor finish and the bottom edge of the door, it is in addition to the 3/32" clearance between the head jamb and the top edge of the door.

If the door must be trimmed at the bottom edge to fit the floor or because of clearance for some type of floor finish, the amount to be removed should be marked lightly with a pencil. Remove the door and carefully place it on carpet-covered finishing sawhorses.

The bottom edge of a hollow-core or solid-core door with veneer faces must be cut carefully so the veneer is not damaged. The damage that can occur is the feathering of the surface veneer away from the base veneers when sawing. To prevent this damage from happening, place a straightedge across the door at the pencil mark indicating the amount to be cut off. Using the straightedge as a guide, run a sharp knife across the grain of the veneer. This should be done on both sides of the door being trimmed. The knife cut will prevent the veneer from feathering or tearing. If the veneer is scored only on one side of the door, the scoring should be on the top side, with the door in a flat position when cutting with a power saw. Masking tape can also be used to reduce splintering.

Another method of cutting off the bottom edge of a door with veneer faces is with the use of a template and a power saw, as shown in *Figure 108*. Use a template if tape is not used to keep the saw off the door. The template is made of 1/4" or 1/8"

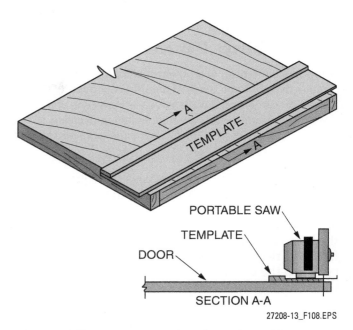

Figure 108 Door-sawing template shown in position.

tempered hardboard. The template saw guide on top of the hardboard is 3/8" thick and may be constructed of a scrap piece of lumber.

Clamp the template to the door in the correct position with the edge up from the bottom of the door by the amount to be trimmed. Run a sharp knife across the bottom of the template, scoring the veneer prior to sawing.

After sawing, place the door in a carpet-covered door jack for the planing operation if needed. Plane the edge of the door to the correct width. Put the door in the opening, and with a shim at the bottom, move the door up against the head jamb.

If the door head is square with the head jamb, nothing more must be done. If the top rail or door head is not square with the head jamb, it must be made square by first scribing it to the head jamb, then sawing and/or planing it to fit. When the door has 3/32" clearance at the head and both sides, as well as the proper clearance at the bottom edge and the floor, remove it from the opening and place it in the door jack with the hinge stile down.

Bevel the edge of the lock stile of the door by planing it to an angle of 1/8" in 2", as shown in *Figure 109*. As shown in *Figure 110*, the door swings on an arc created by a radius that has its center located in the center of the pin of the butt hinge. This center is well forward of the face of the door stop, so the lock edge of the door must be beveled to clear the face edge of the door jamb.

5.2.3 Door-Hinge Installation

The door hinge locations are determined either by the architect or the contractor on the job. The

Reducing Power- and Hand-Plane Surface Drag

To reduce the drag of a power plane or hand plane against the wood of the door, lightly coat the bottom of the plane by rubbing a block of paraffin wax over its surface. This method of reducing friction on surfaces has been used by carpenters for many years. However, use the wax sparingly if the doors must be painted or stained. The wax can interfere with the final finishing process. Paraffin wax can be obtained in the canned-goods department of any grocery store. Paraffin is inexpensive and should be a standard item carried in a carpenter's tool box.

exact hinge location varies across the country and from one prehung door manufacturer to another. Many carpenters, however, position the top of the top door hinge 7" down from the top of the door, the bottom of the bottom door hinge 11" above the finished floor, and the third door hinge centered between the other two. The clearance at the top of the door may be 1/16" or 1/8".

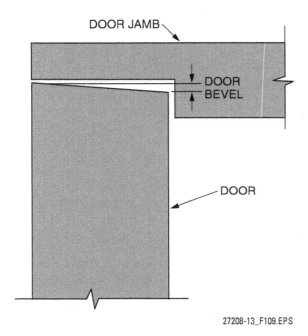

Figure 109 Section of door showing door bevel.

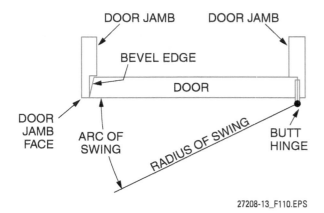

Figure 110 Plan view of door showing swing.

Step 1 Place the door in the opening, and using a wedge at the bottom, move the door upwards until a 3/32" gap is at the top. A thin piece of wood, usually a scrap piece that has been planed to the correct thickness in advance, can be used as a spacer. Next, wedge the door up against the jamb on which the hinges will be located.

Step 2 Measure and then mark the hinge locations on the jamb with a knife, and transfer those marks to the edge of the door stile. Be careful not to mark the face of the door or the jamb any deeper than the thickness of one leaf of the butts being used. Remove the door from the opening, and place it in the door jack with the stile to receive the hinges in the Up position.

> **NOTE**
> Hinge mortises may be hand-cut and -fitted or they may be cut using a template.

There are three measurements to be marked when laying out the mortises:
- Location of the door hinge on the jamb
- Location of the hinge on the door
- Thickness of the hinge leaf on both the door and the jamb

The door hinge backset should be 1/4" for doors up to 2 1/4" thick and 3/8" for doors over 2 1/4" thick, as shown in *Figure 111*. Allow 1/16" clearance for the door when nailing the stop in place.

Sealing the Bottom and Edges of a Door

Always seal the bottom and edges of the door with varnish or paint, especially after cutting it to length. Be sure not to get varnish on the door surface. The bottom must be sealed before the door is hung. Otherwise, the door will have to be removed and sealed.

Consistent Hinge Location

The hinge locations for all doors on the same job should be the same. If there are both prehung door units and doors that must be hung by hand, make sure that the hinge locations on the prehung units are duplicated throughout the job.

The door-hinge leaf location may be marked using a butt marker (*Figure 112*). The butt marker marks a perfect mortise with a minimum of time and effort. The three sides have ground edges and are heat treated. Butt markers are normally available in three sizes: 3", 3½", and 4".

Step 3 Place the butt marker on the pencil marks indicating the correct butt location, and strike it on the top with a soft-faced hammer or mallet until the cutting edges reach the correct mortise depth.

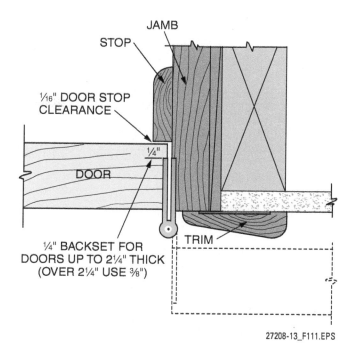

Figure 111 Hinge backset.

> **CAUTION**
> Verify what you are doing before you mortise for the hinges.

Step 4 Place the cutting edge of the chisel across the width of the wood to be removed, and score to the hinge leaf depth about every ⅛", as shown in *Figure 113*.

Step 5 Place the cutting edge of the chisel along the lower edge of the chisel cuts and remove all excess wood, as shown in *Figure 114*.

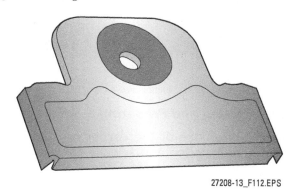

Figure 112 Butt marker.

5.2.4 Butt-Hinge Template for Doors and Jambs

Some tool manufacturers make special templates for routing butt hinges. *Figure 115* shows a template kit, along with the router and power planer needed for hinge installation. *Figure 116* shows the template in use. Accessories are available to increase the template to allow routing for three or four hinges on wood doors and jambs of 7'-0" to 8'-0" or 8'-0" to 9'-0".

> **NOTE**
> In addition to manufactured templates, job-built templates made of plywood are often used.

Reversible Hinges

Most hinges are reversible, meaning they can be used on either a left-hand or a right-hand door. If it is an NRP-type of hinge, it can be used with either end mounted in an upright position. There are, however, some hinges that are specifically manufactured for either a right-hand or left-hand door. They cannot be reversed, and caution must be exercised in selecting a hinge for either a right-hand or left-hand opening.

Module 27208   Doors and Door Hardware   59

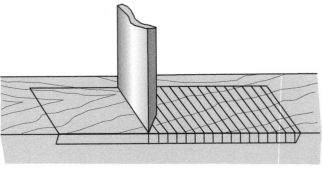

Figure 113 Scoring the wood in the mortise.

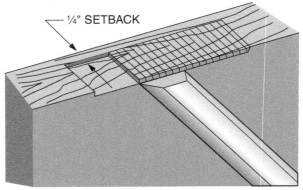

Figure 114 Removing the excess wood from the mortise.

5.2.5 Selecting and Preparing a Router

Determine the size of the corner radius on the hinges to be installed and select a corresponding bit and template guide from those listed in the template manual. To ensure that the correct-size mortise is cut, it is important to use the template guide specified for the size of the bit.

If hinges have square corners, the ⅝"-diameter bit and corresponding template guide are recommended. A corner chisel, which is used to square the corners of the cut mortise, is available as an accessory.

> **NOTE:** The following instructions for preparing and using the router are general. Read and follow the instructions in the owner's manual included with your router.

Step 1 Assemble the template guide to the router base with a locknut and tighten the lock nut securely.

Step 2 With the router unplugged, install the selected bit in the router collet and tighten it securely.

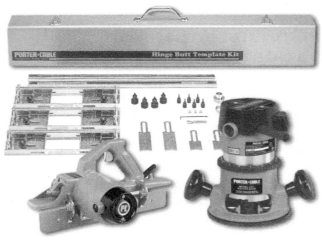

Figure 115 Template kit and tools for door hanging: router, planer, and butt-hinge templates.

Figure 116 Routing the door for hinges.

Step 3 Set the router on the butt-hinge template and adjust the depth of the cut so the bit just touches the door.

Step 4 Set the router depth-adjusting ring to the zero position.

Step 5 Lift the router from the template, and adjust the depth of the cut so that it is equal to the thickness of the hinge to be installed.

Step 6 Firmly tighten the motor locking device.

Step 7 Rout the area to receive the hinge.

5.2.6 Installing Hinges and Hanging the Door

Step 1 Remove the loose pin and place one hinge leaf in the mortise on the door. If the mortise has been properly cut, the hinge leaf may require a few light taps with a soft-faced hammer to seat it to a flush depth. Place the other half of the hinge leaf on the mortised jamb, making sure the hinges on both the door and jamb are in the correct position so the loose pins can be inserted from the top.

Step 2 Place a self-centering punch or self-centering bit in the countersunk holes of each hinge leaf. If a punch is used, gently tap the pin to mark the location of the pilot hole, as shown in *Figure 117*. The pilot hole will be drilled to receive the hinge leaf screw.

Step 3 Carefully position the door in the opening, engage the door hinge leaves, insert the hinge pins, and check the door for proper operation and clearance. The door should have a clearance of $\frac{1}{8}$" to $\frac{3}{32}$" at the top and $\frac{3}{32}$" at both sides. If a slight adjustment at the sides is required, the hinge mortises can be cut slightly deeper or the hinge can be adjusted using cardboard strips, as shown in *Figure 118*. All of this adjustment takes time, and time costs money. If care is taken from the start of the door installation, this adjustment will not be necessary. Ensure that all screws are tightened in the door hinges after adjustments are made.

5.2.7 Lockset Installation

All locksets, regardless of the manufacturer, are packaged with installation instructions and templates to mark drill locations. The installation of most exterior and interior locksets are identical, and most locks require only the drilling of two holes. The entrance lockset, the passage latch, and the privacy lockset are all shown in *Figure 119*.

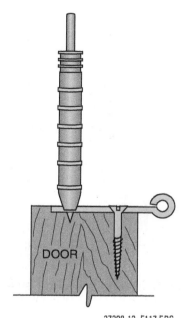

Figure 117 Self-centering screw-hole punch.

Figure 120 shows an exploded view of a typical residential tubular lockset. Most locksets consist of three interlocking parts that contribute to rapid installation and positive alignment. The primary part consists of a pre-assembled outside knob, rose, and spindle assembly. The secondary part consists of a pre-assembled inside knob and rose assembly. The last part is the latch unit that is reversible for doors opening in or out.

Step 1 Remove the lockset from the packing carton and make sure all the parts are there. Check and make sure the lockset is the correct one for the door on which you are working. Read the installation instructions carefully. Open the door to a position that will allow you to work comfortably on both sides. Place two wedges under the lock side of the door and at right angles to the face of the door, as shown in *Figure 121*. These wedges will hold the door firmly and allow safe work.

Lubrication of Hinge Screws

If hinge screws are made from a soft metal such as brass, and the door or the jambs are a hardwood such as oak or birch, some of the screw heads may twist off when installing the screws. To prevent this, rub the screw threads across a block of paraffin or beeswax, and allow some of the material to collect between the threads. The paraffin or beeswax will act as a lubricant and prevent the screws from binding while being turned in and tightened.

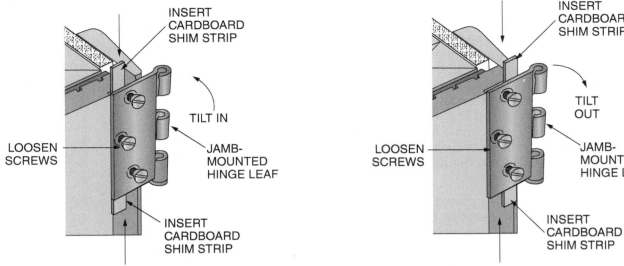

Figure 118 Adjusting the hinge-side jamb clearance using thin cardboard shim strips.

Figure 119 Entrance lockset, passage latch, and privacy lockset.

Lever Locksets

Many doors today are being equipped with lever locksets instead of knob locksets. Lever locksets comply with Americans with Disabilities Act (ADA) requirements for the physically challenged. Most of the basic locksets can be equipped with either a knob or a lever handle and function in the same manner. Like knob locksets, lever locksets are available in a variety of styles and metal finishes including chrome, brass, bronze, antique bronze or silver, and brushed-satin aluminum. Curved or stylized lever locksets must be specified based on the door hand.

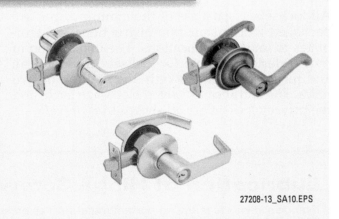

62   NCCER – *Carpentry*

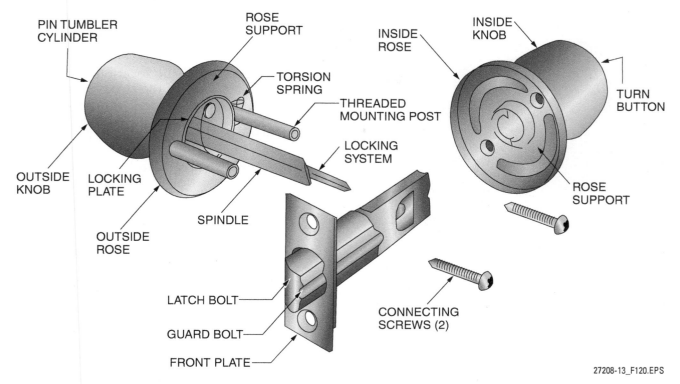

Figure 120 Exploded view of a typical tubular lockset.

Deadbolts

Most residential entrance doors are equipped with single-cylinder key-operated deadbolts in addition to an entrance lockset. In some cases, the primary lockset is not key-operated and only the deadbolt can be used to secure the door. Also available are electronically locked residential deadbolts that are remote-controlled from a key-fob transmitter and manually unlockable from inside the residence. Entrance doors with large door lights or side lights at the strike allow easy access to the turn-button release of a single-cylinder deadbolt. Such doors can be equipped with a double-cylinder key-operated deadbolt. However, some jurisdictions prohibit their use because they are a safety hazard and do not permit a quick exit from the building. Always check local codes before installing a double-cylinder deadbolt.

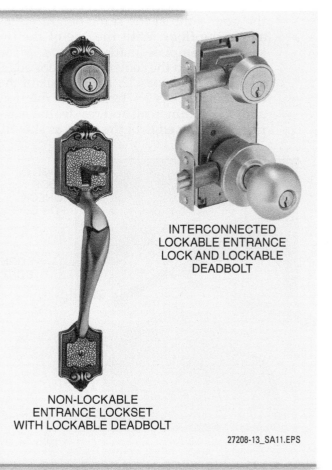

Module 27208 — Doors and Door Hardware 63

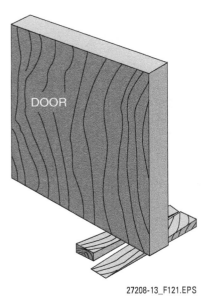

Figure 121 Wedges under the lockset edge of a door.

Step 2 Measure up from the floor. The standard height to the center of the doorknob is 38", but this may vary. Place a light pencil mark on the edge of the door. Using a square, draw a horizontal light pencil line on the edge and inside surface of the door. Also, place a light pencil mark on the door jamb at the same height of 38" above the floor. With the use of the installation template, mark the center of the cross bore and the center of the door edge, as shown in *Figure 122*. The setback of the cross bore hole is 2⅜" (residential interior) and 2¾" (commercial and residential exterior) from the edge of the door, as shown.

Both may be shown on the template. Keep in mind to adjust the setback if the door edge has a bevel (usually ¹⁄₁₆" difference).

Step 3 To determine the location of the vertical center line of the strike, take one-half of the door thickness as measured from the door stop. Using the strike as a template, mark the outline with a sharp knife or pencil. The location of the attaching screws should also be marked at this time, as shown in *Figure 123*.

Step 4 Drill the holes as directed on the installation template. The cross bore hole should be drilled first, and it should be drilled from both sides of the door to avoid splintering the wood. Each side of the hole should have a clean, sharp edge. A hole saw of the correct size will usually give excellent results. Next, using the center mark on the edge of the door, drill the hole for the latch bolt. The hole for the latch bolt should be 3" to 3½" deep from the edge of the door and will extend past the cross bore hole. Insert the latch bolt into the hole in the edge of the door, and using the front trim as a template, mark the outline using a sharp knife or pencil. The location of the attaching screws should also be marked at this time, and pilot holes drilled for them. Chisel out the area marked for the latch bolt to a depth of ⁵⁄₃₂" to allow the front trim to fit flush with the door edge. On the door jamb, drill a 1"-diameter hole ½" deep at the center point of the strike. Drill the pilot holes for the attaching screws and chisel out the jamb to a depth of ¹⁄₁₆", so the face of the strike is flush with the face of the jamb.

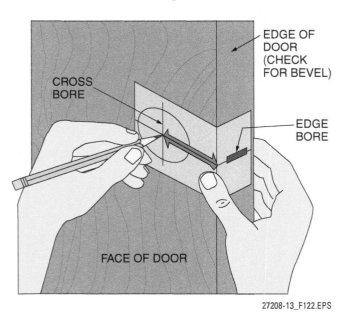

Figure 122 Use of an installation template.

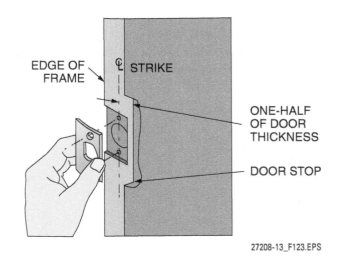

Figure 123 Center-line location of the strike.

Door Movement during Lockset Installation

If the door moves when the pressure of drilling is applied during lockset installation, it will result in damage to the door and possible injury to the individual operating the drill. Always secure the door to prevent movement.

Step 5 Insert the latch bolt into the hole in the door edge, with the beveled edge of the latch bolt facing toward the door jamb. Fasten the latch bolt to the door with the two attaching screws.

Step 6 Install the outside-knob assembly by inserting the spindle and the threaded posts through the holes in the latch bolt. Rotate the locking stem so the knob is unlocked.

Step 7 Install the inside-knob assembly by aligning it with the tip of the spindle of the outside-knob assembly (*Figure 124*). Make sure the turn button is in a horizontal position, then slide the inside-knob assembly up to the door surface. Rotate the rose so that the screw holes line up with the threaded posts. Insert the connecting screws through the holes in the rose and into the posts. Tighten the connecting screws to obtain a firm attachment, but be careful not to overtighten. Attach the strike to the jamb with the two attaching screws and the job should be complete.

Step 8 Check the operation of the lockset with the thumb turn, making sure it operates

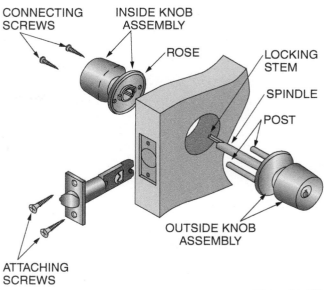

Figure 124 Disassembled lockset.

properly. If the lock functions properly, tighten the connecting screws, but do not overtighten as this may cause a bind in the latch mechanism.

Cylindrical locksets (*Figure 125*) are used in industrial, commercial, and institutional construction. They are ruggedly constructed and designed for unlimited use. To simplify architectural specifications and installation, all locks in all functions are usually uniform in size, self-aligning, and simple to install. While both knob and lever locksets are essentially the same mechanically, the lever locksets are compliant with the requirements for disabled persons. Both types of locksets have a pin tumbler cylinder to offer a wide range of keying and control options.

Cylindrical Lockset Jig

Drilling jigs simplify the installation of cylindrical locksets. These jigs eliminate the use of hand-marked templates and ensure accurate, perpendicular boring of lockset holes in the door and jamb.

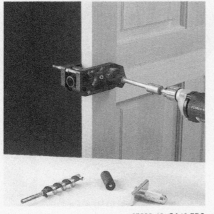

Figure 125 Heavy-duty cylindrical locksets.

A comparison may be made between the internal mechanism of the tubular lockset, as shown earlier, and the internal mechanism of the cylindrical lockset, as shown in *Figure 126*.

Most manufacturers of heavy-duty locksets also offer a knurled knob, or a knob with an abrasive surface. When called for in the specifications, this type of knob is installed as part of the lockset on doors leading to dangerous areas so these areas can be identifiable by touch for blind persons. A typical knurled knob is shown in *Figure 127*.

Tubular and cylindrical locksets are installed in much the same manner by drilling two holes in the door and mortising for the latch bolt and the strike.

5.2.8 Installation of Mortise Locksets

Use the following procedure when installing a mortise lockset in a wood door:

Step 1 Remove the lockset from the packing carton and make sure all the parts are there. Check and make sure the lockset is the correct one for the door on which you are working. Read the installation instructions carefully so they are thoroughly understood. Open the door to a position where it can be worked on and place two wedges at the bottom on the lock side of the door.

Step 2 Measure the recommended height up from the floor (usually 34" to 38") and place a light pencil mark on both the door and the jamb.

Figure 127 Typical knurled knob.

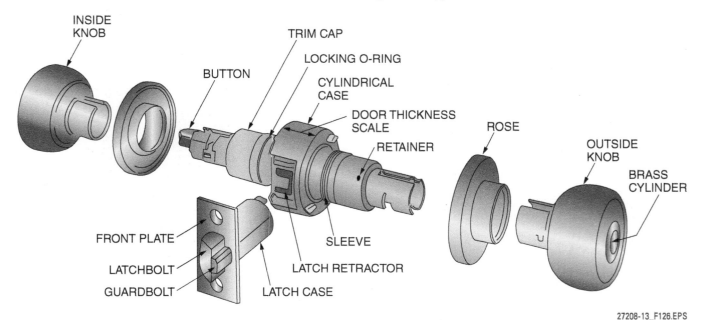

Figure 126 Exploded view of a heavy-duty cylindrical lockset.

Step 3 Using the template supplied, mark the door. This template will locate all the centers of holes that must be drilled on the face of the door and the mortise location and outline on the edge of the door.

Step 4 Using the correct-diameter bits, drill the holes through the face of the door first. Drill from both sides of the door to avoid splintering the wood. Using the correct-size bits, drill a series of holes in the edge of the door within the outline of the mortise, and clean out the remaining wood in the mortise with a sharp chisel.

Step 5 Place the mortise lock in the hole and mark the outline of the lock face on the edge of the door with a sharp knife or pencil. The location of the attaching screws should also be marked at this time and pilot holes drilled for them. Carefully chisel out the area marked for the lock face. Insert the lock in the mortise and fasten it to the door with the two attaching screws. Install the lock cylinder, handles, and strike on the face of the door according to the manufacturer's instructions.

5.2.9 Installation of Door Closers

For installation procedures, the installer will need to refer to the manufacturer's instructions pertaining to the specific unit being installed. Generally, the installation for a door-mounted closer will proceed as follows:

Step 1 Using a supplied template such as the one shown in *Figure 128*, select the angle of the opening desired. Locate and make holes on the door for the closer body and holes on the frame for the arm shoe.

> **NOTE:** The template is shown for illustration purposes only. Use the template that is included with the closer.

Step 2 Install the closer body on the door.

Step 3 Disassemble the secondary arm and shoe assembly from the main arm by removing the elbow screw. Fasten the secondary arm and shoe assembly to the frame face.

Step 4 Place the main arm into the closer pinion shaft, and install and tighten the main arm screw with a ½" wrench.

Mortise Lockset Jig

Mortise lockset jigs greatly simplify the mortising of wood doors for the installation of mortise locksets/deadbolts. These jigs ensure the accurate routing of the mortise in the door and eliminate the need for hand mortising.

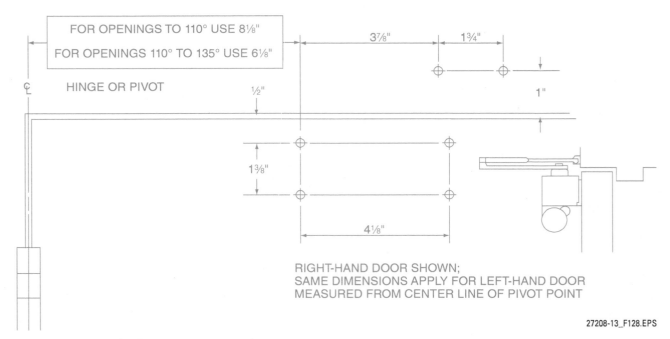

Figure 128 Example of a closer mounting template.

Step 5 Close the door and adjust the secondary arm assembly so the main arm is perpendicular to the face of the door. Reassemble the secondary arm to the main arm and tighten the screw securely.

Step 6 Adjust the closing tension by using the wrench packed with the door closer on the ratchet. Swing the wrench away from the hinge to wind the spring between 3 and 10 notches, then engage the dog on the ratchet. Increase or decrease the swing power to suit the closing conditions of the door (*Figure 129*).

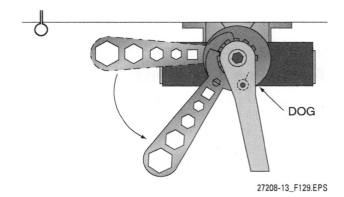

Figure 129 Adjusting closer tension.

5.0.0 Section Review

1. You need to make a door jack to help you to hang some doors, so you are careful to _____.
    a. space the uprights to the exact width of the door
    b. line the jack with carpet scraps to protect the door
    c. diagonally brace one side, but not the other side
    d. make the base length at least three times the height

2. To reduce drag when using a hand or power plane to plane a door edge, lightly coat the bottom of the tool with _____.
    a. lubricating oil
    b. paraffin wax
    c. silicone spray
    d. cooking oil

Section Six

6.0.0 Door Schedules

Objective

List and identify specific items included on a typical door schedule.

a. Describe the hardware finish classifications.
b. Describe the information included in a typical door schedule.

Trade Terms

Finish hardware: The exposed hardware in a building, such as doorknobs, door hinges, door locks, door closers, window hardware, shelf and clothing storage hangers, and bathroom hardware.

Head: The horizontal member at the top of a door or window opening.

Table 3 Common Hardware Symbols and Finishes

Symbol	Finish
USP	Primed for painting
US 3	Bright brass
US 4	Dull brass
US 10	Dull bronze
US 10B	Dull bronze, oxidized, and oil rubbed
US 14	Nickel plated, bright
US 26	Chrome plated, bright
US 26D	Chromium plated, dull
US 27	Satin aluminum, lacquered
US 28	Satin aluminum, anodized
US 32	Stainless steel, bright
US 32D	Stainless steel, dull

Each door must be located and identified on the construction drawings of the building. One way to locate a door is on the door schedule. Hardware and hardware finish are often included in a door schedule.

6.1.0 Hardware Finish

Hardware is divided into two basic classifications: finish hardware and rough hardware. Rough hardware includes items such as bolts, screws, nails, or any similar items that are not usually exposed to view in the completed structure. Finish hardware includes hinges, door lock/passage sets, cabinet handles, pulls, and other such items that are exposed to view in the completed structure.

The term *hardware finish* is used to describe the surface finish on finish hardware. Examples of this include a bright brass finish on door hinges or a satin chromium-plated finish on a kitchen cabinet drawer pull.

A classification system of finish symbols, called US Finish Symbols, has been established. Some of the more common symbols should be familiar to the experienced finish carpenter (see *Table 3*).

6.2.0 Door Schedules

Most sets of commercial drawings include a door schedule. The one shown in *Figure 130* is one example.

> **NOTE**
> The door schedule shown is one of many variations used and should not be thought of as standard.

The first column of the schedule (MARK) indicates the location of the door by number. One method of door numbering is to number all doors on the first floor in the one-hundred series, the second floor in the two-hundred series, and so forth. If it is a large building that has wings shown on the floor plan and these wings are identified by letters, such as Wing A, Wing B, and so on, then a letter may appear with the mark number identifying a door, such as 104-A.

The second column on the schedule (NO. REQ) indicates the number of doors required at the numbered mark indicated in the first column. For example, look at door mark 103; two doors are required at that location. This is indicative of a pair of doors located side by side.

The third column (SIZE) indicates the size of each door with the width shown first and the height last. In the case of a multiple door opening, the complete opening size may be given, and it is up to the individual reading the door schedule to mathematically reduce that opening size to an individual door size. In some schedules and/

DOOR SCHEDULE										
MARK	NO. REQ	SIZE	THICKNESS	CORE	HINGE OR SWING	TYPE	MATERIAL	FINISH	REMARKS	LABEL
101	1	3'0" × 6'8"	1⅜"	H.C.	L	A	WD.	STN.	WOOD	20 MIN.
102	1	2'8" × 6'8"	1⅜"	H.C.	L	B	WD.	STN.	WOOD LOUVER	20 MIN.
103	2	6'0" × 7'0"	1¾"	H.C.	R	C	MET.	PAINT	WOOD LOUVER	90 MIN.
104	1	3'0" × 7'0"	1¾"	S.C.	R	D	WD.	PAINT	¼" P.P. GLASS	45 MIN.
105	1	3'0" × 6'8"	1¾"	H.C.	R	B	WD.	PAINT	SIGHT PROOF LOUVER	60 MIN.
106	1	3'0" × 6'8"	1¾"	S.C.	R	E	WD.	STN.	¼" P.P. GLASS	20 MIN.

Figure 130 Typical door schedule.

or drawings, the sizes may be given without the foot and inch markings and without the "×" or a space between the numbers; for example, 3'-0" × 6'-8" would be 3068. The door swing may also be indicated here, such as 3068R or 3068L.

The fourth column (THICKNESS) indicates the thickness of each door. Door thicknesses may range from ¾" up to 2½" and more in some instances. Take care to ensure that all doors delivered to the job site are of the correct thickness.

The fifth column (CORE) indicates whether the door has a hollow core (HC) or a solid core (SC). Many times, the core may be the only differentiating factor between doors. A mistake in placement is often made as a result of not carefully checking the core requirement for a particular opening.

The sixth column (HINGE OR SWING), if present, indicates the swing of the door. As mentioned previously, this information may be included as part of the size information.

The seventh column (TYPE) indicates the type of door required for that particular opening. The type is usually indicated by a capital letter and that letter is placed at the bottom of a scaled elevation drawing of that particular door. In most cases, the drawing is found on the same page as the door schedule.

The eighth column (MATERIAL) indicates the type of material used in the door construction. Metal or wood is the most common, but plastic or fiberglass may occasionally be indicated on the schedule. The drawings will define any abbreviations used.

The ninth column (FINISH) indicates the final door finish. The finish may be paint, stain, or varnish. On occasion, a door may be indicated as prefinished, and the door will usually be delivered to the job site with a protective paper wrapping.

Other items that may be indicated on the door schedule are the types of jambs required for each door opening. This indication is then tied into a drawing of that jamb in elevation, showing sections cut through at the head and the side jamb. The type of threshold may also be indicated, and a drawing of each type will be shown corresponding to the schedule. Hardware finish is another item that may occasionally be on a schedule.

BHMA Finish Numbering System

The Builders Hardware Manufacturer's Association (BHMA) has established a sectional classification system for hardware finishes. This standard for materials and finishes contains a description of types of finishes and divides them into categories. A numbering system has been established, which identifies the base material and finish. This system can be more readily used in a computer than the US Finish Symbols system. As an example, the BHMA number 605 is the equivalent of US 3, bright brass. Some hardware manufacturers are already using the BHMA numbering system. Any finish carpenter who installs a lot of hardware should contact BHMA for a copy of their finish standards.

The door schedule will typically include a REMARKS column. This column is reserved for any special information about a door. The schedule might also include a LABEL column, which lists fire-rating information. This will only be included if fire-rated doors are to be used. It shows the number of minutes or the duration of the fire test the door type is capable of enduring before allowing the entry of fire into a protected area.

6.0.0 Section Review

1. Hinges, cabinet handles, and similar items are referred to as _____.
   a. rough hardware
   b. appearance items
   c. finish hardware
   d. accessory materials

2. In a door schedule, a door mark designates the _____.
   a. location
   b. core type
   c. thickness
   d. finish

SUMMARY

All carpenters should be familiar with the various steps in a typical door installation, the many kinds of doors used for various openings, and the correct hardware required for each particular situation. Attention to detail and carefully following the manufacturer's instructions are essential to a professional installation.

Review Questions

1. When starting an electric tool, be sure that _____.
   a. keys and adjusting wrenches have been removed
   b. it is plugged in
   c. you are wearing earplugs
   d. the owner's manual is readily available

2. To protect yourself while working, you should *not* _____.
   a. lean against a grounded pipe while operating power tools
   b. carefully put away any tool after you have used it
   c. use clamps to hold work so that both of your hands are free
   d. unplug tools before you change blades, cutters, or bits

3. The two main types of flush doors are _____.
   a. hollow core and solid core
   b. vision lights and door lights
   c. lead lined and panel
   d. stile and rail

4. A staved-wood door is a _____.
   a. hollow-core door
   b. panel door
   c. solid-core door
   d. mineral-core door

5. Solid-core doors have a tendency to _____.
   a. be less expensive than hollow-core doors
   b. be less soundproof than hollow-core doors
   c. warp less than hollow-core doors
   d. be lighter than hollow-core doors

6. The term *swing* refers to the _____.
   a. direction that the door opens
   b. side that the hinges are on
   c. side that the doorknob is on
   d. type of door hinges used

7. The rough-opening size normally used for conventional job-site-installed wood jambs and doors is the door size plus _____.
   a. 1" for height and 2" for width
   b. 1½" for height and 2" for width
   c. 2" for height and 3" for width
   d. 3" for height and 2" for width

8. When installing a loose door stop, the stop on the lock-jamb side of the door must _____.
   a. fit flush with the door
   b. have a ¹⁄₁₆" clearance between it and the door
   c. have a ⅛" clearance between it and the door
   d. have a ¼" clearance between it and the door

9. Adjustable metal door frames usually accommodate wall thicknesses of _____.
   a. 2½" to 7½"
   b. 3" to 8"
   c. 3¾" to 9¼"
   d. 4½" to 10½"

10. To provide fireproofing and sound attenuation, metal door frames may be _____.
    a. painted
    b. locked
    c. grouted
    d. stripped

11. A lockset with a turn button on one knob and an emergency release on the other knob is a(n) _____.
    a. entrance lock
    b. privacy lock
    c. passage latch
    d. patio lock

12. A hospital hinge is a _____.
    a. hardware device used to hold doors open for wheelchairs
    b. type of hinge that allows a door to swing both ways
    c. ball-bearing hinge used with wide, heavy doors
    d. hinge with parts sealed to prevent germ infestation

Trade Terms Introduced in This Module

Access: A passageway or a corridor between rooms.

Astragal: A piece of molding attached to the edge of an inactive door on a pair of double doors. It serves as a stop for the active door.

Butt: Any kind of hinge, except a strap or T-hinge.

Casing: The exposed finish material around the edge of a door or window opening.

Catches: Spring bolts used to secure a door when shut.

Coordinator: A device for use with exit features to hold the active door open until the inactive door is closed.

Cylindrical lockset: A lock in which the keyhole and tumbler mechanism are in a cylinder that is separate from the lock case and can be removed to change the keying of the lock.

Deadbolt: A square-head bolt in a door lock that requires a key to move it in either direction.

Door closer: A mechanical device used to check a door and prevent it from slamming when it is being closed. It also ensures the closing of the door.

Door frame: The surrounding case of a door into which a door closes. It consists of two upright pieces called jambs and a horizontal top piece called the head.

Door stop: The strip against which a door closes on the inside face of the frame or jamb. It can also be a hardware device used to hold the door open to any desired position or a hardware device placed against the baseboard to keep the door from marring the wall.

Dust-proof strike: A metal piece used to receive the latch bolt when a door is closed. The piece has a spring-action cover to keep particles out of the strike when the door is open.

Finish hardware: The exposed hardware in a building such as doorknobs, door hinges, door locks, door closers, window hardware, shelf and clothing storage hangers, and bathroom hardware.

Finishing sawhorse: A pair of trestles used to support lumber, molding, doors, stair parts, and other materials while they are being fitted and shaped for installation.

Flush bolt: A sliding bolt mechanism that is mortised into a door at the top and bottom edge. It is used to hold an inactive door in a fixed position on a pair of double doors.

Flush door: A door of any size that has a totally flat surface.

Hanging stile: The door stile to which the hinges are fastened.

Hardware: Components, such as hinges, locksets, and closers, used to attach a door to its frame or operate the door.

Head: The horizontal member at the top of a door or window opening.

Hinge: The hardware fastened to the edge of a door that allows the frame to pivot around a steel pin, permitting the operation of the door.

Jamb: One of the vertical members on either side of a door or window opening.

Kerf: A slot or cut made with a saw.

Knob lockset: A lock for a door, with the locking cylinder located in the center of the knob.

Latch bolt: A spring-loaded bolt in a lock, with a beveled head that is retracted when hitting the strike.

Lockset: The entire lock unit, including locks, strike plate, and trim pieces.

Molding: Long strips of material used for finishing and decorative trim.

Mortise: A measured portion of wood removed to receive a piece of hardware such as a lock or butt.

Mortise lockset: A rectangular metal box that houses a lock. It usually has a latch and deadbolt as part of the unit. Options consist of a locking cylinder and/or thumb turn by which it can be secured. It is used on residential entry doors and commercial doors.

Panic hardware: Hardware that provides an emergency escape exit.

Plumb: A true vertical position.

Prefinished: Material such as molding, doors, cabinets, and paneling that has been stained, varnished, or painted at the factory.

Prehung door: A door that is delivered to the job site from the mill already hung in the frame or jamb. In some instances, the trim may be applied on one side.

Rabbet: A rectangular groove cut in the edge of a board.

Rail: A horizontal member of a door or window sash.

Rough opening: Any unfinished door or window opening in a building.

Sanitary stop: Door stop with a 45-degree angle cut at the bottom on the vertical jambs.

Shim: A thin, tapered piece of wood such as a wooden shingle used to fill in gaps and to level or plumb structural components.

Sill: The lowest member at the bottom of a door or window opening. It also refers to the lowest member of wood framing supporting the framing of a building.

Smoke gasket: A rubber strip that goes all the way around the door to keep smoke from penetrating an area in case of fire.

Sound attenuation: The reduction of sound as it passes through a material.

Sound transmission class (STC): The rating by which the sound attenuation of a door is determined. The higher or greater the number, the better the sound reduction.

Stile: The vertical edge of a door.

Strike plate: A metal plate screwed to the jamb of a door so that when the door is closed, the bolt of the lock strikes against it. The bolt is then retracted and slides along the metal plate. When the door is fully closed, the bolt inserts itself into a hole in the plate to hold the door securely in place.

Sweep: A type of weather stripping. A sweep is a felt or rubber flap mounted in a metal channel to seal door bottoms to prevent air infiltration.

Template: A thin piece of material such as plastic or heavy paper with a shape cut out of it, or the whole of it cut to a shape on its perimeter. A template is used to transfer that shape to another object by tracing it with a pencil or scribe.

Threshold: A piece of wood, metal, or stone that is set between the door jamb and the bottom of a door opening.

Transom: A panel above a door that lets light and/or air into a room or is used to fill the space above a door when the ceiling heights on both sides of the door opening allow.

Weather stripping: Strips of metal or plastic used to keep out air or moisture that would otherwise enter through the spaces between the outer edges of doors and windows and the finish frames.

Additional Resources

This module presents thorough resources for task training. The following resource material is suggested for further study.

Finish Carpentry. 1997. Newtown, CT: Taunton Press.

Figure Credits

Courtesy of Pella® Windows and Doors., CO01

Marlite Decorative Wall System, Figure 2

Pioneer Industries, Inc., Figure 20, Figure 21, Figures 38, 39, 40, 41, 42, 43, 44, 45, 46, 47, 48, 49, 50, 51, 52, 53, 54, 55, 56, 57, 58, 59, 60, 61, 62, 63, 64, 65, 66, 67, 68, 69, 70, 71

Photo courtesy of Crane Revolving Doors, Figure 23

Raynor, Figure 24

HMMA Division of NAAMM, National Association of Architectural Metal Manufacturers, Figure 36

Timely Prefinished Steel Door Frames, Figure 37A, Figure 37B

The drawings of Yale® locks used in these materials have been used with the consent of Yale Security Inc. Nothing in these materials is intended to describe or depict the actual performance of YALE locks or other products of Yale Security Inc., Figure 74, Figures 120, 121, 122, 123, 124, Figures 126, 127, 128, 129

Ingersoll Rand Safety & Security, Figures 77, 78A, 78B, 79, 80, 81, 82, 83, 84, Figure 95, Figure 96, Figure 97, SA08, SA10, SA11, Figure 119, Figure 125

Photo Courtesy of DORMA Architectural Hardware, Figure 78B

SDC Security Door Controls, Figure 85, Figure 88 (bottom)

Courtesy of Security Door Controls, Figure 86, Figure 87, Figure 89

Securiton Magnalock Corp. An ASSA ABLOY Group Company, SA07

Detex Corporation, SA09

Porter Cable Corporation, SA12, Figure 115, Figure 116

Section Review Answer Key

Answer	Section Reference
Section One	
1. b	1.0.0
Section Two	
1. c	2.0.0
2. c	2.2.1
Section Three	
1. d	3.1.0
2. b	3.2.2
Section Four	
1. a	4.1.2
2. c	4.2.4
Section Five	
1. b	5.1.1
2. b	5.2.2
Section Six	
1. c	6.1.0
2. a	6.2.0

NCCER CURRICULA — USER UPDATE

NCCER makes every effort to keep its textbooks up-to-date and free of technical errors. We appreciate your help in this process. If you find an error, a typographical mistake, or an inaccuracy in NCCER's curricula, please fill out this form (or a photocopy), or complete the online form at **www.nccer.org/olf**. Be sure to include the exact module ID number, page number, a detailed description, and your recommended correction. Your input will be brought to the attention of the Authoring Team. Thank you for your assistance.

Instructors – If you have an idea for improving this textbook, or have found that additional materials were necessary to teach this module effectively, please let us know so that we may present your suggestions to the Authoring Team.

NCCER Product Development and Revision
13614 Progress Blvd., Alachua, FL 32615

Email: curriculum@nccer.org
Online: www.nccer.org/olf

❏ Trainee Guide ❏ Lesson Plans ❏ Exam ❏ PowerPoints Other _____

Craft / Level: _____ Copyright Date: _____

Module ID Number / Title: _____

Section Number(s): _____

Description: _____

_____

_____

_____

_____

Recommended Correction: _____

_____

_____

_____

_____

Your Name: _____

Address: _____

_____

Email: _____ Phone: _____

This page is intentionally left blank.

Drywall Installation

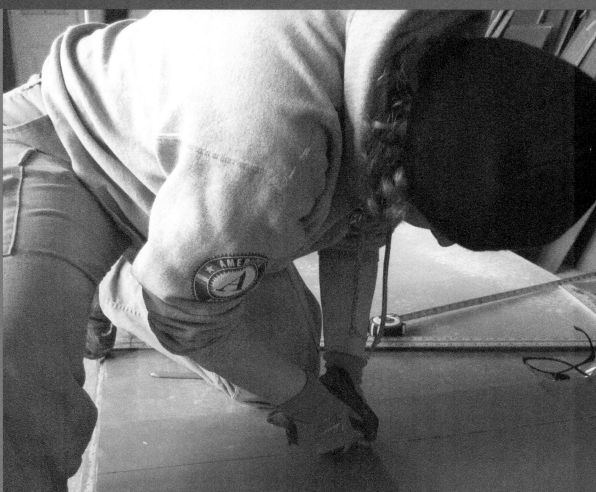

OVERVIEW

Gypsum drywall is the most common wall finish materials used in residential and commercial construction. There are a variety of drywall materials used for different applications, as well as a number of construction methods used to build walls to meet building codes for fire resistance and sound transmission. There are also many different types of fasteners used in drywall installation. The selection of materials, fasteners, and construction methods is controlled by building codes, and therefore must be carefully considered.

Module 27206

Trainees with successful module completions may be eligible for credentialing through NCCER's National Registry. To learn more, go to **www.nccer.org** or contact us at **1.888.622.3720**. Our website has information on the latest product releases and training, as well as online versions of our *Cornerstone* magazine and Pearson's product catalog.

Your feedback is welcome. You may email your comments to **curriculum@nccer.org**, send general comments and inquiries to **info@nccer.org**, or fill in the User Update form at the back of this module.

This information is general in nature and intended for training purposes only. Actual performance of activities described in this manual requires compliance with all applicable operating, service, maintenance, and safety procedures under the direction of qualified personnel. References in this manual to patented or proprietary devices do not constitute a recommendation of their use.

Copyright © 2013 by NCCER, Alachua, FL 32615, and published by Pearson, New York, NY 10013. All rights reserved. Printed in the United States of America. This publication is protected by Copyright, and permission should be obtained from NCCER prior to any prohibited reproduction, storage in a retrieval system, or transmission in any form or by any means, electronic, mechanical, photocopying, recording, or likewise. To obtain permission(s) to use material from this work, please submit a written request to NCCER Product Development, 13614 Progress Blvd., Alachua, FL 32615.

27206 V5

From *Carpentry, Trainee Guide*. NCCER.
Copyright © 2013 by NCCER. Published by Pearson. All rights reserved.

27206
DRYWALL INSTALLATION

Objectives

When you have completed this module, you will be able to do the following:

1. Identify components of a drywall assembly.
    a. List the types of gypsum products.
    b. Identify drywall fasteners and list their uses.
    c. Identify drywall accessories and state their applications.
2. Describe the installation of drywall.
    a. Describe the purpose of a finish schedule.
    b. List the tools used for drywall application.
    c. Identify methods of sound-isolation construction.
    d. Describe the procedure for drywall construction.
    e. List special applications for drywall.
3. Contrast rated assemblies to nonrated assemblies.
    a. Describe single-ply drywall application.
    b. Describe how fire-rated walls are constructed.
    c. List multi-ply drywall applications.
    d. Describe how to prioritize walls.
4. Identify how to calculate a quantity takeoff for proper drywall installation.
    a. Explain how to perform a material takeoff for drywall.
    b. Explain how to perform a material takeoff for drywall fasteners.

Performance Tasks

Under the supervision of your instructor, you should be able to do the following:

1. Select the type and thickness of drywall required for an installation.
2. Install gypsum drywall panels on a stud wall and a ceiling using any or all of the following fastening systems:
    - Nails
    - Screws
    - Adhesives
3. Estimate material quantities for an installation.

Trade Terms

Corner bead
Drywall
Floating interior angle construction
Joint
Nail pop
Slurry
Substrate

Code Note

Codes vary among jurisdictions. Because of the variations in code, consult the applicable code whenever regulations are in question. Referring to an incorrect set of codes can cause as much trouble as failing to reference codes altogether. Obtain, review, and familiarize yourself with your local adopted code.

Industry-Recognized Credentials

If you're training through an NCCER-accredited sponsor, you may be eligible for credentials from NCCER's Registry. The ID number for this module is 27206. Note that this module may have been used in other NCCER curricula and may apply to other level completions. Contact NCCER's Registry at 888.622.3720 or go to **www.nccer.org** for more information.

Contents

Topics to be presented in this module include:

1.0.0 Drywall Assembly Components 1
  1.1.0 Types of Gypsum Products 2
  1.2.0 Drywall Fasteners 5
    1.2.1 Nails 5
    1.2.2 Screws 7
    1.2.3 Adhesives 8
    1.2.4 Application of Adhesives 10
    1.2.5 Adhesive Application to Metal Framing 11
    1.2.6 Adhesive Application to Concrete and Masonry 11
  1.3.0 Drywall Accessories 12
    1.3.1 Corner Beads and Casings 12
    1.3.2 Control Joints 13
2.0.0 Drywall Installation 16
  2.1.0 Fastening Schedules 16
  2.2.0 Tools Used for Drywall Application 16
  2.3.0 Sound-Isolation Construction 18
    2.3.1 Separated Partitions 20
    2.3.2 Resilient Mountings 22
    2.3.3 Sound-Isolating Materials 23
  2.4.0 Installing Drywall 24
    2.4.1 Preparing the Job Site 24
    2.4.2 Cutting and Fitting Procedures 25
    2.4.3 Floating Interior Angle Construction 26
    2.4.4 Moisture-Resistant Construction 26
    2.4.5 Resurfacing Existing Construction 29
  2.5.0 Special Applications 29
3.0.0 Rated vs. Nonrated Assemblies 31
  3.1.0 Single-Ply Application 31
  3.2.0 Fire-Rated Walls 31
    3.2.1 Fire-Rated Construction 33
    3.2.2 Fire-Stopping 34
  3.3.0 Multi-Ply Application 37
    3.3.1 Attaching the Base Ply to Metal Framing or Furring 37
    3.3.2 Face-Ply Attachment 38
    3.3.3 Adhesive Attachment 38
    3.3.4 Supplemental Fasteners 38
  3.4.0 Prioritizing Walls 39
4.0.0 Estimating Drywall 40
  4.1.0 Drywall 40
  4.2.0 Fasteners 40

Figures and Tables

Figure 1 Typical board application..1
Figure 2 Standard edges of drywall..4
Figure 3 Architectural drywall 4
Figure 4 Uniform depression or dimple..5
Figure 5 Nails..5
Figure 6 Single nail spacing 6
Figure 7 Double-nail spacing 6
Figure 8 Correct and incorrect nailing 6
Figure 9 Incorrect and correct alignment 7
Figure 10 Drywall screws...7
Figure 11 Adhesive applicators...9
Figure 12 Adhesive applied to the edges of a stud 11
Figure 13 Supplemental wall and ceiling fasteners 11
Figure 14 Prebowing of gypsum panels...11
Figure 15 Adjustable wall furring..12
Figure 16 Examples of protective beads 13
Figure 17 Moldings for prefinished drywall...13
Figure 18 Typical control joint..13
Figure 19 Designs for perimeter relief 14
Figure 20 Carbide cutting tool 16
Figure 21 T-square..17
Figure 22 Utility knife 17
Figure 23 Hook-bill knife 17
Figure 24 Drywall rasp...17
Figure 25 Circle cutter..17
Figure 26 Power cutout tool and jab saw 18
Figure 27 Drywall saw..18
Figure 28 Light box cutter..18
Figure 29 Drywall lifter...18
Figure 30 Drywall hammer 18
Figure 31 Screw guns 18
Figure 32 Drywall lift 19
Figure 33 T-brace 20
Figure 34 Caulking of sound-isolation construction (1 of 2).....................21
Figure 34 Caulking of sound-isolation construction (2 of 2).....................22
Figure 35 Attached resilient furring channel 23
Figure 36 Nailing patterns for installation over rigid foam insulation..........24
Figure 37 Drywall storage..25
Figure 38 Floating interior angle construction..27
Figure 39 Fastener patterns for floating interior angle construction 28
Figure 40 Pan is on the same plane as the face of the board........................28
Figure 41 Sound-rated construction 28
Figure 42 Single-ply construction 32
Figure 43 Multi-ply construction 33

Figure 44 High fire/noise resistance partition 33
Figure 45 Partition-wall examples 34
Figure 46 An example of a fire-rated wall abutting a nonrated wall 34
Figure 47 Mechanical fire-stop device 35
Figure 48 Fire barrier of moldable putty installed around electrical cables 36
Figure 49 Floating-corner treatment 37
Figure 50 Spreading adhesive using a notched spreader 38
Figure 51 Properly spacing adhesive ribbons 38

Table 1   Types and Uses of Drywall 3
Table 2   Maximum Spacing of Ceiling Framing 12
Table 3   Fastener Spacing 37
Table 4   Fastening Materials Required for 1,000 Square Feet of Drywall 40

This page is intentionally left blank.

SECTION ONE

1.0.0 DRYWALL ASSEMBLY COMPONENTS

Objective

Identify components of a drywall assembly.
a. List the types of gypsum products.
b. Identify drywall fasteners and list their uses.
c. Identify drywall accessories and state their applications.

Performance Task 1

Select the type and thickness of drywall required for an installation.

Trade Terms

Corner bead: A metal or plastic angle used to protect outside corners where drywall panels meet.

Drywall: A generic term for paper-covered gypsum-core panels; also known as *gypsum drywall*.

Joint: A place where two pieces of material meet.

Nail pop: The protrusion of a nail above the wallboard surface that is usually caused by shrinkage of the framing or by incorrect installation. Also applies to screws.

Slurry: A thin mixture of water or other liquid with any of several substances such as cement, plaster, calcined gypsum, or clay.

Substrate: The underlying material to which a finish is applied.

Drywall, also known as gypsum board or gypsum drywall, is one of the most popular and economical methods of finishing the interior walls and ceilings of wood-framed and metal-framed buildings. When properly installed and finished, gypsum drywall can give a wall or ceiling made from many panels the appearance of being made from one continuous sheet.

The responsibility for drywall installation and finishing varies from job to job and from one locale to another. In some situations, carpenters install the drywall and painters finish it. In others, professional drywall workers do the entire job. The smaller the project, the more likely it is that the carpenter will install and finish the drywall.

At a minimum, it is important for the carpenter to understand the framing techniques that are necessary for the proper installation of drywall.

Drywall is a generic name for products consisting of a noncombustible core. This product is made primarily of gypsum with a paper covering on the face, back, and long edges. A typical board application is shown in *Figure 1*.

Drywall differs from products such as plywood, hardboard, and fiberboard because of its noncombustible core. Gypsum is a mineral found in sedimentary rock formations in a crystalline form known as calcium sulfate dihydrate.

One hundred pounds of gypsum rock contains approximately 21 pounds (10 quarts) of chemically combined water. The gypsum rock is mined and then crushed. The crushed rock is heated to about 350°F, driving out or evaporating three-fourths of the chemically combined water in a process called calcining. The calcined gypsum is then ground into a fine powder used in plaster, wallboard, and other gypsum products.

To produce drywall, the calcined gypsum is mixed with water and additives to form a slurry, which is fed between continuous layers of paper on a board machine. As the board automatically moves down a conveyor line, the calcium sulfate

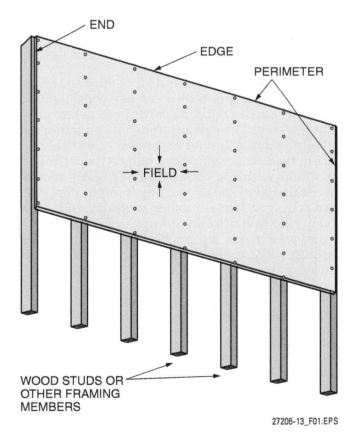

Figure 1 Typical board application.

recrystallizes or rehydrates, reverting to its original rock state.

The paper becomes chemically and mechanically bonded to the core. The board is then cut to length and conveyed through dryers to remove any free moisture.

Walls and ceilings finished with drywall have a number of outstanding advantages:

- Fire resistance
- Sound insulation
- Durability
- Economy
- Versatility

Drywall is an excellent fire-resistive material. It is the most commonly used interior finish where fire resistance classifications are required. Its noncombustible core contains chemically combined water, which is slowly released under high heat as steam, effectively retarding heat transfer. Even after complete calcination, when all of the water has been released, it continues to act as a heat-insulating barrier.

In addition, tests conducted in accordance with the standard *ASTM E84* of the American Society for Testing and Materials (ASTM) International, show that it has low flame spread and low fuel and smoke contribution factors. When installed in combination with other materials, it serves to effectively protect building elements from fire for prescribed time periods. Type X board is most often used in fire-rated assemblies. Be sure all local codes and regulations are met.

Control of unwanted sound that might be transmitted to adjoining rooms is a key consideration in the design of a building. It has been determined that low-density paneling transmits an annoying amount of noise. Sound-absorbing acoustical surfacing materials, while they reduce the reflection of sound within a room, do not greatly reduce transmission of sound into adjoining rooms. Wall and ceiling systems finished with drywall effectively help to control sound transmission.

Drywall makes strong, high-quality walls and ceilings with excellent dimensional stability. Their surfaces are easily decorated and refinished.

Drywall products are easy to apply. They are the least expensive of the wall surfacing materials that offer a fire-resistant interior finish. Both regular and architectural wallboard may be installed at relatively low cost. When architectural board is used, further decorative treatment is unnecessary.

Drywall products satisfy a wide range of architectural requirements for design. Ease of application, performance, availability, ease of repair, and adaptability to all forms of decoration combine to make drywall unmatched by any other surfacing product.

1.1.0 Types of Gypsum Products

Many types of drywall are available for a variety of building needs (see *Table 1*). Drywall panels are mainly used as the surface layer for interior walls and ceilings; as a base for ceramic, plastic, and metal tile; for exterior soffits; for elevators and other shaft enclosures; and to provide fire protection for architectural elements.

Foil-backed drywall reduces radiant heat loss in the cold season and radiant heat gain in the warm season. However, foil-backed drywall should not be used as a backing material for tile, as a second face-ply on a two-ply system, in conjunction with heating cables, or when laminating directly to masonry, ceiling, and roof assemblies.

Various thicknesses of drywall are available in regular, Type X, water-resistant, and architectural boards:

- $1/4"$ *drywall* – A lightweight, low-cost board used as a base in a multilayer application for improving sound control, to cover existing walls and ceilings in remodeling work, and for curved walls and barrel ceilings.
- $5/16"$ *drywall* – A lightweight board developed for use in manufactured construction, primarily mobile homes.
- $3/8"$ *drywall* – A lightweight board principally applied in a double-layer system over wood framing and as a face layer in repair and remodeling.
- $1/2"$ *drywall* – A board generally used for single-layer wall and ceiling construction in residential work and in double-layer systems for

The Way It Was

Until the 1930s, walls were typically finished by installing thin, narrow strips of wood or metal, known as lath, between studs, and then coating the lath with wet plaster. Skilled plasterers could produce a very smooth wall finish, but the process was time consuming and messy. In the early 1930s, paper-bound drywall was introduced and soon came into widespread use as a replacement for the tedious lath and plaster process.

Table 1 Types and Uses of Drywall

Type	Thickness	Sizes	Use
Regular, paper faced	¼"	4' × 8' to 10'	Re-covering old gypsum walls
	⅜"	4' × 8' to 14'	Double-layer installation
	½", ⅝"	4' × 8' to 16'	Standard single-ply installation
Regular with foil back	½", ⅝"	4' × 8' to 14'	Use as a vapor barrier or radiant-heat retarder
Type X, fire retardant	⅜", ½", ⅝"	4' × 8' to 16'	Use in garages, workshops, and kitchens, as well as around furnaces, fireplaces, and chimney walls; ⅝" is ¾-hour fire rated
Moisture resistant	½", ⅝"	4' × 6' to 16'	For tile backing in areas not exposed to constant moisture
Architectural panels	⁵⁄₁₆"	4' × 8'	Any room in the house
Gypsum lath	⅜", ½", ⅝"	16" × 4'	Use as a base for plaster
		2' × 8' to 12'	Use ⅜" for 16" on center (OC) stud spacing; ½" or ⅝" for 24" OC Stud spacing
Gypsum coreboard	1"	2' × 8' to 12'	Shaft liner Laminated partitions

greater sound and fire ratings. These panels are also available in 54" widths for use with 9' ceilings.

- *⅝" drywall* – A board used in quality single-layer and double-layer wall systems. The greater thickness provides additional fire resistance, higher rigidity, and better impact resistance. It is also used to separate occupied and unoccupied areas, such as a house from a garage or an office from a warehouse.
- *1" drywall* – A special board also known as coreboard. Either a single 1" board or two ½" factory-laminated boards may be used as a liner or core in shaft walls and in semisolid or solid drywall partitions.

Standard drywalls are 4' wide and 8', 10', 12', or 14' long. The width is compatible with the standard framing of studs or joists spaced 16" or 24" on center (OC). Other lengths and widths are available from the manufacturers on special order. The standard edges are rounded, tapered, beveled, square edge, and tongue-and-groove, as shown in *Figure 2*.

- *Regular drywall* – Regular drywall is used as a surface layer on walls and ceilings. Type X drywall is available in ½" and ⅝" thicknesses and has an improved fire resistance made possible by the use of special core additions. It is also available with a predecorated finish. Type X drywall is used in most fire-rated assemblies.

Special-Use Drywall

Regular ½" and ⅝" drywall are the most common types. There are, however, several types of drywall designed for special applications. These include:

- Type X gypsum board provides improved fire ratings because its core material is mixed with fire-retardant additives. Type X is often used on walls that separate occupancies. Examples are walls and ceilings between apartments or a wall separating a garage from the living area of a house. Use of Type X is normally specified by local building codes for protection of occupants.
- Flexible ¼" drywall panels have a heavy paper face and are designed to bend around curved surfaces.
- Special high-strength drywall panels are made for ceiling applications. The core of these panels is specially treated to resist sagging.
- A weather-resistant drywall panel is available for installation on soffits, porch ceilings, and carport ceilings.

Gypsum sheathing panels are used in cases where the required fire rating of exterior walls exceeds that available with oriented strand board (OSB), plywood, or other types of sheathing. Gypsum sheathing panels have a water-resistant core covered on both sides with water-repellent paper. Gypsum sheathing panels are widely used in commercial construction.

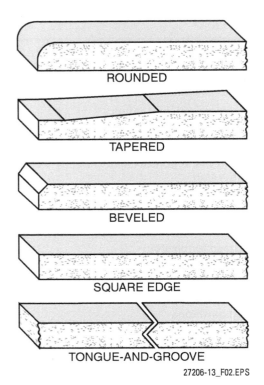

Figure 2 Standard edges of drywall.

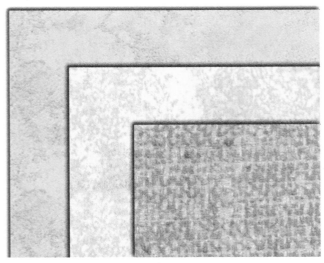

Figure 3 Architectural drywall.

- *Architectural drywall* – Architectural drywall (*Figure 3*) has a decorated surface that does not require further treatment. The surfaces may be coated, printed, or have a vinyl film. Textured patterns are also available. It requires additional trim, divider, and corner pieces. It is also known as predecorated drywall.
- *Water-resistant drywall* – Water-resistant drywall, also known as green board, has a water-resistant gypsum core and water-repellent paper. The facing typically has a light green color. It serves as a base for the application of ceramic or plastic wall tile and plastic finish panels in kitchen and laundry areas. It is available with a regular or Type X core and in ½" or ⅝" thicknesses. Water-resistant drywall is not recommended for use in tub and shower enclosures and other areas exposed to water; tile backer is now preferred for these high-moisture applications.
- *Tile backer* – This type of drywall is replacing water-resistant gypsum board as a backing for tile in damp areas such as baths and shower stalls. One type of this board is known as cement board, which is made from a slurry of portland cement mixed with glass fibers. It is colored light blue for easy recognition. This backing, which is available under a variety of trade names, such as Hardibacker™, Durock™, and Denshield™, is very versatile. In addition to its use as a tile backer, it can be used as a floor underlayment, countertop base, heat shield for stoves, and a base for exterior finishes such as stucco and brick veneer. It is available in 4' × 8' and 3' × 5' panels. Common thicknesses are ¼", ⁷⁄₁₆", and ½".
- *Gypsum backing board* – Gypsum backing board is designed to be used as a base layer or backing material in multilayer systems. It is available with aluminum foil backing and with regular or Type X cores.
- *Gypsum form board* – Gypsum form board has a fungus-resistant paper, and is used as a permanent form and support for poured-in-place reinforced gypsum concrete roof decks.
- *Gypsum coreboard* – Gypsum coreboard is available as a 1"-thick solid coreboard or as a factory-laminated board composed of two ½" boards. It is used in shaft walls and laminated gypsum partitions with additional layers of drywall applied to the coreboard to complete the wall assembly. It is available in a width of 24" and with a variety of edges, of which square and tongue-and-groove are the most common.
- *Gypsum sheathing* – Gypsum sheathing is used as a protective, fire-resistive membrane under exterior wall surfacing materials such as wood siding, masonry veneer, stucco, and shingles. It also provides protection against the passage of water and wind, and adds structural rigidity to the framing system. The noncombustible core is surfaced with firmly bonded water-repellent paper. In addition, a water-repellent material may be incorporated in the core. It is available in 2' and 4' widths and ½" and ⅝" thicknesses. The latter is also available with a Type X core.
- *Drywall substrate* – Drywall substrate for floor or roof assemblies has a Type X core that is ½" thick. It is available in 24" or 48" widths. It is

used under combustible roof coverings to protect the structure from fires originating on the roof. It can also serve as an underlayment when applied to the top surfaces of floor joists and under subflooring. It may also be used as a base for built-up roofing applied over steel decks.

- *Gypsum base for veneer plaster* – Gypsum base for veneer plaster is used as a base for thin coats of hard, high-strength gypsum veneer plaster.
- *Gypsum lath* – Gypsum lath is a board product used as a base to receive hand-applied or machine-applied plaster. It is available in ⅜" or ½" thicknesses and in widths of 16" or 24". Gypsum lath comes in 48" lengths. Other lengths are available on special order.
- *Flexible drywall* – Flexible drywall is specially designed for radius applications. A ¼" sheet can be laminated on a ¼" sheet of Masonite™ for a tight radius application.
- *Abuse-resistant panels* – Some drywall panels are designed specifically for use in areas where they might be subject to impact or vandalism. These types of panels have heavy-duty face paper and either heavy-duty backing or internal reinforcement.

1.2.0 Drywall Fasteners

Nails and screws are commonly used to attach drywall in both single-ply and multi-ply installations. Clips and staples are used only to attach the base layers in multi-ply construction. Special drywall adhesives can be used to secure single-ply drywall to framing, furring, masonry, and concrete, or to laminate a face ply to a base layer of drywall or other base material. Adhesives must be supplemented with mechanical fasteners.

Where fasteners are used at the board perimeter, they should be placed at least ⅜" from the board edges and ends. Fastening should start in the middle of the board and proceed outward toward the board perimeter. Fasteners must be driven as near to perpendicular as possible while the board is held firmly against the supporting construction.

Special Fasteners

Application of drywall requires special fasteners. Ordinary wood or sheet metal screws and common nails are not designed to penetrate the board without damage, to hold it tightly against the framing, or to permit correct countersinking for proper concealment. For high-end installations, sheets can be connected between studs using butt clips.

Also, fasteners must be used with the correct shield, guard, or attachment recommended by the manufacturer. Nails should be driven with a crown-headed hammer, which forms a uniform depression or dimple that is not more than ¹⁄₃₂" deep around the nail head. See *Figure 4*.

1.2.1 Nails

Both annular and cupped-head nails are acceptable for drywall application (*Figure 5*). Preferably, the nails should have heads that are flat or concave and thin at the rim. The heads should be between ¼" and ⁵⁄₁₆" in diameter to provide adequate holding power without cutting the face paper when the nail is dimpled. Casing and common nails have heads that are too small in relation to the shank; they easily cut into the face paper and should not be used.

Nail heads that are too large are also likely to cut the paper surface if the nail is driven incorrectly at a slight angle. The nails should be long enough to go through the wallboard layers and far enough into the supporting construction to provide adequate holding power. The nail penetration into the framing member should be ⅞" for smooth-shank nails and ¾" for annular-ring nails, which provide more withdrawal resistance and require less penetration. For fire-rated assemblies, greater penetration is required (generally 1⅛" to 1¼" for one-hour assemblies).

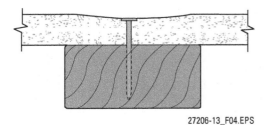

Figure 4 Uniform depression or dimple.

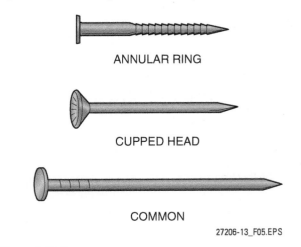

Figure 5 Nails.

Particular care should be taken not to break the face paper or crush the core by striking it too hard with the hammer.

Drywall can be attached by either a single-nailing or a double-nailing method. Double nailing produces a tighter board-to-stud contact. Whenever fire-resistive construction is required, the nail spacing specified in the fire test should be followed. Always check local codes for nailing requirements.

Single nails should be spaced at a maximum of 7" on center on ceilings and 8" on center on walls along framing members. See *Figure 6*.

Nails are first driven into the center, or field, of the board and then outward to the edges and ends. In single-ply installations, all ends and edges of drywall are placed over framing members or other solid backing, except where treated joints are at right angles to framing members.

In double nailing, the spacing of the first set of nails is 12" on center, with the second nailing 2" to 2½" from the first. See *Figure 7*.

The second set of nails is applied in the same sequence as the first set, but not on the perimeter of the board. The first nails driven should be reseated as necessary following the application of the second set. The general attachment procedure is as follows:

Step 1 Carefully measure and cut the board.

Step 2 Prior to nailing, mark the drywall to indicate the location of the framing.

Step 3 To avoid nail pops or protrusions, hold the board firmly against the framing when nailing.

Step 4 Drive the nails straight into the framing member. Nails that miss the framing member should be removed, and the nail hole dimpled and covered with joint compound.

Step 5 Damage to the board caused by overdriving nails may be corrected by driving a new nail 2" away to provide firm attachment. Examples of correct and incorrect nailing are shown in *Figure 8*.

Other common causes of face-paper fractures are misaligned or twisted supporting framing members and projections of improperly installed blocking or bracing, as shown in *Figure 9*.

Framing faults prevent solid contact between the drywall and framing members, and hammer impact causes the board to rebound and rupture the paper. Defective supports should

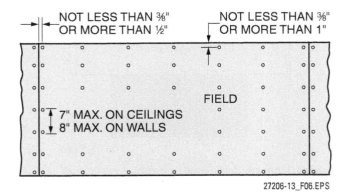

Figure 6 Single nail spacing.

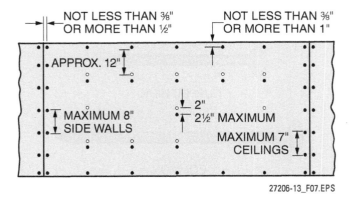

Figure 7 Double-nail spacing.

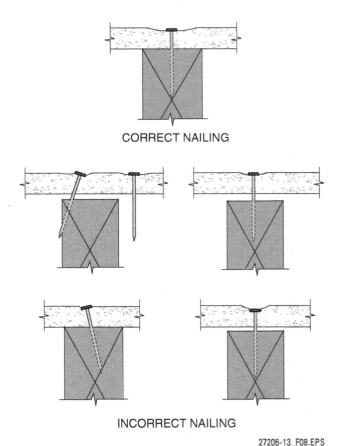

Figure 8 Correct and incorrect nailing.

be corrected prior to the application of the drywall. Protruding framing members should be trimmed or reinstalled. Shims can be used, if necessary, for receiving framing members. The use of screws, adhesives, or two-ply construction will minimize problems resulting from these defects.

1.2.2 Screws

Drywall screws (*Figure 10*) are used to attach drywall to wood or steel framing or to other drywall. They have a cupped Phillips-head design that is intended to be used with a drywall power screwdriver. These screws pull the board tightly to the supports without damaging the board, and minimize fastener and surface defects due to loose boards. The specially contoured head, when properly driven, makes a uniform depression that is free of ragged edges and fuzz.

Type W gypsum drywall screws are designed for fastening drywall to wood framing or furring. The Type W screw points are diamond shaped to provide efficient drilling action through both gypsum and wood, and their specially designed

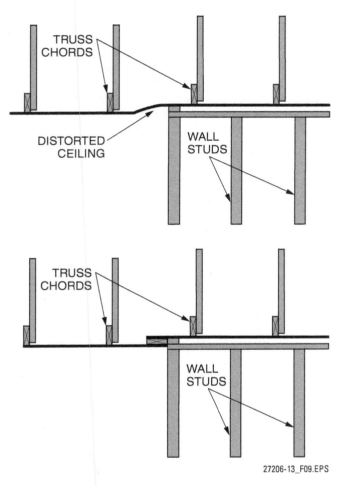

Figure 9 Incorrect and correct alignment.

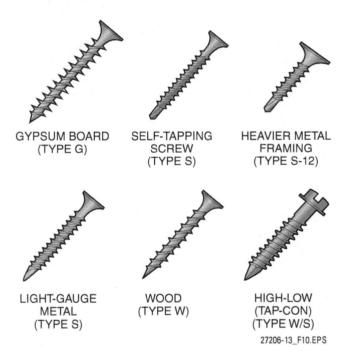

Figure 10 Drywall screws.

Selecting Drywall Screws

If you go to a building supply store to get drywall screws, you will find that there are many different types and sizes. It is important to determine exactly what you need before you make the trip.

Module 27206   Drywall Installation 7

threads provide both quick penetration and increased holding power.

The recommended minimum penetration into supporting construction is ⅝". However, in two-ply construction where the face layer is screw attached, additional holding power is developed in the base ply, which permits reduced penetration into supports down to ½". Type S screws may be substituted for Type W screws in two-ply construction.

Type S gypsum drywall screws are designed for fastening drywalls to steel studs or furring. They are self-drilling with a self-tapping thread and generally a mill slot or hardened drill point that is designed to penetrate sheet metal with little pressure. Easy penetration is important because steel studs are often flexible. They tend to bend away from the screws, and the screws tend to strip easily.

Type G gypsum drywall screws are used for fastening drywall panels to gypsum backing boards. They are similar to Type W screws, but have a deeper, special-thread design. They are generally 1½" long, but other lengths are available. Gypsum drywall screws require a penetration of at least ½" of the threaded portion into the supporting board. Allowing approximately ¼" for the point results in the minimum penetration of ¾".

Gypsum drywall screws should not be used to attach wallboard to ⅜" backing board because they do not provide sufficient holding strength. Nails or longer screws should be driven through both the surface layer and the ⅜" backing board (base ply) to provide the proper penetration in the supporting wood or metal construction.

For best results, the screw gun should be kept perpendicular to the work surface. Adequate pressure must be exerted to engage the clutch and prevent the screw from slipping (also known as walking). The tool should be triggered continuously until each fastener is seated. A one-piece socket makes driving easier and more efficient than separate socket and extension pieces because it provides a more rigid base and firmer control. Depth gauges are useful to ensure proper penetration.

Because fewer fasteners are required when screws are used to attach drywall, the number of fasteners to be finished is reduced and possible application defects are minimized. Screws should be placed 12" on center on ceilings and 16" on center on walls where the framing members are 16" on center. Screws should be placed at a maximum of 12" on center on walls and ceilings where the framing members are 24" on center. Double screws are recommended in the latter case.

The required penetration for screws is as mentioned earlier in this section. Drywall should be attached to steel framing and furring with Type S screws spaced no more than 12" on center along supports for both walls and ceilings. Type S-12 screws are required for steel framing that is 20 gauge or heavier. A 12" screw spacing is also appropriate when drywall is mounted on resilient furring channels over wood framing.

1.2.3 Adhesives

Adhesives are used to bond single layers of drywall directly to the framing, furring, masonry, or concrete. They can be used to laminate drywall to base layers of backer boards, sound-deadening boards, rigid foam, and other rigid insulating boards. The adhesive must be used in combina-

Electric Screwdriver Attachment

This attachment holds a strip of 50 drywall screws. It has a depth control that allows the operator to set and lock in the correct depth for drywall screws.

27206-13_SA02.EPS

tion with nails or screws, which provide supplemental support.

The adhesives used for applying wallboard finishes are classified as follows:

- Stud adhesives
- Laminating adhesives such as dry powder (including joint tape compound), special drywall laminating adhesives, and drywall contact and modified contact adhesives

Stud adhesives are specially prepared to attach single-ply wallboard to steel or wood studs and are generally used in conjunction with nails. Some permit a significant reduction in the use of mechanical fasteners, but they still require some fastening, at least at the board perimeters. These adhesives should be of caulking consistency so that they bridge framing irregularities. Stud adhesives should meet the requirements of the *Standard Specification for Adhesives for Fastening Gypsum Wallboard to Wood Framing* (*ASTM C557*). This specification covers workability, consistency, open time, wetting characteristics, strength, bridging ability, aging, and freeze/thaw resistance. These adhesives are applied with an electric, pneumatic, or hand-operated gun (*Figure 11*) in a continuous or semicontinuous bead.

If the stud adhesive has a solvent base, it should not be used near an open flame, in poorly ventilated areas, or for laminating drywall. Special adhesives are available for architectural drywall.

> NOTE: Check the temperature range for the adhesive you plan to use, to make sure it is compatible with the expected operating temperatures of the building.

Dry-powder drywall joint compounds are sometimes used to laminate drywall panels to each other or to suitable masonry or concrete surfaces. Dry-powder adhesives are not intended for use in bonding drywall to wood framing or furring, although special laminating adhesives, as well as some stud adhesives, can be used when recommended by the manufacturer.

Only as much laminating adhesive should be mixed as can be used within the working time specified by the manufacturer. The water used should be at room temperature and clean enough to drink. The adhesive may be applied over the entire board area with a suitable spreader, applied in spaced parallel ribbons, or applied in a pattern of spots, as recommended by the manufacturer. All dry-powder laminating adhesives require permanent mechanical fasteners at the board perimeters.

CORDLESS ELECTRIC

PNEUMATIC

HAND OPERATED

Figure 11 Adhesive applicators.

If the boards are applied vertically on side walls, fasteners are placed at the top and bottom. Face boards may require temporary support or supplemental fasteners until the full bond strength is developed.

Drywall contact adhesives require permanent mechanical fasteners at least at the perimeters of all boards applied to walls, ceilings, and soffits. If used to apply architectural drywall vertically on side walls, permanent fasteners are required only at the top and bottom of the boards, where they can be hidden by base and ceiling moldings or other decorative trim.

Contact adhesives may be used to laminate drywall panels to each other or to steel studs. The adhesive is applied by roller, spray gun, or brush in a thin, uniform coating to both surfaces to be bonded. For most contact adhesives, some drying time is usually required before surfaces can be joined and the bond can be developed.

To ensure proper adhesion between surfaces, the face board should be impacted over its entire surface with a suitable tool, such as a rubber mallet. No temporary supports are needed while a contact adhesive sets and the bond forms.

One disadvantage of contact adhesives is their inability to fill irregularities between surfaces, which leaves some areas without an adhesive bond. Another disadvantage is that most of these adhesives do not permit moving of the board once contact has been made. A sheet of polyethylene film or tough building paper can be slipped between the surfaces so gradual bonding of surfaces will occur as the slip sheet is withdrawn. Extra care should be taken when contact adhesives are used. The manufacturer's recommendations should always be followed.

Modified contact adhesives provide a longer placement time. They have an open time (up to a half hour) during which the board can be repositioned, if necessary. They combine good long-term strength with a sufficient immediate bond to permit erection with a minimum of temporary fasteners.

In addition, these adhesives have enough bridging ability to cover up minor framing irregularities. Modified contact adhesives are intended for attaching wallboard to all types of supporting construction, such as solid walls, other drywall panels, and various insulating boards, including rigid foam insulation.

Adhesives are also used for securing drywall materials and paneling to steel studs. The use of adhesives will eliminate some of the fasteners required. Use only the adhesives specified by the manufacturers or suppliers of the particular steel framing and sheathing material being used. Improper adhesives will not only fail to add to the structural integrity, but may be detrimental to system performance.

> **WARNING!**
> Observe all safety data sheet (SDS) precautions for adhesives. Extreme caution must be taken when using contact cement, as it is highly flammable. It should be used in a well-ventilated area, as the fumes can quickly overcome a worker.

1.2.4 Application of Adhesives

Stud adhesives should be applied with a caulking gun in accordance with the manufacturer's recommendations. A straight bead, approximately ¼" in diameter, is applied to the face of the studs in the field (center) of the panel. See *Figure 12*. Where two gypsum panels join over a stud, two parallel beads of adhesive should be applied, one near each edge of the stud.

Single-ply drywall systems attached with stud adhesives require supplemental perimeter fasteners. The fasteners should be placed 16" on center along the edges or ends on boards that are perpendicular or parallel to the supports.

Ceiling installations require supplemental fasteners in the field as well as on the perimeter. They should be placed 24" on center. See *Figure 13*.

Using Adhesives on Drywall

When drywall is applied using adhesive, either drywall adhesive or construction adhesive may be used if it meets the requirements of *ASTM C557*. The adhesive should be allowed to dry for 48 hours before finishing the joints. Using adhesive does not eliminate the need for fasteners; it simply reduces the number of fasteners needed. Check the job specifications or local codes.

Adhesives cannot be used to attach drywall panels to studs if the building has an inside moisture barrier.

When laminating drywall panels for multilayer installation, either lightweight or standard setting-type joint compound may be used in place of adhesive to laminate the panels.

27206-13_SA03.EPS

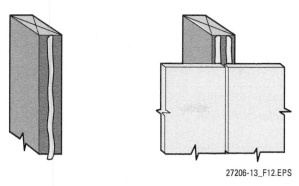

Figure 12 Adhesive applied to the edges of a stud.

Adhesive is not required at top or bottom plates, bridging, bracing, or fire stops. Where fasteners at vertical joints are undesirable, gypsum panels may be prebowed, as shown in *Figure 14*.

Prebowing puts an arc in the drywall, which keeps it in tight contact with the adhesive after the board is applied. Supplemental fasteners (placed 16" on center) are then used at the top and bottom plates.

Drywall may be prebowed by stacking it, face up, with the ends resting on 2 × 4 lumber or other blocks, and with the center of the boards resting on the floor. Allow it to remain overnight or until the boards have a permanent bow.

Architectural boards can also be installed using adhesive, but care should be taken to avoid adhesive contact with the decorated face. Position the boards within the open time specified for the adhesive, and use a rubber mallet to tap the boards along the studs to ensure a continuous bond with the framing. Follow the manufacturer's specifications for architectural drywall.

1.2.5 Adhesive Application to Metal Framing

Some stud adhesives, such as those used with steel framing, require fasteners on intermediate supports as well as at the perimeters of gypsum panels. The framing spacing varies both according to the load and the type of board being used. See *Table 2*.

1.2.6 Adhesive Application to Concrete and Masonry

Drywall panels can be laminated directly to above-grade interior masonry and concrete wall surfaces if the surface is dry, smooth, clean, and flat. Drywall can be laminated directly to exterior cavity walls if the cavities are properly insulated to prevent condensation and the inside face of the cavity is properly waterproofed.

Prefinished drywall with a surface that is highly resistant to water vapor should not be laminated to concrete or masonry, because moisture may be-

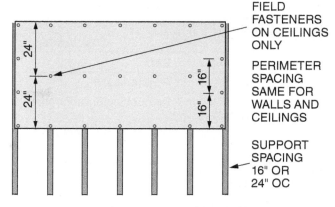

HORIZONTAL APPLICATION

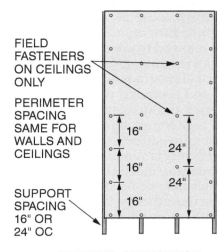

VERTICAL APPLICATION

Figure 13 Supplemental wall and ceiling fasteners.

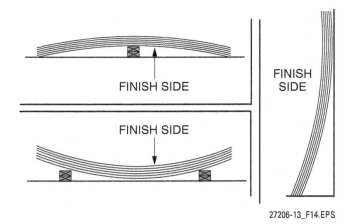

Figure 14 Prebowing of gypsum panels.

come trapped within the gypsum core of the board. The base surface must be made as level as possible. Rough or protruding edges and excess joint mortar should be removed and any depressions filled with mortar to make the wall surface level.

Table 2 Maximum Spacing of Ceiling Framing

Gypsum Board (Thickness)		Application to Framing		Maximum OC Spacing of Framing
Base	Face	Base	Face	
⅜"*	⅜"	Perpendicular	Perpendicular or Parallel	16"
½"**	⅜" or ½"	Perpendicular or Parallel	Perpendicular or Parallel	16"
⅝"**	½"	Perpendicular or Parallel	Perpendicular or Parallel	16"
⅝"**	⅝"	Perpendicular or Parallel	Perpendicular or Parallel	24"

* Adhesive between plies should be dried or cured prior to any decorative treatment. This is especially important when a spray-applied, water-based texture finish is to be used.

** Sidewalls – For two-layer application with adhesive between the plies ⅜", ½", or ⅝" gypsum board may be applied perpendicularly (horizontally) or parallel (vertically) on framing spaced a maximum of 24" OC.

Base surfaces should be cleaned of all oil, curing compound, loose particles, dust, and grease in order to ensure an adequate bond. Concrete should be allowed to cure for at least 28 days before drywall is laminated directly to it.

Exterior below-grade walls or surfaces should be furred and protected with the installation of a vapor barrier and insulation in order to provide a suitable base for attaching the drywall. This is also true for any surface that cannot be prepared readily for direct lamination.

Supplemental mechanical fasteners spaced 16" on center may be used to hold the drywall in place while the adhesive is developing a bond.

A variety of clips, runners, and adjustable brackets are available with furring systems to facilitate installation over irregular masonry walls, as shown in *Figure 15*. When special clips are used, the manufacturer's instructions for their use must be followed.

> **NOTE**
> An alternative to anchoring furring to a masonry wall is to build a 1⅝" metal stud wall.

1.3.0 Drywall Accessories

A variety of drywall accessories, such as corner beads, casings, and control joints, are used in drywall construction. Corner beads and casings are installed to protect and cover the ends of drywall panels. Control joints provide room for expansion and contraction of the base material and drywall panels.

1.3.1 Corner Beads and Casings

Figure 16 shows common beads and casings used around doors, windows, and other openings. They are also used when drywall is butted against a different surfacing material. Casing moldings cover the unfinished ends of drywall to protect it, while corner beads protect outside corners. L-beads protect the drywall where it abuts dissimilar materials, but they require finishing; J-beads serve the same purpose, but do not require finishing.

Figure 17 shows examples of decorative moldings, also called trim, used with architectural drywall panels.

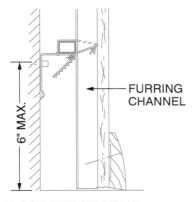
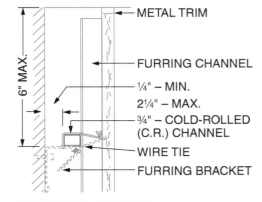

Figure 15 Adjustable wall furring.

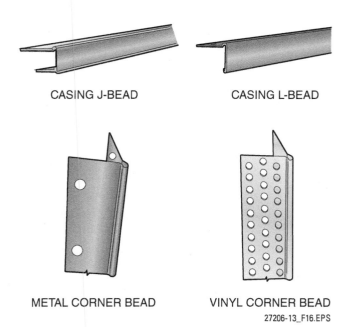

Figure 16 Examples of protective beads.

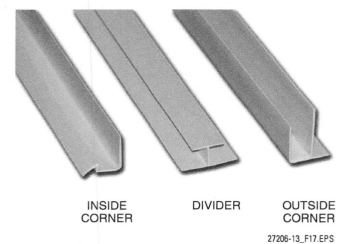

Figure 17 Moldings for prefinished drywall.

1.3.2 Control Joints

Control joints (*Figure 18*) should be installed in drywall systems wherever expansion or control joints occur in the base exterior wall and not more than 30' on center in long wall furring runs. Wall or partition height door or window frames may be considered control joints.

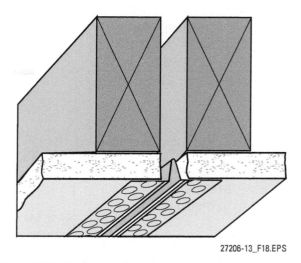

Figure 18 Typical control joint.

Control joints should also be used over window and door openings. The drywall should be isolated from structural elements such as columns, beams, and loadbearing interior walls, and from dissimilar wall or ceiling finishes by control joints, metal trim, or other means. Refer to the specifications and drawings for the locations of control joints.

Movement of the structure can impose severe stresses and cause cracks, either at the joint or in the field of the board. Cracks are more prevalent at an archway or over a door because this is usually the weakest point in the construction. In new construction, it is wise to wait until at least one heating season has passed before repairing or refinishing.

A source of cracking in nonbearing walls of high-rise or commercial buildings is the modern trend toward less-rigid structures. Larger deflections in structural members and greater expansion and contraction of exterior columns can impose unexpected loads on nonbearing walls and lead to cracking. Detail designs for perimeter relief of nonbearing partitions are available to improve this condition. One solution is to use relief runners to attach nonbearing walls to ceiling and column members (*Figure 19*).

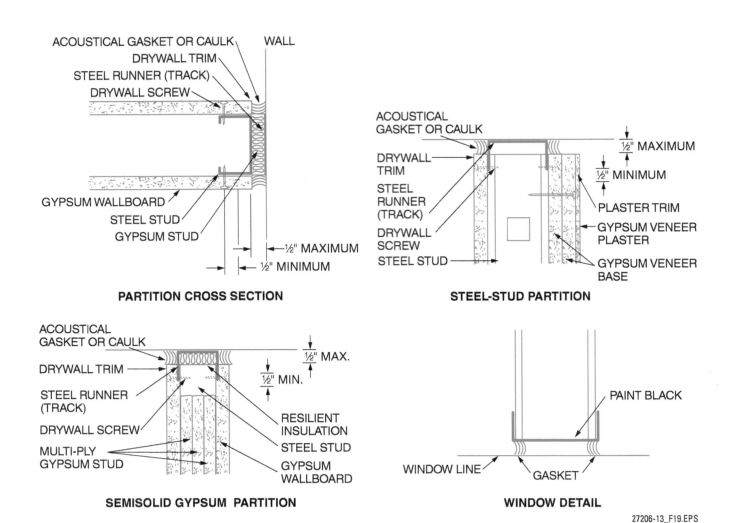

Figure 19 Designs for perimeter relief.

Control Joints

Control (expansion) joints are used in large expanses of wall or ceiling drywall to compensate for the natural expansion and contraction of a building. Control joints help prevent cracking and joint separation. They are common in commercial construction, especially where exterior concrete walls contain expansion joints.

If the control joint is installed in a space where fire rating and/or sound control are important, a seal must be used behind the control joint. The control joint has a ¼" slot that is covered by plastic tape. The tape is removed after the joint is finished, leaving a small recess.

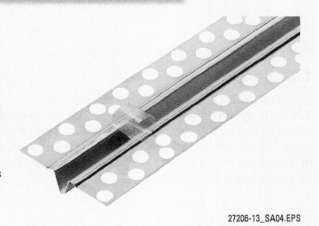

Cutting Small Strips

The drywall stripper shown here is designed to cut narrow strips of wallboard up to 4½". There is a sharp edge on each side of the tool, so it cuts both sides of the panel at once. The drywall stripper makes a cleaner cut than a utility knife when cutting long, narrow strips such as those that might be needed around a window or door.

and other openings can have an adverse effect on the ability of a building to achieve its STC rating, particularly in higher-rated construction. Where a very high STC performance is needed, air conditioning, heating, and ventilating ducts should not be included in the assembly. Failure to observe special construction and design details can destroy the effectiveness of the best assembly.

Improved sound isolation is obtained by the following:

- Separate framing for the two sides of a wall
- Resilient-channel mounting for the drywall
- Using sound-absorbing materials in wall cavities

Figure 31 Screw guns.

Figure 32 Drywall lift.

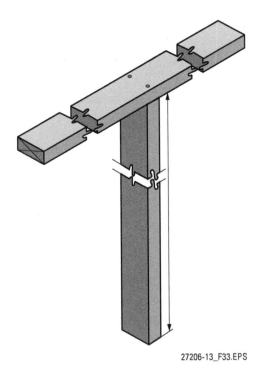

Figure 33 T-brace.

- Using adhesive-applied drywall of varying thicknesses in multilayer construction
- Caulking the perimeter of drywall partitions, openings in walls and ceilings, partition/mullion intersections, and outlet box openings
- Locating recessed wall fixtures in different stud cavities

The entire perimeter of sound-isolating partitions should be caulked around the drywall edges to make it airtight, as detailed in *Figure 34*. The caulking should be a nonhardening, nonshrinking, nonbleeding, nonstaining resilient sealant.

Sound-control sealing must be covered in the specifications, understood by all related tradespeople, supervised by the appropriate party, and inspected carefully as the construction progresses.

2.3.1 Separated Partitions

A staggered wood-stud gypsum partition placed on separate plates will provide an STC between 40 and 42. The addition of a sound-absorbing material between the studs of one partition side can increase the STC by as much as 8 points. With $\frac{5}{8}$" Type X drywall on each side, an assembly has a fire resistance classification of one hour. Separated walls without framing can also be constructed by using an all-gypsum, double-solid, or semisolid partition. Steel or wood tracks fastened to the floor and ceiling hold the partitions in place.

For the attachment of kitchen cabinets, lavatories, ceramic tile, medicine cabinets, and other fixtures, a staggered stud wall rather than a resilient wall is recommended. The added weight and fastenings may short-circuit the construction acoustically.

Locating Electrical Boxes

The Blind Mark™ system uses magnets to find electrical boxes under drywall. A target magnet is placed in the electrical box before the drywall is installed. After the drywall is up, a locator magnet is used to find the box containing the target.

(1)

(2)

(3)

(4)

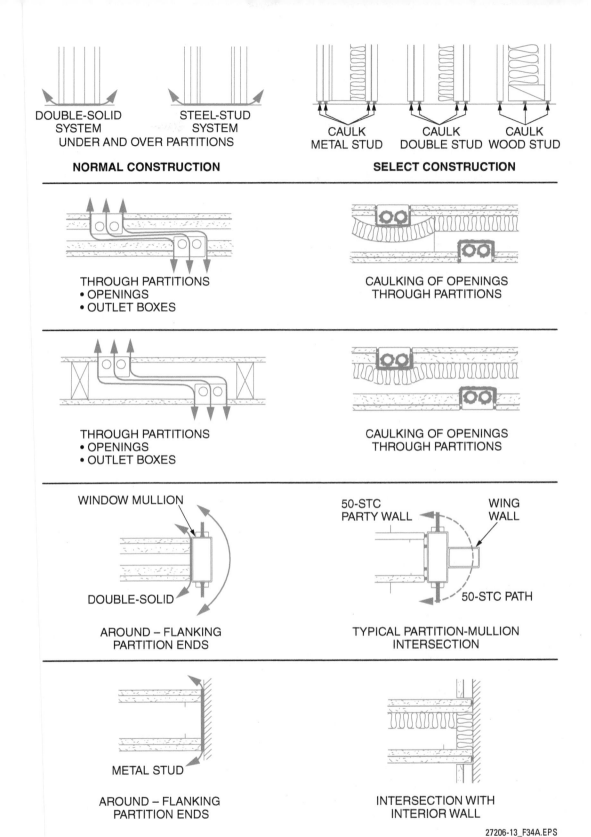

Figure 34 Caulking of sound-isolation construction (1 of 2).

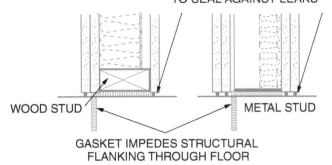

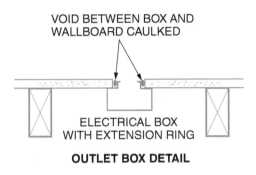

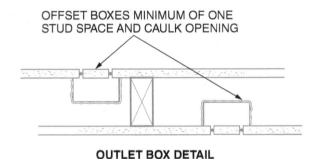

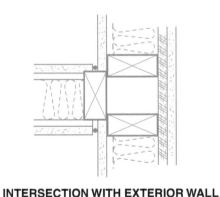

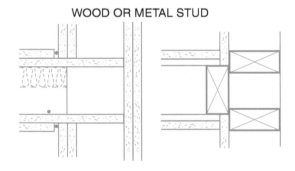

Figure 34 Caulking of sound-isolation construction (2 of 2).

2.3.2 Resilient Mountings

Resilient attachments acting as shock absorbers reduce the passage of sound through the wall or ceiling and increase the STC rating. Further STC increases can result from more complex construction methods incorporating multiple layers of drywall and building insulation in the wall cavities.

Resilient furring channels are attached with the nailing flange down and at right angles to the wood stud, as shown in *Figure 35*.

To install furring channels, drive 1¼" Type W screws or 6d coated nails through the prepunched holes in the channel flange. With extremely hard lumber, ⅞" or 1" Type S screws may be used. Locate the channels 24" from the floor, within 6" of the ceiling line, and no more than 24" on center. Extend the channels into all corners and fasten them to the corner framing. Attach ½" × 3" drywall filler strips to the bottom plate directly over the studs by overlapping the ends and fastening both flanges to the stud. Apply the drywall hori-

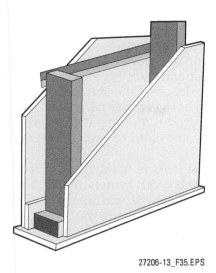

Figure 35 Attached resilient furring channel.

zontally with the long dimension parallel to the resilient channels using 1" Type S screws spaced 12" on center along the channels. The abutting edges of boards should be centered over the channel flange and securely fastened.

2.3.3 Sound-Isolating Materials

Sound-isolating materials include:

- Mineral fiber (including glass) blankets and batts used in wood stud assemblies
- Semirigid mineral or glass-fiber blankets for use with steel studs and laminated gypsum partitions
- Mineral (including glass) fiberboard
- Gypsum-core sound-insulating board used behind drywall, applied with adhesive or mechanically fastened
- Rigid plastic foam furring systems
- Lead or other special shielding materials

Mineral-wool or glass-fiber insulating batts and blankets may be used in assembly cavities to absorb airborne sound within the cavity. They should be placed in the cavity and carefully fitted behind electrical outlets and around any cutouts necessary for plumbing lines. Insulating batts and blankets may be faced with paper or another vapor barrier, and may have flanges or be of the unfaced friction-fitted type.

Drywall may be applied over rigid plastic foam insulation. It is applied on the interior side of exterior masonry and concrete walls to provide a finished wall, and to protect the insulation from early exposure to fire originating within the building. Additionally, these systems provide the high insulating values needed for energy conservation.

In new construction or for remodeling, these systems can be installed with as little as 1" dimension from the inside face of the framing or masonry (½" insulation and ½" Type X drywall).

When applying drywall over rigid foam insulation, the entire insulated wall surface should be protected with the drywall, including the surface above ceilings and in closed, unoccupied spaces.

Single-ply or double-ply, ½" or ⅝" drywall should either be screw-attached to steel wall furring members attached to the masonry or nailed directly into wood framing, as shown in *Figure 36*. Follow the insulation manufacturer's instructions.

Furring members should be designed to minimize thermal transfer through the member and to provide a 1¼" minimum-width face or flange for screw application of the drywall.

Furring members should be installed vertically and spaced 24" on center. Blocking or other backing as required for attachment and support of fixtures and furnishings should be provided. Furring members should also be attached at floor/wall and wall/ceiling angles (or at the termination of the drywall above suspended ceilings) and around door, window, and other openings. Single-ply drywall should be applied vertically, with the long edges of the board located over furring members. The installation should be planned carefully to avoid end joints.

In double-ply applications, the base ply should be applied vertically. In horizontal face-ply applications, the face ply and end joints should be offset by at least one framing or furring member space from the base-ply edge joints.

In wallboard applications, mechanical fasteners should be of such a length that they do not penetrate completely to the masonry or concrete. In single-layer applications, all joints between drywall panels should be reinforced with tape. In

Avoid Back-to-Back Fixtures

Medicine cabinets; electrical, telephone, television, and intercom outlets; and plumbing, heating, and air conditioning ducts should not be installed back-to-back. Any opening for such fixtures, piping, and electrical outlets should be carefully cut to the proper size and fire-caulked.

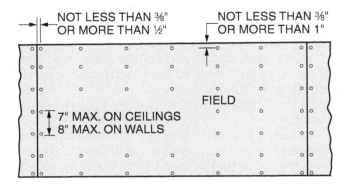

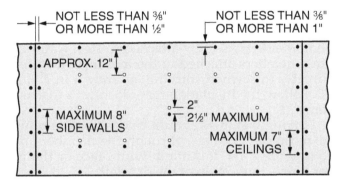

Figure 36 Nailing patterns for installation over rigid foam insulation.

addition, drywall joints should be finished with joint compound. In two-ply applications, the base-layer joints may be concealed or left exposed.

Adhesive should not be used to apply vinyl-faced drywall face layers over a wall insulated with rigid foam.

2.4.0 Installing Drywall

Drywall panels can be applied over any firm, flat base such as wood or steel framing and furring. Gypsum can also be applied to masonry and concrete surfaces, either directly or to steel or wood furring strips. If the board is applied directly, any irregularities in the masonry or concrete surfaces must be smoothed or filled. Furring is a means to provide a flat surface for standard fastener application. It also provides a separation to overcome dampness in exterior walls.

The most common type of residential interior wall construction is the standard drywall system with joints between the panels and internal corners reinforced with tape and covered with joint treatment compound to prepare them for decoration.

The term *joint* is used to describe any point where two drywall panels meet. A butt joint is where two sheets of wallboard with untapered sides meet. A flat joint is the intersection of two bevel-edged wallboards. External corners are normally reinforced with corner bead, which, in turn, is covered with joint compound. Exposed edges are covered with metal or plastic trim. The result is a smooth, unbroken surface ready for final decoration. When architectural board is used, no further decoration is necessary, but trim moldings or battens can be used to cover the joints.

2.4.1 Preparing the Job Site

Job conditions such as temperature and humidity can affect the performance of joint treatment materials and the appearance of the joint. These conditions may also affect adhesive materials and their ability to develop an adequate bond. During the cold season, interior finishes should not be installed unless the building is maintained between 50°F and 80°F. These temperatures should also be maintained for at least 48 hours after the installation. All materials must be protected from the weather.

When ceilings are to receive water-based spray texture finishes, special attention must be given to the spacing of framing members, the thickness of the board used, ventilation, vapor barriers, insulation, and other factors that can affect the performance of the system and cause problems, particularly sag of the drywall between framing members.

Lumber must be kept dry during storage and installation at the job site. Its moisture content should not exceed 15 percent at the time of drywall application. Green lumber should not be

Rigid Foam Insulation

Drywall applied over rigid plastic foam insulation in the manner described may not necessarily provide the finish ratings required by local building codes. Many building codes require a minimum fire protection for rigid foam on interior surfaces equal to that provided by ½" Type X drywall when tested over wood framing. The flammability characteristics of rigid foam insulation products vary widely, and the manufacturer's literature should be reviewed.

used for framing. Since lumber shrinks across the grain as it dries, it tends to expose the shanks of nails driven into the edges of the framing members. If shrinkage is substantial or the nails are too long, separation between the drywall and its framing member can result in nail pops.

The delivery of drywall should coincide with the installation schedule. Boards should be placed for convenience at the work location. Boards should be stored flat and under cover. The materials used as storage supports should be at least 4" wide. As the units are tiered, the supports should be carefully aligned from bottom to top so that each tier rests on a solid bearing, as shown in *Figure 37*. Be careful to avoid excessive weight.

Stacking long lengths on short lengths should be avoided to prevent the longer boards from breaking. Leaning boards against the framing members for a prolonged period of time with the long edges horizontal is not recommended. You should avoid leaning boards during periods of high humidity, as the boards could be subject to warping. All materials should remain stored in their original wrappers or containers until ready to use on the job site. When boards are moved, they should be carried, not dragged, so the edges are not damaged.

Figure 37 Drywall storage.

2.4.2 Cutting and Fitting Procedures

Any drywall installation should be carefully planned. Accurate measuring, cutting, and fitting are very important. In residential buildings with less than 8'-1" ceiling heights, it is preferred that the wallboard be installed at right angles to the supporting members because there are usually fewer joints to finish. On long walls, boards of maximum practical lengths should be used to minimize the number of end joints. Scored, scratched, broken, or otherwise damaged boards should not be used.

Measurements should be done accurately at the correct ceiling or wall location for each edge or end of the board. Accurate measuring will usually reveal any irregularities in framing or furring so corrective allowances can be made in cutting. Poorly aligned framing should be corrected before applying drywall.

Drywall should be cut by first scoring through the paper down to the core with a sharp utility knife, working from the face side. The board is then snapped back away from the cut face.

The back of the paper is broken by cutting it with a utility knife. Drywall may also be cut by sawing. All cut edges and ends of the drywall should be smoothed to form neat, tight-fitting joints when installed. Ragged cut ends or broken edges can be smoothed with a rasp or sandpaper, or trimmed with a sharp knife. If burrs on the cut ends are not removed, they will form a visible ridge in the finished surface.

The practices listed below should be followed to ensure a sound application:

- Install the ceiling boards first, then the wall panels.

Moving Drywall

A drywall dolly like the one shown here is specially designed to transport drywall panels at the job site.

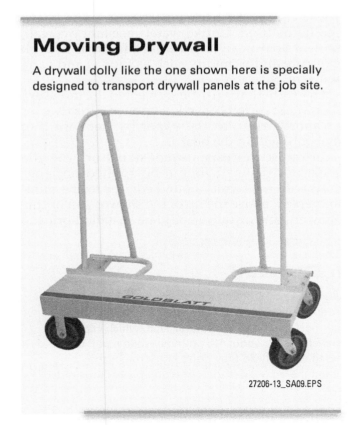

- The panels should fit easily into place without force.
- Always match edges and ends. For example, tapered end to tapered end and square-cut end to square-cut end.
- Plan to span the entire length of ceilings or walls with single boards, if possible, to reduce the number of end joints, which are more difficult to finish.
- Stagger end joints and locate them as far from the center of the ceiling or wall as possible, so they will be inconspicuous.
- In a single-ply application, the board ends and edges parallel to the supporting members (framing) should fall on these members to reinforce the joint.
- Mechanical and electrical equipment, such as cover plates, registers, and grilles, should be installed to provide for the final wall thickness when applying the trim.
- Place a shim under the wallboard to keep it from absorbing moisture from the floor.

NOTE

The depth of electrical boxes should not exceed the framing depth, and boxes should not be placed back-to-back on the same stud. Electrical boxes and other devices should not be allowed to penetrate completely through the walls. This is detrimental to sound isolation and fire resistance. Make sure the wires are pushed back into the box.

2.4.3 Floating Interior Angle Construction

To minimize the possibility of fastener popping in areas adjacent to wall and ceiling intersections and to minimize cracking due to structural stresses, the floating-angle method may be used for either the single-layer or double-layer application of drywall to wood framing.

Floating interior angle construction helps to eliminate nail popping and corner cracking by omitting fasteners at the intersections of walls and ceilings. This is applicable for single nailing, double nailing, and screw attachment. *Figure 38* shows a typical single-layer application. The same nail-free clearances at corners should be maintained in double nailing.

In floating interior angle construction where the ceiling framing members are perpendicular to the wall/ceiling intersection, the ceiling fasteners should be located 7" from the intersection for single nailing and 11" to 12" for double nailing or screw applications.

On ceilings where the joists are parallel to the wall intersection, nailing should start at the intersection. Drywall should be applied to the ceiling first and then to the walls. See *Figure 39*.

Drywall on side walls should be applied to provide a firm, level support for the floating edges of the ceiling board.

Apply the overlapping board firmly against the underlying board to bring the underlying board into firm contact with the face of the framing member behind it. The overlapping board should be nailed or screwed, and the fasteners should be omitted from the underlying board at the vertical intersection.

2.4.4 Moisture-Resistant Construction

Special consideration must be given when finishing bathrooms, laundries, kitchens, and other areas subject to moisture. Although water-resistant gypsum board can be used in these applications, a waterproof tile backer such as cement board should be used. Unlike water-resistant drywall, cement board can also be used in areas of high moisture and humidity such as saunas and gang showers.

Drywall that will be subjected to moisture should not be foil backed or applied directly over a vapor barrier, as the vapor barrier will trap moisture within the board.

In moisture-resistant construction, the tile backer or drywall should be applied horizontally, with the factory-bound edge spaced a minimum of ¼" above the lip of the shower pan or tub. Shower pans or tubs should be installed prior to

Storing and Handling Gypsum Drywall

Figure 37 shows gypsum drywall as it would be stored in a warehouse or building supply store. Just before installation, the drywall panels would be distributed along interior walls and stood on edge.

Drywall panels are sold in pairs. Two sheets of drywall are connected by a strip of paper tape, which can be stripped off to separate the panels. Drywall is heavy. Two ½" panels weigh about 110 pounds, while a pair of ⅝" panels weigh close to 150 pounds. The panels need to be handled carefully so they don't break under their own weight. They should be lifted and carried by the edges rather than the ends. Also, proper lifting procedures must be used by people handling drywall, in order to prevent injury.

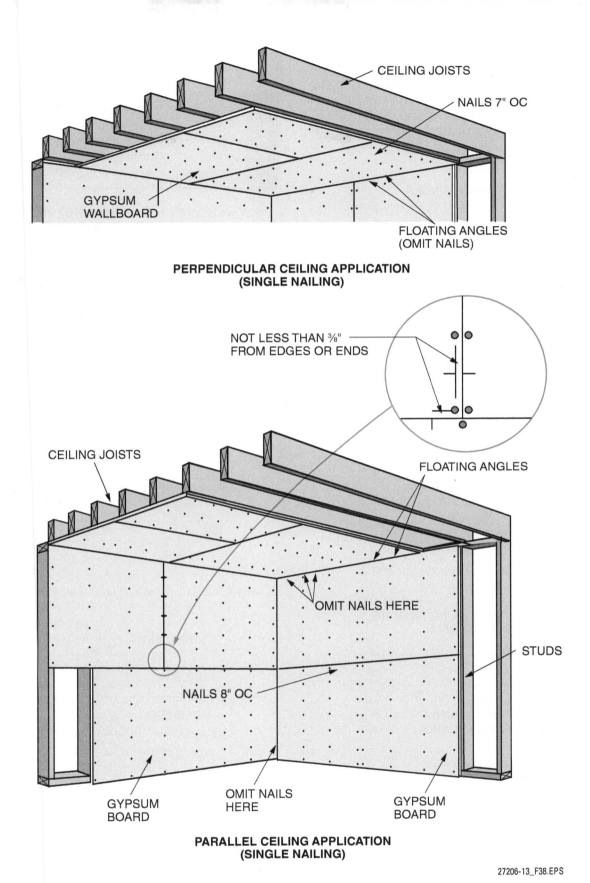

Figure 38 Floating interior angle construction.

the installation of the board. Shower pans should have an upstanding lip or flange located at a minimum of 1" higher than the entry wall to the shower. It is recommended that the tub be supported. If necessary, the board should be furred away from the framing members so the upstanding leg of the pan (*Figure 40*) will be on the same plane as the face of the board.

> **NOTE**
> Different types of waterproof boards have different applications and limitations, so it is always necessary to check the manufacturer's product data sheets and installation instructions for the type of board being used.

An additional board extending the full height from floor to ceiling is required for a fire-rated or sound-rated construction (*Figure 41*).

Suitable blocking should be provided approximately 1" above the top of the tub or pan. Between-stud blocking should be placed behind the horizontal joint of the board above the tub or shower pan. For ceramic tile applications, use studs that are at least 3½" deep and placed 16" on

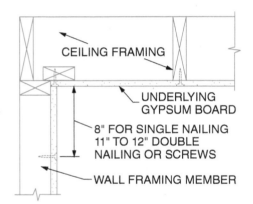

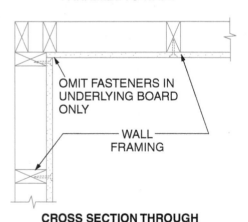

Figure 39 Fastener patterns for floating interior angle construction.

Figure 40 Pan is on the same plane as the face of the board.

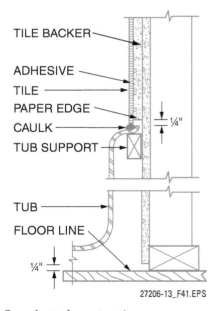

Figure 41 Sound-rated construction.

center. Appropriate blocking, headers, or supports should be provided for tub plumbing fixtures and to receive soap dishes, grab bars, towel racks, and similar items.

Tile backer boards should be applied with nails or screws spaced not more than 8" on center. When ceramic tile more than ⅜" thick is to be applied, the nail or screw spacing should not exceed 4" on center. When it is necessary for joints and nail heads to be treated with joint compound and tape, either use waterproof, nonhardening caulking compound, or seal-treat joints and nail heads with a compatible sealer prior to the installation.

Interior angles should be reinforced with supports to provide rigid corners. The cut edges and openings around pipes and fixtures should be caulked flush with a waterproof, nonhardening, silicone caulking compound or adhesive complying with the *USA Standard for Organic Adhesives for Installation of Ceramic Tile*. The directions of the manufacturer of the tile, wall panel, or other surfacing material should also be followed.

The surfacing material should be applied down to the top surface or edge of the finished shower floor, return, or tub, and installed to overlap the top lip of the receptor, subpan, or tub.

> **NOTE**
> The caulking compound or sealer must be compatible with the adhesive to be used for the application of the tile. Follow the adhesive manufacturer's instructions.

2.4.5 Resurfacing Existing Construction

Drywall may be used to provide a new finish on existing walls and ceilings of wood, plaster, masonry, or wallboard. If the existing surface is structurally sound and provides a sufficiently smooth and solid backing without shimming, ¼" drywall can be applied with adhesives, nails, or screws. Drywall nails should penetrate the framing by ⅞". When power-driven screws are to be used, the threaded portion of the screw must penetrate the framing by at least ⅝".

Tile Application

Ceramic wall tile application to drywall should meet the *USA Standard for Organic Adhesives for Installation of Ceramic Tile* for installation of ceramic tile with water-resistant organic adhesive. The adhesives used should meet or exceed the information from the same standard.

Existing surfaces that are too irregular to receive drywall directly should be furred and shimmed to provide a suitable fastening surface. The minimum drywall thickness for various support spacing and installation methods should be the same as new construction over furring. Any surface trim for mechanical and electrical equipment, such as switch plates, outlet covers, and ventilating grilles, should be removed and saved for reinstallation. Electrical boxes should be reset prior to the installation of new drywall.

2.5.0 Special Applications

Drywall is also available for specialty applications. Abuse-resistant drywall panels are manufactured with a strong paper face and a heavy-duty backing sheet to improve the integrity of the panels. Abuse-resistant panels offer greater resistance to indentation and through-penetration than standard gypsum drywall panels.

Unlike standard drywall panels that have a paper facing, paperless drywall is faced with a fiberglass netting, which is moisture resistant. One of the disadvantages of standard drywall panels is that the paper facing helps spread mold if the panels are exposed to moisture. While the fiberglass netting is not mold proof, it does minimize the spread of mold.

Lead-lined drywall, also known as leaded drywall, is a standard drywall panel with a sheet of rolled lead laminated to it. Lead-lined drywall is available in ½" and ⅝" thicknesses, with the lead sheet ranging from ½₂" to ⅛" in thickness. Lead-lined gypsum drywall panels can be ordered

Selecting the Right-Size Panel

Drywall panels come in two common sizes: 8' and 12'. The 12' size is heavier and costs more to buy than the 8' size. However, the 12' size is likely to be more economical for some spaces because, even though the initial cost may be greater, 12' panels result in fewer joints. Therefore, less finishing labor and finishing materials are required. For example, if you have an 11' wide, 8' high wall, two 12' panels would span the entire wall, leaving only one wall joint plus the corner joints to finish. With 8' panels, the same wall would have three wall joints to finish.

with a standard butt edge or with the lead sheet extending approximately 1" beyond the drywall, for overlapping adjacent pieces.

Drywall products are also commonly used as exterior (nonappearance) sheathing on commercial buildings. These exterior sheathing panels are faced with fiberglass mats to provide moisture and mold resistance. These drywall products provide a rigid substrate for a wide variety of air and moisture barriers, and are commonly used behind brick, siding, exterior insulation finish systems (EIFS), and stucco.

2.0.0 Section Review

1. Fastening schedules provide information about _____.
   a. the estimated arrival time of fastners on the job site
   b. the size of the needed fastners
   c. the proper spacing of fastners on the drywall
   d. the type of fastner material needed

2. Exact hole spaces for electrical boxes are cut in drywall using a _____.
   a. jab saw
   b. light box cutter
   c. drywall saw
   d. rasp

3. Each of the following is a construction method used to control noise *except* _____.
   a. caulking around outlet box openings
   b. placing air conditioning ducts back-to-back
   c. using separate framing for the two sides of a wall
   d. mounting gypsum board in resilient channels

4. The intersection of two bevel-edged wallboards is called a _____.
   a. flat joint
   b. taper joint
   c. butt joint
   d. bevel joint

5. Gypsum-board materials used behind brick, stucco, or siding are referred to as _____.
   a. paperless drywall
   b. cement board
   c. abuse-resistant drywall
   d. sheathing panels

SECTION THREE

3.0.0 RATED VS. NONRATED ASSEMBLIES

Objective

Contrast rated assemblies to nonrated assemblies.
  a. Describe single-ply drywall application.
  b. Describe how fire-rated walls are constructed.
  c. List multi-ply drywall applications.
  d. Describe how to prioritize walls.

Rated and nonrated construction refers to the fire rating of walls, which is dictated by the sound and fireproofing requirements of the local building code. A carpenter should have a good understanding of the differences between rated and nonrated walls, and their construction techniques.

3.1.0 Single-Ply Application

In light commercial and residential construction, single-ply drywall systems are commonly used (*Figure 42*). Generally, they are adequate to meet fire-resistance and sound-control requirements. Multi-ply systems, as shown in *Figure 43*, have two or more layers of drywall to increase sound isolation and fire-resistive performance. They also provide better surface quality because face layers are often laminated over base layers, thereby reducing the number of fasteners. As a result, the surface joints of the face layer are reinforced by the continuous base layers of drywall. Nail pop and ridging problems are less frequent, and imperfectly aligned supports have less effect on the finished surface.

Satisfactory results can be ensured with either single-ply or multi-ply assemblies by requiring the following:

- Proper framing details, consisting of straight, correctly spaced, and properly cured lumber
- Proper job conditions, including controlled temperatures and adequate ventilation during application
- Proper measuring, cutting, aligning, and fastening of the board
- Proper joint and fastener treatment
- Special requirements for proper sound isolation, fire resistance, thermal properties, or moisture resistance

3.2.0 Fire-Rated Walls

The construction of walls and partitions is driven by the fire and soundproofing requirements specified in local building codes. In some cases, a frame wall with ½" gypsum drywall on either side is satisfactory. In extreme cases, such as the separation between the offices and the manufacturing spaces in a factory, it may be necessary to have a concrete block (concrete masonry unit, CMU) wall combined with fire-resistant drywall, along with rigid and/or fiberglass insulation, as shown in *Figure 44*. This is especially true if there is any explosion or fire hazard.

While they are only occasionally used in residential construction, steel studs are the standard for framing walls and partitions in commercial construction.

Once the studs are installed, one or more layers of drywall and insulation are applied. The type and thickness of the wallboard and insula-

Parallel vs. Perpendicular Installation

Perpendicular installation (at right angles to the framing members) is generally preferred when ceilings are normal height—8'-1" or less. A 54" board is available for use with 9' ceilings. Perpendicular installation has several advantages over parallel installation, including:

- There are fewer joints.
- There is less measuring and cutting required.
- Joints are at a convenient height for finishing.
- A single panel ties together more framing members to increase strength and hide framing irregularities.

For rooms with taller ceilings, parallel application is more practical. Parallel application may also be required for normal-height ceilings in order to meet fire ratings.

For ceiling application, select the method that results in the fewest joints.

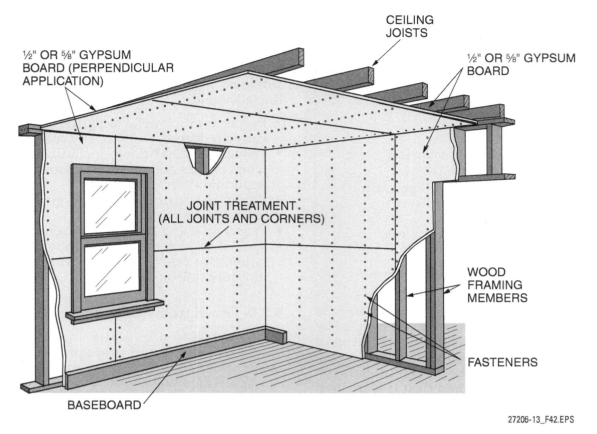

Figure 42 Single-ply construction.

tion depend on the fire rating and soundproofing requirements. Soundproofing needs vary from one use to another and are often based on the amount of privacy required for the intended use. For example, executive offices and medical examination rooms may require more privacy than general offices.

The requirements for sound reduction and fire resistance can significantly affect the thickness of a wall. For example, a steel stud wall with a high

Achieving a Rounded Appearance

A smooth, rounded finish appearance can be obtained by using a bull-nose corner molding and cap such as the ones shown here.

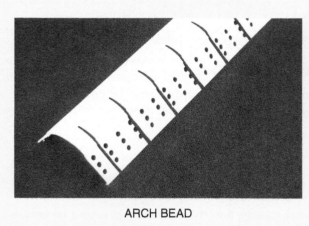

ARCH BEAD

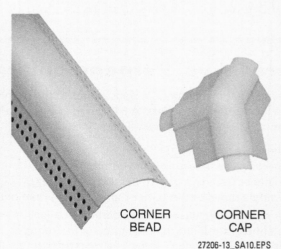

CORNER BEAD   CORNER CAP

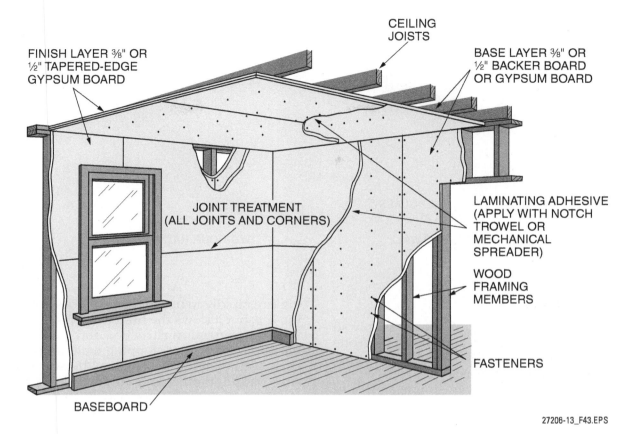

Figure 43 Multi-ply construction.

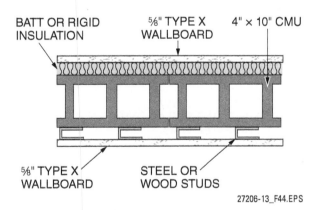

Figure 44 High fire/noise resistance partition.

sound transmission class (STC) and fire resistance might have a total thickness of nearly 6", while a low-rated wall might have a thickness of only 3" using steel studs and 2¼" using wooden studs.

3.2.1 Fire-Rated Construction

Every wall, floor, and ceiling in a building is rated for its fire resistance, as established by building codes. The fire rating is stated in terms of hours, such as one-hour wall or two-hour wall. The rating denotes the length of time an assembly can withstand fire and provide protection from it, as determined under laboratory conditions (*Figure 45*). The greater the fire rating, the thicker the wall is likely to be.

In multifamily residential construction, such as apartments and town houses, the walls and ceilings dividing the occupancies must meet special fire and soundproofing requirements. The code requirements will vary from one location to another and may even vary within areas of a jurisdiction. For example, dwellings in high-risk areas may have stricter standards than those in other areas of the same city or county.

In some cases, the code may require a masonry wall between occupancies. This masonry wall may even be required to penetrate the roof of the building so that if a fire occurs, it is contained within the unit in which it started because it is unable to travel through the walls or across the attic space.

There are many different construction methods for so-called party walls. Each is designed to meet different fire and soundproofing standards. The wall is likely to be more than 3" thick and contain several layers of drywall and insulation. A fire-rated wall may abut a nonrated partition or wall. In this case, the rated wall must be carried through to maintain the fire rating. *Figure 46* shows an example of how this can be done.

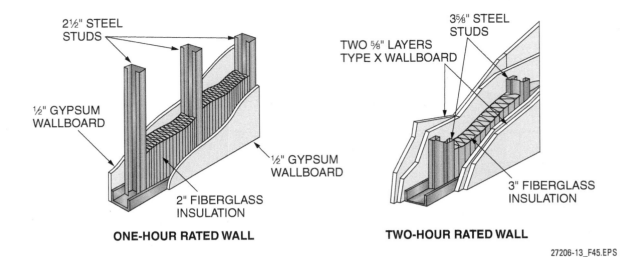

ONE-HOUR RATED WALL

TWO-HOUR RATED WALL

Figure 45 Partition-wall examples.

3.2.2 Fire-Stopping

Fire-stopping means cutting off the air supply so that fire and smoke cannot readily move from one location to another. You will hear the term *fire-stop* used in two different ways.

In frame construction, a fire-stop is a piece of wood or fire-resistant material inserted into an opening such as the space between studs. This fire-stop acts as a barrier to block airflow that would allow the space to act as a chimney, carrying fire rapidly to upper floors. It does not put out the fire, but it slows the fire's progress.

In commercial construction and some residential applications, fire-stopping material is used to close wall penetrations such as those created to run conduit, piping, and air conditioning ducts. If such openings are not sealed, fire will travel through the openings in its search for oxygen.

In order to meet the fire rating standards established by the building and fire codes, the openings must be sealed. The fire-stopping methods

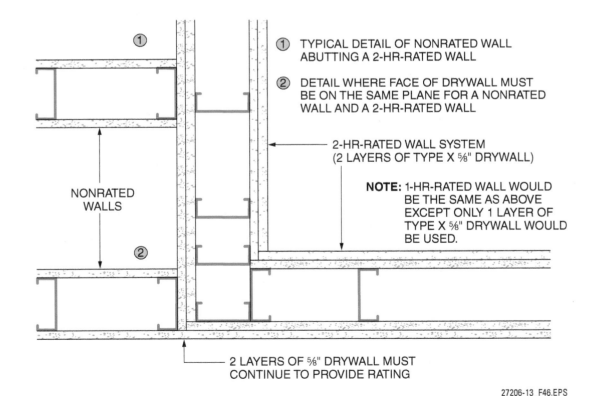

Figure 46 An example of a fire-rated wall abutting a nonrated wall.

Code Compliance

Local codes specify the fire ratings that must be achieved in different occupancies and uses. They may also specify the nailing pattern to be used on drywall panels. Check the local codes before proceeding. Also, keep in mind that electrical and plumbing installations must be inspected in order for the building to receive a certificate of occupancy. Before covering these installations with drywall, make sure the inspection has been performed. An inspection sheet should be at the site.

used for this purpose are classified as mechanical and nonmechanical.

Mechanical fire-stops are devices such as the one shown in *Figure 47* that mechanically seal the opening.

Nonmechanical fire-stops are fire-resistant materials, such as caulks and putties that are used to fill the space around the conduit or piping. You may be required to install various nonmechanical fire-stopping materials when working with fire-rated walls and floors. Holes or gaps affect the fire rating of a floor or wall. Properly filling these penetrations with fire-stopping materials maintains the rating. Fire-stopping materials are typically applied around all types of piping, electrical conduit, ductwork, electrical and communication cables (*Figure 48*), and similar devices that run through openings in floor slabs, walls, and other fire-rated building partitions and assemblies.

Nonmechanical fire-stopping materials are classified as intumescent or endothermic. Both are formulated to help control the spread of fire before, during, and after exposure to open flames. When subjected to the extreme heat of a fire, intumescent materials expand (typically up to three

Figure 47 Mechanical fire-stop device.

Fire Ratings

When gypsum drywall is used in a party wall, the architect's plans must be followed precisely. If the inspector finds flaws in the construction, a certificate of occupancy will not be issued. This photo illustrates why building codes place so much emphasis on properly constructed party walls.

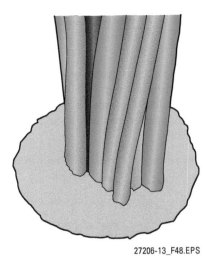

Figure 48 Fire barrier of moldable putty installed around electrical cables.

times their original size) to form a strong insulating material that seals the opening for three to four hours. Should the insulation on the cables, pipes, etc., passing through the penetration become consumed by the fire, the expansion of the fire-stopping material also acts to fill the void in the floor or wall in order to help stop the spread of smoke and other toxic products of combustion.

Endothermic materials block heat by releasing chemically bound water, which causes them to absorb heat.

Fire-stopping materials are formulated in such a way that when activated, they are free of corrosive gases, reducing the risks to building occupants and sensitive equipment.

Fire-stopping materials are made in a variety of forms including composite sheeting, caulks, silicone sealants, foams, moldable putty, wrap strips, and spray coatings. They come in both one-part and two-part formulations. The installation of these materials must always be done in accordance with the applicable building codes and the manufacturer's instructions for the product being used. Depending on the product, fire-stopping materials can be applied via spray equipment, conventional caulking guns, pneumatic pumping equipment, or a putty knife.

Any fire-stopping materials used must meet the criteria of standard *ASTM E814, Fire Test*, as tested under positive pressure. They must also have an hourly rating that is equal to or greater than the hourly rating of the floor or wall being penetrated. Based on *ASTM E814/UL 1479* tests, one of four ratings, measured in time, may be applied to fire-stopping materials and systems. These ratings are as follows:

- *F rating* – A fire-stopping system meets the requirements of an F rating if it remains in the opening during the fire test for the rating period without permitting the passage of flames through the opening or the occurrence of flaming on any element of the unexposed side of the assembly.
- *FT rating* – A fire-stopping system meets the requirements of an FT rating if it remains in the opening during the fire test within the limitations as specified for an F rating. In addition, the transmission of heat through the fire-stopping system during the rated period shall not have been such as to raise the temperature of any thermocouple on the unexposed surface of

Drilling Fire-Rated Walls

There are many wall variations, so you must first establish the type of construction before undertaking any invasive action, such as drilling. If drilling is permitted, you may have to use fire-stopping materials to seal off the opening.

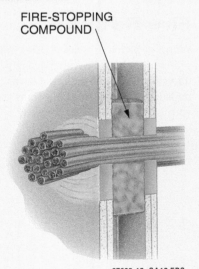

the fire-stopping system by more than 347.8°F (181°C) above its initial temperature.
- *FH rating* – A fire-stopping system meets the requirements of an FH rating if it remains in the opening during the fire test within the limitations for an F rating. In addition, during a hose stream test, the fire-stopping system shall not develop any opening that would permit a projection of water from the stream to the unexposed side.
- *FTH rating* – A fire-stopping system shall be considered as meeting the requirements of an FTH rating if it remains in the opening during the fire test and hose stream test within the limitations as described for FT and FH ratings.

3.3.0 Multi-Ply Application

Multi-ply construction consists of one or more layers of drywall applied over a base layer. This results in an improved surface finish, greater strength, and higher fire resistance and sound classifications. The base layer can consist of gypsum backer board (with or without foil), regular drywall, or another base material.

The maximum support spacing for multi-ply systems depends upon the location, the type of face ply, and the type of fastener to be used. See *Table 3*.

When a multi-ply system with a laminated face ply is to be used over wood supports, it should be fastened as recommended for single-ply construction. Double nailing is not needed because the fasteners used on a two-ply application will produce a firmly fastened system.

In a normal wall installation, the base ply may be installed with the long edges either parallel or perpendicular to the framing. End joints should occur on framing members. If foil-backed wallboard is used, apply the foil side against the framing.

The face ply is installed perpendicular to the base ply, with the joints offset from the base ply joints. The exception occurs when the face ply consists of decorated gypsum panels. In that case, the face ply is always installed parallel to the framing, regardless of how the base ply is installed. The wallboard should not be in contact with the floor, as it can wick up moisture that will damage the wallboard.

At inside corners, it is recommended that only the overlapping base ply be nailed or screwed and that the fasteners be omitted from the face ply. The floating-corner treatment is better able to resist structural stresses, as shown in *Figure 49*.

A floating-angle construction for multi-ply systems has the overlapping side of the base ply nailed only at the interior corners.

3.3.1 Attaching the Base Ply to Metal Framing or Furring

Base-ply drywall is normally attached to steel framing and furring with screws that are at least ⁵⁄₁₆" longer than the thickness of the board. The board application may be either perpendicular or parallel. In a perpendicular application where no adhesive is used between the plies, the base ply should be fastened with a single screw in each stud or furring channel at the board edges and with one screw at the middle of the board at each stud or channel.

In parallel applications with no adhesive between the plies, the base ply is fastened with

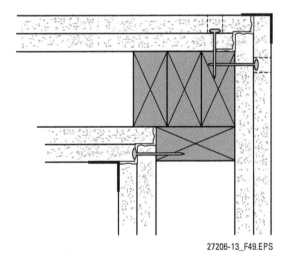

Figure 49 Floating-corner treatment.

Table 3 Fastener Spacing

	Nail Spacing		Screw Spacing	
Location	Laminated Face Ply	Nailed Face Ply[1]	Laminated Face Ply[2]	Screwed Face Ply[1]
Walls	8" OC	16" OC	16" OC	24" OC
Ceilings	7" OC	16" OC	16" OC	24" OC

Note: Fastener size and spacing for sound-deadening boards vary for different types of fire-rated and sound-rated construction. Follow the manufacturer's recommendations.
[1]Fastener spacing for face ply shall be the same as for single-layer application.
[2]12" OC for both ceilings and walls when supports are spaced 24" OC.

screws located 12" on center along the edges of the board and 24" on center in each stud in the field of the board. When the base ply is to be attached perpendicular or parallel to steel framing that is 16" on center and with adhesive between the plies, the screw spacing should be 12" on center for walls. The maximum screw spacing on steel framing at 24" on center is 12" on center for both walls and ceilings.

3.3.2 Face-Ply Attachment

The joints in the face ply should be offset at least 10" from the joints in the base ply. Drywall can be applied to framing in either a perpendicular or parallel configuration, whichever results in the least waste of materials. A perpendicular application is preferred on walls, because it usually results in fewer joints. Architectural board is usually installed parallel and does not require joint treatment. Some systems are designed to use decorative battens over the joints. When the face ply is attached with fasteners and with no adhesive between the plies, the maximum spacing and minimum penetration recommended for screws should be the same as for single-ply applications.

3.3.3 Adhesive Attachment

Typical multi-ply construction may employ sheet lamination, strip lamination, or spot lamination to attach the face ply to the base ply. Sheet lamination involves covering the entire back with a laminating adhesive using a notched spreader (*Figure 50*), box spreader, or other suitable tool.

Notched spreaders are used to apply adhesive for sheet lamination. The size and spacing of the notches are determined by the type of adhesive being used. The drywall should be applied using moderate pressure. Any adhesive squeezed out at the joints must be promptly removed.

Strip and spot lamination are preferred to sheet lamination in sound-rated partitions. In strip lamination, the adhesive is applied in ribbons using a special spreader. The ribbons are normally spaced 16" to 24" on center, as shown in *Figure 51*. In spot lamination, spots of adhesive are brushed or daubed on in a regular pattern.

3.3.4 Supplemental Fasteners

In order to ensure a satisfactory adhesive bond, it is necessary to hold the face ply firmly against the base ply with supplemental fasteners, shoring, or bracing while the adhesive is setting up. Generally, the fasteners are applied in the field of each laminated face ply in ceiling applications. On side walls, these fasteners can be placed at the perimeters of the boards where they will be concealed by the joint treatment or trim.

It should be remembered that in fire-rated assemblies, the specific fastener spacing is given for the particular assembly tested and may not be related to whether or not adhesive is used between the plies. Fastener spacing details for fire-rated assemblies are available from the sponsor of the test and, in many cases, from the Gypsum Association.

In sound-rated partitions where fire resistance is not a consideration, an acceptable practice is to attach the face ply vertically over sound insulation board or backing board with permanent fasteners in the back ends. Intermediate fasteners are omit-

Figure 50 Spreading adhesive using a notched spreader.

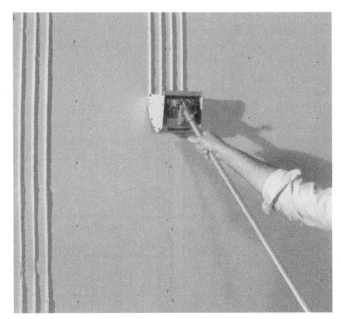

Figure 51 Properly spacing adhesive ribbons.

ted, and the panels are temporarily braced until the adhesive has developed a sufficient bond.

3.4.0 Prioritizing Walls

For some commercial structures, the order in which walls are constructed is critical to the integrity of the fire rating of the walls and overall fire rating of the building. In general, the overall priority of wall construction is as follows, from highest priority to lowest priority:

- Two-hour fire wall and smoke wall
- Two-hour fire wall and two-hour shaft wall
- One-hour fire wall and smoke wall
- One-hour fire wall
- Nonrated wall

Always refer to the local building code, project specifications, and construction drawings for information on wall priorities.

Did You Know?
Multi-Ply Wallboard

The most common reason for using multi-ply wallboard construction is to meet fire-rating requirements.

3.0.0 Section Review

1. Multi-ply drywall systems provide _____.
    a. fewer joints
    b. moisture control
    c. increased sound isolation
    d. poor surface quality

2. Joint-filling material that expands when exposed to heat of a fire is called _____.
    a. intumescent fire-stopping
    b. exothermic fire-stopping
    c. endothermic fire-stopping
    d. coalescent fire-stopping

3. When attaching base-ply gypsum board to steel studs, the screw length should exceed the board thickness by _____.
    a. $3/16"$
    b. $1/4"$
    c. $1/2"$
    d. $5/16"$

4. In a commercial building, the order in which interior walls are constructed may affect the building's fire rating.
    a. True
    b. False

Section Four

4.0.0 Estimating Drywall

Objective

Identify how to calculate a quantity takeoff for proper drywall installation.
   a. Explain how to perform a material takeoff for drywall.
   b. Explain how to perform a material takeoff for drywall fasteners.

Performance Task 3

Estimate material quantities for an installation.

This section covers the guidelines for performing a quantity takeoff for wallboard and fasteners. *Table 4* shows rules of thumb for ordering different types of drywall nails and screws.

4.1.0 Drywall

To estimate how many sheets of wallboard you will need for a job, first determine how many square feet of space the room contains. Multiply the length of each wall by two, add the results, then multiply the total wall space by the wall height to get the total square footage of wall space.

Table 4 Fastening Materials Required for 1,000 Square Feet of Drywall

Material	Amount Required for 1,000 Square Feet of Drywall
1¼" annular-ring nails	6¼ lb
1⅜" annular-ring nails	6¾ lb
1" drywall screws	3 lb
1¼" drywall screws	4¼ lb
1⅝" drywall screw	5½ lb

If the ceiling is to be covered with wallboard, its square footage would be included as well. Then divide the total square footage by 32, which is the square footage of a 4' × 8' sheet, or 36, which is the square footage of a 4'-6" × 8' panel. For example, a 10' × 12' room with 8' high walls would contain 352 square feet of wall space. Converting this number into sheets (352 ÷ 32) yields 11. Door and window openings are usually figured solid unless there is a large picture window or door. This creates a built-in allowance for waste.

4.2.0 Fasteners

The types and amounts of fasteners will vary from job to job. It is important to consult the job specifications as well as the manufacturer's guidelines for this information. The following will serve as a rule-of-thumb guide to estimating fasteners.

For single-layer application 16" OC, approximately 1,000 screws are needed per 1,000 square feet of wallboard. If the framing is 24" OC, about 850 screws per 1,000 square feet will be needed.

If nails are used, you need to calculate the amount required in pounds. To install a single layer of ⅝" wallboard, you will need about 5 pounds of nails for every 1,000 square feet of wallboard. The size of the nail, and therefore the total weight, depends on the thickness of the wallboard. It can range from 4½ pounds of nails for ¼" wallboard to 7 pounds of nails for 1¼" (two ⅝" sheets laminated) wallboard.

Think About It

Estimating Drywall Needs

How many square feet of drywall are required for the walls and ceiling for a room that is 12' × 16' with 8' ceilings? How many pounds of 1¼" annular ring nails would be required for a single-layer installation?

4.0.0 Section Review

1. For each 1,000 square feet of ⅝" drywall, the amount of nails needed would be _____.
   a. 2 lb
   b. 3½ lb
   c. 5 lb
   d. 7 lb

Summary

This module covered gypsum drywall and its installation. Once the drywall is properly installed, the joints must be finished to create a smooth surface for application of the final decorative finish, such as paint or wall covering.

Gypsum drywall panels installed over wood or steel framing are the most common method of finishing walls. In residential applications, drywall panels are commonly used to finish ceilings as well.

There are different types and sizes of drywall designed for different applications. It is therefore important that the carpenter know the different types, their applications, and their methods of installation.

Review Questions

1. A common use of ¼" gypsum board is as a _____.
   a. coreboard in shaft walls
   b. backing for tile in showers
   c. new single-layer wall system
   d. base layer in a multilayer system

2. Regarding the use of water-resistant gypsum board, the true statement below is: _____.
   a. it is recommended for use in saunas and steam rooms
   b. it is not recommended for showers and tubs
   c. it should have a foil backing
   d. it should be applied directly over a vapor barrier

3. An attachment method that is *not* recommended for securing drywall to framing in single-layer applications is _____.
   a. stapling
   b. screws
   c. nails
   d. adhesives

4. Type G drywall screws are used to secure drywall to _____.
   a. wood studs
   b. gypsum backing boards
   c. steel studs
   d. concrete masonry units

5. Fasteners should be applied to wallboard working from _____.
   a. top to bottom
   b. edge to edge
   c. center to edge
   d. corner to corner

6. In a double-nailing system, the second set of nails is driven about _____ from the first.
   a. 2"
   b. 4"
   c. 8"
   d. 12"

7. When ceiling drywall is attached with screws and the framing members are 16" OC, the screws should be spaced _____.
   a. 7" OC
   b. 8" OC
   c. 12" OC
   d. 16" OC

8. When you are installing fire-rated wallboard, the nails should penetrate the studs by at least _____.
   a. ½"
   b. ¾"
   c. ⅞"
   d. 1⅛"

9. The tool commonly used to cut gypsum-board panels to size is a _____.
   a. circular saw
   b. crosscut saw
   c. utility knife
   d. hook-bill knife

10. The abbreviation STC stands for _____.
    a. sound transmission class
    b. sound transmission capture
    c. sound transmission control
    d. sound transmission containment

11. To control noise, a construction practice that should be avoided is _____.
    a. caulking around outlet box openings
    b. locating recessed wall features in the same stud cavity
    c. using separate framing for the two sides of a wall
    d. mounting gypsum board in resilient channels

12. In semisolid partitions, gypsum panels are applied to _____.
    a. metal or wood tracks
    b. gypsum studs
    c. plywood
    d. a foam core

13. The point at which the edges of two panels of drywall meet is known as a _____.
    a. bedding seam
    b. joint
    c. gypsum lath
    d. dimple

14. In floating interior angle construction where the framing is perpendicular to the wall/ceiling intersection, the distance that ceiling fasteners should be located from the intersection when using single nailing is _____.
    a. 1"
    b. 7"
    c. 10"
    d. 12"

15. When you are installing cement board, the factory-bound edge should be placed above the lip of the shower pan or tub by at least _____.
    a. ¼"
    b. ½"
    c. ¾"
    d. 1"

16. When covering existing walls that have structurally sound surfaces, use gypsum board with a thickness of _____.
    a. ⅝"
    b. ⅜"
    c. ¼"
    d. ⅛"

17. The fire resistance (fire rating) required for a building's walls, floors, and ceilings is established by _____.
    a. building codes
    b. OSHA
    c. the architect
    d. testing laboratories

18. Fire-stopping materials are used to _____.
    a. extinguish fires
    b. seal wall openings to prevent fire from spreading
    c. prevent fires from starting
    d. lower the ignition point of flammable materials

19. In a multi-ply application, the face-ply joints should be separated from the base-ply joints by at least _____.
    a. 6"
    b. 8"
    c. 10"
    d. 12"

20. When adhesive is spread over the entire back surface of the face ply in a multi-ply system, the method is referred to as _____.
    a. complete lamination
    b. blanket lamination
    c. full back lamination
    d. sheet lamination

21. When single-ply construction on 16" centers is used, drywall screw requirements can be calculated on the basis of _____.
    a. one screw per two square feet
    b. one screw per square foot
    c. two screws per square foot
    d. four screws per square yard

Trade Terms Quiz

Fill in the blank with the correct term that you learned from your study of this module.

1. Two pieces of material meet at a _____.
2. A thin mixture of cement, plaster, or a similar dry material with water is called a _____.
3. To allow for structural stresses, _____ is an installation technique for drywall that does not use fasteners at the edge of the panel.
4. Caused by either framing shrinkage or poor installation technique, _____ results in fasteners protruding above the surface of the drywall.
5. When drywall panels meet at an outside corner, a metal or plastic _____ is installed for protection.
6. A finish is applied to an underlying material, such as drywall, that is referred to as the _____.
7. The generic term for gypsum drywall (panels with a core of gypsum and a paper covering) is _____.

Trade Terms

Corner bead
Floating interior angle construction
Drywall
Joint

Nail pop
Slurry
Substrate

Trade Terms Introduced in This Module

Corner bead: A metal or plastic angle used to protect outside corners where drywall panels meet.

Floating interior angle construction: A drywall installation technique in which no fasteners are used at the edge of the panel in order to allow for structural stresses.

Drywall: A generic term for paper-covered gypsum-core panels; also known as *gypsum drywall*.

Joint: A place where two pieces of material meet.

Nail pop: The protrusion of a nail above the wallboard surface that is usually caused by shrinkage of the framing or by incorrect installation. Also applies to screws.

Slurry: A thin mixture of water or other liquid with any of several substances such as cement, plaster, calcined gypsum, or clay.

Substrate: The underlying material to which a finish is applied.

Additional Resources

This module presents thorough resources for task training. The following resource material is suggested for further study.

Installing and Finishing Drywall. 2008. William Spence. New York: Sterling Publishing Company, Inc.
The Gypsum Construction Handbook. 2009. Chicago, IL: R.S. Means.

Figure Credits

FEMA Photo/Christopher Mardorf, CO01
Gypsum Association, Figures 1, 2, 3, 4, 5, 6, 7, 8, 9, 10, 12, 13, 14, 15, 34A, 34B, 35, 36, 38, 39, 40, 41, 42, 43, SA 11, SA 12, 49
Senco Products, Inc., SA02
Z-PRO INTERNATIONAL, Inc., Figure 11
National Gypsum Company, SA04
USG Corporation, SA05, SA06

Courtesy of Irwin Tools, Figures 21, 22, 26B, 27
The Stanley Works, Figures 25, 29, SA07, 30, SA09
ROTOZIP® by Bosch, Figure 26A
Corey Builders, Inc., SA08
Telpro, Inc., Figure 32
AMICO, SA10
USG Corporation, Figures 51, 52

Section Review Answer Key

Answer	Section Reference	Objective
Section One		
1. d	1.0.0	1a
2. b	1.2.2	1b
3. d	1.2.3	1c
Section Two		
1. c	2.1.0	2a
2. b	2.2.0	2b
3. b	2.3.0	2c
4. a	2.4.0	2d
5. d	2.5.0	2e
Section Three		
1. b	3.1.0	3a
2. a	3.2.2	3b
3. d	3.3.1	3c
4. a	3.4.0	3d
Section Four		
1. c	4.2.0	4b

This page is intentionally left blank.

NCCER CURRICULA — USER UPDATE

NCCER makes every effort to keep its textbooks up-to-date and free of technical errors. We appreciate your help in this process. If you find an error, a typographical mistake, or an inaccuracy in NCCER's curricula, please fill out this form (or a photocopy), or complete the online form at **www.nccer.org/olf**. Be sure to include the exact module ID number, page number, a detailed description, and your recommended correction. Your input will be brought to the attention of the Authoring Team. Thank you for your assistance.

Instructors – If you have an idea for improving this textbook, or have found that additional materials were necessary to teach this module effectively, please let us know so that we may present your suggestions to the Authoring Team.

NCCER Product Development and Revision
13614 Progress Blvd., Alachua, FL 32615

Email: curriculum@nccer.org
Online: www.nccer.org/olf

❏ Trainee Guide ❏ Lesson Plans ❏ Exam ❏ PowerPoints Other _____

Craft / Level: _____ Copyright Date: _____

Module ID Number / Title: _____

Section Number(s): _____

Description: _____

_____

_____

_____

_____

Recommended Correction: _____

_____

_____

_____

_____

Your Name: _____

Address: _____

_____

Email: _____ Phone: _____

This page is intentionally left blank.

Drywall Finishing

OVERVIEW

When gypsum drywall is first installed, there are visible seams between the drywall sheets and at corners. These seams must be closed in a way that makes them invisible to anyone looking at the finished wall. Look around your home. Although the walls were most likely finished with 4 × 8 sheets of drywall, it should look as if it was done with one solid sheet. The seams are covered with paper or fiberglass tape, which is embedded with a compound commonly called mud. When the mud has dried, the joint is sanded flat. It sounds simple, but like other areas of construction, it can only be learned with practice.

Module 27207

Trainees with successful module completions may be eligible for credentialing through NCCER's National Registry. To learn more, go to **www.nccer.org** or contact us at **1.888.622.3720**. Our website has information on the latest product releases and training, as well as online versions of our *Cornerstone* magazine and Pearson's product catalog.

Your feedback is welcome. You may email your comments to **curriculum@nccer.org**, send general comments and inquiries to **info@nccer.org**, or fill in the User Update form at the back of this module.

This information is general in nature and intended for training purposes only. Actual performance of activities described in this manual requires compliance with all applicable operating, service, maintenance, and safety procedures under the direction of qualified personnel. References in this manual to patented or proprietary devices do not constitute a recommendation of their use.

Copyright © 2013 by NCCER, Alachua, FL 32615, and published by Pearson, New York, NY 10013. All rights reserved. Printed in the United States of America. This publication is protected by Copyright, and permission should be obtained from NCCER prior to any prohibited reproduction, storage in a retrieval system, or transmission in any form or by any means, electronic, mechanical, photocopying, recording, or likewise. To obtain permission(s) to use material from this work, please submit a written request to NCCER Product Development, 13614 Progress Blvd., Alachua, FL 32615.

27207 V5

From *Carpentry, Trainee Guide*. NCCER.
Copyright © 2013 by NCCER. Published by Pearson. All rights reserved.

27207
DRYWALL FINISHING

Objectives

When you have completed this module, you will be able to do the following:

1. Identify differences between the six levels of finish established by industry standards.
2. Identify the different materials for proper drywall finishing.
   a. Describe how to select the proper trim.
   b. Describe the purposes of tapes, compounds, coatings, and sanding materials.
3. Identify the proper tools used in drywall finishing.
   a. Identify the hand tools used in drywall finishing.
   b. Identify the automatic tools used in drywall finishing.
4. Describe proper drywall finishing procedures.
   a. Identify ideal site conditions for drywall finishing.
   b. Describe the process for finishing drywall.
   c. Describe the hand-finishing procedures involved in drywall finishing.
   d. Describe the automatic taping and finishing procedures involved in drywall finishing.
   e. Identify common joint problems when finishing drywall.
   f. Identify common compound problems when finishing drywall.
   g. Identify common fastener problems when finishing drywall.
   h. Identify common problems when finishing drywall.
5. Explain how to estimate the proper amount of drywall finishing materials.

Performance Tasks

Under the supervision of your instructor, you should be able to do the following:

1. State the differences between the six levels of finish established by industry standards and distinguish a finish level by observation.
2. Properly apply a corner bead, tape, and finish to a drywall panel.
3. Patch damaged drywall.

Trade Terms

All-purpose compound	Joint compound	Ridges	Tapered joint
Bull nose	Lightweight compound	Skim coat	Taping compound
Feathering	Mud	Tape	Topping compound

Code Note

Codes vary among jurisdictions. Because of the variations in code, consult the applicable code whenever regulations are in question. Referring to an incorrect set of codes can cause as much trouble as failing to reference codes altogether. Obtain, review, and familiarize yourself with your local adopted code.

Industry-Recognized Credentials

If you're training through an NCCER-accredited sponsor, you may be eligible for credentials from NCCER's Registry. The ID number for this module is 27207. Note that this module may have been used in other NCCER curricula and may apply to other level completions. Contact NCCER's Registry at 888.622.3720 or go to **www.nccer.org** for more information.

Contents

Topics to be presented in this module include:

1.0.0 Finishing Standards 1
2.0.0 Drywall Finishing Materials 3
  2.1.0 Trim 3
    2.1.1 Trim Materials 3
  2.2.0 Tapes, Compounds, Coatings, and Sanding Materials 4
    2.2.1 Tape 4
    2.2.2 Drywall Finishing Compounds 5
    2.2.3 Powder Compounds 6
    2.2.4 Premix Compounds 7
    2.2.5 Quick-Setting Compounds 8
    2.2.6 Texture Materials 9
    2.2.7 Dry-Mix Safety 9
    2.2.8 Sanding Materials 10
3.0.0 Drywall Finishing Tools 12
  3.1.0 Hand Tools 12
    3.1.1 Mixing Tools 15
    3.1.2 Tape Dispensers 15
    3.1.3 Texture Tools 16
  3.2.0 Automatic Finishing Tools 17
    3.2.1 Automatic Taping Tools 17
    3.2.2 Nail Spotters 19
    3.2.3 Flat Finishers 19
    3.2.4 Corner Applicators and Finishers 20
    3.2.5 Automatic Loading Pumps 21
    3.2.6 Vacuum Sanders 21
4.0.0 Drywall Finishing Procedures 23
  4.1.0 Site Conditions 23
    4.1.1 Drywall Inspection 23
  4.2.0 Taping and Finishing Process Overview 24
  4.3.0 Hand Finishing Procedures 24
    4.3.1 Sanding 26
    4.3.2 Spotting Fastener Heads 26
    4.3.3 Outside Corners 27
    4.3.4 Inside Corners 27
    4.3.5 Safety and Good Housekeeping 27
  4.4.0 Automatic Taping and Finishing Procedures 28
    4.4.1 Pump Loading Procedures 28
    4.4.2 Pumping 29
    4.4.3 Automatic Nail Spotter Procedures 29
    4.4.4 Using the Automatic Taping Tool on Ceiling Joints 30
    4.4.5 Using the Automatic Taping Tool on Wall Joints 31
    4.4.6 Using the Corner Roller 32
    4.4.7 Corner-Plow Operation 33
    4.4.8 Flat Finisher Procedures 34

4.4.9 Flat Finishing for Other Joints 35
4.4.10 Finishing Coats with the Flat Finisher 35
4.4.11 Corner Applicator Operation 35
4.5.0 Finished Joint Problems 36
  4.5.1 Ridging 36
  4.5.2 Photographing 37
  4.5.3 Joint Depressions 37
  4.5.4 High Joints 37
  4.5.5 Discoloration 37
  4.5.6 Tape Blisters 37
  4.5.7 Cracking 38
4.6.0 Problems with Compound 38
  4.6.1 Debonding, Flaking, or Chipping 38
  4.6.2 Moldy or Contaminated Compound 39
  4.6.3 Pitting 39
  4.6.4 Sagging 39
  4.6.5 Excessive Shrinkage 39
  4.6.6 Delayed Shrinkage 40
4.7.0 Fastener Problems 40
  4.7.1 Nail Pops 40
  4.7.2 Preventing Nail Pops 40
  4.7.3 Repairing Nail Pops 42
  4.7.4 Fastener Depressions 42
4.8.0 Problems with Wallboard 42
  4.8.1 Board Blisters 43
  4.8.2 Damaged Edges 43
  4.8.3 Water Damage 43
  4.8.4 Board Bowing 43
  4.8.5 Board Cracks and Fractures 43
  4.8.6 Loose Drywall Panels 44
  4.8.7 Patching Drywall 45
5.0.0 Estimating Drywall Finishing Materials 47

Figures

Figure 1	Corner bead with mesh flanges	38
Figure 2	L-bead	3
Figure 3	J-bead	3
Figure 4	Expansion (control) joint	4
Figure 5	Joint-reinforcing tape	4
Figure 6	Metal edge tape	5
Figure 7	Joint compound	5
Figure 8	All-purpose compound	6
Figure 9	Lightweight compound	7
Figure 10	Quick-setting compound	9
Figure 11	Drywall saw with thin, pointed blade for detail cuts	13
Figure 12	Putty knife	13
Figure 13	Finishing knives	13
Figure 14	Finishing trowel	14
Figure 15	Corner trowels	14
Figure 16	Corner-clinching tool with a rubber mallet	14
Figure 17	Mud pan	14
Figure 18	Hawk	15
Figure 19	Pole sander	15
Figure 20	Commercial hand sander	15
Figure 21	Hand-operated mud mashers	15
Figure 22	Mud mixers used with a power drill	16
Figure 23	Tape dispenser used with paper or fiberglass mesh tape	16
Figure 24	Banjo	16
Figure 25	Examples of texture patterns	17
Figure 26	Texture rollers	18
Figure 27	Typical texture-finishing equipment	18
Figure 28	Automatic taping tool	18
Figure 29	Nail spotter	19
Figure 30	Flat finishers	20
Figure 31	Corner applicator	20
Figure 32	Corner finisher (plow)	20
Figure 33	Corner roller	21
Figure 34	Automatic joint-compound loading pump and gooseneck	21
Figure 35	Power vacuum sander	22
Figure 36	Dimple or uniform depression	23
Figure 37	Butt-joint misalignment (low side on right)	24
Figure 38	Applying tape	25
Figure 39	Finishing a butt joint using the double-wide method	26
Figure 40	Using a corner taping tool	28
Figure 41	An automatic taping tool being loaded with a loading pump	29
Figure 42	Nail spotter	29
Figure 43	Using an automatic taping tool	31
Figure 44	Using a corner roller	33
Figure 45	Using a corner plow	33

Figure 46 Using a corner plow with a corner applicator 33
Figure 47 Using a flat finisher 34
Figure 48 Coating with topping compound 37
Figure 49 Shrinkage contributes to nail pops 40
Figure 50 Nonaligned framing 40
Figure 51 Force of gravity and vibrations can cause nail pops 41
Figure 52 Improper drywall installation 41
Figure 53 Veneer L-trim casing bead 45
Figure 54 Applying an expansion joint 45
Figure 55 Patching a large hole 45
Figure 56 Patching a hole in drywall using the hot-patch (blowout) patching technique 46

This page is intentionally left blank.

Section One

1.0.0 Finishing Standards

Objective

Identify differences between the six levels of finish established by industry standards.

Performance Task 1

State the differences between the six levels of finish established by industry standards and distinguish a finish level by observation.

Trade Terms

Joint compound: Patching compound used to finish drywall joints, conceal fasteners, and repair irregularities in the drywall. It dries hard and has a strong bond. Sometimes called *mud* or *taping compound*.

Mud: See *joint compound*.

Ridges: Slight protrusions in the center of a finished drywall joint that are usually caused by insufficient drying time. Also known as *beads*.

Skim coat: A thin coat of joint or topping compound that is applied over the entire drywall surface. Sometimes required under a high-gloss finish.

Tape: A strong paper or fiberglass tape used to cover the joint between two sheets of drywall.

Taping compound: See *joint compound*.

In some situations, the carpenter may have to install and finish gypsum drywall. A remodeling project is one example. In other cases, it may be up to the carpenter to repair damaged drywall or drywall that was improperly installed by someone else. Therefore, it is essential to be thoroughly familiar with the tools, materials, and procedures used in drywall finishing and repair.

Drywall requires different levels of finish depending on its location, lighting conditions, and decorative treatment. A hidden surface, such as an attic area, requires far less finishing than one in full view, such as a living room wall. In addition, even minor flaws are apt to be more evident when the surface is exposed to strong lighting conditions or decorated with certain finishes, such as gloss or semigloss paints or thin wall coverings. Generally, the more visible a surface, the more likely its lighting or decoration are to show surface defects, and the more finishing work it requires.

These factors are addressed in *A Recommended Specification for Levels of Gypsum Board Finish*, which was jointly developed by the Painting and Decorating Contractors of America, the Association of the Wall & Ceiling Industries–International, the Gypsum Association, and the Ceilings & Interior Systems Construction Association. This specification is designed to serve as a standard reference for architects, specification writers, contractors, building owners, and others. It provides them with a specific description of the final appearance of gypsum walls and ceilings finished to different levels before the application of a decorative coating of paint, texture material, or wall covering.

The specification describes the following six levels of finish and typical applications for each of them:

- *Level 0* – No taping, finishing, or accessories required. This level might be used for temporary construction or where final decoration is undetermined.
- *Level 1* – All joints and interior angles shall have tape embedded in joint compound (also referred to as mud or taping compound): Surface shall be free of excess joint compound. Tool marks and ridges are acceptable. A Level 1 finish might be specified for attics, areas above ceilings, service corridors, and other areas not generally seen by the public. It provides some degree of smoke and sound control. In some areas, it is called fire taping.
- *Level 2* – One separate coat of joint compound shall be applied over all joints, angles, fastener heads, and accessories. The surface shall be free of excess joint compound. Tool marks and ridges are acceptable. All joints and interior angles shall have tape embedded in joint compound. Joint compound applied over the body of the tape at the time of tape embedment shall be considered a separate coat of joint compound and shall satisfy the conditions of this level. A Level 2 finish might be recommended for garages, warehouses, and other areas where surface appearance is not critical. It is specified where water-resistant gypsum backing board (*ASTM C630*) is used as a substrate for tile.
- *Level 3* – All joints and interior angles shall have tape embedded in joint compound and one separate coat of joint compound applied over all joints and interior angles. Fastener heads and accessories shall be covered with two separate coats of joint compound. All joint compound surfaces shall be smooth and free of tool marks

and ridges. A Level 3 finish might be specified for surfaces to be finished with a medium or heavy texture before painting or with heavy-grade wall covering. It is not recommended for light- or medium-weight wall coverings or for smooth painted surfaces.

- *Level 4* – All joints and interior angles shall have tape embedded in joint compound and two separate coats of joint compound applied over all flat joints and one separate coat of joint compound applied over interior angles. Three separate coats of joint compound shall be applied over all fastener heads and accessories. All joints shall be smooth and free of tool marks and ridges. Light textures or wall coverings require this level of finish. The specification notes that in critical lighting areas, flat paint over light textures reduces shadowing of finished joints through the surface decoration, but that gloss, semigloss, and enamel paints are not recommended for this level of finish. It also notes that the type of wall covering applied over this level should be chosen carefully to properly conceal joints and fasteners.
- *Level 5* – All joints and interior angles shall have tape embedded in joint compound, two separate coats of joint compound applied over all flat joints, and one separate coat of joint compound applied over interior angles. Three separate coats of joint compound shall be applied over all fastener heads and accessories. A thin skim coat of joint compound, or a material manufactured especially for this purpose, shall be applied to the entire surface. The surface shall be smooth and free of tool marks and ridges. Level 5 is recommended for gloss, semigloss, enamel, or nontextured flat paints or severe lighting conditions. This is the highest level of finish and provides the best coverage against joints or fasteners being visible through the decorative coating.

The specification, which is available from any of the associations that developed it, notes that for Levels 3, 4, and 5, it is recommended that the prepared surface be coated with a drywall primer prior to the application of finish paint. It also notes that the effects of severe lighting on a surface can be minimized by skim coating the drywall, by decorating it with medium to heavy textures, or by using window coverings to soften shadows.

Finishing Steps

The application of paper tape and compound to a drywall joint is done in four basic steps:

Step 1 A coat of taping compound is applied to the joint.
Step 2 The paper tape is pressed into the base layer and is smoothed with a 5" to 6" knife. (When automatic taping equipment is used, the compound and tape are applied in one step.)
Step 3 When the first coat of compound is dry, a coat of topping or all-purpose compound is applied with a 6" taping knife and feathered out 7" to 10" on either side of the joint with a beveled trowel or finishing knife.
Step 4 When the second coat is dry, a third coat is applied and feathered out at least 2" wider than the second coat.

1.0.0 Section Review

1. *A Recommended Specification for Levels of Gypsum Board Finish* specifies gypsum wall and ceiling finish requirements for _____.
   a. three levels
   b. four levels
   c. five levels
   d. six levels

Section Two

2.0.0 Drywall Finishing Materials

Objective

Identify the different materials for proper drywall finishing.
  a. Describe how to select the proper trim.
  b. Describe the purposes of tapes, compounds, coatings, and sanding materials.

Trade Terms

All-purpose compound: Combines the features of taping and topping compounds. It does not bond as well as taping compound but finishes better.

Bull nose: A metal corner bead with rounded edges.

Lightweight compound: An all-purpose compound having less weight than standard compounds.

Topping compound: A joint compound used for second and third coats. It dries soft and smooth and is easier to sand than taping compound.

Only a few materials are required to finish gypsum drywall. These materials include trim pieces, tapes, compounds, coatings, and sanding materials. This section covers various types of drywall finishing materials.

2.1.0 Trim

Trim is available in a variety of shapes and sizes, each one having a particular function. Since gypsum drywall has unfinished edges, they must be protected to minimize damage and to make the intersection of drywall panels look good. Trim pieces are made of metal or vinyl.

2.1.1 Trim Materials

Corner trim, or corner bead as it is usually called, serves as reinforcement for outside corner joints. It consists of edging that strengthens drywall corners and provides a straight guide for finishing.

One type of corner bead is nailed or clinched onto the framing members through the drywall panels. Other types might have a bull nose (a metal corner bead with rounded edges) and/or metal mesh flanges, either in regular or expanded widths (*Figure 1*). Still another type has paper flanges attached to the corner bead, which is applied with joint compound. Corner beads usually receive a three-coat finishing process to obtain a smooth surface and to conceal any fasteners. The exposed nose of outside corners or the edge of the inside corner bead provides a guide for making the finish a flush surface.

L-bead and J-bead (sometimes called casings) are metal or plastic pieces shaped like Ls or Js (*Figures 2* and *3*). They provide maximum protection and neat finished edges for drywall at window and door jambs and other abutments. They are available in sizes to accommodate all thicknesses

Figure 1 Corner bead with mesh flanges.

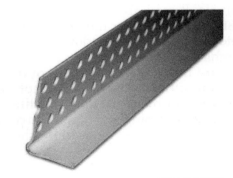

Figure 2 L-bead.

Figure 3 J-bead.

of gypsum board. They are especially useful and attractive when gypsum board abuts dissimilar objects.

Metal L- and J-bead may be either finished or unfinished. Unfinished L- and J-beads are treated like regular corner beads and covered with compound and sanded. Finished trim is not covered with compound or texture.

Flexible tape is another type of drywall trim used to reinforce edges and corners that serve as decoration. This material might be used as border or edging on arches, splayed angles, door and window frames, or to reinforce in a decorative way various odd intersections of ceiling or wall panels.

Expansion joints are control joints used to relieve stresses caused by expansion and contraction in large ceiling or wall expanses in interior drywall systems. They are made of roll-formed zinc, making them corrosion resistant. *Figure 4* shows one type of expansion joint.

Expansion joints are similar to J-bead in that part of the expansion joint, the ¼" open slot, remains visible after finishing. In a room, it simply looks like another panel dividing line, but the joint contracts and expands as the large drywall sections expand and contract. It helps prevent large expanses of drywall from cracking.

Reveal trim is another type of finishing or decorating material often classified as decorative trim or architectural molding. In general, reveal trim works like decorative spacing between drywall panels. It may also function as decorative batten strips or borders or moldings around room perimeters, door frames, windows, archways, and other architectural features.

There are many different types of decorative trim, but all of them generally attempt to eliminate the need for the usual joint-finishing procedures. In place of taping and topping, a decorative piece can be installed by drywall finishers. However, some reveal trims need to be installed at the time the drywall is installed. Such trim pieces usually have a design feature that allows them to stay covered up until the rest of the room or project is finished. Then the protective covering is removed to show off the reveal trim between the drywall panels.

2.2.0 Tapes, Compounds, Coatings, and Sanding Materials

After the drywall and trim is properly installed, various materials are used to cover and reinforce seams, and make the overall appearance of the drywall pleasing. Skilled craftworkers, such as drywall finishers, apply the tape, compounds, and/or other coatings, and sand the exposed surfaces to achieve a true surface.

2.2.1 Tape

Three kinds of tape may be used in drywall finishing: paper tape, fiberglass mesh tape, and metal edge tape. Each kind may further be divided into those precoated with adhesive and those without adhesive. The tape most frequently used is paper tape without adhesive.

Standard paper tape is used to cover and reinforce seams, joints, and patchwork. It is a strong paper with feathered edges. There are two types of paper tape available: plain paper and perforated paper.

Paper tape (*Figure 5*, left) can vary in width, but it is generally about 2" wide. The specific width required for most automatic taping tools is 2¹⁄₁₆". The tape is usually available in rolls that range in length from 60' to 500'.

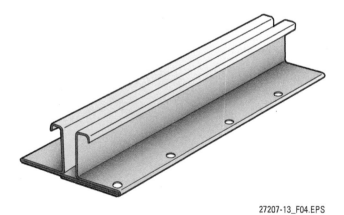

Figure 4 Expansion (control) joint.

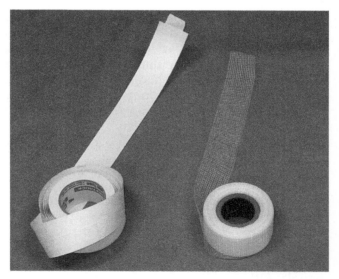

Figure 5 Joint-reinforcing tape.

The paper tape may also contain fibers, crisscrossed or woven into the material, and it often comes tapered or feathered at the edges. The tape surface may also be scuffed or roughened to allow better bonding with the taping compound. Paper tape is considered to be superior to fiberglass tape for many applications because paper tape resists stretching and distortion. It also provides a more consistent bond between the face papers of the drywall on each side of most joints.

Perforated paper tape has larger, more visible holes to allow more compound to penetrate, producing a better bond once dry. These holes may vary in size, depending upon the manufacturer, but they are usually about 1/16" in diameter.

Fiberglass mesh tape (*Figure 5*, right) is generally self-sticking fiberglass joint tape made of fabric-woven filaments that do not decay. This type of tape is also available without adhesive. Fiberglass mesh makes a strong tape and may be more durable than paper tape under certain conditions, such as in high-moisture areas and when used with moisture-resistant drywall. Fiberglass mesh tape is much more costly than paper tape, but is good for repair work, veneer taping, and other specialized applications. Its use can increase productivity enough to offset the extra cost of the tape.

Metal edge tape (*Figure 6*) is a type of paper tape with two strips of galvanized steel down the center, with a small gap between them to allow for the crease. The metal strips are typically 1/2" wide. This tape is sometimes referred to as flexible metal corner-reinforcing tape. However, the tape is still applied and finished just like regular paper tape.

Metal edge tape is best used for corners with other than 90-degree angles. It is also used for corners formed by radius-wall and ceiling intersections, arches, drops, splays, and wherever drywall panels need to join in unusual configurations. Metal edge tape makes any outside angled corner straight and sharp, with some reinforcing qualities. The edges beyond the metal strips are feathered like those of standard paper tape.

Corner bead is not very flexible, so something else is needed for odd-angle corners made by curved or angled wallboard intersections. The principal idea of metal edge tape is to provide, as

Figure 6 Metal edge tape.

much as possible, the same strength and finishing quality that corner bead provides.

Metal edge tape is generally 2 1/8" wide and is available in 100' rolls. The gap between the metal strips, allowing for the paper crease, is usually only 1/16" wide. The tape is designed so that the metal side is applied to the wall.

2.2.2 Drywall Finishing Compounds

Joint compound, commonly known as mud and also sometimes called taping compound, is available in both liquid and powder forms. Liquid compound has been premixed by the manufacturer, while powder compound (*Figure 7*) is mixed

Figure 7 Joint compound.

Taping Corners

The center of all paper tape is scored (running the length of the tape) to allow the tape to be easily creased for inside corners. The crease side is always the side you press into the corner. In other words, if you fold the tape like a V, the point of the V always goes against the wall.

on the job site by adding the proper amount of cold, clean water. Each of these products is designed to accomplish a specific result, depending on the job. Their ingredients are often different, so do not mix liquid compound with dry compound.

Several types of compounds are used in finishing drywall. One popular approach is the two-step system in which the tape is embedded in the joint using a taping compound to obtain a strong bond. The joint is then finished with a topping compound, which is much easier to sand.

Taping compound is designed for its bonding qualities and strength in bedding and reinforcing taped joints. It is also used as a first coat on metal corner beads or other trim, nail or screw heads, and other fasteners. Taping compound is also used as prefill and fill coats, and for repairing surface drywall cracks and cracks in plaster. Taping compound is generally the most likely to shrink. It is also the strongest bonding and most difficult to sand.

Topping compound is used for the second and third coats of the finishing process. This type of compound is softer drying and easier to sand. It also produces less shrinkage. You must never use topping compound for taping because topping compound is not designed to embed and bond a taped joint. A joint bedded with topping compound will crack with the first slight movement or settling of the wall. Topping compound is designed to be molded and sanded flat on top of a joint that has already been fastened together.

All-purpose compounds combine the characteristics of the two-step system into a single compound. All-purpose compound (*Figure 8*) combines the features of both taping and topping compounds, but in so doing, it gives up some of the qualities of each. For example, it loses some of the bonding qualities of taping compound and some of the soft and smooth drying capabilities of topping compound. All-purpose compound is also excellent for use in textured finish applications. This type of compound is often used to finish walls with various interesting effects. In that respect, you are almost practicing the art of plastering.

Lightweight compound (*Figure 9*) has the advantages of an all-purpose compound and is 25 to 35 percent lighter. It also has less shrinkage and sands as easily as topping compound. Lightweight compound can be used to laminate drywall panels, coat interior concrete ceilings and above-grade columns, and patch cracks in plaster. It can also be used for texturing.

2.2.3 Powder Compounds

Compounds packaged in powder form store better than other forms of compound. Powder can be stored at any temperature and in any storage area

Figure 8 All-purpose compound.

Using Fiberglass Tape

Fiberglass mesh tape precoated with adhesive is designed to be installed without first applying a bedding coat of taping compound. In other words, adhesive-backed tape is applied directly to the joint and pressed into place with a taping knife or trowel. After the tape is applied, it is covered with layers of compound. Fiberglass mesh tape without adhesive backing is sometimes installed using staples.

Some manufacturers recommend that you do not use fiberglass mesh tape with the usual ready-mixed or powder joint compounds. Instead, special powder compounds, such as quickset compounds, have been developed that work much better with fiberglass tape. Quickset compounds are discussed in more detail later in this module.

An automatic taping tool specifically designed for placing fiberglass tape is shown here.

Figure 9 Lightweight compound.

or warehouse that is kept dry and free from moisture. Since warehouses are rarely heated, powder compounds are best for winter storage. However, it is recommended that powder compounds be moved to a warm mixing room a full day before they are mixed.

Powder compounds must be mixed with clean water in exactly the proportions specified on packaging instructions. Generally, these proportions depend upon the amounts of compound you need for the job. The amounts needed are also determined by the square footage of the joint areas you intend to cover. Once mixed, the compound may be kept in tightly covered containers in storage for many days, as long as the storage area is kept at room temperature. If the powder has been properly mixed to start with, it will not clump up or harden in stored closed containers, but you should always stir or mash it again before use.

Both powder and liquid forms of compound shrink by drying out. As water evaporates from the compound, the compound shrinks to fill in space vacated by the water. This process is different from the chemical hardening process used by special quickset compounds (discussed later, which is more like glue drying.

Taping, topping, and all-purpose compounds are generally available in powder form. They may be packaged in bags or cartons. The bags generally contain 25 pounds of powder. Cartons may be measured in gallons or pounds and usually contain more powder than bags.

Current building codes and standards generally forbid any use of asbestos in construction materials. Always make sure that the powder you are going to use is right for the job and complies with local regulations.

2.2.4 Premix Compounds

In general, premix (ready-mix) compounds are formulated, mixed, and packaged by their manufacturers. They are vinyl based and require little or no mixing. This feature reduces the need for readily available water at the job site. Premix compounds also reduce the waiting or soak times after application, and they offer good crack resistance after drying.

These products will freeze, however, so precautions need to be taken in cold weather. If vinyl premix compound does freeze, thaw it only at room temperature and do not apply any additional heat. Always use premix products in their packaged state of consistency as much as possible,

Working with Powder Compounds

Here are a few other practical rules you should follow regarding powders:

- Always use clean, cold water for mixing.
- Never vary from the correct proportions of water-to-powder specified for the product you are mixing.
- Always label and date the container you use for mixing. Some carpenters write a description of the contents and the date on a piece of masking tape and attach it to the lid. This is generally better than using ink or magic marker directly on the lid because you can pull the tape off and use the lid again later.
- Never confuse the different mixes. Keep taping compounds separate from topping compounds and so forth. If everything is labeled, you will not have this problem. (Hint: Mix powder taping compound only in empty premix taping compound pails and mix powder topping compound only in empty premix topping compound pails.)
- Be sure to use the oldest dated compound first before mixing a new batch.
- Always avoid mixing one brand of powder with a different brand of powder. Even if they are supposed to be the same kind of powder, different manufacturers make their products differently. They are usually not compatible with similar products made by other companies.
- Follow the dry-mix safety instructions given later in this module.

to minimize shrinkage. Follow the specific instructions on the label.

Premix compounds are available in both plastic pails and sealed cartons. They are also quite heavy because, as the name implies, the water is already mixed in. A full pail weighs over 60 pounds; full cartons might be 50 or 60 pounds.

The advantage of premix compounds is that they can be used at job sites where there is no supply of fresh water, which must be available in order to use powders. Of course, powders can be mixed where there is fresh water and then transported in closed pails to sites without water. Generally, however, most contractors simply use the premix compounds in these situations.

The disadvantage of premix compounds is that they must be stored where they will not freeze during the winter. Finishing compounds must be used only in room-temperature environments. In colder climates in winter, job-site interiors must be closed up and heat must be installed and working before drywall finishing can proceed.

Even though premix compound is already mixed at the factory and ready to use on the job, mix or mash it again before using it. Hand mashing is done by forcing the masher down repeatedly in the center and around the sides of the pail. Use smooth, complete downward strokes. On the upward stroke, scrape the masher along the sides of the pail to loosen any compound sticking to the edge.

Two or three minutes of mixing is usually enough to ensure a smooth, even consistency all through the compound.

Premix compound will need to be diluted slightly for use with automatic finishing tools. Generally, add ½ cup of clean, cold water to 4½ gallons of premix compound. For the automatic taper, mix in 2 cups of clean, cold water per 4½ gallons of compound. Note that some manufacturers have compounds specifically designed for use with their line of taping and finishing tools.

The full range of compounds is available in premix form: taping, topping, all-purpose, and specialty compounds. Specialty compounds are all-purpose compounds that offer enhanced bonding, shrinking, and sanding capabilities. They often eliminate the need for a third topping coat, and they are good for laminated applications. Specialty compounds are also useful for texturing applications.

2.2.5 Quick-Setting Compounds

Quick-setting compound, commonly called quickset, 20-minute mud, 30-minute mud, or hot mud, is a compound that sets up very quickly in comparison to other compounds because it hardens chemically instead of by water evaporation. Shrinkage is reduced considerably, so quick-setting compounds make an excellent filler for metal trim, repairs, and around pipes.

Quick-setting compound is available only in powder form; it needs to be mixed with water on the job. Clean and cold water must be used. It is also very important to mix only as much as you can apply in the allotted setup time.

Quick-setting compound is good for small jobs, corner beads, and finishing bathrooms and other high-moisture areas. It is especially good in humid weather. Because the compound sets so quickly, you need to wipe off excess compound immediately. Sanding the dried compound is also difficult, so take care to apply and wipe it as smooth as possible before it dries. Accelerators are available to make quick-setting compounds set even faster for special needs.

Quick-setting compound is packaged and sold by its setup time, which generally ranges from 20 minutes to 360 minutes (6 hours). (Note that the compound shown in *Figure 10* sets up in 90 minutes.) Because it works chemically, once it is hardened it will not shrink, even though it is not dry. This allows a strong bond to form and remain in high-humidity environments. It also enables successive coats to be applied even before previous coats have completely dried, allowing faster finishing and greater cost savings.

Quick-setting compounds are also very good for laminating applications, especially for laminating drywall layers together. Quick-setting compounds can be used for coating concrete walls and ceilings (above ground), for filling in cracks and holes, for skim coating, and even for surface texturing. They are also preferred for finishing exterior ceiling boards and for presetting joints of veneer finish systems.

When Is 5 Gallons Not 5 Gallons?

The traditional 5-gallon pail is no longer 5 gallons. Currently, these full plastic pails contain only 4½ gallons of premixed compound.

2.2.6 Texture Materials

Texture materials may be manufactured in both powder and ready-mixed forms. The four types are described as follows:

- *Powder textures* – These textures may be either aggregated or unaggregated, which means there may or may not be other particles mixed into the powder, respectively. In aggregated products, particles of such substances as perlite, vermiculite, and polystyrene are used to make textured effects on primed surfaces, particularly ceilings. Aggregated powder products are designed for spray application. They also have a good solution time, only minimum-to-moderate fallout, and good bonding power and crack resistance. When properly sprayed on, these textures hide substrate imperfections very well. Unaggregated powder products may either be sprayed or hand-applied to primed walls or ceilings. Several of these products are limited to hand applications only so that you can produce crow's foot, stipple, or other pattern texture effects. Crow's foot is a design produced by a roller, which makes a kind of random bird-track pattern across the textured surface.
- *Powder joint-compound textures* – Generally, powder joint-compound textures are the same as the topping or all-purpose powder compounds used for normal joint finishing. For texturing with these products, use a brush, roller, or trowel to produce light and medium hand-formed textures on walls or ceilings. Typically, swirling motions are used to make random patterns. Powder joint compounds are easily mixed in the usual way. They are smooth working and easy to texture by hand. They also hide surface imperfections very well, produce good bonding, and resist cracking. The color is white after drying, which can be left as is or painted another color, depending upon the finish specifications.
- *Premixed textures* – These include thick, heavy-bodied, vinyl-based materials, which are able to produce smooth, very deep textures. They generally dry to hard white finishes that are often left unpainted, especially on ceilings. These textures may be sprayed, troweled, or applied with a roller or brush. They go well over concrete and are able to fill voids and cracks and cover surface blemishes. Premixed texture offers good resistance against cracking on walls and ceilings. These textures are factory mixed to a smooth consistency. They are easy and fast to apply, and generally produce favorable results.
- *Premixed joint-compound textures* – These textures consist of topping or all-purpose compounds, which are able to produce light to medium textures on ceilings and walls. These compounds are well-suited for small jobs that need only brushes, rollers, or trowels. However, they can also be spray-applied. A large variety of patterns and designs can be created. These textures dry white, offer good crack resistance, and can be painted when dry. They are factory mixed to be smooth and free from lumps. These compounds go on quickly and easily for low-cost, yet good-quality, results.

Figure 10 Quick-setting compound.

2.2.7 Dry-Mix Safety

Whenever mixing dry compounds, be aware of the dust level and try to keep it to a minimum. Wear the appropriate respiratory protection when mixing dry powder compounds. In addition:

Quick-Setting Compounds

A disadvantage of quick-setting compound is that it can be difficult to sand. That is why it is generally used as a prefill or tape-and-bed coat, while regular topping compound is used for second and third finishing coats. Quick-setting compound should not be used for the final coat because it might bleed through the decorated surface. Sandable quick-setting compounds are available.

- Make sure all mixing containers are clean and free of residue.
- Use only clean, drinkable water for mixing.
- Make sure all tools and mixing blades are clean.
- Mix compounds only according to the directions on their labels.
- If using a power mixer or mixers operated by a power drill, use a slow speed such as 300 rpm to 500 rpm.
- Do not try to mix different types of compounds together. Their chemical makeup often differs from manufacturer to manufacturer. Even with the same manufacturer, the different types of compound are not compatible with each other.

2.2.8 Sanding Materials

Sanding operations may be done with sponge sanders, hand sanders, pole sanders, or power sanders. The sanders require sheets or strips of sandpaper to be fitted to them. Sanding is an important part of the finishing process, and tools are available to help with this task. Use a hand sander to sand dried finished joints and nail spots that are within normal reach. With the pole or power sander, the same job can be performed over your head without using scaffolds, ladders, or stilts.

Sandpaper is rated by coarseness, called grit, which varies by degree of fineness. The lower the grit number, the coarser the paper. Coarse paper is used for first sanding jobs where the surface is very rough and a craftsperson is trying to smooth it out fairly quickly. Fine paper is used for finish sanding to make the surface as smooth as possible, usually in preparation for painting.

Sanding is typically done with 180-grit sandpaper. Sandpaper is packaged in sheets or rolls for hand sanders and in discs for power sanders. The precut sheets are designed to fit specialized tools such as angle sanders and wall sanders (sanding poles). Precut sheets are usually sold in packages of 100.

Sandpaper rolls are designed to be cut off to whatever length is needed to fit onto a hand sander or pole sander. These rolls may vary in total length from 30' to 150' and in width from $3\frac{1}{3}$" to 12".

Other sanding materials include sanding cloth, abrasive mesh cloth, and open mesh cloth, which are primarily used for cutting joint mud. These materials may also be packaged in rolls or as individual sheets.

Mesh cloth, commonly called sanding screen, is preferred by many finishers for use on pole sanders. For first sanding of a joint or seam, mesh cloth prevents raising the nap (tiny hair-like fibers) on the face paper of the drywall. Do not raise this nap at all, if possible, because it makes for better painting if the nap is smooth. An advantage of mesh cloth is that it does not load up with excess finishing material.

Film-backed drywall abrasives are also effective. A major advantage is that they do not raise the face of recycled paper.

Rules for Quick-Setting Compounds

Observe the following rules when using quick-setting compounds:

- It is best to mix with power mixers.
- Mix only enough to be used in the time stated in the product instructions.
- Clean mixing equipment and tools immediately after using them because hardening will occur even when they are submerged in water.
- Never attempt to remix quick-setting compound.
- Never try to use quick-setting compound in an automatic finishing tool, such as an automatic taper.

Texture Materials

Texture materials are similar to joint compounds in that they are made in both powder and premixed forms. In fact, some joint compounds can be easily used for texturing.

Spray-Texturing Machines

Spray-texturing machines are available for large jobs. They range from a hopper with a pneumatic spray nozzle to a self-contained machine like this one with its own built-in compressor.

Wet Sanding

If only minimal sanding is required, a wet sponge can be used. This method produces no dust and will not scuff the paper. Use a polyethylene sponge, which looks something like carpet padding. Wet the sponge with clean water that is cool to lukewarm. Wring out the sponge enough to prevent dripping, and then rub the joints to remove high spots, using as few strokes as possible. Clean the sponge frequently during use.

2.0.0 Section Review

1. Finished appearance and protection from damage for drywall around windows and doors are provided by L-bead and J-bead trim pieces, which are also known as _____.

    a. moldings
    b. casings
    c. bull noses
    d. caps

2. Paper joint tape is more easily stretched and distorted than fiberglass tape.

    a. True
    b. False

SECTION THREE

3.0.0 DRYWALL FINISHING TOOLS

Objective

Identify the proper tools used in drywall finishing.
a. Identify the hand tools used in drywall finishing.
b. Identify the automatic tools used in drywall finishing.

Trade Terms

Feathering: Tapering joint compound at the edges of a drywall joint to provide a uniform finish.

Proper finishing of walls and ceilings would be nearly impossible without a variety of finishing tools. These hand-operated and mechanical tools not only speed up the finishing process, but also help create walls and ceilings that are smooth and flat.

Although they appear simple and harmless, finishing tools can be as dangerous as any other kind of tool when handled improperly. Finishing knives have very sharp edges and corners. Automatic tapers have gears and chains that can catch fingers or hair. Automatic finishers have sharp blades, and some have spring-loaded hatches that can catch fingers, hair, and clothing.

The best way to avoid accidents and tool-related injuries is to treat the tools with respect. Do not handle any tool unless you have been thoroughly trained in its use. If a tool is unfamiliar, ask an instructor, job supervisor, or another carpenter for information.

Always keep finishing tools clean and free of rust or excess compound, both of which can be poisonous if they get into a cut. Do not use knives with chipped blades or broken handles. Not only can unsafe tools hurt you, but they can also ruin finishing work.

Treat automatic tools with special care. They must be cleaned and maintained according to the manufacturer's directions. Do not try to force a mechanical tool to work when the tool is jammed with dried joint compound. Inspect automatic tools before every use; if a tool looks damaged or in bad repair, report the problem to your supervisor.

3.1.0 Hand Tools

The following hand tools are used to cut, install, and finish drywall:

- 4' straightedge or T-square
- Utility knife with plenty of blades
- Drywall saw
- Circle cutter
- Drywall hammer
- Caulking gun
- Screwdriver
- Broad knife
- Joint trowel
- Corner tool
- Mud pan or hawk
- Sandpaper/drywall screen
- Sanding block, pole sander, or electric sander
- Sponge sander

An easy way to cut drywall is to place the straightedge on the finished side of the panel and score the drywall with a utility knife, cutting through the paper facing and into the gypsum core. Then snap the panel along the score line by applying pressure from the back of the board. If the backing paper is still intact, use the knife to finish cutting through it.

An alternative method of cutting through drywall is to use a drywall saw. It is good for making straight as well as curved cuts. A saw with a long, thin, pointed blade (*Figure 11*) works well for limited or detail cuts such as when cutting out a hole for an electrical box. The saw can cut in a straight line or in an arc. Use a drill to make a starter hole for the blade.

Finishing knives range in type and size from the 1¼" putty knife (*Figure 12*) to the extremely wide 24" taping knife (*Figure 13*). They are all similar in function, however, because they bed and feather the taping or topping compound. Each

---

Dress Properly

Always wear the proper clothing and appropriate personal protective equipment (PPE) when finishing drywall. Wear goggles and a dust mask to keep dripping compound and dust out of your eyes, nose, and mouth. Adequate clothing can help protect you from cuts and falling objects.

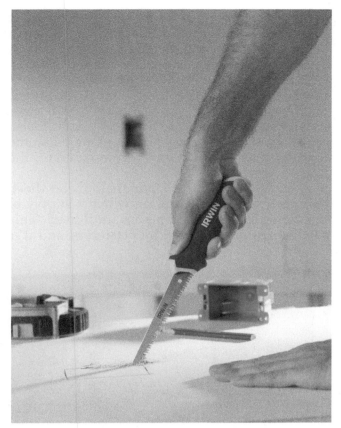

Figure 11 Drywall saw with thin, pointed blade for detail cuts.

Figure 12 Putty knife.

Figure 13 Finishing knives.

knife is designed for use in a different situation. For example, the smallest putty knife is used in hard-to-reach spaces, for patching, and for working the compound around windows, cabinets, and doors. The 4", 6", 8", 10", and 12" widths are the most common for drywall finishing.

Taping knives generally have blades made of blue steel, which flexes under the pressure of bedding and feathering taped joints. They range in size according to blade width.

A long-handled broad knife is used to wipe excess compound from freshly taped joints. The broad knife's blade is made of stainless steel or blue steel and may range in width from 7" to 9". Typically, a broad knife's handle is about 10" to 12" long, but handles up to 28" long are available. These long knives are handy tools for cleaning up messy joints and spatters on high walls and ceilings.

Finishing trowels (*Figure 14*) are similar to cement and plaster trowels, but they are available with either a flat or curved finishing surface (also called a blade). The concave blade is very useful for shearing the joint tape. Curved blade trowels are often preferred over taping knives for finishing drywall joints. Blades range from 10" to 18" in length and are usually 4½" wide. Blades maintain their bow shape due to the spring steel or flexible stainless steel from which they are made. Flat-blade trowels may be used for applying textured finishes.

Corner tools enable you to finish both inside and outside corners. They include not only trowels, but sanding and bead-attaching devices as well. Corner trowels are made for finishing inside and outside corners. While inside and outside corner trowels are shaped alike, their handles are on opposite sides (*Figure 15*). These tools are generally not used by professional drywall finishers.

Corner beads can be screwed on or applied with staples. Some corner systems have a paper-faced corner bead that can be coated with joint compound for direct application to a corner.

For attaching corner bead and metal trim to outside corners, the corner-clinching tool may also be used. The clincher is struck with a rubber mallet once the corner bead is positioned and the tool is set over it. When struck, the corner-clinching tool centers the bead and clinches it to the corner by crimping each side into the drywall at four points. The entire length of corner bead can be attached by moving the clincher up or down the corner and striking it several times (*Figure 16*).

A mud pan (*Figure 17*) is simply a long container that holds joint compound to be knifed or troweled over the wallboard. The pan is commonly made of steel, aluminum, or plastic. It also may have replaceable steel edges mounted on the long sides for scraping excess compound off the knife or trowel.

A hawk (*Figure 18*) is a handheld metal sheet, similar to a mortarboard, on which a supply of plaster or compound is placed until used. This tool was originally used by plasterers, and it is

CURVED BLADE TROWEL

TROWEL IN USE

Figure 14 Finishing trowel.

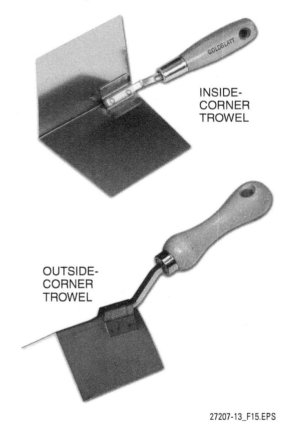

INSIDE-CORNER TROWEL

OUTSIDE-CORNER TROWEL

Figure 15 Corner trowels.

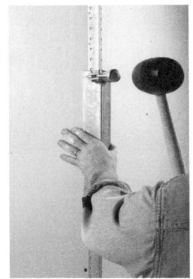

Figure 16 Corner-clinching tool with a rubber mallet.

Figure 17 Mud pan.

still known by the name given to it by workers in that trade.

Sanding is an important part of the finishing process, and tools such as sanders are available to help you with this task. There are two basic types of sanders: the pole sander (*Figure 19*) and the hand sander (*Figure 20*). Both use the same principle of fastening down a strip of sandpaper by means of clamps and wing nuts.

With the hand sander, you are better able to sand dried finished joints and fastener heads that are within normal reach. With a pole sander, you can do the same job over your head without relying on a scaffold, ladder, or stilts.

Hand sanders, also known as sanding blocks, may be purchased or made on the job site. A commercial sanding block consists of a wooden or metal base, around which a sheet of sandpaper is wrapped. A second block is pressed against one side of the base and tightened down with a wing nut to hold the sandpaper in place around the base. The sander allows you to apply even pressure over the sandpaper's entire face while saving your fingers from abrasions and other injuries.

Power vacuum sanders, which will be discussed later, are also available.

3.1.1 Mixing Tools

Both hand tools and attachments to power drills can be used to mix joint compound and other liquefied materials at the job site. For hand mixing, you can use a mud masher, which is a lot like a potato masher (*Figure 21*). The mud is mashed in a pail until it is mixed to a smooth consistency.

If you need to mix a full bucket of compound, it is faster and easier to use a power drill equipped with a long-stemmed mud mixer (*Figure 22*). There are several types of mud mixers. The mixer end of the device is placed in the mud once its shaft has been secured in the power drill chuck. The drill is used like a kitchen hand mixer to stir the mud.

3.1.2 Tape Dispensers

A 500' roll of joint tape can be awkward to carry around while trying to apply compound or tape a long joint. Simple tape dispensers and tape holders solve the problem. Many of these lightweight tape-holding devices are designed to hang from belts or shoulder slings, leaving your hands free (*Figure 23*). Some dispensers even crease the tape for application in corners.

Figure 18 Hawk.

Figure 19 Pole sander.

Figure 20 Commercial hand sander.

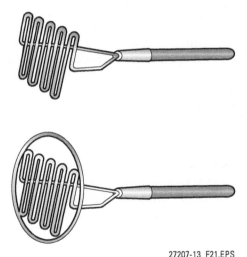

Figure 21 Hand-operated mud mashers.

Making a Sanding Block

At the job site, you can make a sanding block by wrapping a sheet of sandpaper around a 4"-long piece of 2 × 4 lumber and stapling the edges down.

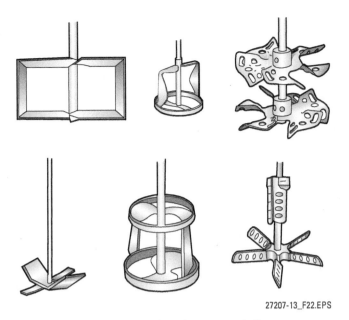

Figure 22 Mud mixers used with a power drill.

A banjo (*Figure 24*) is a large tape dispenser. It is loaded with a full 500' roll of joint tape and a supply of joint compound. The banjo applies the compound to the tape as the tape is pulled out. The banjo can be adjusted to change the amount of compound applied to the tape. The banjo may get its name from its similarity in shape to the musical instrument of the same name.

3.1.3 Texture Tools

A texture is any wall or ceiling coating that serves as its own finish. It may also serve to hide tape and other finishing imperfections so that the surface need not be sanded smooth or otherwise prepared for painting, wallpapering, or other treatments. *Figure 25* shows examples of texture finishes.

Decorative textures are very popular in both residential and commercial construction. Interesting patterns, simulated acoustical effects, and light or heavy finishes may be applied with rollers, brushes, stencils, sprayers, and other tools and equipment. *Figure 26* shows some of the many roller patterns available. Some of the more common equipment used in texture finishing is shown in *Figure 27*.

1. *Glitter gun* – Used to embed glitter in wet texture ceilings. The manually operated model shown is most economical, but not as efficient as an air-powered type.
2. *Drywall mud paddle* – Used with an electric drill at less than 400 rpm to mix drywall mud.

Figure 23 Tape dispenser used with paper or fiberglass mesh tape.

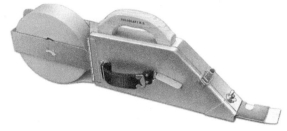

Figure 24 Banjo.

It is designed to reduce the entrapment of air bubbles in the mixture.

3. *Stucco brush* – Used to create a variety of textures from stipple to swirl. Other variations can be achieved with thicker application and deeper texturing.
4. *Texture brush* – Available in many sizes and styles; tandem-mounted brushes cover a large area to speed a texturing job.

LIGHT STIPPLE TEXTURE

MEDIUM-LIGHT FINISH APPLIED BY SPRAY OF MULTIPURPOSE TEXTURE FINISH

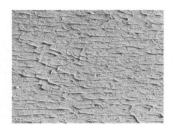
BOLD SHADOWING WITH ROLLER APPLICATION

MEDIUM STIPPLE TEXTURE

SWIRL FINISH

CROW'S FOOT

Figure 25 Examples of texture patterns.

5. *Wipe-down blade* – Has a hardened steel blade and long handle to speed cleaning of walls and floors after application of texture materials. The blade has rounded corners to avoid gouging.

6. *Long-handled roller* – A standard paint roller adapted to the particular type of finish required. Available roller sleeves include short nap, long nap, looped, foam stipple, and carpet types in professional widths.

7. *Texture roller pan* – Used with rollers. Some models can hold up to 25 pounds of mixed texture.

8. *Flat blade knife* – Used to apply texture material and for troweled finishes.

9. *Circular patterned sponge* – Used to achieve patterned swirl finishes.

10. *Sea sponge* – Used to achieve free-form texture finishes.

11. *Whisk broom* – Similar to a stucco brush but stiffer. It can be used to produce a bolder brush pattern.

12. *Short-handle roller* – Same as the long-handled roller with the same sleeves. A looped texture roller sleeve is shown.

13. *Trowel* – Used to apply texture materials and for achieving troweled or knockdown finishes.

14. *Texture paddle* – Similar to the drywall mud paddle, but designed for mixing texture materials at 300 rpm to 600 rpm.

3.2.0 Automatic Finishing Tools

Automatic finishing tools are operated manually, but they are referred to as automatic finishing tools or mechanical finishing tools because they use intricate mechanisms to do work that otherwise would have to be done with hand tools and fingers. Some of these tools combine the compound application and taping and finishing joints into one step. Although they are operated by hand, the hand operation usually involves simply pushing the tool along a path.

3.2.1 Automatic Taping Tools

One of the most popular automatic finishing tools is the automatic taping tool, which applies mud and tape to joints (*Figure 28*).

The automatic taping tools, commonly known by the brand name of BAZOOKA®, quickly guide the tape, coat it with a measured layer of mud, and dispense the tape along the joint. The automatic taping tools even cut the tape at the end of a pass and can crease the tape for application in corners.

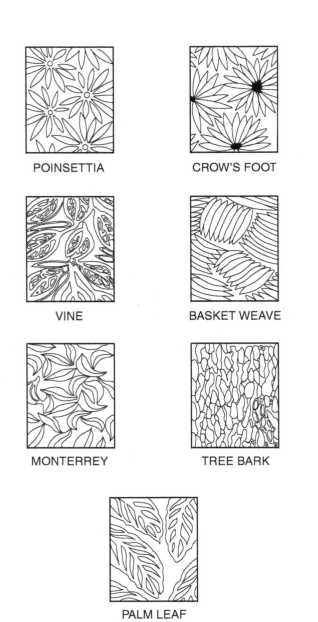

Figure 26 Texture rollers.

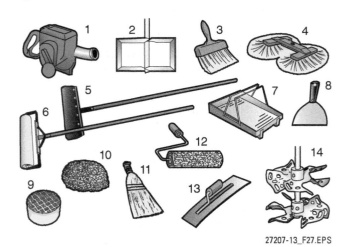

Figure 27 Typical texture-finishing equipment.

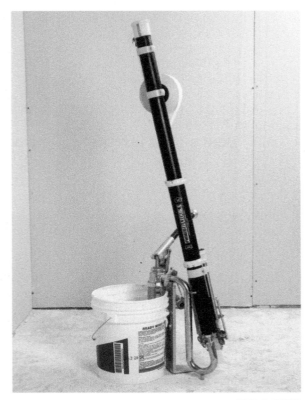

Figure 28 Automatic taping tool.

Using an automatic taping tool, ceilings up to 10' high can be finished without using a ladder, scaffold, or stilts. An extension can be mounted to the taping tool's base to reach even higher ceilings. For closets and other tight spaces, miniature tapers are also available.

A good deal of practice may be required to master the use of an automatic taping tool, but it is likely time well spent. Using an automatic taping tool, an experienced carpenter can tape an entire room faster than several carpenters using only hand tools.

Automatic taping tools are made primarily of aluminum, plastic, and other rust-resistant materials. This makes it easy to keep the unit clean, which is necessary to guarantee proper operation.

An automatic taping tool is cleaned by filling the empty joint-compound chamber with water. Water is then forced out of the unit through the valve that distributes the compound by moving the floating piston up and down inside the compound chamber. The gears and valve openings at the head of the taper should be sprayed clean with a high-pressure water hose.

Automatic taping tools use a small razor blade to cut the joint tape at the end of a pass. This blade must be changed occasionally, as must the cable that operates the unit's piston. The piston moves up and down inside the taper's tube, forcing out compound. You can change the blade or cable in just a few minutes, following the instructions that come with the taping unit. Always follow the manufacturer's instructions for maintaining an automatic taping tool, to avoid breakdowns and lost time on the job.

3.2.2 Nail Spotters

A much simpler automatic finishing tool is the nail spotter (*Figure 29*). A nail spotter quickly applies compound over nail and screw dimples in fastened wallboard.

The spotter is simply a metal box on a swiveling pole. The pole's swivel allows you to use the spotter at any angle and to push the device along a wall or ceiling while standing still.

The pole is attached to a hinged plate on top of the spotter. As you push the spotter along, the pressure you exert on the pole is transferred to the plate. As the plate sinks into the metal box, it forces compound out through a small opening in the bottom of the box. The compound fills the dimples and excess mud is automatically scraped off by a blade mounted in the trailing edge of the box.

Nail spotters are commonly made in 2" and 3" widths. The pole lets you reach high ceilings and walls without a ladder or stilts. The mechanism is small and light enough to be used in closets and other cramped spaces.

A nail spotter is much easier to use than the automatic taping tool. Once it is mastered, an entire room can be spotted in just a few minutes. Since the tool automatically scrapes off excess mud and feathers the spot, each dimple must be gone over only once.

The spotter's blade is bowed slightly to leave a small crown of compound over each dimple. The blade is very hard, but can be damaged or broken by an exposed nail or screw head. Broken or worn blades must be replaced. The blade is usually mounted to the unit with a clamp and one or two wing nuts. By removing the nuts, you can back off the clamp and slide out the blade. The new blade is simply slid into place and held in place by the wing nuts.

The nail spotter should be cleaned thoroughly after every use by flushing it with a high-pressure water hose. The units are usually made of rust-resistant aluminum or stainless steel.

3.2.3 Flat Finishers

The flat finisher is also known as a box. You will generally use more than one box because they are available in different widths (*Figure 30*). The idea of a flat finisher is to apply topping and finishing coats to taped drywall joints. As each successive layer is applied, you use a wider box. The flat finisher works on the same principle as the nail spotter and dispenses an even, measured strip of compound along any flat joint. Each size of box applies topping coats of those dimensions. Use the smallest box for the first topping coat after taping; use the 10" or 12" box for the second topping coat if two topping coats are all that are required; or use the 10" for the second coat and the 12" for a third coat.

Figure 29 Nail spotter.

Keeping Your Compound Fresh

To prevent the compound from drying out in the head of an automatic taping tool during use, always stand the unit headfirst in a pail of clean water whenever you must stop working for a short time. At the end of the workday, pump all compound out of the tool before cleaning it out with water.

Like the nail spotter, the flat finisher is a metal box with a hinged lid. When you push the box along the joint, pressure on the handle forces the hinged lid down, squeezing mud out of the box through a small opening at the bottom. The opening can be adjusted to change the amount of mud released. Automatic-feed versions supply compound to the finisher, reducing the amount of effort required.

Once flat joints have been taped either by hand or with an automatic taping tool, they can be finished using an automatic-feed flat finisher. This tool applies a smooth, even coat of mud over the taped joint, automatically feathering the edges and crowning the center.

Flat finishing tools are available in three widths: 7" for applying a topping coat to a taped joint; 10" for applying a third or finish coat; and 12" for applying a fourth or skim coat.

Flat finishers can be used on wall and ceiling joints running in any direction. The tool is not only easy to operate, but allows you to get the job done in a small fraction of the time it would take to finish joints by hand.

Adjustable handles enable the finisher to reach ceilings up to 12' high. The device is compact enough to use in closets and other cramped spaces. Like most other automatic finishing tools, the flat finisher must be kept clean in order to work properly. These applicators must be cleaned thoroughly after each use by squeezing the remaining compound out of the reservoir and flushing the unit with water from a high-pressure water hose.

3.2.4 Corner Applicators and Finishers

The corner applicator (*Figure 31*) works in conjunction with a corner finisher, or plow (*Figure 32*). By attaching the plow to the ball/cone end of the corner applicator with a locking retainer clip, the operator can put a finish coat on both sides of an angle at the same time. The plow operates on the same principle as the flat finisher or nail spotter, but its dispensing surface is shaped in a 90-degree angle to fit into interior corners. The applicator is available in 2" or 3" widths and can be used to apply bedding coats of compound prior to taping or to apply finish coats that are feathered and smoothed with other tools.

Once a bed of compound and tape has been applied in interior corners, the tape is smoothed and embedded into the mud using a corner roller (*Figure 33*). This device consists of four stainless steel rollers mounted on a swiveling pole.

Figure 31 Corner applicator.

Figure 30 Flat finishers.

Figure 32 Corner finisher (plow).

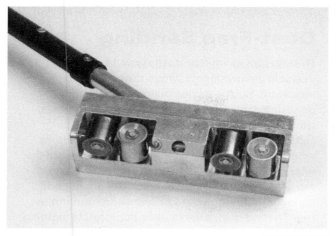

Figure 33 Corner roller.

Figure 34 Automatic joint-compound loading pump and gooseneck.

For best results, start from the middle of the angle joint and use light pressure to roll toward both ends. Make a second pass, again from the middle, working toward both ends with firm pressure. This will force excess compound from under the tape. To maintain the corner roller, spray it with a high-pressure water hose after use.

3.2.5 Automatic Loading Pumps

Automatic taping and finishing tools are filled with compound through automatic loading pumps (*Figure 34*). These pumps work on the same principle as the old-fashioned hand-cranked water pumps.

The pump's intake nozzle is placed in a 5-gallon pail of compound. To stabilize the device, place a foot on the foot plate on the pump's main body, which sticks out of the bucket and sits on the floor.

As the pump's handle is operated, compound is forced out through a J-shaped outlet nozzle. Several attachments can be used to adapt the nozzle to feed various automatic tools. The attachments allow the tools to be filled quickly without any mess.

Like all automatic tools, the pump must be maintained properly. At the end of the day, remove the device from the pail and force out any compound remaining inside the pump. Flush the pump with water from a high-pressure water hose or by placing the intake nozzle in a bucket of fresh water and pumping the water through the compound.

To prevent clogging the pump, make sure the joint compound is properly mixed and free of lumps. Remix the compound periodically if it must stand unused for a while.

The pump's intake nozzle features a small screen that stops lumps or debris from being drawn into the pump. Following the manufacturer's directions, remove and clean the screen after each use. Replace the screen occasionally.

3.2.6 Vacuum Sanders

A manually operated vacuum sander works like a hand-operated pole sander. The vacuum sander's pole, however, is actually a rigid hose connected to a powerful vacuum cleaner. A piece of screen-backed sandpaper or a tough sanding mesh is stretched across the hose's flat, hollow head. As the sanding head is pushed back and forth across the dried joint compound, the vacuum sander pulls dust and chunks through the sanding mesh and into the hose. The dust is collected in a large filter bag, as in an ordinary household vacuum cleaner.

Powered versions of vacuum sanders are also available (*Figure 35*). They provide a fast, clean way of finishing drywall jobs. The vacuum hose is connected to a shop vacuum. The sander uses foam-backed sanding pads that are specially designed for drywall work. The machines are safe, effective, and easy to use. Hoses and attachments enable the finisher to sand normal-height ceilings and walls without stilts or ladders.

Priming the Pump

A new or fresh compound pump should be primed by pouring ½ cup of water into the outlet.

Figure 35 Power vacuum sander.

Dust-Free Sanding

The vacuum sander, or dustless sander, is very useful in environments where dust control is important. Such a device may be required for drywall finishing jobs in hospitals, supermarkets, clean rooms, and food preparation areas.

Be sure to follow the manufacturer's directions when using a vacuum sander. Empty or replace the filter bag as often as required, and see that hoses are kept clean and sanding materials are in good condition. Although the vacuum sander minimizes free-floating dust, always wear appropriate clothing and PPE, including goggles and a dust mask.

3.0.0 Section Review

1. A finishing tool with a thin flexible blade made of blue steel is the _____.
   a. putty knife
   b. corner trowel
   c. taping knife
   d. flat-blade trowel

2. An automatic taping tool is cleaned by _____.
   a. careful scrubbing with steel wool
   b. chemcial cleaning solutions
   c. scraping out excess compound with a putty knife
   d. forcing water through the unit from an empty compound chamber

Section Four

4.0.0 Drywall Finishing Procedures

Objective

Describe proper drywall finishing procedures.
a. Identify ideal site conditions for drywall finishing.
b. Describe the process for finishing drywall.
c. Describe the hand-finishing procedures involved in drywall finishing.
d. Describe the automatic taping and finishing procedures involved in drywall finishing.
e. Identify common joint problems when finishing drywall.
f. Identify common compound problems when finishing drywall.
g. Identify common fastener problems when finishing drywall.
h. Identify common problems when finishing drywall.

Performance Tasks 2 and 3

Properly apply a corner bead, tape, and finish to a drywall panel.

Patch damaged drywall.

Trade Terms

Tapered joint: A joint where tapered edges of drywall meet.

Drywall finishing involves taping, topping (also known as buttering), and sanding the drywall seams and joints, whether on walls or ceilings, so these surfaces can be made ready for final decorating. Some jobs use texturing and have a reduced need for taping and sanding. Other jobs require a large amount of butt and seam taping, three or more topping or skim coats, and a lot of sanding to produce the expected results.

4.1.0 Site Conditions

Job-site conditions such as temperature and humidity affect the performance of most finishing materials. During winter conditions, drywall finishing should not be attempted unless the building is heated to between 50°F and 80°F. In addition, finishing materials must be protected from the cold temperatures at all times. If the humidity is excessive, ventilation must be provided. Windows should be kept open to provide air circulation. In enclosed areas without natural ventilation, fans should be used to move the air. When drying is slow, additional drying time should be allowed between applications of joint compound. During hot, dry weather, drafts should be avoided so the joint compound will not dry too rapidly.

4.1.1 Drywall Inspection

A professional always inspects installed drywall before finishing it to determine if the drywall is ready for joint treatment. Improperly installed drywall is difficult to finish.

Examine the installed drywall. Nail and screw heads should all be dimpled. This means that each fastener head should have been driven into the drywall so that a slight depression is made (*Figure 36*). No part of the fastener head should be above the rest of the drywall surface. Drag an 8" drywall knife over the rows of fasteners to check for flatness.

Determine if the fasteners are holding the drywall panels tightly against the framing members. Detect loose fasteners by placing a finger over the fastener head and press the adjoining drywall area in toward the framing with your other hand. When loose, movement will be felt through the fastener head. If a loose fastener is discovered, install an extra fastener above and below the original fastener to make the drywall more secure.

Check the butted joints and outside corners. Remove loose paper and broken drywall from the edges. Cut the paper back to where it still adheres; do not pull it.

Examine the drywall fields and the butted joints. Check for torn areas and large gaps. Mark all damaged areas with a lead pencil only; ink or crayon will bleed through. These damaged areas must be repaired.

Examine the inside and outside corners. How well do the drywall panels butt together? Make

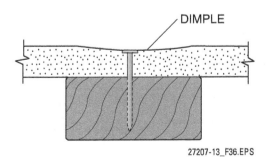

Figure 36 Dimple or uniform depression.

sure the panels are properly aligned. When one panel sticks out farther than the other, the joint is difficult to tape and finish smooth. Another way of saying this, especially for butt joints, is that there might be a high side and a low side. A high side is produced by a butted panel that sticks out too far from the framing. A low side is then produced in the other abutting panel, which does not stick out as far. This is shown in *Figure 37*. This condition might be caused by a twisted stud. Part of the panel rests against a part of the stud that is not even with the part of the stud to which the other board is attached.

Since at this stage it is too late to correct the framing, the best you will be able to do is hide the offset. You can hide it by applying a little more compound and finishing the joint a little wider than usual. In fact, all butt joints are finished wider than a tapered joint because there is no tapered depression to hold the compound.

All these inspections help to determine which procedures to use first in any given situation. If a major repair is required, inform a supervisor. If drywall needs serious correction, such as reframing or rehanging, it is much better to get it done before taping and trying to hide everything with compound.

4.2.0 Taping and Finishing Process Overview

Taping and finishing is a multistep process that varies between three and five different stages of finishing of each board joint or seam. Joints are generally considered to be any place where two edges of drywall abut one another. Seams are places where tapered board edges meet each other. Butt joints are places where square (nontapered) board edges meet each other. Flat joints are places where two beveled edges meet.

Generally speaking, joint taping includes prefilling the joint, taping and bedding, topping and skim coats, and sanding. After each step, the compound is allowed to dry, usually overnight.

Joint compound and tape shrink as they dry. This shrinkage results in slight depressions that need to be filled out again by applying topping, skim, or finishing coats of compound. Using an automatic taping tool system, the actual processes at each joint or seam are greatly speeded up. The whole idea is to make every joint or seam as flat as the rest of the wall or ceiling. The idea is also to make these joints and seams undetectable once the decorating is complete.

As a general statement, the ideal drywall finishing process requires five steps. Depending on the drying time, it may take five different trips to the job site:

Step 1 Complete repairs, cutouts, and prefill. Apply bedding tape at all joints and seams. Complete corner bead and trim installation. Top outside corners. Use an 8" knife on headers and spot fastener heads.

Step 2 Apply first topping coat over taped joints and seams. Use a 10" box. Make double-wide topping coats at all butt joints. Top all corners and angles.

Step 3 Sand lightly, and scuff angles and flats. Apply a second (flash) topping coat to fill in all pits, gaps, depressions, or shrinkage. Apply a straddle coat on the butt joints. Apply another coat on the fastener heads.

Step 4 Perform light sanding and scuffing. Apply a third (skim) topping coat. Use an 18" knife on flat seams and butts. Use a 3" plow for angle topping. Apply a flash coat on headers, seams, and outside corners.

Step 5 Complete pole and hand sanding.

4.3.0 Hand Finishing Procedures

Drywall finishing procedures start with gathering all the proper tools, equipment, and materials necessary to do the job. Decisions are then made about prioritizing the various tasks: what is done first, second, and so on. These tasks include:

- Inspecting, repairing, and prefilling
- Taping flat joints, corners, and other angles
- Installing bead and trim pieces
- Spotting fastener heads
- Topping, scuffing, and sanding

To finish drywall, proceed as follows:

Step 1 Dimple the nail and screw heads, and cover them with joint compound. A dimple can sometimes be created by hitting the nail with the butt end of your knife handle. Damaged drywall must be patched. Do only a minimum of scuff-

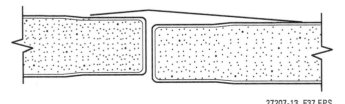

Figure 37 Butt-joint misalignment (low side on right).

ing or sanding on the paper so as not to roughen it or raise the nap of the drywall paper.

Step 2 Once the drywall has been properly installed, carefully inspected, and fixed where necessary, the next step is to prepare for spotting and taping. The usual finishing sequence is to spot the fastener heads; prefill gaps, damaged areas, and butt joints; tape the ceiling joints; and then tape the corners, other angles, and finally the flat joints.

Step 3 Place a suitable sheet of scrap drywall on the floor and mix all materials over the scrap board. Prepare the joint compound according to directions and the job-site conditions.

Step 4 For hand taping, load the joint compound into the mud pan or onto the hawk. Obtain the proper compound consistency. It should generally be the consistency of soft putty. Properly mixed compound does not fall off the hawk when it is tilted for a short time.

Step 5 Apply the compound to the bare joint with a broad knife. (If self-adhesive mesh tape is used, it is applied without a bedding coat.) While the joint compound is still wet, apply the joint-reinforcing tape. Press the tape into the compound (*Figure 38*). Smooth the tape with a broad knife as it is applied. Force the excess compound out from under the tape and remove it with the knife. This bonds the tape to the compound. For perforated tape, force the excess compound up and out through the holes. Again, wipe away the excess with the knife. Make sure there is enough compound under the tape or bubbles will result.

Step 6 Spread a thin coat of joint compound over the top of the tape. Allow these bedding coats of joint compound to dry overnight. Special precautions need to be taken when finishing butt joints. For hand finishing, it is critically important to look at the taped butt before you finish it. Follow these guidelines when finishing butt joints:
- If the butt joint has a high side and a low side, coat the low side.
- If the wallboard butt forms a hollow, fill it back to the normal plane by adding much more compound than usual.
- If the butt joint is regular, it still needs to be finished with the double-wide method (*Figure 39*). For the first topping coat, do not cover the tape. On day one, make a crown of compound on each side of the tape. Then, on day two, apply the straddle coat to cover the tape between the crowns, thereby making one slightly larger crown in the middle of the seam.

Step 7 Apply a thin coat of joint topping compound. Feather the compound's outer edges. Feathering spreads the compound thinly from the center of the taped joint outwards beyond each edge of the tape, causing a slight crown over the center. However, the smaller the crown and the

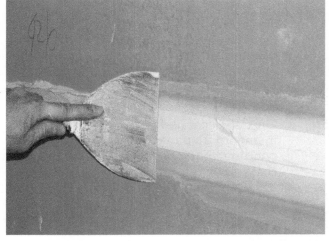

Figure 38 Applying tape.

Maintaining the Proper Temperature

If drywall finishing is done during cold weather, the building must be heated to 50°F minimum, and the heat must be maintained during the entire finishing process and until the finish is dry. Avoid sudden changes in temperature, which can cause cracking due to thermal expansion or contraction.

Finishing compounds lose strength if they are subjected to freeze-thaw cycles. If a compound has been frozen, it may have to be discarded.

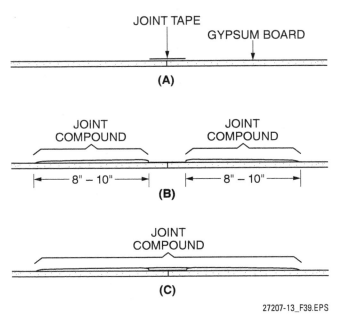

Figure 39 Finishing a butt joint using the double-wide method.

finer the feather, the less sanding will be required.

Step 8 Allow this coat to dry. Some compounds require 24 hours to dry; some take even longer. Drying times also depend upon atmospheric conditions within the structure. If the building is only partially closed off, the finishing work will be affected by the outside weather. If it is too cold or wet, the compound might not dry out at all until the weather changes. Finishing work depends upon a controlled interior.

Step 9 Once the coat has dried thoroughly, sand it smooth. Remove the sanding dust. Depending on the situation, another topping coat or one or more skim coats may need to be applied. These determinations are almost always made by a supervisor.

4.3.1 Sanding

Always wipe down after any light or heavy sanding to remove sanding dust and tiny particles of compound or other debris. Always check the coated surface to see if it is straight and smooth. Use a straightedge or level to do this.

Always have a hawk or mud pan available filled with topping compound, no matter what finishing procedure you are doing. This way, whenever you see something that needs a little touch-up, like a pit or scratch, you can take care of it immediately to avoid poor-quality finishing.

For general sanding and scuffing procedures, 100-grit, 120-grit, or 150-grit sandpaper is recommended. Sand screen is also used for scuffing as well as for final sanding, because it does not raise the nap on drywall face paper. Sandpaper of less than 100 grit should be avoided, and any sandpaper coarser than 80 grit is unacceptable.

Remember, you are sanding a dried, porous wall joint covering material in order to smooth out tiny bumps and spaces so it will hold a final decoration as well as wallboard face paper. Guard against oversanding, which tends to grind out hollows.

> **WARNING!**
> Be sure to wear proper PPE, including eye protection and appropriate respiratory equipment, when sanding. Check the safety data sheets (SDSs) for the applicable drywall to learn about the safety hazards associated with that material.

4.3.2 Spotting Fastener Heads

Check the drywall nails. Be sure they have been properly dimpled. Apply the first coat of joint compound on top of the nail heads. Allow it to dry. Sand the dried coat with an abrasive cloth or sand screen. Apply a second coat of joint compound. Allow this coat to dry. Sand it smooth and apply a third coat. The covered area should be smooth and level with the substrate.

Many contractors prefer hand spotting and will not allow the use of a nail spotter for this task. Their reasoning is based on several factors:

- Fasteners may not always be below the substrate level, so the nail spotter's blade is frequently damaged. This causes downtime to change blades or tools. When there are no more

Correcting Oversanding

Excessive sanding or use of coarse sandpaper can cause the paper fibers of the drywall to stand up. If the problem is not too severe, light sanding with a very fine sandpaper or wiping the panel down with a damp sponge or cloth can correct it. Otherwise, use a light skim coat of topping or all-purpose compound to correct the problem.

replacement blades or tools, hand spotting will be the only remaining option.

- Spotting by hand is a reasonably fast method when done by an experienced craftsperson.
- Hand spotting is thorough. It forces the compound more effectively onto the head and completely fills the dimple. It also packs compound down into the crossed indentations of Phillips-head screws better than the automatic tool usually does.
- Hand spotting is something even the newest apprentice can master almost at once. It helps them appreciate the nature of finishing work faster than any other process. It can also build confidence, speed, and an eye for detail.

All fastener heads normally need to receive three coats of topping so they are undetectable when the surface is finished. Use a 5" knife for the first two coats and a 6" or 8" knife for the third coat. Make sure the compound fills in the crossed indentations in Phillips-head drywall screws.

4.3.3 Outside Corners

To finish outside corners, proceed as follows:

Step 1 Attach the metal corner beads to the outside corner angles. Fasten them with drywall nails or a clinch-on tool or by applying tape with compound. Staples are sometimes used, although many contractors try to avoid them. Dimple the nail heads. Apply joint compound to each flange with a broad knife right after fastening.

Step 2 Spread the compound 7" from either side of the nose (center outside corner). Cover the metal edges with compound. Allow the compound to dry. Sand it lightly, then apply the next coat. Feather the coat out 2" from the first coat. Let this dry and sand it smooth. Also apply and smooth a third coat of topping at outside corners. Use an 8" knife for all three coats, a 6" knife for the first coat and an 8" knife for the others, or an 8" knife for the first two coats and a 10" knife for the third coat.

The goal with each coat is to fill the corner so it is flat, not concave. Too much compound will make the corner concave. Finish each outside corner so that the corner bead is completely invisible.

4.3.4 Inside Corners

To finish inside corners, proceed as follows:

Step 1 Cut the joint tape to the length of the corner angle. Apply recommended amounts of compound to both sides of the corner angle. This prevents thin cracks from occurring at the angle. Crease the tape length along the center.

Step 2 Use a 4" broad knife to press and embed the tape in the compound. Apply enough pressure to wipe the compound from under the tape. Feather the compound 2" beyond the tape edges. Let it dry and then sand. A corner tool is available as an alternative (*Figure 40*).

Step 3 Apply and feather the next coat about 2" beyond the first coat. Dry and sand. Then apply a third coat where needed.

Step 4 For inside corners, it is better when hand finishing to apply compound first to one side, wait overnight, and then apply compound to the other side. With the automatic corner applicator/finisher tool, you only need apply one coat of topping to inside corners. This is called plowing or glazing the angles.

You might find using trowels more comfortable than taping knives. There are many craftspersons in the trade who were trained on trowels. They find taping knives too stiff and awkward. Other carpenters, however, cannot imagine finishing a taped joint with a trowel. Either tool can be used to produce equally attractive results. Try both types of tools, if you like, until you decide which is best for you.

4.3.5 Safety and Good Housekeeping

Follow all recommended safety practices for drywall finishing. Maintain tools and equipment.

Drying Time

Atmospheric conditions and other factors always play a part in how fast the taping and topping coats dry. Another factor might even be the wallboard face paper itself. New and recycled paper might well have different drying rates, and different compound materials will vary as to how fast they dry. Therefore, drying times might be longer than just overnight.

Figure 40 Using a corner taping tool.

If stilts are permitted in your jurisdiction, practice safety; always put stilts on and take them off when you are able to lean against a wall, preferably in a corner. Never climb stairs or try to pick up something off the floor while wearing stilts; ask for help from someone who is not wearing stilts.

Be aware that taping and topping tools, knives, and trowels carry with them a certain degree of hazard. For example, even a dull blade can cut or injure an eye or face. Any tool, if mishandled, can cause an accident or injury. If tools are allowed to clutter up a work area, they can also cause an accident or injury. Keep unused tools in a safe place out of everyone's way.

One of the most common hazards when finishing are slips, often caused by wet compound carelessly spilled on the floor. This is especially hazardous for someone on stilts. The best rule is if you spill something, clean it up—no matter what it is. If wearing stilts, either remove them and clean up the spill or ask someone who is not on stilts to clean it up for you; do not attempt to bend over on stilts.

Store finishing materials in a cool, dry, protected location. Provide adequate ventilation during dry sanding. Always wear a face mask or respirator to prevent inhalation of sanding dust. Wash hands after applying joint compound as well as after sanding. Proper safety and housekeeping procedures minimize illness and injury.

Final cleanup is always your responsibility. This means a complete sanding and wipe-down of all ceilings and floors, and a thorough scraping and sweeping of all floors.

Keep in mind that sanding dust travels. If you are doing remodeling or repair in a finished building, secure the area in which you are working by covering doors and other wall openings with plastic. If you are working in a room that contains furniture, equipment, or other items, cover them.

4.4.0 Automatic Taping and Finishing Procedures

Large jobs may require drywall finishing equipment. These specialized tools, which were discussed earlier in this module, enable drywall finishers to tape and finish drywall uniformly and efficiently. Basic tool components of such a system are the loading pump, nail spotter, automatic taping tool, corner roller, flat finisher, corner finisher, and corner applicator. The finishing process using automatic tools is outlined as follows:

Step 1 Apply the tape using the automatic taping tool.

Step 2 Press the tape into corners using the corner roller.

Step 3 Wipe down excess compound and embed the tape using a broad knife or taping knife.

Step 4 After the bedding coat has dried, apply a topping coat using a flat finisher for seams and butts. Use a corner applicator/finisher for angles.

Step 5 After the first topping coat has dried, apply a second topping coat using a wider flat finisher for seams and butts, and a corner applicator and finisher for angles.

Neatness Counts

Always set up your mixing pails and other equipment on a large scrap sheet of drywall. This helps keep splashes off the floor and will greatly simplify your cleanup efforts. You should also keep at least one full bucket of water handy in this area to soak automatic tool heads when not in use.

Step 6 After this topping coat has dried, another skim coat may be applied using taping knives.

Step 7 After all coats have dried, sand the areas treated with compound. Wipe them down with a damp sponge after sanding. This helps the paper fibers to lie down.

4.4.1 Pump Loading Procedures

The loading pump (*Figure 41*) has nozzles of different sizes. Nozzle selection depends upon the equipment and material being used. The pump has a replaceable screen at the loading pump intake. This screen prevents large particles from passing through the pump. The pail holder is designed for a standard 5-gallon pail. A gooseneck attachment mounts on the pump to fill the automatic taping tool. The loading pump without the gooseneck attachment fills the nail spotter, flat finisher, and corner applicator.

Be sure to mix all compound thoroughly before using and especially before pumping it into any automatic tool. Be sure to mix out any lumps, especially when using powder compounds. Use a power drill with mixer attachment, if available, or a mud masher. With a mud masher, you should plunge it down into the center of the pail and bring it up, scraping against the sides. Rotate around the pail as you mix. This keeps lumpy residues from forming on the sides of the pail. Also, be sure to remix any compound that has been left standing for a period of time.

Keep the pump screen clean and free of lumps or dried compound. It may need to be replaced quite often, along with the O-ring on the gooseneck.

A final recommendation is to have two loading pumps at each job site, one with the gooseneck attachment, and the other with a fill adapter. This will speed up the operations of filling the different types of automatic tools.

4.4.2 Pumping

Set the pump into a full standard-sized premix compound pail, step on the pump's foot plate outside the pail, and pump the compound using the pump handle. Before pumping any compound into an automatic tool, however, water may need to be added to improve the consistency of the premixed compound. Taping compound should have a thinner consistency for use with the automatic taping tool. A thicker consistency is needed for topping applications so less water is added. Always follow the recommended mixing instructions on the labels of the products.

Also, before pumping any compound into an automatic tool, be sure to pump the handle a few

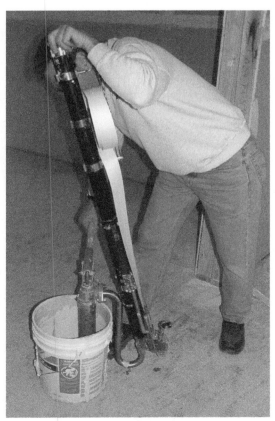

Figure 41 An automatic taping tool being loaded with a loading pump.

Figure 42 Nail spotter.

times to clear out any air. Without the gooseneck attachment, simply pump the compound back into its own pail. With the gooseneck attached, use another container to catch these first few pumps. Never attach a taping tool to the gooseneck until you are sure all the air is out. Then attach the tool upside down to the fitting and pump it full.

4.4.3 Automatic Nail Spotter Procedures

You may use a nail spotter (*Figure 42*) to fill countersunk nail and screw heads with joint compound. A complete row of fastener heads can be filled in one pass with this tool, which is normally available in 2" and 3" widths. A nail spotter allows you to fill rows of fastener head dimples with compound, on both walls and ceilings, while working from the floor.

Use the loading pump to fill the nail spotter with compound. Set the pump into the pail, making sure the compound is well mixed and free of lumps. The pump needs only to be pumped full of compound; no priming is needed. Use the adapter spout (not the gooseneck) to fill the nail spotter at its opening.

Each row of dimples can be filled in one pass. Be sure to make positive contact with the drywall surface at the beginning of each row. Draw the tool smoothly along the entire row, applying some pressure to force the compound out into the surface. The blade skims off excess compound and leaves a slight crown over each dimple as the tool floats along on the rocking skid.

After passing over the last dimple, gradually break contact with the surface by using a sweeping motion. This procedure will fill the dimples without leaving excess compound that needs to be removed.

4.4.4 Using the Automatic Taping Tool on Ceiling Joints

Fill the automatic taping tool with joint compound using the gooseneck adapter with the loading pump. Close the gate valve and turn the tool upside down to fit its opening over the gooseneck opening. Place one to two fingers of your free hand into the end of the tool. To avoid overfilling, pump in the compound and until your fingers feel the piston. If the tool is overfilled, relieve pressure by depressing the filler valve stem with a nail.

> **WARNING!** When filling the tool, start slowing down at about six or seven pumps so your fingers will not be injured if the piston rises too quickly.

You will likely find that it takes about 9½ pumps to fill an empty taping tool.

Using Stilts

Do not try to use stilts unless you have been properly trained to use them. If you plan to use stilts, ensure they are not prohibited by local codes.

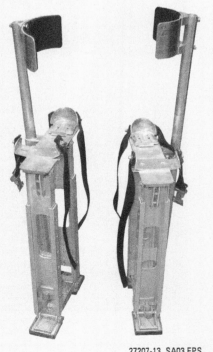

27207-13_SA03.EPS

Step 1 After loading the taping tool, open the gate valve and turn the key counterclockwise until joint compound covers the leading edge of the tape. This is only necessary the first time the tool is used for taping after each filling.

Step 2 Hold the taping tool with one hand on the control tube, called the slide, which is similar in operation to a shotgun pump (see *Figure 43*). Put your other hand on the bottom of the mud tube so the creaser control lever can be operated. Several fingers can be placed into the end of the mud tube if this method provides greater control.

Step 3 Start by taping ceiling butt joints, and then tape the ceiling flat joints. Use both drive wheels for the first 4" to 6" in order to secure the tape to the ceiling. Start at one end of the seam and work to the other end. After taping the first 6", tilt the taping tool at about a 20- or 30-degree angle away from the plane being worked in so that only one drive wheel is pressed to the board surface. This helps limit the amount of compound and prevents air bubbles. Walk backwards as rapidly as possible, leading with the head of the tool.

Step 4 When approaching the end of a joint, gradually bring the tool back to alignment with the vertical working plane. At about 3" to 4" from the end, stop and pull down sharply on the control tube, which will cut the tape. You will need to slow down and stop completely in order to do this because the tape or blade will jam if you do not stop to make this cut.

Step 5 Return the slide to its neutral position and bring back the other drive wheel so both wheels press against the surface once again. Keep both wheels rolling to maintain the continuous buttering of compound needed to press the last bit of tape onto the end of the joint. At this ending sequence (when both wheels are rolling), push the slide forward to eject

Figure 43 Using an automatic taping tool.

the end of a new piece of tape. It will be buttered with the correct amount of compound that allows the next run to be immediately started.

Step 6 To finish the ceiling taping procedure, close the gate valve on the automatic taping tool and put it headfirst into a pail of clean water. Use a long-handled broad knife to wipe down all the ceiling butt and flat joints. The bedding joint compound should be wiped down while it is still wet, so that it is easily workable and excess moisture does not soak into the wallboard.

Step 7 Use firm pressure and hold the knife blade at about a 45-degree angle to the surface. Wipe along each taped joint, laying the tape flat and forcing out excess compound from underneath. Start in the middle of each seam and work first toward one end, then the other. This helps to avoid wrinkles and bunching up of the tape. Be sure to catch all the excess compound squeezed out on your knife and scrape it off on your hawk or into a mud pan. The whole process is meant to embed the tape, fill the joint, and leave a generally flat surface.

Thinning the Compound

Experience has shown that best results come from adding ½ cup of clean, cold water to 4½ gallons of compound for all automatic tools except the automatic taping tool. For the automatic taper, mix in 2 full cups of clean, cold water per 4½ gallons of compound. Use the compound full strength for all hand-tool applications, nail spotting, prefill, and skim coating.

4.4.5 Using the Automatic Taping Tool on Wall Joints

After the ceiling is wiped down, the next process is to tape the horizontal and vertical wall joints. Again, start with the butt joints. Remember to open the gate valve on the taping tool.

Step 1 For vertical wall joints, place the taping tool at the bottom of the joint, about parallel to the floor. Push the control tube forward 1" or 2" to make a tape leader. Start the vertical taping with the leader overlapping the floor a little. While proceeding upwards, the tape will be drawn up. As soon as possible, when moving upwards off the floor, maneuver the taping tool to lead with the head. Also, shift your position to track with just one wheel in contact with the wall.

Step 2 At 3" or 4" from the top of the wall, pull back the control tube to cut the tape and continue rolling to the ceiling intersection on both wheels. To start the next joint, roll the wheels slightly against the surface, starting the flow of compound while ejecting a new leader with the control tube.

Step 3 When taping horizontal joints, push forward on the control tube to produce a 2" or 3" tape leader. Place the leader, again with a slight overlap, at the beginning of the horizontal seam. Except for the start and end of a joint, always hold the taping tool at an angle to the wall (base of the tool angled downward) so only the bottom drive wheel is rolling against the surface. At the beginning of each seam, however, push both wheels against the wall for about 6".

Step 4 When approaching within 3" or 4" of the end, pull back on the slide tube to cut the tape, and continue on both drive wheels to the end of the joint while pushing forward on the slide. This applies the last several inches of tape while feeding out another leader for the next joint.

Step 5 For outside corners (where you are not using corner bead), simply follow the same procedure as explained above for vertical wall joints, but this time only apply tape to one side of the corner. Let the other side remain exposed to the air. When the vertical run has been completed, close the gate valve and fold the tape over the corner using a broad knife.

Step 6 At ceiling angle (or wall intersection) joints, use the creaser wheel, which is extended by pulling on the trigger near the end of the automatic taping tool. The creaser wheel can also be used to help roll the tape against a flat seam, which is critically important when taping inside corners. Bisect the angle with the tape and make sure both wheels press equally on each side of the angle while rolling. Be sure to track in a straight line. Avoid twisting the automatic taping tool while moving it. Again, start with a 1" or 2" leader to allow for the tape to be pulled toward the joint end (sometimes called creeping). Push the leader into position in the angle using your fingers, if needed, before you are able to proceed. Otherwise, taping inside corners and ceiling intersections is the same process that is used for vertical wall joints.

Step 7 If taping alone, stop using the taping tool after about 10 minutes, close the gate valve, and place the head into the pail of water. Then proceed to wipe down all the tape just installed, to embed it and remove the excess compound.

4.4.6 Using the Corner Roller

After the tape and joint compound have been placed in ceiling and corner angles, use the corner-roller tool (*Figure 44*) to embed the tape in the joint compound at inside corners and force out excess compound from the tape. At the corners, remove excess compound with a broad knife. Then, after the angles are rolled, go back over the full length of the angle with a corner finisher called a plow. The general sequence at all corners is as follows:

- Taping
- Rolling
- Plowing
- Corner-applicator finishing

Four metal rollers in the head of the corner roller will embed and smooth the tape while forming a sharp corner crease. Work it from the middle of the taped joint out toward the ends of the joint. This will force any overlap of tape due to stretching out to the end of the joint, where it can be trimmed off. This method stretches the tape in place. It also helps to prevent bunching up, which can easily happen if rolling is started at one end and applied going toward the middle (instead of the other way around).

Figure 44 Using a corner roller.

4.4.7 Corner-Plow Operation

The corner finisher, or plow, is used to take out excess compound from angles after the corner roller has embedded the tape. This tool is available in widths of 2" and 3". Either size can be used to wipe down the excess compound. Simply snap one of these tools onto the associated ball on the end of the corner applicator handle and use it like a plow (*Figure 45*). Another nickname for a plow is a butterfly, probably due to its shape. It has four skimming blades, and is designed to wipe down and feather both sides of inside corners and other angles at once.

With the arrow end leading, work from top to bottom for vertical angles, and from one end to the other for ceiling intersections. Wipe down these angles further on both sides with a 6" taping knife.

The plow is also used for topping, together with the corner applicator (*Figure 46*). It smooths and finishes both taped and topped corners. The corner finisher feathers the joint compound out from the corner and onto the drywall. It finishes both sides of the corner at once.

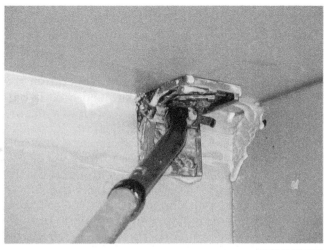

Figure 45 Using a corner plow.

Figure 46 Using a corner plow with a corner applicator.

The plow has a spring action that compensates for corners slightly greater or less than 90 degrees. The blade design produces a smoothly feathered joint. When using this tool, be sure the three tips, or arrows, are pointing in the forward direction of travel.

Using the Automatic Taping Tool

Except for beginnings and ends of tape runs, always hold the automatic taper at about a 20-degree or 30-degree angle to the wall or ceiling, and operate on one wheel only. It is important to have at least one wheel pressing against the drywall surface at all times while you are taping. These wheels control compound flow onto the tape. If the tape is pushed along without engaging a wheel, air pockets or bubbles will be produced, which are gaps in the tape where compound is missing. Correct this immediately by going back, tearing out the unbuttered tape, and retaping. If this is not done, the topping will dry over the air pocket, and will eventually crack apart and crumble or fall off the surface completely.

4.4.8 Flat Finisher Procedures

The flat finisher (box) adjusts the amount of topping compound applied to the joint (*Figure 47*). The compound is automatically feathered out from crown to edges. The crown runs down the center of the taped joint. This raised area is balanced out by the shrinkage that normally occurs when the compound dries in the joint. A box can be used to apply topping to both tapered and butt joints.

A small dial is located under the box by the handle connection. This dial controls the size of the crown left by the box on the surface. It also controls the amount of topping applied to the surface. As this dial is turned, the blade that controls the crown raises or lowers. Establish the setting needed for a particular application. In general, more crown is needed for earlier coats and less crown for later coats.

A flat finisher is available with handles of various sizes, usually from around 3' to 6'. Longer handles allow a craftsperson to finish most higher wall and ceiling levels right from the floor, without stilts or scaffolds.

Always run the box with the wheels leading and the blade trailing. Adjust your grip in relation to your body to lead the box with the handle, except at the joint ends. Also note that the end of the handle has a special gripping lever. This grip locks the box into the desired position in relation to the handle while moving across the surface.

To finish flat joints, proceed as follows:

Step 1 Load the box through its opening behind the blade using the adapter spout, or nozzle, on the loading pump. The box loads in the same manner as the nail spotter. In general, the topping compound should be a little thicker than the taping compound.

Step 2 Apply topping compound on wall or ceiling taped joints by drawing the box steadily along the joint while applying pressure to the back of the box with the handle. This forces out the compound evenly through the opening, depending on the crown you set on the dial. The blade also serves to feather the compound thinly out to the edges, leaving the crown in the center. Always start from one end of the seam and go straight across to the other end without stopping. It should always be a smooth process. Always ensure there is enough compound in the box before beginning the run.

Step 3 Ceiling joints should be topped first. The first ceiling joints to receive the topping are the butt joints. Butt joints receive a first topping coat, one coat on each side of the butt, using the 12" box on each side. Do not put this coating across the center of the butt; leave the center alone for the first coat. Set the dial on the 12" box to #1 (fullest) crown. Think of this as giving each butt joint a double-wide treatment. The reason for this is that you deliberately finish butt joints much wider than regular tapered (flat) seams to hide the very slight crown already caused by the butted wallboard panels. The finished crown is so gradual and so slight that it will hardly be noticeable.

Figure 47 Using a flat finisher.

Teamwork

On large finishing crews, one carpenter operates the taping tool, another follows with a broad knife to wipe down excess compound, another spots fastener heads, another comes with the corner roller, and so on. When operating the automatic taping tool alone, however, put down the tool after taping a joint in order to wipe down excess compound.

Step 4 The flat joints receive topping after the butt joints. Again, start with the ceiling. Use the 7" flat finisher set at #3 (medium) crown. Start at one end of the joint and apply even pressure to the middle of the joint. Lock onto the grip as soon as possible after starting the run. Lead with the handle.

Step 5 When approaching the middle of a run, keep the lock on the handle and gradually release the pressure. Then remove the box in a sweeping motion from the surface.

Step 6 Next, reverse hand positions and start at the other end of the run. Repeat the process described above for beginning the run. Lead with the handle toward the middle of the joint. Again, keeping the lock on, slightly overlap the stopping point and remove the box with a sweeping motion.

4.4.9 Flat Finishing for Other Joints

For wall vertical flat joints, start at the bottom and lock onto the handle grip right away; then remove the box in a sweeping motion about 2' to 3' above the floor line. Start at the top of the joint and apply pressure down to the previous stopping point. Again, with the handle locked, slightly overlap that point and then sweep off the surface, which should neatly finish the two topped sections.

When applying topping to joints near doors, windows, and other openings, always work from the corner and move towards the opening. Just before the wheels reach the opening, keep the handle locked and lift the wheels. Then sweep away out into the opening so that topping compound is applied all the way to the edge.

4.4.10 Finishing Coats with the Flat Finisher

Second and third topping coats are applied in the same way as the first topping coat, but in thinner and wider layers. These finishing coats are applied to fill out minor shrinkage and unevenness that could produce shadows and other imperfections after painting. Always allow each coat to dry overnight. In cold or humid conditions, each coat might take even longer to dry.

Before starting each finishing coat, scuff the surfaces. This means lightly sanding the dried compound areas to remove any crumbs, burrs, globs, and so forth. This also prevents any of these things from producing scratches in the surface when you make another pass over them with a box. Be sure to wipe down all sanded areas.

Fill the 10" box, set the dial to #3 (medium), and then apply a straddle coat on all butt joints, starting with the ceilings. Apply this coat of topping compound right down the center (on top of the taped seam) between each of the previous double-wide coats. A correctly finished butt joint should always have a final width of at least 25".

Do the flat seams next. Use the 12" box if only one coat is required, or use the 10" box for the second coat, and the 12" box for the third. Reset the dial to #5 (least) crown, and cover the ceiling and wall flat joints in the same way as for the first topping coat.

4.4.11 Corner Applicator Operation

In topping operations, the corner applicator provides a final finish for ceiling and inside corner angles. Use it after the bedding coat has dried. Attach the corner finisher to the corner applicator by snapping it into place. The plow becomes the skimming or troweling blade for the tool. This runs in the topping coat at the corner angles. The corner applicator smooths and feathers the final coat. The corner applicator does for angles and corners what the flat finisher does for flat and butt joints.

To use a corner applicator, proceed as follows:

Step 1 To load the corner applicator, first remove the nozzle end from the chain-slung filler adapter that goes on the loading pump. Inside the nozzle housing is a rubber O-ring that prevents leakage during filling. The filler valve on the corner applicator is inserted into the housing against the O-ring, which seals it. Pressure from the pumped compound pushes the tool's filler valve open.

Step 2 Once the corner applicator is full of topping compound, attach the 3" plow over the round opening. Place the tool at one end of the corner angle. Then with the nose of the plow leading, draw the tool along the angle, applying steady pressure with the handle on the back of the box. This forces the compound out where the plow distributes and feathers it along the run.

Step 3 As end of the run is approached, sweep the tool away from the surface in the same way as with a flat finisher. Reverse hand and body positions and start again at the unfinished end. Draw the tool back to the previous stopping point. Overlap just a little and then sweep away from the surface. This should neatly join both sections of the run. Use the plow to neatly bisect the angle. Also, keep the tool at as close to a right angle to the corner as possible.

Step 4 To apply compound to vertical angles, start at the top of the wall and draw the tool downward, sweeping away from the surface at about knee height. Then place the head in the bottom of the angle seam, and draw the tool back upward to barely overlap the previous stopping point. Gradually sweep away from the surface, neatly joining both sections of the run.

Step 5 Detail out the corner intersections with a broad knife. Feather any excess compound away from the angle on both sides, also with a broad knife.

When the plow is removed from the corner applicator and a ribbon dispenser is attached, it becomes a tool known as a flat applicator. The flat applicator is an alternative tool to the automatic taping tool. It applies a bedding coat of taping compound to which the tape can be attached to the surface by hand. This semi-automatic taping method is useful in places where an automatic taping tool cannot fit. There is also a taping mini-tool that might work just as well in confined areas. The flat applicator is convenient for emergencies, for hand operations, or for use by one craftsperson when another craftsperson is using the only taping tool on the job.

All these automatic taping and topping tools must be kept clean. Keep the tool heads submerged in a bucket of clean water whenever they are not being used.

4.5.0 Finished Joint Problems

The true test of finished work will come not from how well mistakes were avoided, but rather how well the mistakes are repaired. The common joint problems found in drywall work include:

- Ridging
- Tape photographing
- Joint depressions
- High joints
- Discoloration
- Tape blisters
- Cracks in the joint

4.5.1 Ridging

When a ridge occurs along a joint between two drywall panels, it is often because there has been movement at the joint. Ridging is also sometimes known as beading or picture framing because a visible ridge that surrounds a drywall panel resembles a frame surrounding a picture. There are three probable causes of ridging:

- *High humidity, poor heat distribution, or not enough ventilation* – These result in expansion and contraction of the framing and drywall panels.
- *Drywall that is not properly installed* – Improper installation includes misaligned framing and butt joints that do not fit well together. If two panels are forced together, joint compound may be squeezed out, forming a ridge. Make sure the joint is not overstressed by too tight a fit.
- *Too much joint compound* – To correct ridging caused by too much mud, first sand the ridge smooth, then apply a finishing coat of joint compound. Hold a light at an angle to the area to make sure you have eliminated the ridge and left a smooth surface.

Drywall may need to be cut away at the joint to make a little gap between the panels. Do this by cutting with a knife, making several passes, or use a handsaw or chisel to remove a sliver of one of the boards. If the space left between the panels is too wide, the joint will be weaker than a properly fitted joint. Ideally, the entire width of the joint tape is bonded to a drywall surface. The tape itself only needs to span a small space between the boards. If the space between the boards is too wide, less of the tape is bonded to the drywall, which can weaken the joint and promote ridging.

Two pieces of tape may need to be overlapped to strengthen a joint. If the joint is very wide, it may be because one board, or both, is not securely fastened to the framing at the joint. Perhaps neither board is directly over the framing member. If needed, install fasteners at an angle through the drywall into the framing to make the joint tight against the framing.

This presents a problem; one of the most basic rules is always to drive the fasteners straight so that the edges of the heads will not protrude. In this case, however, they will. Repair the problem by using a hammer to straighten the head (by bending the shank) after the fastener is installed. You may also need to cut and install a thin strip of gypsum board to bridge a space that is too wide.

4.5.2 Photographing

If the joint is still visible even after the wall is finished and painted, this condition is known as photographing. The joint may show through as a slightly different color than the finished wall. Or it may be the same color as the finished wall but have a higher or lower gloss (shine) to it. Photographing can also occur over fasteners if there was insufficient joint or topping compound spotted on the heads.

The usual causes of joint photographing are:

- The installer failed to force excess joint compound out from under the joint.
- High-humidity conditions delayed drying of the second and third topping coats of compound.
- The tape absorbed too much moisture from the compound, causing the mud to shrink and conform to the shape of the joint tape. Avoid this by wetting the tape before installation.

To correct photographing, sand the tape edges to feather them into the surface of the wall or ceiling. Then cover the tape with thin coats of joint or topping compound. Use thin coats so as not to rewet the tape too much, or seal the tape with a primer after sanding and before applying the final coats of mud. This keeps the tape from drawing too much moisture out of the mud.

4.5.3 Joint Depressions

A joint depression is a valley that occurs at a seam or joint. It will be most obvious when a light strikes the drywall at an angle. (Hint: Use a 10" or 12" knife to help find high and low spots.) There are two common causes of joint depressions:

- There may not be enough joint compound over the joint. This can happen when the joint compound mixture is too thin or when not enough joint mud is applied to the joint.
- The joint may be sanded too deeply.

The cure for joint depressions is to add more joint or topping compound to the joint. Then smooth it and sand again to get a flat surface at the joint. Make sure the joint is flush with the surface of the wall or ceiling.

4.5.4 High Joints

A high joint is the opposite of a joint depression. It occurs when a wide section of joint is raised above the rest of the drywall surface. Like the joint depression, a high joint is most noticeable when a light strikes it from an angle. High joints are the result of too much mud built up underneath or on top of the tape and/or improper feathering of each coat. The edge of the coat must be feathered into the drywall surface. When this is not done, high joints will result.

To repair a high joint, sand the area as flush as possible without sanding into the tape. Then apply one or two final skim coats of topping compound. Feather each coat into the drywall surface. Make each coat wider than the previous coat to conceal the area of the joint (*Figure 48*).

4.5.5 Discoloration

Joints may discolor or turn lighter or darker than the rest of the finished surface. There are several common reasons for joint discoloration:

- Moisture may be trapped inside the joint. Until a joint is sealed, it can absorb water. If a joint is sealed before it is dry, water is sealed inside the joint. Trapped water will degrade the finish and discolor the surface. Be sure the joint is dry before sealing it.
- The joint was painted in conditions of excess humidity. To prevent this, reduce the room humidity before any painting is done.
- A poor-quality paint was used. Always be sure to use a good-quality paint. Cheap paint often gives uneven coverage and sealing. This increases discoloration.

4.5.6 Tape Blisters

A tape blister is really an air bubble in the surface of a joint. It can be several inches long or as small as a dime. Tape blisters occur when the bond fails between the tape and the first bedding coat of joint compound. One way a bond can fail is when there is no bond to start with. This may happen if

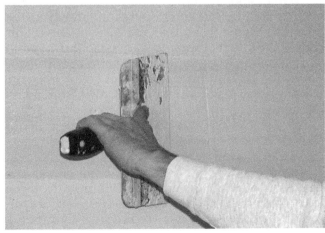

Figure 48 Coating with topping compound.

care is not taken when using an automatic taping tool and sections of tape come out without a mud coat underneath.

Tape blisters can also occur if the joint is too wide, because the tape was not properly embedded in the joint compound, or because the tape draws moisture too quickly from the mud. Another cause of a blister is when topping compound is used instead of joint compound to embed the tape.

To repair a tape blister, proceed as follows:

Step 1 Slit the blister with a knife. If the blister is large, cut and remove the entire section of tape that came unbonded.

Step 2 Sand or scrape out enough of the dried mud so a new section of tape can be embedded.

Step 3 Work joint compound underneath the tape, smoothing the slit in the old or new section of tape into the joint compound. This embeds the blistered section. This is a hand procedure only. Do not attempt to do this with another run of the automatic taping tool.

Step 4 Apply a skim coat of mud over the tape. When this coat is dry, apply the required number of topping coats, always allowing enough drying time in between coats. Sand enough to produce a smooth finish that is flush with the surrounding surface.

4.5.7 Cracking

The two common types of joint cracks are those that run along the edges of a joint and those that run along the center of a joint. Each has its own causes.

Edge cracks are cracks along joint edges that occur when the air temperature is high and the humidity is low when the joints are finished. This causes the joint compound to dry too quickly and unevenly, resulting in uneven shrinkage. To slow down the drying rate, run a wet roller over the joint or spray it with a fine water mist from an atomizer.

Edge cracks can also be caused by tape that has a thick edge or by joint compounds applied in coats that are too thick.

The procedure for repairing edge cracks depends on whether the crack is thin or wide. If the crack is small and thin, coat it with a latex emulsion or a thin coat of joint or topping compound. Then sand it as needed.

If the crack is wide, gouge out some of the joint compound if needed to prepare the surface. Paint the gouged-out crack with a primer, then fill with joint compound and sand smooth.

Center cracks are cracks running along the center of a joint that occur for several reasons:

- If the tape is still intact, the crack is probably the result of applying joint compound too thickly. Also, low humidity may have caused the mud to dry too quickly.
- If the tape under the crack has been torn, it is possible that the structure is settling or another type of movement caused the crack.

To repair cracks along the center of the joint, follow these procedures:

- If the tape is still intact and the crack is narrow, apply latex emulsion to the crack. If the crack is wide, use joint or topping compound to bridge the space.
- If the tape is torn, remove a section of tape and old joint compound before making repairs. Then retape the joint following the usual finishing procedures.

4.6.0 Problems with Compound

Compound has its own special set of problems. It can debond, grow mold, become pitted, sag, and shrink.

4.6.1 Debonding, Flaking, or Chipping

When the joint or topping compound will not bond to the tape or the drywall or becomes unbonded from either one, the condition is known as compound debonding, flaking, or chipping. Common causes of compound debonding include:

- A foreign substance was on the gypsum drywall surface or on the surface of the tape when the mud was applied. Examples of foreign substances include dirt, oil, sanding dust, and incompatible paint.
- The mud was mixed improperly, or the wrong ratio of water to powder was used to mix the compound.
- Too much water was added during mixing, or incompatible compounds were mixed with each other in order to add to the working supply, combine containers for storage, etc.
- Dirty water was used to mix the compound, or dirty tools were used to mix or apply it.
- Hot or heated water was used to mix the compound. One reason for avoiding hot water is

because of possible sediment problems associated with water heaters.
- The installer used old or expired compound.

Most compound debonding problems can be avoided by following the manufacturer's mixing and usage instructions exactly. Some manufacturers request that you let the compound sit for a while after mixing it. There is a good reason for this, so do not take any shortcuts. They will end up costing excessive time and waste in the long run.

Be sure to always use only clean, cold water to mix any compound. Also use clean mixing and application tools and equipment. Remember that automatic taping tools need to be kept in pails of clean water between uses. Always make sure the gypsum drywall surface, tape, and all mixing pails are clean, too.

Repairing compound debonding is very much like repairing tape blisters, only on a larger scale. First, separate the debonded section of tape from the dried mud. Then remove enough of the old mud to allow you to apply a new layer in which to embed the tape. If the old compound crumbles easily, remove all of it. You will also have to remove whatever mud was used to feather the joint. Apply new compound and tape as you would for a new joint.

4.6.2 Moldy or Contaminated Compound

Using contaminated water, dirty containers or tools, or letting the compound stand too long can result in mold, bacteria, and bad odors in the compound. Hot and humid weather also contributes to the growth of mold and bacteria.

Always be careful to examine every pail before mixing anything in it. If it is discovered that a batch of compound has become moldy or contaminated, throw it out. Then be sure to soak tools and containers in a solution of chlorine bleach and clean water at least overnight.

Be sure to clean and wash all tools and equipment components at the end of every working day. The mud intended for use the next day must be stored in covered containers and kept at room temperature overnight.

4.6.3 Pitting

Small pits may appear in the finish of the mud after it dries. Pitting has several common causes:

- Air escaped that was trapped in the mud mixture. This can happen if the compound is mixed too vigorously or for too long.
- The mud mixture was too thin.
- Not enough pressure was used to apply the mud to the joint; that is, it was not embedded or wiped down properly.
- Mud was not adequately mixed prior to application.

To prevent pitting, mix the compound thoroughly using a slow, steady motion. Set power mixers, if used, on slow speed and try to achieve a smooth mixture that is free of lumps. When you apply mud, use enough force to establish a good bond to the surface, smooth it out, and feather the edges.

Repair a section of pitted compound by simply skim-coating with another topping layer in order to fill the pits. Sand a little to form a smooth base for applying the new mud. Then apply the new compound as if applying a topping coat to the joint. Feather it out to conceal the joint area. The joint may need to be feathered wider than the original topping coat in order to completely hide the joint.

4.6.4 Sagging

When compound sags or shows evidence of runs, these conditions are usually present:

- The mud was too thin. When mixed properly, compound is thick and smooth. Be sure to follow the mixing instructions exactly.
- Water added to the mud or to the powder compound was too cold to mix completely. Again, cold water is essential, but do not use ice-cold water. Remember that finishing is a room-temperature process. Anything colder than what normally comes out of a faucet in a warm room is just too cold.

To repair sags and runs, sand them very smooth after they dry. Then, recoat with layers of joint or topping compound as needed.

4.6.5 Excessive Shrinkage

If the mud shrinks too much when it dries, it is probably the result of one of the following:

- Mud mixed too thin
- Insufficient drying time between coats
- Too much mud applied at any one time

To prevent this problem, use lightweight mud, which tends to shrink less. This problem is similar to the joint depression problem discussed earlier. As in that case, remedy excessive compound shrinkage by applying more mud. However, ensure that each previous coat is thoroughly dry before adding more compound.

4.6.6 Delayed Shrinkage

Delayed shrinkage is caused when too much time elapses before the correct amount of shrinkage occurs. The mud is not shrinking enough and tends to resist drying out. Delayed shrinkage has several common causes:

- Atmospheric conditions (slow drying capabilities and very high humidity)
- Insufficient drying time between coats of compound (trying to rush the job before it is actually ready for each finishing procedure)
- Excess water added to the mud mixture
- Heavy fills (adding too much mud as a prefill, or trying to fill large spaces in the drywall with compound instead of slivers or strips of drywall)

One way to prevent delayed shrinkage is to use a faster-drying compound, perhaps a quick-setting compound, which sets up chemically and does not depend on water evaporation. Quick-setting compounds were discussed earlier in this module.

A remedy for this condition is to allow extra drying time, and then to reapply a full cover coat of a heavy-mixed mud over the tape. Most shrinkage will generally take place on this heavy topping coat. With the right mud, the coat will dry faster and allow finishing procedures to continue in the usual way.

The best defense against delayed shrinkage is to use a faster-drying compound in the first place. There is very little you can do to mud that needs more drying time, except to give it more time to dry.

4.7.0 Fastener Problems

Two common fastener problems that may be encountered are nail pops and fastener depressions.

4.7.1 Nail Pops

When drywall nail heads work up from under the finished surface after the installation is complete, the job is said to have nail pops. Nail pops are unsightly, protruding fastener heads.

If enough nails pop out, the drywall will loosen and sag. The nails can be driven in again and the hole refinished, but the best remedy is preventing nail pops before they happen. Here are the primary reasons for nail pops:

- Wood framing with relatively high moisture content will shrink as the lumber dries out. As the wood shrinks, the nails lose their tight holding power (*Figure 49*). When the drywall is

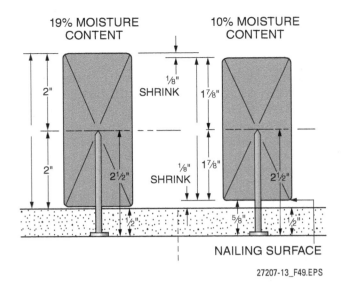

Figure 49 Shrinkage contributes to nail pops.

no longer securely attached, a space develops between the panel and the stud; the nail shank is exposed at that point. Then almost anything that puts pressure against the drywall will push it against the stud and the nail—which does not move—will actually pop right out of the panel along with the compound covering it.
- When drywall is fastened to framing that is out of alignment, stress on the drywall causes fasteners to work up above the surface (*Figure 50*).
- Gravity acting on ceilings and vibrations acting on walls will tend to work the nails loose (*Figure 51*).
- The drywall may not have been installed properly (*Figure 52*).
- If a building has poor ventilation or an inadequate heating system, large temperature fluctuations will cause expansion and contraction of the framing and drywall. If there is too much of that, the fasteners will begin to loosen.

4.7.2 Preventing Nail Pops

Nail pops may show up days or weeks after installation is complete or gradually over a period of many months. How soon they appear depends on the degree of misalignment, the type of fasteners used, the moisture content of the framing at

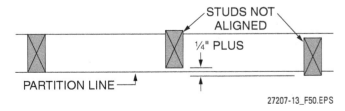

Figure 50 Nonaligned framing.

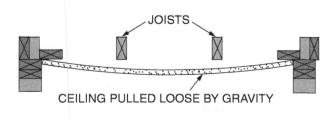

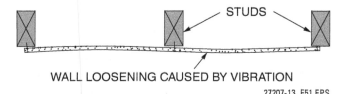

Figure 51 Force of gravity and vibrations can cause nail pops.

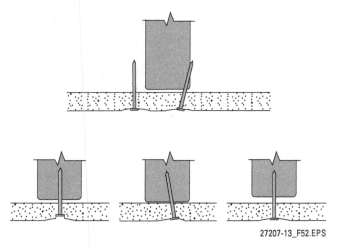

Figure 52 Improper drywall installation.

the time of installation, the amount of vibration present, and the temperature cycles to which the drywall and framing are exposed. Prevention is easier once the causes of nail pops are known. To prevent nail pops, follow these key rules:

- Make sure framing lumber is dry before fastening drywall to it. The builder should provide enough ventilation to speed up the drying process. When working in cold or humid weather, use portable heaters or blowers to warm or circulate the air. It may take only a few days to reduce the moisture content of lumber to an acceptable level, depending on temperature and humidity. Lumber is too green for hanging drywall if any wet spot appears when the lumber is hit sharply with the head of a hammer. Test several lengths before forming an opinion. The amount of moisture can vary from one piece to the next.
- Make sure the framing members are aligned in the same plane. Sight along the edges of the studs and joists to see that they are in a straight line. Alignment can also be checked by holding a long straightedge up against the studs and joists.
- If framing is out of alignment, repair it. Applying drywall to framing members that are out of alignment will prove to be a mistake. The framing will eventually spring back to its original position and the nails will pull out from the framing.
- Use ring-shank drywall nails or screws to fasten the drywall to the framing. They have more gripping power than plain-shank nails and offer more protection against nail pops. Also, consider using double rather than single fasteners. Double nailing provides more holding power than single nailing. Use floating corner angles to reduce stress on the drywall. Finally, use adhesive in addition to nails or screws to fasten the drywall panels.
- Always nail from the center of the drywall panel toward the edges. If nailing from one edge of the panel to the other, the drywall may not rest firmly against the framing for the entire length of the panel.

For example, assume that the final drywall panel is being installed in a wall. The other panels along the wall are already installed, and so are the panels that cover the wall that forms the other half of the corner. On this final panel, if the fasteners are first installed at one edge instead of at the center, the drywall may move slightly toward the opposite edge. This particular panel is trapped by a corner and other drywall panel. Any movement after nailing is started will stress the board, eventually causing it to bow and pop the fasteners.

A similar problem can occur if working from the other edge toward the center of the board. The center may actually spring away from the framing. If the center of the panel is fastened first with work moving toward the edges, the panel will not be able to move once the first fasteners are in place. From the moment the first fastener is driven home, the panel is forced flat against the framing in the center and along the edges. Be sure to hold the drywall panel tightly

Nail Pops

Nail pops that occur after the building has been heated for more than a month are usually caused by lumber shrinkage. Once this begins to occur, it is better to wait until the end of the heating season to repair the nail pops.

against the framing as fasteners are installed, to ensure the panel stays flat. Do not worry about installing a panel with a slight bow. With proper installation, any stress put on the board by flattening its bow will relax in a short time. Before you cover the fastener heads on any drywall finishing job, check to be sure the nails or screws are tight. Redrive any loose nails or screws. It is also a good idea to drive another nail on each side of a nail that has worked loose. Drive them about 1½" away from the old nail. After you redrive any loose nails or screws and add extra nails, go back and check all the nail heads again. The vibration caused by hammering may have loosened more nails. A few seconds spent now can save a few hours later. Also, if drywall panels are installed on both sides of a wall, driving the nails on one side may have loosened the nails on the other side. Be sure to check the first side again.

Attention to detail should prevent most nail pops. Take the time to check framing lumber carefully and install panels properly.

4.7.3 Repairing Nail Pops

When nail pops occur after a job has been finished, perhaps even after the texturing and painting are done, fixing them takes a little more time. When the nail head works out of the framing, it will show above the surface of the drywall. It may even lift the compound from the depression around the nail head. Repair it as follows:

Step 1 If the nail head has worked loose and become visible, just driving it back may not solve the problem. It is better to pull out the nail and replace it. The best and most permanent replacement is a drywall screw; otherwise, use a longer nail and/or another nail within 1½" of the first.

Step 2 Once the fasteners are in, fill the depression with mud and let it dry. If necessary, use a second coat of compound.

Step 3 If necessary, repair the texture, depending on the surface's original texture. Be sure the texture of the patch matches all the surrounding texture.

Step 4 When the new texture has dried, paint it with an oil-based primer. If this step is skipped, the paint may soak into the repaired spot when it is later repainted, which will make the repaired area very evident.

4.7.4 Fastener Depressions

A fastener depression is a depressed area over the fastener head. This is the opposite of a nail pop. The joint compound over a nail or screw has sunk lower than the surface of the surrounding drywall.

Fastener depressions can be caused by several problems:

- Nails were dimpled too deeply or screws were driven in too far.
- Not enough mud was applied to the fastener heads to cover them properly.
- The framing lumber was extremely dry. Dry lumber will absorb moisture, squeezing the board between the nail head and the edge of the stud or joist, and pulling the fastener head deeper into the drywall.
- The installer used too few fasteners to hold the drywall firmly against the framing, allowing the drywall to flex independently of the framing and forcing the fastener heads deeper into the surface.

To prevent fastener depressions, avoid driving fasteners through the face paper. Install the correct number of fasteners and space them properly. Spot the fastener heads with two coats of compound, sanding lightly between coats, if necessary.

Repairing fastener depressions is a simple matter. First, make sure enough fasteners have been installed. If more nails or screws are needed to hold the drywall firmly against the framing, add them. Second, spot the fastener heads with joint compound to bring the surface flush with the surrounding drywall.

4.8.0 Problems with Drywall Panels

Common problems with gypsum drywall panels include blisters, damaged edges, water damage, oard bowing, board cracks, fractures, and brittleness.

Repairing Nail Pops

Another way to repair a nail pop is to drive a Type W screw about 1½" from the popped nail. Then, place a broad knife over the popped nail and hit the knife with a hammer to drive the nail back down.

4.8.1 Board Blisters

When the facing paper becomes unbonded from the surface of a piece of drywall panel, it is known as a board blister. It may be caused by a manufacturing defect, or it may be the result of careless handling or improper storage. The gypsum filler tends to break apart inside the wrapped panel, causing the facing paper to loosen.

There are two common ways to repair board blisters:

- Inject an aliphatic resin glue, such as yellow or white carpenter's or wood glue, into the blister, and then press the paper flat. This is the best remedy when the blister is small or when the blister is not discovered until after the wall has been textured and/or painted.

WARNING! Before using any adhesive, check the manufacturer's instructions and applicable SDS to identify any hazards. Wear protective equipment and apparel as specified by the manufacturer.

- Cut out the entire blistered area and finish it with tape and joint compound. Follow the usual procedure for embedding tape and finishing joints. If one width of tape is not going to be enough to cover the blistered area, add as many other strips as necessary.

4.8.2 Damaged Edges

Improper handling of gypsum drywall panels is what generally causes damaged surfaces and edges. Such carelessness may cause the facing paper to tear or the gypsum core to crumble.

The only way to repair such damage is to cut off the damaged area back to sound gypsum areas prior to installation.

If a panel has already been installed and you detect damage along an edge or joint, cut out the damaged area back to sound drywall and prefill with mud. If this produces too large an area, install a filler strip of good drywall either laminated to a board layer beneath or attached with screws to the framing. Prefill around the strip and finish the joints as usual.

4.8.3 Water Damage

When drywall becomes wet, the core becomes soft and is easily deformed. Also, the facing paper may come unbonded (blistered) from the core.

If a drywall panel has been exposed to water, let it dry thoroughly before using it. Be very sure it is completely dry before installing it on the framing. If it is so badly warped that even screw attaching will not straighten it, then, after it is dry, put it under a stack of new panels lying flat on the floor.

If a panel is already installed and then becomes so wet that it warps away from the framing, drive in some additional screws to hold it. If additional screws do not help, remove that panel from the framing and replace it.

4.8.4 Board Bowing

Board bowing is similar to the warping problem discussed above. In this case, a board may have been forced into too small a space on the framing, causing the board to bow or warp.

Whenever this problem is discovered, the best remedy is to trim the panel edges to relieve the stress that caused the bowing. You may have to remove the panel to do a proper trim job on the edges. Re-attach the panel when it has been trimmed to fit properly, so that it doesn't need to be forced into place.

4.8.5 Board Cracks and Fractures

A gypsum board can crack along its face, or it may fracture all the way through to the other side. There are various causes and cures for cracks and fractures.

Panel cracks may occur along the face of any drywall panel, but they are most likely to show up over a doorway, where there is a smaller and weaker section of drywall. If a crack is over $\frac{1}{8}$" wide, treat it just as you would a regular joint. Repair it by taping and feathering the joint compound and topping compound until the crack does not show.

Manufacturing Defects

If you suspect that the drywall panels you are installing are defective, immediately stop work and get instructions from your supervisor. Suppliers and manufacturers will usually replace defective material. There is no sense in putting up defective material only to rip it out later.

This type of cracking is often caused by movement or settling of the building. Many larger buildings, such as skyscrapers, have a built-in flexibility that may contribute to the cracking of interior drywall. In a building with a flexible frame, the best choice is nonbearing interior partitions that have a clearance at the top of every wall. The tracks are fastened to the ceiling to hold the tops of metal studs, which may or may not be actually fastened to the track.

There can be up to ½" clearance between the wallboards and the ceiling. Fill this space with caulk or a specialty gasket or trim such as the type shown in *Figure 53*. A control joint or expansion joint (*Figure 54*) might also be used. Place such metal or plastic trim around the appropriate panel edges to give a finished appearance to the room and add protection to the walls.

Any one of three possible reasons can contribute to drywall panel fractures:

- The panel was attached across the wide face of the structural framing members, such as the headers. If the framing is wood and the lumber shrinks, the panel is compressed and it will crack. If the framing is steel and not very adequate, loads put on it may stress and crack some of the drywall panels attached in this way.
- The drywall was improperly handled or stored.
- The face paper was scored past the edge of a cutout.

To repair broken or fractured panels, completely cut out the damaged sections and replace them. If the damage was produced by scoring the facing paper beyond the cutout edges, simply repair this score with tape as you would any other joint.

4.8.6 Loose Drywall Panels

Loose drywall panels might be caused by any of the following:

- The panels were improperly fastened.
- The framing members were misaligned, uneven, or warped (in the case of wood).
- The screws or nails were not driven in all the way or else (with lumber) some shrinkage has occurred, pulling the framing away from the board, and thereby making it loose.

Improper fastening may be due to using incorrect types of screws or an improperly adjusted screwgun. This can result in screws being stripped or not seated properly, contributing to drywall panel looseness.

The remedy for fastener problems is generally to remove all faulty fasteners. Replace them with correct fasteners and properly drive them all the way in, so that they are well fixed into the framing and produce a good dimple in the surface.

When redriving fasteners, make sure your free hand is pushing solidly against the board near the fastening point. It is important that the board be perfectly flat against the framing member while you are driving the fastener.

Double-check to make sure you are using the correct type of drywall screw. Also, readjust the clutch on your screwgun to give you the proper depth into the board. You do not want to tear the face paper, but you need a dimple of about $\frac{1}{16}$" to allow proper spotting and finishing. If nails are used, use the double-nailing method.

If the cause of loose panels is poor framing, which may be out of alignment, twisted, or warped, redriven fasteners alone may not pull the board flush to where it should be. The only way to fix the problem may be to remove all the drywall panels and correct the framing.

Another way to make a better drywall panel attachment is to use adhesive as well as additional screws to hold the board to the framing. However, if the framing is badly warped and you succeed in firmly fixing the drywall panel, the wall or ceiling might be just as warped as the framing. It is better to fix the framing.

One other possibility is to laminate an entire new layer of drywall over the warped layer using adhesive or other material. Not only will this provide new drywall laminate, but it will also fill in any spaces caused by the first layer's warping.

Finally and most easily, if loose panels are caused by loose nails or screws, drive them in farther. Check this before finishing any wall or ceiling. Pushing with your hands against the panel (even while you are spotting with mud) will indicate if any panel is loose. If it is, stop and redrive the fasteners. Fasteners can often be driven by simply using the butt end of a broad knife. Add other fasteners if necessary, then continue spotting and finishing. The time taken to interrupt the finishing process and fix the panel-hanging problem will prevent anyone from having to do the job over again.

As in all repairs, the fix should stay fixed. Do not settle for shortcut methods. If you have to rip off the drywall panels and reset the framing, it is better to do it now than to have the problem reported later. If the general contractor or customer discovers the poor framing, it could cost an employer their business and could result in job loss. Fix these problems right from the start.

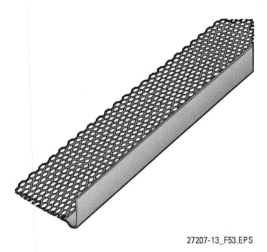

Figure 53 Veneer L-trim casing bead.

4.8.7 Patching Drywall

Drywall defects such as holes and dents require patching. For holes 2 inches or less in diameter, apply joint compound and reinforcing tape over the hole. An additional tape layer may also be needed. Once the bedding coat and tape have dried, apply a topping coat, feathering the edges. Apply a finish coat, if necessary.

Large holes require a different method of repair. Square off and cut out the defective area. Bevel the edges of the squared opening so that the bevels face you. Measure and cut out a patch of new drywall to fit this opening. Bevel the patch edges to mate with the opening's edges. Use joint compound to cement the patch in place, then tape and finish the edges like butt joints.

For holes 12" or larger, square off and cut out a whole drywall section back to the framing members (*Figure 55*). Cut a fresh patch to fit, cement the patch in place, and use fasteners through the patch into the framing members. Only one drywall screw in each corner of the patch should be necessary. Tape and finish the patch edges like butt joints.

To repair a drywall dent, first sand the dented section. This raises the nap, but it also permits the joint compound to grip the drywall face paper. Fill the dent with one or more layers of compound. Allow each layer to dry before lightly sanding

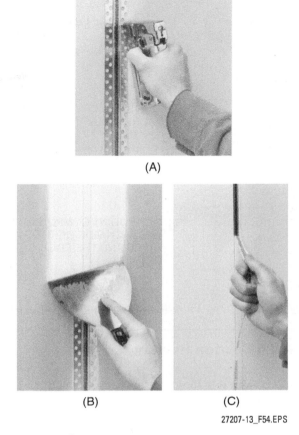

Figure 54 Applying an expansion joint.

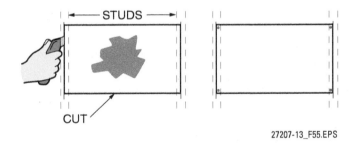

Figure 55 Patching a large hole.

and then applying the next layer. Finally, sand the filled dent smooth and level with the surrounding drywall.

Another patching technique is the hot patch or blowout patch, as shown in *Figure 56*.

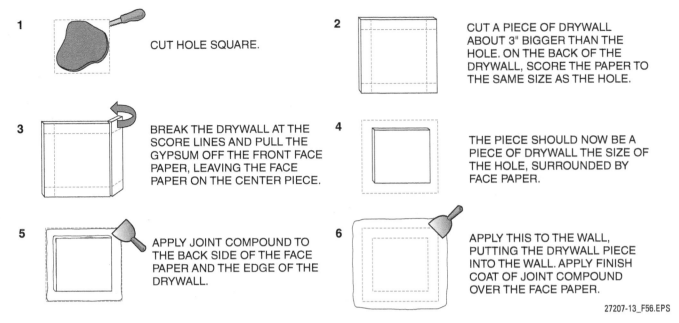

Figure 56 Patching a hole in drywall using the hot-patch (blowout) patching technique.

4.0.0 Section Review

1. When finishing work is being done during cold weather, the building interior must be heated to at least _____.
   a. 32°F
   b. 45°F
   c. 50°F
   d. 65°F

2. The ideal drywall finishing process requires _____.
   a. the fastener heads to remain uncovered
   b. the compound to dry between steps
   c. an automatic taping tool
   d. an automatic nail spotter

3. When sanding drywall, *never* use sandpaper that is _____.
   a. 80 grit
   b. 100 grit
   c. 120 grit
   d. 150 grit

4. When using an automatic taping tool, a taping coat should be applied with a flat finisher for seams and butts and a corner finisher for angles.
   a. True
   b. False

5. Which of the following would *not* a cause a tape blister?
   a. use of topping compound instead of joint compound
   b. a joint that is too wide
   c. too much joint compound
   d. no joint compound beneath the tape

6. If the joint compound fails to adhere to the tape or drywall, the condition is called _____.
   a. mud failure
   b. compound debonding
   c. layer peeling
   d. compound extrusion

7. Nail pops occurring after a building has been heated for more than a month are usually a result of _____.
   a. lumber shrinkage
   b. high humidity
   c. incorrect fastener installation
   d. building settling

8. Cracks in drywall are most likely to appear _____.
   a. below a window
   b. in a ceiling
   c. over a doorway
   d. at the bottom edge of a panel

Section Five

5.0.0 Estimating Drywall Finishing Materials

Objective

Explain how to estimate the proper amount of drywall finishing materials.

The approximate quantities of materials needed for 1,000 square feet of drywall are as follows:

- *Joint tape* – 370'
- *Joint compound* – 83 pounds of conventional powder or 9.4 gallons of lightweight ready-mix compound

These figures can be used to calculate the requirements for other square footages by reducing the square footage to a decimal percentage of 1,000 and multiplying the area required by the above amounts. For example, 2,000 square feet would require twice the amount listed above and 1,400 square feet would require 1.4 times the amount listed above.

To determine the area of walls and ceiling for a room, first add the wall lengths and multiply by the wall height. For this example, assume that a room measures 20' wide by 45' long and has 8' high walls.

Wall area = (Wall length 1 + Wall length 2 + Wall length 3 + Wall length 4) × Wall height
Wall area = (20' + 45' + 20' + 45') × 8'
Wall area = 130' × 8'
Wall area = 1,040 sq ft
Ceiling area = Ceiling width × Ceiling length
Ceiling area = 20' × 45'
Ceiling area = 900 sq ft
Total area = Wall area + Ceiling area
Total area = 1,040 sq ft + 900 sq ft
Total area = 1,940 sq ft

To determine the amounts of materials needed, multiply the coverage per 1,000 sq ft by the total area and then divide by 1,000. For example, 717.8' of tape is required for 1,940 sq ft of drywall ([370' × 1,940 sq ft] ÷ 1,000). For the same room, 161.02 lb of conventional powder joint compound ([83 lb × 1,940 sq ft] ÷ 1,000) or 18.25 gallons of lightweight ready-mix compound ([9.4 × 1,940] ÷ 1,000) is required. Each of these values would be rounded up.

5.0.0 Section Review

1. To finish a room with 1,500 square feet of drywall, the amount of lightweight ready-mix joint compound needed would be _____.

    a. 9.4 gallons
    b. 13.16 gallons
    c. 14.1 gallons
    d. 18.8 gallons

Summary

Drywall finishing is the difference between craft and mechanics. Once it is finished and painted, the properly finished seam cannot be distinguished from the wall surface itself. Modern finishing equipment, if correctly used, can allow a crew to complete rooms much more quickly than a crew with hand equipment. However, you must use and maintain the equipment correctly in order to get good results. You also need to be able to finish by hand as well as with automatic tools as there will be places where manual work is the only way to get good results.

The textured surface, more and more popular, eliminates finish sanding, but a poorly taped or fastened joint will still show through. Different kinds of corner beads and tape require different mud mixes and techniques. Drywall that is properly hung is much easier to finish well, preventing nail pops, alignment problems, and gaps. Remember, the walls are out in plain sight; any craftsman or customer will see and recognize shoddy work.

Review Questions

1. A drywall finishing job with tape embedded in joint compound, one separate coat of compound on joints and interior angles, and two separate coats of compound over fastener heads and accessories meets part of the requirements for _____.
   a. Level 1
   b. Level 2
   c. Level 3
   d. Level 4

2. A drywall finish meeting the specifications for Level 1 would most likely be found in a(n) _____.
   a. kitchen
   b. attic
   c. office
   d. living room

3. Finishing with gloss, semigloss, or enamel paints is recommended only for _____.
   a. Level 2
   b. Level 3
   c. Level 4
   d. Level 5

4. To obtain a smooth surface and conceal fasteners, the finishing of a corner bead normally consists of _____.
   a. two coats
   b. three coats
   c. four coats
   d. five coats

5. Fiberglass mesh tape may be preferred over paper tape in applications that involve _____.
   a. high moisture
   b. low humidity
   c. high temperature
   d. low temperature

6. A location in which you would be unlikely to use metal edge tape is a(n) _____.
   a. outside angled corner
   b. 90-degree inside corner
   c. ceiling and radius-wall intersection
   d. arch

7. The *incorrect* statement below regarding topping compound is _____.
   a. topping compound is used for the second and third finishing coats
   b. topping compound is easier to sand because it dries softer than joint compound
   c. topping compound and joint compound are the same material—you just add more water to joint compound to make topping compound
   d. topping compound shrinks less than joint compound

8. Texture compounds that have good solution time, minimum-to-moderate fallout, and good bonding power are _____.
   a. premixed textures
   b. powder joint-compound textures
   c. aggregated powder textures
   d. unaggregated powder textures

9. When mixing dry-mix compounds, you are required to wear _____.
   a. gloves
   b. respiratory protection
   c. a hairnet
   d. protective coveralls

10. A safety hazard posed by automatic taping machines is _____.
    a. gears and chains that can catch hair or fingers
    b. heated surfaces
    c. toxic fumes
    d. exposed electrical connections

11. Drywall is commonly cut using a _____.
    a. hacksaw
    b. band saw
    c. circular saw
    d. utility knife

12. A tool that is often preferred over taping knives for finishing drywall joints is the _____.
    a. flat-blade finishing trowel
    b. broad knife
    c. curved-blade finishing trowel
    d. putty knife

13. A rubber mallet is used to strike a _____.
    a. flat finisher
    b. corner-clinching tool
    c. corner plow
    d. bull nose

14. Small quantities of joint compound are often hand mixed, using tool called a _____.
    a. compound crusher
    b. blending trowel
    c. puddler
    d. mud masher

15. The nail spotter is a tool used for _____.
    a. applying joint compound to cover fastener heads
    b. locating studs before driving fasteners
    c. removing incorrectly driven fasteners
    d. locating fasteners that have been covered with joint compound

16. The automatic tool that applies a smooth finish with feathered edges and a center crown to a taped seam is the _____.
    a. flat applicator
    b. banjo
    c. flat finisher
    d. nail spotter

17. If an area of paper covering is torn or pulled away during installation, it should be prepared for finishing by _____.
    a. gluing it back in place
    b. cutting it back to the point where it still adheres
    c. fastening it down with ring-shank nails
    d. tearing it loose

18. Drywall fasteners should be installed so that _____.
    a. the head penetrates the paper
    b. there is a slight depression, or dimple, in the drywall
    c. the head protrudes 1/64 of an inch from the drywall surface
    d. the head is exactly flush with the drywall surface

19. When sanding drywall joints, you should never use sandpaper that is coarser than _____.
    a. 200 grit
    b. 150 grit
    c. 120 grit
    d. 100 grit

20. The number of coats of topping that fasteners normally receive is _____.
    a. two
    b. three
    c. four
    d. five

21. When you are putting on stilts or taking them off, you should _____.
    a. sit on the floor
    b. lean against a wall
    c. ask another person to hold you upright
    d. hold onto a railing

22. If you find bubbles in tape joints after using the automatic taper, it probably means that _____.
    a. only one wheel was pressing on the drywall surface
    b. neither wheel was pressing on the drywall surface
    c. both wheels were pressing on the drywall surface
    d. the taper was upside down

23. When you are able to see a taped joint after the wall has been painted, it is known as _____.
    a. ridging
    b. photographing
    c. high joints
    d. discoloration

24. To avoid bonding problems when mixing joint compound, you should use clean water that is _____.
    a. room temperature
    b. hot
    c. cold
    d. distilled

25. A finishing defect that is the opposite of a nail pop is known as a _____.
    a. sinking fastener
    b. surface pit
    c. fastener depression
    d. pothole

26. A 2,500 sq ft area will require approximately how many feet of tape?
    a. 370 feet
    b. 700 feet
    c. 925 feet
    d. 1,500 feet

Trade Terms Quiz

Fill in the blank with the correct term that you learned from your study of this module.

1. The thin, strong fiberglass mesh or paper strips used to cover joints between drywall panels is generally referred to as _____.

2. Insufficient drying time can cause slight protrusions called _____ in the center of a finished drywall joint.

3. _____ is a technique that provides a uniform finish by tapering the edges of the joint compound.

4. A mixture used to cover drywall joints and fasteners, _____ dries hard and can be sanded to a smooth finish.

5. A joint where sheets of drywall with tapered edges meet is called a _____.

6. A common term for joint compound or taping compound is _____.

7. Used for second and third coats over joints, _____ is easier to sand than joint compound, since it dries softer.

8. Because of its rounded edges, this type of corner bead is known as a _____.

9. Also called "mud," _____ is another term for joint compound.

10. _____ does not bond as well as taping compound but finishes better.

11. Sometimes specified for use under a high-gloss finish, a _____ is a thin layer applied over the entire drywall surface.

12. _____ is a version of all-purpose compound that is 25 to 35 percent lighter than standard joint compound.

Trade Terms

All-purpose compound
Bull nose
Feathering
Joint compound
Lightweight compound
Mud
Ridges
Skim coat
Tape
Tapered joint
Taping compound
Topping compound

Trade Terms Introduced in This Module

All-purpose compound: Combines the features of taping and topping compounds. It does not bond as well as taping compound but finishes better.

Bull nose: A metal corner bead with rounded edges.

Feathering: Tapering joint compound at the edges of a drywall joint to provide a uniform finish.

Joint compound: Patching compound used to finish drywall joints, conceal fasteners, and repair irregularities in the drywall. It dries hard and has a strong bond. Sometimes called *mud* or *taping compound*.

Lightweight compound: An all-purpose compound having less weight than standard compounds.

Mud: See *joint compound*.

Ridges: Slight protrusions in the center of a finished drywall joint that are usually caused by insufficient drying time. Also known as *beads*.

Skim coat: A thin coat of joint or topping compound that is applied over the entire drywall surface. Sometimes required under a high-gloss finish.

Tape: A strong paper or fiberglass tape used to cover the joint between two sheets of drywall.

Tapered joint: A joint where tapered edges of drywall meet.

Taping compound: See *joint compound*.

Topping compound: A joint compound used for second and third coats. It dries soft and smooth and is easier to sand than taping compound.

Additional Resources

This module presents thorough resources for task training. The following resource material is suggested for further study.

A Recommended Specification for Levels of Gypsum Board. 2010. Gypsum Association.

Gypsum Construction Guide. 2011. Charlotte, NC: National Gypsum Company.

The Gypsum Construction Handbook. 2009. Chicago, IL: R.S. Means.

Installing and Finishing Drywall. 2008. William Spence. New York: Sterling Publishing Company.

Painting and Decorating Craftsman's Manual and Textbook. 1995. Fairfax, VA: Painting and Decorating Contractors of America.

Figure Credits

FEMA Photo/Marty Bahamonde, CO01
National Gypsum Company, Figure 1
Courtesy of Irwin Tools, Figure 11
The Stanley Works, Figures 13, 14, 15, 16, 19, 24, 40
USG Corporation, Figures 25, 27, 54
Kraft Tool Company, Figure 26
Ames Taping Tool Systems, Inc., Figures 28, 31, 32, 33, 34
Porter Cable Corporation, Figure 35

Section Review Answer Key

Answer	Section Reference	Objective
Section One		
1. d	1.0.0	1
Section Two		
1. b	2.1.1	2a
2. b	2.2.1	2b
Section Three		
1. c	3.0.0	3a
2. d	3.2.1	3b
Section Four		
1. c	4.1.0	4a
2. b	4.2.0	4b
3. c	4.3.1	4c
4. a	4.4.0	4d
5. c	4.5.6	4e
6. b	4.6.1	4f
7. a	4.7.1	4g
8. c	4.8.5	4h
Section Five		
1. c	5.0.0	5

NCCER CURRICULA — USER UPDATE

NCCER makes every effort to keep its textbooks up-to-date and free of technical errors. We appreciate your help in this process. If you find an error, a typographical mistake, or an inaccuracy in NCCER's curricula, please fill out this form (or a photocopy), or complete the online form at **www.nccer.org/olf**. Be sure to include the exact module ID number, page number, a detailed description, and your recommended correction. Your input will be brought to the attention of the Authoring Team. Thank you for your assistance.

Instructors – If you have an idea for improving this textbook, or have found that additional materials were necessary to teach this module effectively, please let us know so that we may present your suggestions to the Authoring Team.

NCCER Product Development and Revision
13614 Progress Blvd., Alachua, FL 32615

Email: curriculum@nccer.org
Online: www.nccer.org/olf

❏ Trainee Guide ❏ Lesson Plans ❏ Exam ❏ PowerPoints Other _____

Craft / Level: _____ Copyright Date: _____

Module ID Number / Title: _____

Section Number(s): _____

Description: _____

_____

_____

_____

_____

Recommended Correction: _____

_____

_____

_____

_____

Your Name: _____

Address: _____

_____

Email: _____ Phone: _____

This page is intentionally left blank.

Suspended Ceilings

OVERVIEW

Suspended ceilings are found in most commercial buildings. Unlike fixed ceilings, they provide easy access to wiring, cabling, and air conditioning equipment located in the area between the ceiling and the overhead deck. The ceiling tiles are designed to suppress sound transmission. Suspended ceilings are built by first installing a grid that is suspended from the overhead deck. Ceiling tiles are then installed in the grid. Proper installation of these ceilings is a skill that takes training and practice.

Module 27209

Trainees with successful module completions may be eligible for credentialing through NCCER's National Registry. To learn more, go to **www.nccer.org** or contact us at **1.888.622.3720**. Our website has information on the latest product releases and training, as well as online versions of our *Cornerstone* magazine and Pearson's product catalog.

Your feedback is welcome. You may email your comments to **curriculum@nccer.org**, send general comments and inquiries to **info@nccer.org**, or fill in the User Update form at the back of this module.

This information is general in nature and intended for training purposes only. Actual performance of activities described in this manual requires compliance with all applicable operating, service, maintenance, and safety procedures under the direction of qualified personnel. References in this manual to patented or proprietary devices do not constitute a recommendation of their use.

Copyright © 2013 by NCCER, Alachua, FL 32615, and published by Pearson, New York, NY 10013. All rights reserved. Printed in the United States of America. This publication is protected by Copyright, and permission should be obtained from NCCER prior to any prohibited reproduction, storage in a retrieval system, or transmission in any form or by any means, electronic, mechanical, photocopying, recording, or likewise. To obtain permission(s) to use material from this work, please submit a written request to NCCER Product Development, 13614 Progress Blvd., Alachua, FL 32615.

27209 V5

From *Carpentry, Trainee Guide*. NCCER.
Copyright © 2013 by NCCER. Published by Pearson. All rights reserved.

27209
SUSPENDED CEILINGS

Objectives

When you have completed this module, you will be able to do the following:

1. Identify the components necessary to properly install a suspended ceiling system.
   a. Identify the suspension systems and hardware necessary to properly install a suspended ceiling system.
   b. Identify the system components necessary to properly frame a suspended ceiling system.
   c. Identify the safe material handling and storage procedures required when installing a suspended ceiling system.
2. Interpret a reflected ceiling plan.
   a. Interpret the layout information.
   b. Interpret the MEP locations.
3. Identify the procedures to lay out and install a suspended ceiling system.
   a. Identify the layout and takeoff procedures to procure materials to lay out and install a suspended ceiling system.
   b. Identify the tools and equipment to lay out and install a suspended ceiling system.
   c. Identify the installation methods and procedures for a suspended ceiling system.

Performance Tasks

Under the supervision of your instructor, you should be able to do the following:

1. Estimate the quantities of materials needed to install a lay-in suspended ceiling system in a typical room from an instructor-supplied drawing.
2. Establish a level line at ceiling level such as is required when installing the wall angle for a suspended ceiling.
3. Lay out and install a lay-in suspended ceiling system according to an instructor-supplied drawing.

Trade Terms

Acoustical materials	Ceiling panels	Dry lines	Plenum
Acoustics	Ceiling tiles	Fissured	Striated
A-weighted decibel (dBA)	Decibel (dB)	Frequency	
	Diffuser	Hertz (Hz)	

Code Note

Codes vary among jurisdictions. Because of the variations in code, consult the applicable code whenever regulations are in question. Referring to an incorrect set of codes can cause as much trouble as failing to reference codes altogether. Obtain, review, and familiarize yourself with your local adopted code.

Industry-Recognized Credentials

If you're training through an NCCER-accredited sponsor, you may be eligible for credentials from NCCER's Registry. The ID number for this module is 27209. Note that this module may have been used in other NCCER curricula and may apply to other level completions. Contact NCCER's Registry at 888.622.3720 or go to **www.nccer.org** for more information.

Contents

Topics to be presented in this module include:

1.0.0 Suspended Ceiling System Components 1
   1.1.0 Ceiling Systems 2
      1.1.1 Exposed Grid Systems 2
      1.1.2 Metal Pan Systems 2
      1.1.3 Direct-Hung Systems 3
      1.1.4 Integrated Ceiling Systems 3
      1.1.5 Luminous Ceiling Systems 4
      1.1.6 Suspended Drywall Furring Ceiling System 5
      1.1.7 Special Ceiling Systems 5
   1.2.0 Ceiling System Components 5
      1.2.1 Exposed Grid Components 6
      1.2.2 Ceiling Panels and Tiles 8
      1.2.3 Metal-Pan System Components 9
      1.2.4 Direct-Hung System Components 9
   1.3.0 Material Handling and Storage 10
      1.3.1 Handling and Storing Ceiling Materials 10
      1.3.2 Safely Working with Ceiling Materials 10
2.0.0 Reflected Ceiling Plan 12
   2.1.0 Interpreting Ceiling Plans 12
   2.2.0 Interpreting the Mechanical, Electrical,
          and Plumbing (MEP) Drawings 13
3.0.0 Laying Out and Installing Suspended Ceiling Systems 17
   3.1.0 Laying Out and Estimating Materials for a Suspended Ceiling 17
      3.1.1 Alternate Method for Laying Out a Suspended
          Ceiling Grid System 18
      3.1.2 Establishing Room Center Lines 19
   3.2.0 Suspended Ceiling Tools and Equipment 20
      3.2.1 General Tools 20
      3.2.2 Ceiling Leveling Equipment 20
   3.3.0 Installing Suspended Ceiling Systems 22
      3.3.1 Exposed Grid Systems 23
      3.3.2 Metal Pan Systems 28
      3.3.3 Direct-Hung Concealed Grid Systems 30
      3.3.4 Installing Standard Luminous Systems 31
      3.3.5 Installing Nonstandard Luminous Systems 31
      3.3.6 Installing Suspended Drywall Ceilings 33
      3.3.7 Installing Acoustical Panel 34
      3.3.8 Installing Furring Channels Directly to Structural Members 34
      3.3.9 Drywall Suspension Systems 35
      3.3.10 Ceiling Cleaning 35

Figures and Tables

Figure 1   Sound-wave reflection 2
Figure 2   Integrated grid system 3
Figure 3   Integrated ceiling schematic.. 4
Figure 4   Luminous ceiling system 4
Figure 5   Special metallic system 4
Figure 6   Special wood system 5
Figure 7   Planar system 5
Figure 8   Reflective system 6
Figure 9   Transparent tile system.. 6
Figure 10   Exposed-grid system components 7
Figure 11   Metal-pan ceiling components.. 9
Figure 12   Concealed grid system 10
Figure 13   Direct-hung concealed-grid system components........................ 10
Figure 14   Example of a reflected ceiling plan... 13
Figure 15   Mechanical, electrical, and plumbing drawings 14
Figure 16   Completed sketch of a suspended-ceiling layout........................ 19
Figure 17   Method for laying out the center lines of a room......................... 20
Figure 18   Level 21
Figure 19   Water level 21
Figure 20   Laser-level wall/ceiling mount... 22
Figure 21   Wall/ceiling mount attached to wall-angle flange........................ 22
Figure 22   Transferring room center lines to the height of
            the suspended ceiling grid 25
Figure 23   Fitting the tiles to the room 25
Figure 24   Locating the positions of the first main runner and
            cross runner... 26
Figure 25   Bending hanger wire.. 27
Figure 26   Inserting hanger wires in the main runner................................... 28
Figure 27   End of cross runner resting on wall angle 28
Figure 28   Main runner (tee bar) clipped to a furring channel..................... 29
Figure 29   Tee bar splice 29
Figure 30   Tile layout 30
Figure 31   Pan-removal tool 30
Figure 32   Metal-pan ceiling.. 31
Figure 33   Ceiling-panel spline.. 31
Figure 34   Access for concealed grid ceilings.. 32
Figure 35   Carrying channel supported by hanger wire................................ 33
Figure 36   Splicing carrying channels.. 33
Figure 37   Hat furring channel.. 33
Figure 38   Splicing (lapping) furring channels... 34
Figure 39   Drywall screwed to a furring channel 34
Figure 40   Examples of drywall suspension system hangers 35

Table 1   Sound Levels of Some Common Noises.. 3

This page is intentionally left blank.

SECTION ONE

1.0.0 SUSPENDED CEILING SYSTEM COMPONENTS

Objective

Identify the components necessary to properly install a suspended ceiling system.
a. Identify the suspension systems and hardware necessary to properly install a suspended ceiling system.
b. Identify the system components necessary to properly frame a suspended ceiling system.
c. Identify the safe material handling and storage procedures required when installing a suspended ceiling system.

Trade Terms

Acoustical materials: Types of ceiling panel, plaster, and other materials that have high absorption characteristics for sound waves.

Acoustics: A science involving the production, transmission, reception, and effects of sound. In a room or other location, it refers to those characteristics that control reflections of sound waves and thus the sound reception in the area.

A-weighted decibel (dBA): A single number measurement based on the decibel but weighted to approximate the response of the human ear with respect to frequencies.

Ceiling panels: Acoustical ceiling boards that are suspended by a concealed grid mounting system. The edges are often kerfed and cut back.

Ceiling tiles: Any lay-in acoustical board that is designed for use with an exposed grid mounting system. Ceiling tiles normally do not have finished edges or precise dimensional tolerances because the exposed grid mounting system provides the trim-out.

Decibel (dBa): An expression of the relative loudness of sounds in air as perceived by the human ear.

Diffuser: An attachment for duct openings in air distribution systems that distributes the air in wide flow patterns. In lighting systems, it is an attachment used to redirect or scatter the light from a light source.

Fissured: A ceiling-panel or ceiling-tile surface design that has the appearance of splits or cracks.

Frequency: Cycles per unit of time, usually expressed in hertz (Hz).

Hertz (Hz): A unit of frequency equal to one cycle per second.

Plenum: A chamber or container for moving air under a slight pressure. In commercial construction, the area between the suspended ceiling and the floor or roof above is often used as the HVAC return air plenum.

Striated: A ceiling panel or ceiling tile surface design that has the appearance of fine parallel grooves.

Suspended ceilings are widely used in commercial construction and to some extent in residential construction. Modern suspended ceilings serve many purposes. They are designed to help keep outside noise from entering the room and to reduce noise levels occurring within the room itself. In some cases, ceilings are integrated with the electrical and HVAC (heating, ventilating, and air conditioning) functions to provide for correct lighting and temperature control. Use of attractive ceiling panels or ceiling tiles helps give a warm, relaxed feeling to a room. Complete ceiling systems offer a wide variety of options, both functional and visual. The selection of acoustical materials, plans for the acoustical ceiling, and the method of ceiling installation depend on the intended use of the room.

Sounds travel through air in a room as a series of pressure waves. These sound pressure waves travel outward in all directions. When sound waves strike a wall or ceiling, some of the sound-wave energy is absorbed and some is reflected in wave patterns moving in the opposite direction (*Figure 1*). The result is that you will hear both the original sound and its reflected image. The sound is also transmitted through the air in the wall or ceiling cavity to the opposite surface, causing the surface to vibrate and transmit the sound to any adjoining room(s).

Sound waves have a frequency. The frequency, or pitch, measured in hertz (Hz), is the number of vibrations or cycles that occur in the wave in one second. The greater the number of cycles per second (Hz), the higher the frequency and the higher the pitch.

The intensity of sound refers to its degree of loudness or softness. An A-weighted decibel (dBA) is a unit of measure used for establishing and comparing the intensity of sound sources. It is used to express the value of all sounds in a range from 0 dBA to 140 dBA and higher.

Table 1 shows some typical examples of noise situations and their relative decibel (dBa) levels.

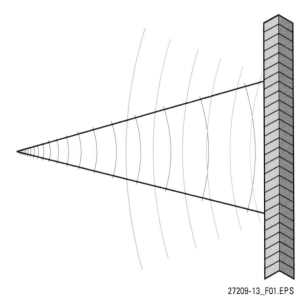

Figure 1 Sound-wave reflection.

Note that any sounds greater than 120 dBA can actually produce a physical sensation. Sounds above 130 dBA can cause pain and/or deafness. Continued exposure to sound levels above 85 dBA can cause hearing loss over time.

Some terms you may encounter and should understand when selecting or working with acoustics and acoustical ceiling materials include the following:

- *Reflection* – The bouncing back of sound waves after hitting some obstacle or surface such as a ceiling or wall.
- *Reverberation* – The prolonging of a sound through multiple reflections of that sound as it travels back and forth across a room. These multiple reflections of sound occur so fast that they are usually not heard as distinct repetitions of that sound. However, they can cause a higher noise level than the original sound source.
- *Noise reduction coefficient (NRC)* – Used by manufacturers to compare the noise absorbencies of acoustical products. The higher the number, the better the absorbency. The NRC measures the average percentage of noise a material absorbs at four selected frequencies.
- *Sound transmission loss* – The amount of sound lost as a noise travels through a material. Acoustical ceiling assemblies are rated in terms of sound transmission class (STC). An STC value of 20 to 25 indicates that even normal speech can be easily understood in an adjoining room. On the other hand, an STC value of 50 to 60 indicates that loud sounds will be heard only faintly or not at all. Acceptable STC ratings range from approximately 39 to 65.
- *Articulation class (AC)* – Rates a ceiling's ability to achieve normal privacy in open office spaces by absorbing noise reflected at an angle off the ceiling into adjacent areas (cubicles). Per the American Society for Testing and Materials (ASTM) International *E1110* and *E1111* standards, the generally accepted AC ratings for normal privacy in open-plan offices is a minimum of 170, with 190 to 210 preferred.
- *Ceiling attenuation class (CAC)* – Rates a ceiling's efficiency as a barrier to airborne sound transmission between adjacent work areas, where sound can penetrate plenum spaces and travel to other spaces. CAC is stated as a minimum value. Per *ASTM E1264*, CAC minimum 25 is acceptable in an open-plan office, while a rating of CAC minimum 35 to 40 is preferred for closed offices.
- *Absorption* – The energy of sound waves being taken in (entering) and absorbed by a surface of any material rather than being bounced off or reflected.

1.1.0 Ceiling Systems

There is a wide variety of suspended ceiling systems, with each system being somewhat different from the others. They use the same basic materials, but their appearances are completely different.

The focus of this module is on the following ceiling systems:

- Exposed grid systems
- Metal pan systems
- Direct-hung concealed grid systems
- Integrated ceiling systems
- Luminous ceiling systems
- Suspended drywall furring ceiling system
- Special ceiling systems

Also covered in this module is background information relevant to acoustics and acoustical ceilings, including information on the propagation of sound waves, acoustical ceiling product terminology, and ceiling-related drawings.

1.1.1 Exposed Grid Systems

An exposed grid system is a suspension system for lay-in ceiling tiles. The factory-finished supporting members are exposed to view.

1.1.2 Metal Pan Systems

The metal pan system is similar to a conventional exposed grid ceiling system except that metal panels or pans are used in place of the conventional

Table 1 Sound Levels of Some Common Noises

Sound Level (dBA)	Intensity Level	Environment	
		Outdoor	Indoor
140	Deafening	Jet aircraft, artillery fire	Gunshot
130	Threshold of pain	—	Loud rock band
120	Threshold of feeling	Elevated train	Portable stereo headset on high setting
110	Extremely loud	Overhead jet aircraft at 1,000'	Loud nightclub
100	Very loud	Chainsaw, motorcycle at 25', auto horn at 10'	—
90	Loud	Lawn mower, noisy city street	Full symphony band, noisy factory
80	Moderately band	Diesel truck at 50'	Garbage disposal, dishwasher
70	Average	—	Face-to-face conversation, vaccum cleaner, printers and copiers
60	Moderately quiet	Air conditioning condenser at 15', auto traffic near an interstate highway	Normal conversation, general office
50	Quiet	Large transformer at 50'	—
40	Very quiet	Bird calls	Private office, soft radio music
30	Extremely quiet	Quiet residential neighborhood	Average residence
20	Nearly silent	Rustling leaves	Quiet theater, whisper
10	Just audible	—	Human breathing
0	Threshold of human hearing	—	One's own heartbeat in a silent room

sound-absorbing tile. In some cases, the panels or pans are snapped into place from below rather than being laid-in from above the ceiling frame.

1.1.3 Direct-Hung Systems

A direct-hung ceiling system is used if the grid system is to be concealed from view. A mechanical clip or tongue-and-groove joint is used to connect the tiles together. The tiles are then tied to the suspended grid.

1.1.4 Integrated Ceiling Systems

As noted by its name, the integrated ceiling system incorporates the lighting and/or air supply diffusers as part of the overall ceiling system, as shown in *Figures* 2 and 3.

Some manufacturers of ceiling materials offer specialty ceiling designs. Many integrated ceilings can be used to create distinctive architectural appearances and interior artistic designs. Complete ceiling systems offer dozens of options, both functional and visual.

The functional aspect allows for enhanced lighting and lighting effects. Mechanically, it can incorporate the air supply system through spaced air supply diffusers and return air grilles, all of which have been designed to go beyond function to enhance the artistic appearance of the ceiling.

Integrated ceiling systems are available in units called modules. The common sizes are 30" × 60" and 60" × 60". The dimensions refer to the spacing of the main runners and cross tees.

Courtesy of CeilingsPlus.com

Figure 2 Integrated grid system.

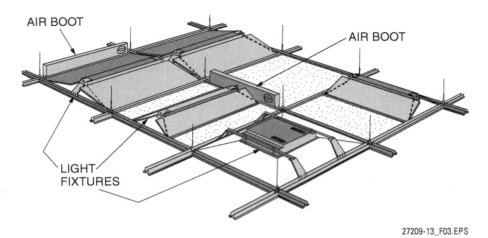

Figure 3 Integrated ceiling schematic.

1.1.5 Luminous Ceiling Systems

Luminous ceiling systems (*Figure 4*) are available in many styles, such as exposed grid systems with drop-in plastic light diffusers or an aluminum or wood framework with translucent acrylic light diffusers.

Fluorescent fixtures are generally installed above the translucent diffusers. Standard modules of 2' × 2' up to sizes of 5' × 5' are available. It is also possible to purchase custom sizes for special fit conditions. There are two types of luminous ceilings: standard and nonstandard. Standard systems are, as their name indicates, those that are available in a series of standard sizes and patterns. Nonstandard systems differ in that they deviate from the normal spacing of main supports and may include unusual tile sizes, shapes, and configurations.

All surfaces in the luminous space, including pipes, ductwork, ceilings, and walls, are painted with a 75 to 90 percent reflective matte white finish. Any surfaces in this area which might tend to flake, such as fireproofing and insulation, should receive an approved hard surface coating prior to painting to prevent flaking onto the ceiling below.

1.1.6 Suspended Drywall Furring Ceiling System

The suspended drywall furring system is used when it is desirable or specified to use a drywall finish or drywall backing for an acoustical panel ceiling.

1.1.7 Special Ceiling Systems

Numerous special ceiling systems differ from those covered in this module. Some of these are special metallic systems, special wood systems, planar systems, and reflective (mirrored) systems (*Figures 5* through *8*).

Transparent tiles produce a subtly refracted light that extends over all areas of a room's interior (*Figure 9*).

1.2.0 Ceiling System Components

A variety of components are combined to create the ceiling grid. Each type of system and each manufacturer's components may be slightly different. Always refer to the manufacturer's instructions before installing a ceiling grid. The grid, composed of main runners, cross runners, and wall angles, is commonly made from light-gauge metal members.

Courtesy of CeilingsPlus.com

Figure 4 Luminous ceiling system.

1.2.1 Exposed Grid Components

For an exposed-grid suspended ceiling, a light-gauge metal grid is hung by wires attached to the original ceiling or structural members. Tiles that usually measure 2' × 2' or 2' × 4' are then placed in the frames of the metal grid. Exposed grid systems are constructed using the components and materials described as follows and shown in *Figure 10*.

Figure 5 Special metallic system.

Figure 6 Special wood system.

Look Up

The next time you are out, take a look at the ceilings in the stores, theaters, malls, and other buildings you enter. You will see that the available styles, designs, materials, and color schemes are more numerous than you ever imagined. The one thing that most of them have in common is that they are some form of suspended ceiling using panels set into or attached to a framework.

Courtesy of CeilingsPlus.com, © Keith Peterson

Figure 7 Planar system.

Courtesy of CeilingsPlus.com, © Benny Chan

Figure 9 Transparent tile system.

Courtesy of CeilingsPlus.com

Figure 8 Reflective system.

- *Main runners* – Primary support members of the grid system for all types of suspended ceiling systems. They are 12' in length and are usually constructed in the form of an inverted T. When it is necessary to lengthen the main runners, they are usually spliced together using extension inserts; however, the method of splicing may vary with the type of system being used.
- *Cross runners (cross ties or cross tees)* – Inserted into the main runners at right angles and spaced an equal distance from each other, forming a complete grid system. They are held in place by either clips or automatic locking devices. Typically, they are either 2' or 4' in length and are usually constructed in the form of an inverted T. Note that 2' cross runners are only required for use when using 2' × 2' ceiling tiles.
- *Wall angles* – Installed on the walls to support the exposed grid system at the outer edges.
- *Ceiling tiles* – Tiles that are laid in place between the main runners and cross ties to provide an acoustical treatment. The acoustical tiles used in suspended ceilings stop sound reflection and reverberation by absorbing sound waves. These tiles are typically designed with numerous tiny sound traps consisting of drilled or punched holes or fissures, or a combination of both. When sound strikes the tile, it is trapped in the holes or fissures. A wide variety of ceiling tile designs, patterns, colors, facings, and sizes are available, allowing most environmental and appearance demands to be met. Tiles are typically made of glass or mineral fiber. Generally, glass-fiber tiles have a higher sound absorbency than mineral-fiber tiles. Tile facings are typically embossed vinyl in a choice of patterns such as fissured, pebbled, or striated. The specific ceiling tiles used must be compatible with the ceiling suspension system due to variations in manufacturers' standards.

USDA-Compliant Tiles

The United States Department of Agriculture (USDA) is responsible for ensuring that the nation's commercial supply of meat, poultry, and egg products is safe through safety guidelines and inspections. USDA-compliant ceiling tiles are designed for use in kitchens and central food-preparation areas. These tiles are washable, waterproof, and bacteria resistant.

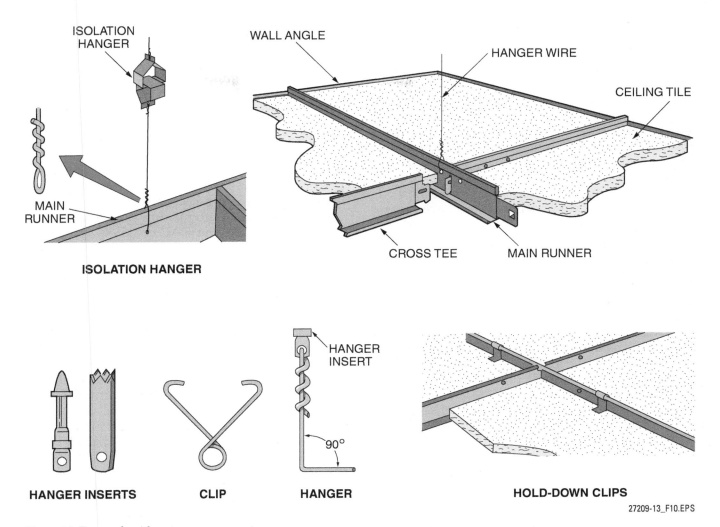

Figure 10 Exposed-grid system components.

- *Hanger inserts and clips* – Many types of fastening devices are used to attach the grid-system hangers or wires to the ceiling or structural members located above the suspended ceiling. Screw eyes and star anchors are commonly used and may require a hammer drill for installation. Powder-actuated fasteners are commonly used when fastening to reinforced concrete. Clips are used where beams are available and are typically installed over the beam flanges. Then the hanger wires are inserted through the loops in the clips and secured.
- *Hangers* – The devices attached to the hanger inserts and used to support the main runners. The hangers can be made of No. 12 wire or heavier rod stock. Ceiling isolation hangers are also available to isolate the ceiling from noise traveling through the building structure.
- *Hold-down clips* – Used in some systems to hold the ceiling tiles in place.
- *Nails, screws, rawls, toppets, and molly bolts* – Used to secure the wall angle to the wall. The specific fastener used depends on the wall construction and material.

What's in a Name?

The terms *ceiling panel* and *ceiling tile* have specific meanings in the trade. Ceiling tiles are typically any lay-in acoustical board that is designed for use with an exposed grid system. They do not have finished edges or precise dimensional tolerances because the grid system provides the trim-out. Ceiling panels are acoustical ceiling boards, usually 12" × 12" or 12" × 24", which are nailed, cemented, or suspended by a concealed grid system. The edges are often kerfed and cut back.

1.2.2 Ceiling Panels and Tiles

Ceiling panels and tiles range in size from 12" × 12" up to 60" × 60". Various colors and designs are available. Most are fabricated from mineral fiber and glass fiber. Depending on their design and purpose, mineral-fiber panels and tiles are made with painted, plastic, aluminum, ceramic, or mineral faces. Glass-fiber panels and tiles are made with painted, film, glass cloth, and molded faces.

There are basically three types of tile/ceiling grid interfaces: lay-in, concealed tee, and profiled edge. Lay-in tiles are widely used and are generally the most cost-effective style. Concealed tee tiles are butt-jointed tiles that provide a monolithic ceiling design with no visible support system. Profiled edge tiles are made in a wide selection of edge designs from soft-edged chamfered or curved tiles to highly articulated edges. Beveled and angular reveal-edged tiles provide a three-dimensional look in a suspended ceiling.

Many interior spaces have specific requirements for ceiling tile materials and characteristics. These can include sound control, fire resistance, thermal insulation, light reflectance, moisture resistance, maintenance, appearance, and cost considerations. High-performance acoustical tiles are used to help prevent noise in open-plan and closed types of offices. Three factors contribute to noise distractions in a workplace:

- *General office noise* – The ability of a ceiling material to absorb general office noise is measured using a value known as the noise reduction coefficient (NRC), which is used by tile manufacturers to compare the noise absorbency of their ceiling tile products. The higher the number, the better the absorbency. The NRC measures the average percentage of noise a tile absorbs at various frequencies.
- *Reflected conversational noise that angles off ceilings into adjacent cubicles* – The ability of a ceiling material to absorb reflected conversational noise is measured using a value known as the articulation class (AC), which rates a ceiling's ability to achieve normal privacy in open office spaces by absorbing noise reflected at an angle off the ceiling into adjacent areas (cubicles). Per *ASTM E1110* and *E1111*, the generally accepted AC rating for normal privacy in open-plan offices is a minimum of 170, with 190 to 210 preferred.
- *Sound transmission through cubicles, partitions, walls, and ceilings* – The ability of a ceiling material to absorb sound transmission is measured using a value known as the ceiling attenuation class (CAC), which rates a ceiling tile's efficiency as a barrier to airborne sound transmission between adjacent work areas, where sound can penetrate plenum spaces and carry to other spaces. The CAC is stated as a minimum value. Per *ASTM E1264*, CAC minimum 25 is acceptable in an open-plan office, while a rating of CAC minimum 35 to 40 is preferred for closed offices.

Fire-resistant ceiling tiles and support systems are specially made of materials that provide increased resistance against flame spread, smoke generation, and/or structural failure in the event of a fire. Two ratings based on ASTM, ANSI (American National Standards Institute), and NFPA (National Fire Protection Association) standards are used to evaluate fire-resistant tiles: the flame-spread rating of the material (*ASTM E84*) and the fire-resistance rating of a ceiling assembly (*ANSI/UL 263, ASTM E119*, and *NFPA 251*).

Basically, the flame-spread rating of a ceiling material is the relative rate at which a flame will spread over the surface of the material. This rate is compared against a rating of 0 (highest rating) for fiber-cement board and a rating of 100 for red oak. Class A ceilings have flame-spread ratings of 25 or less, the required standard for most commercial applications. The fire-resistance rating of a ceiling assembly represents the degree to which the entire assembly, not the individual components, withstands fire and high temperatures (measured in hours). Specifically, it is an assembly's ability to prevent the spread of fire between spaces while retaining structural integrity.

High-humidity-resistant tiles are tiles that have superior resistance to sagging caused by highly

Fire-Rated Applications

Suspended ceiling systems can be used in fire-rated applications. Factors such as resistance to fire and flame spread for the tiles, as well as the suspension system are taken into account. However, the ceiling materials alone do not determine the fire rating. Rather, the materials and construction methods used in the entire system, including the floor/ceiling or ceiling/roof assembly, all enter into the fire-rating determination. If a suspended ceiling is to be part of a fire-rated system, you must consult the manufacturer's product literature to determine the ceiling tile and grid that must be used to meet the rating.

humid conditions. Many also inhibit the growth of mold or mildew that may appear in highly humid conditions. Sagging not only diminishes the attractiveness of a ceiling, but it also causes ceilings to chip and soil more easily, and reduces the high light reflectivity of the tiles. High-humidity-resistant tiles are typically installed in high humidity climates or areas such as kitchens, locker rooms, shower areas, and indoor pools; buildings where the HVAC systems may be shut down for extended periods; or where the ceiling might be installed early in the construction process before the building is fully enclosed.

Ceiling tiles with a high light reflectance (LR) value of 0.83 or greater per *ASTM E1477* help increase effective lighting levels and reduce light fixture costs and energy consumption, especially with indirect lighting systems. Their use also helps to reduce eyestrain. These tiles typically have soil-resistant surfaces that stay cleaner longer than standard ceilings, resulting in a much lower loss in light reflectance over time.

Detailed information regarding the various tile characteristics described above can normally be found in ceiling manufacturers' product catalogs and literature. When selecting tiles for a particular application, it is best to follow the manufacturer's recommendations.

1.2.3 Metal-Pan System Components

The metal pan system is similar to the conventional suspended acoustical ceiling system except that metal panels, or pans, are used in place of the conventional sound-absorbing tile (*Figure 11*). In some cases, the metal pans are snapped in from the bottom of the grid.

The pans are made of steel or aluminum and are generally painted white; however, other colors are available by special order. Pans are also available in a variety of surface patterns. Metal-pan ceiling systems are effective for sound absorption. They are durable and easily cleaned and disinfected. In addition, the finished ceiling has little or no tendency to have sagging joint lines or drooping corners. The metal pans are die-stamped and have crimped edges, which snap into the spring-locking main runner and provide a flush ceiling.

The tools, room layout, and installation of hanger inserts, hangers, and wall angle for the metal pan system are basically the same as for the conventional exposed-grid suspended ceiling.

1.2.4 Direct-Hung System Components

A concealed grid system is advantageous if the support runners need to be hidden from view, resulting in a ceiling that is not broken by the pattern of the runners (*Figure 12*).

The panels used for this system are similar in composition to conventional panels, but are manufactured with a kerf on all four edges. Kerfed and rabbetted 12" × 12" and 12" × 24" panels are used with this system. Splines are inserted in the kerfs to tie the panels together. Panels of various colors and finishes are available. Refer to *Figure 13* for a diagram of the components in a typical concealed grid system.

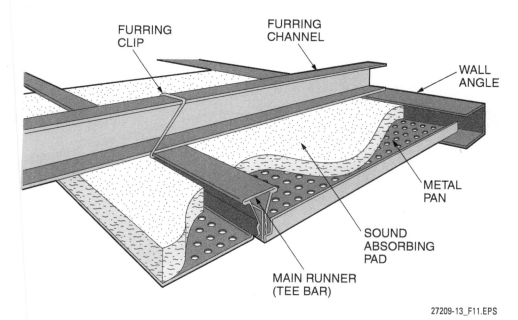

Figure 11 Metal-pan ceiling components.

Figure 12 Concealed grid system.

1.3.0 Material Handling and Storage

Since the ceiling panels or tiles, and in some cases the tracks and runners, are exposed to view they must be handled and stored carefully to prevent damage. In addition, working with ceiling materials often presents safety hazards of its own.

1.3.1 Handling and Storing Ceiling Materials

Finish ceiling materials should be stored in their original unopened packages and be protected from damage and exposure to the elements. By keeping them in their unopened packages, the materials can be easily moved to their installation location. The conditions where the materials are stored should be as close as possible to the place where they will be installed. Ensure there is proper support for the ceiling materials being placed. Materials should be stored at the job site for a minimal amount of time before being installed. Long-term storage should be avoided. Other procedures for proper handling and storage are as follows:

- Excess humidity during storage can cause expansion of material and possible warp, sag, or poor fit after installation.
- Chemical changes in the mat and/or coatings can be aggravated by excess humidity and cause discoloration during storage, even in unopened cartons.
- Cartons should be removed from pallets and stringers to prevent distortion of material.

1.3.2 Safely Working with Ceiling Materials

When installing grid support members or the panels or tiles themselves, carpenters typically work from ladders, movable scaffold, or lifts, depending on the height of the ceiling. Refer to the *Core Curriculum* for specific safety information relating to ladders and lifts. A common type of movable scaffold used for ceiling projects is called a baker's scaffold. Baker's scaffolds are short and

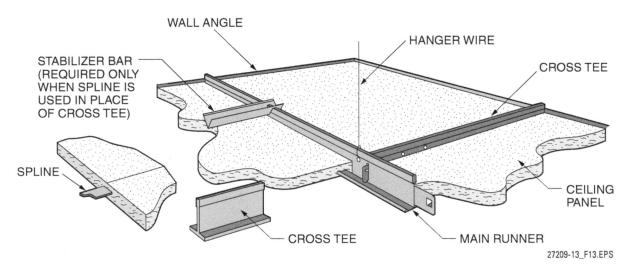

Figure 13 Direct-hung concealed-grid system components.

lightweight scaffolds with wheels that allow scaffolds to roll easily. Safety guidelines for using a baker's scaffold include the following:

- Inspect the scaffold before each use for defects or damage.
- Do not stand on or attach any equipment to cross braces or diagonal braces.
- Do not place boxes or ladders on a scaffold to increase your reach or height.
- Do not sit or stand on guardrails. Ensure that all guardrails are secured, in place, on all four sides.
- Never ride on a moving scaffold.
- Do not attempt to move a scaffold by applying a pushing or pulling force at or near the top of the scaffold.
- When hoisting material up to a scaffold platform, ensure the scaffold is attached to a permanent structure to keep the scaffold from tipping.
- Workloads on the scaffold must not exceed the capacity of the lowest-rated scaffold component.
- When working around scaffold, always wear a hard hat, safety glasses, work gloves, and steel-toe boots.

In addition to standard hand-tool and power-tool safety guidelines that should be followed, eye protection is especially important when installing hanger wire. The loose ends of hanger wire are commonly at eye level when working from a baker's scaffold. Carefully handle the cut ends of wall angle and other support members. The cut ends of these items are very sharp and can cause serious injury.

Using Stilts to Install Ceilings

While stilts are permitted in certain areas, they are not permitted in all jurisdictions. Always refer to the specific safety regulations in effect in the area in which you are working before using stilts.

1.0.0 Section Review

1. Integrated ceiling systems incorporate _____.
    a. both ceiling panels and ceiling tiles
    b. fire-rated panels and sprinkler systems
    c. lighting and/or air supply diffusers
    d. both exposed and concealed grid systems

2. A ceiling grid consists of main runners, cross runners, and wall angles.
    a. True
    b. False

3. Finished ceiling materials should be stored _____.
    a. on pallets
    b. in their original unopened package
    c. in a climate-controlled environment
    d. on edge to prevent warping

Section Two

2.0.0 Reflected Ceiling Plan

Objective

Interpret a reflective ceiling plan.
a. Interpret the layout information.
b. Interpret the MEP locations.

Performance Task 1

- Estimate the quantities of materials needed to install a lay-in suspended ceiling system in a typical room from an instructor-supplied drawing.

On some large construction jobs, there may be a set of reflected ceiling plans. These show the details of the ceiling as though it were reflected onto the floor (*Figure 14*). This view shows features of the ceiling while keeping those features in proper relation to the floor plan. For example, if a pipe runs from floor to ceiling in a room and is drawn in the upper left corner of the floor plan, it is also shown on the upper left corner of the reflected ceiling plan of that same room. Reflected ceiling plans show in detail how the ceiling will be constructed. The plans will indicate the following:

- Layout (direction) of the ceiling panels or tiles
- Location of the center (starting) line
- Size of the borders
- Position of the light fixtures
- Location of the air diffusers

As a rule, reflected ceiling plans also show all items that penetrate the ceiling, including:

- Return grilles
- Diffusers
- Sprinkler heads
- Recessed lights
- Recessed speakers for sound systems

Other information is also used when constructing a ceiling. Your employer may prepare shop drawings that show in detail just how the ceiling should be installed and also indicate the finished appearance. These drawings provide insurance against errors in the details of installation.

2.1.0 Interpreting Ceiling Plans

In order to avoid mistakes and/or omissions when installing ceilings, an organized and systematic approach should be used when reading the related construction drawings. The following is a general procedure for reading construction drawings:

Step 1 Check the room schedule on the construction drawings.
- Identify the type of material to be used.
- Locate the protrusions into the ceiling.

Step 2 Locate the room on the floor plan.
- Find the room dimensions. If none are found, locate the drawing scale.
- Using the given scale, determine the dimensions of the room.

Step 3 Check to see if there is a reflected ceiling plan. If no reflected ceiling plan is found, check to see if a shop drawing is included.

Step 4 Be sure the construction drawings are the final revised set.
- Construction drawings are often revised several times before the ceiling is ready to be installed. To be sure the construction drawings are the final revised set, check the date of issuance against the work order to see if they are the same.
- If work that has already been done is not reflected on the construction drawings, chances are the construction drawings are not the final revised set. This could have a significant impact on the ceiling installation.

Step 5 Read the specifications.
- Be sure the ceiling to be installed is the same as that listed in the specifications.
- If the job conditions do not agree with what is shown in the specifications and/or plans, call your supervisor and ask for instructions on how to proceed.

Step 6 Check the mechanical and electrical plans prior to the layout of the ceiling.
- On the mechanical plans, locate the air diffusers (HVAC air supply outlets), return grilles, ducts, and sprinkler heads.
- On the electrical plans, locate the light fixtures, fans, and other ceiling protrusions.

2.2.0 Interpreting the Mechanical, Electrical, and Plumbing (MEP) Drawings

Along with reading and interpreting the reflected ceiling plan, the mechanical, electrical, and plumbing (MEP) drawings must be referenced to ensure that all ceiling penetrations are accounted for. The MEP drawings will indicate where plumbing and electrical risers penetrate the ceiling, and also will show the routing of mechanical equipment. *Figure 15* shows the mechanical, electrical, and plumbing drawings for similar areas of the same set of construction drawings.

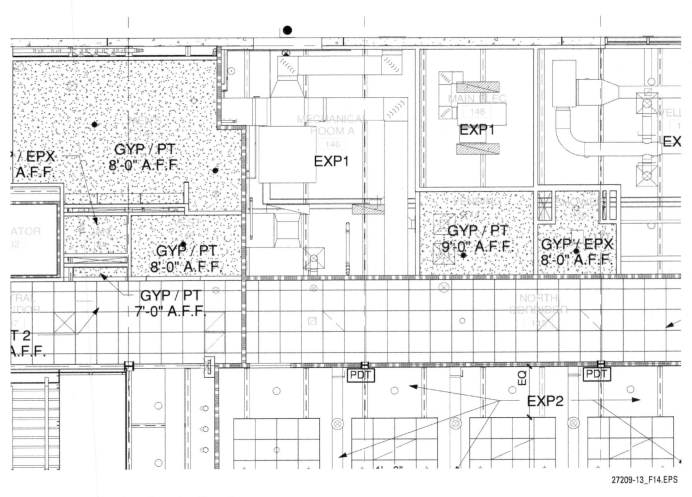

Figure 14 Example of a reflected ceiling plan.

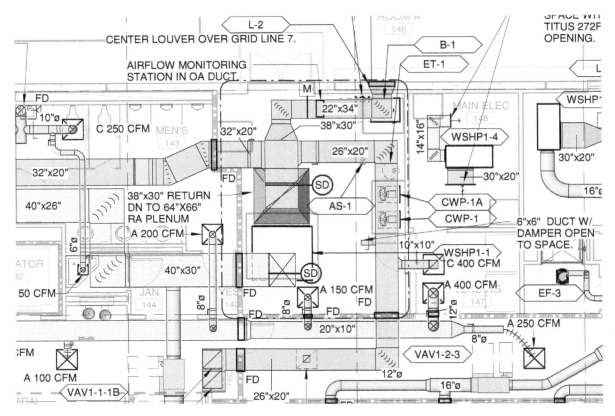

MECHANICAL PLAN

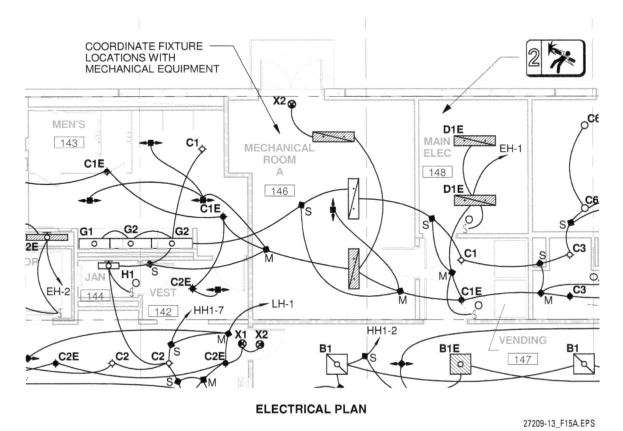

ELECTRICAL PLAN

Figure 15A Mechanical, electrical, and plumbing drawings. (1 of 2)

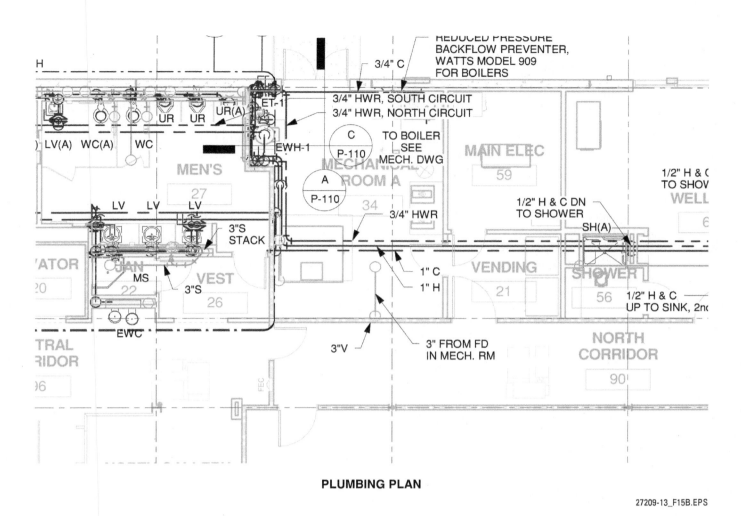

PLUMBING PLAN

Figure 15B Mechanical, electrical, and plumbing drawings. (2 of 2)

Plenum Ceilings

The systems that provide heating and cooling for most commercial buildings are forced-air systems. Blower fans are used to circulate the air. The blower draws air from the space to be conditioned and then forces the air over a heat exchanger, which cools or heats the air. In a cooling system, for example, the air is forced over an evaporator coil that has very cold refrigerant flowing through it. The heat in the air is transferred to the refrigerant, so the air that comes out the other side of the evaporator coil is cold. In homes, the air is delivered to the conditioned space and returned to the air conditioning/heating system through ductwork that is usually made of sheet metal. In commercial buildings with suspended ceilings, the space between the ceiling and the overhead decking is often used as the return air plenum. It is often called an open plenum. (A plenum is a sealed chamber at the inlet or outlet of an air handler.) This approach saves money by eliminating about half the cost of materials and labor associated with ductwork.

One thing to keep in mind is that anything in the plenum space (electrical or telecommunications cable, for example) must be specifically rated for plenum use in order to meet fire ratings. Plastic sheathing used on standard cables gives off toxic fumes when burned. Plenum-rated cable uses nontoxic sheathing.

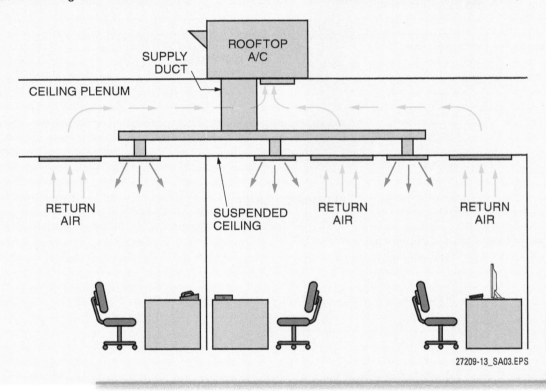

2.0.0 Section Review

1. To be sure that the construction drawings you are using are the final revised set, you should _____.
   a. make sure the date of issuance matches the work order
   b. ask your supervisor
   c. call the architect's office
   d. compare them to the specifications

2. MEP drawings indicate _____.
   a. the direction and location of main runners
   b. where open plenums should be installed
   c. datum lines for ceiling elevations
   d. where plumbing and electrical risers penetrate the ceiling

SECTION THREE

3.0.0 LAYING OUT AND INSTALLING SUSPENDED CEILING SYSTEMS

Objective

Identify the procedures to lay out and install a suspended ceiling system.
  a. Identify the layout and takeoff procedures to install a suspended ceiling system.
  b. Identify the tools and equipment to lay out and install a suspended ceiling system.
  c. Identify the installation methods and procedures for a suspended ceiling system.

Performance Tasks 2 and 3

Establish a level line at ceiling level such as is required when installing the wall angle for a suspended ceiling.

Lay out and install a lay-in suspended ceiling system according to an instructor-supplied drawing.

Trade Term

Dry lines: A string line suspended from two points and used as a guideline when installing a suspended ceiling.

Suspended ceilings must be properly laid out based on the ceiling plan. Proper layout will allow tiles or panels on each side of a room to be equal width to provide a pleasing appearance.

3.1.0 Laying Out and Estimating Materials for a Suspended Ceiling

The estimate of materials for a suspended ceiling should be based on the ceiling plan provided with the construction drawings or on a scaled sketch of the ceiling layout. These drawings should show the direction and location of the main runners, cross tees, light tiles, and border tiles. In a typical suspended ceiling, the main runners are spaced 4' apart and are usually run parallel with the long dimension of the room. For a standard 2' × 4' pattern, 4' cross tees are spaced 2' apart between the main runners. If a 2' × 2' pattern is used, 2' cross tees are installed between the midpoints of the 4' cross tees. Main runners and cross tees should be located in such a way that the border tiles on both sides of the room are equal and as large as possible.

If no ceiling plan or sketch is available, you will need to make one in order to determine the required quantity of materials. Sketch a ceiling plan to scale on graph paper. Use a convenient scale; for example, one square equals one square foot. Measure along the ceiling level, including irregular areas such as bays, alcoves, and beams, noting each dimension on the drawing. Then proceed as follows:

Step 1 Determine and sketch the locations of the main runners.
  - Convert the dimension for the room width to inches.
  - Divide the room width in inches by 48" and add 48" to any remainder.
  - Divide the sum by 2 to find the distance from the wall to install the first main runner. Note that this distance is also the length of the border panels or tiles. For example, assume the room has a width of 12'-8" and a length of 20'-8", as shown in *Figure 16*. Changing the room width to inches equals 152". Dividing 152" by 48" yields 3.166, with a remainder of 8". Adding 48" to 8" equals 56". Dividing 56" by 2 equals 28", the distance from the wall to the first runner. If there is no remainder, the distance from the wall to the first runner is 48".
  - Draw a main runner the calculated distance from, and parallel to, the long dimension of the ceiling. Draw the remaining main runners parallel to the first runner at 4' on center. The distance between the last main runner and the wall should be the same as the distance between the first main runner and the wall.

Step 2 Determine and sketch the locations of the 4' cross tees.
  - Convert the long dimension of the room to inches.
  - Divide the room length in inches by 24" and add 24" to any remainder.
  - Divide the sum by 2 to find the distance in inches from the wall to install the first row of cross tees. Note that this dimension is also the width of the border panels or tiles. For our example room, the length of the room is 20'-8". Changing this dimension to inches equals 248". Dividing 248" by 24" yields 10,

with a remainder of 8". Adding 24" to 8" equals 32". Dividing 32" by 2 equals 16", the distance from the wall to the first row of cross tees.

- Draw the first row of cross tees the calculated distance from, and parallel to, the short wall. Draw the remaining rows of cross tees parallel to each other at 2' on center. The distance from the last row of cross tees to the wall should be the same as the distance from the first row of cross tees to the opposite wall.

> **NOTE**
> Additional 2' cross tees are required if 2' × 2' panels are used. You will also need one hanger device and wire for about every 4' of main runner.

Step 3 Determine the quantity of ceiling materials. From the ceiling plan or sketch, determine the number of pieces required for the wall angle, main runners, cross tees, and ceiling tiles. Normally, main runners come in 12' lengths, cross tees in 4' lengths, and wall angle in 10' lengths.

- Using the ceiling plan or sketch, find the number of main-runner sections needed. For our example room (*Figure 16*), six main-runner sections are required. This is because main runners are made 12' in length and no more than two pieces can be cut from any one 12' runner. Three main runners, each 20'-8" (20.667') in length, total 62.001'. Therefore, 62.001' ÷ 12' = 5.167 lengths, or six lengths when rounded off.
- Find the number of 4' cross tees. For our example room, 40 cross tees (10 per row × 4 rows) are required. Note that the border cross tees must be cut from full-length cross tees.
- If using 2' × 2' panels or tiles, find the number of 2' cross tees. A 2' × 2' grid is made by installing 2' cross tees between the midpoints of the 4' cross tees. The number of 2' cross tees required for our example room is 44 (11 per row × 4 rows).
- Find the number of sections of wall angle needed. Divide the perimeter of the room [perimeter = (2 × length) + (2 × width)] by 10'. For our example room, the perimeter is 66'-8" (66.667'). Therefore, seven sections are needed (66.667' ÷ 10' = 6.667, or 7 when rounded off).
- Find the number of ceiling panels or tiles. One method is to count the total number of ceiling tiles shown on the ceiling plan or sketch. Note that each border tile requires a full-size ceiling tile. Also, subtract one tile for each lighting fixture installed in the ceiling. Assuming the use of 2' × 4' ceiling tiles and six light fixtures, our example ceiling would require 38 tiles.
- Find the approximate number of hanger wires and hangers needed. Assume one hanger device and wire for about every 4' of main runner. For our example ceiling, approximately 16 hangers are required (62.001' ÷ 4' = 15.5, or 16 when rounded off). Multiply the number of hangers needed by the required length of each hanger wire to find the total linear feet of hanger wire needed.

> **NOTE**
> Another method is to determine the total area (in square feet) of the ceiling by multiplying the length by the width (area = length × width). Then, divide the total ceiling area by the coverage (in square feet) printed on the carton of the tiles intended for use. If using tiles that cover 64 sq ft per carton, our example room ceiling would require 4.09 cartons of ceiling tiles (12.667' × 20.667' = 261.789' ÷ 64 = 4.09, or four cartons plus one extra tile).

Reflective Ceilings

GOING GREEN

Reflective ceiling panels can reduce lighting costs because they allow the use of lower-wattage light bulbs.

3.1.1 Alternate Method for Laying Out a Suspended Ceiling Grid System

Another common method for laying out the grid system for a suspended ceiling is given here:

Step 1 Locate the room center line parallel to the long dimension of the room and draw it on the ceiling sketch.

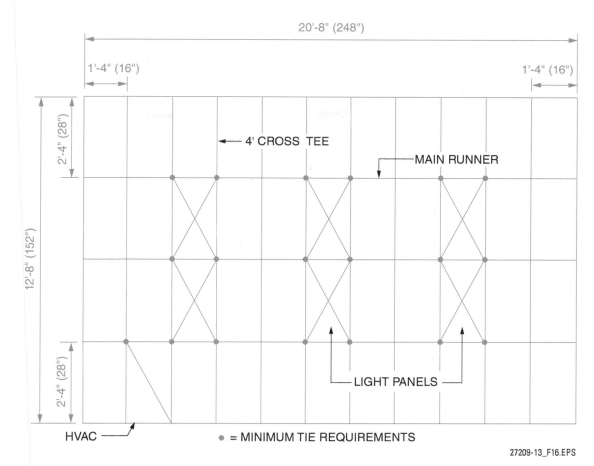

Figure 16 Completed sketch of a suspended-ceiling layout.

Step 2 Beginning at the center line and going toward each side wall, mark off 4' intervals on the sketch. If more than a 2' space remains between the last mark and the side wall, locate the main runners at these marks. If less than a 2' space remains between the last mark and the side wall, locate the main runners at 4' intervals beginning 2' on either side of the center line. This procedure provides for symmetrical border tiles of the largest possible size. Remember to consider the locations of light fixtures and air diffusers in the room.

Step 3 Locate the 4' cross tees by drawing lines 2' on center at right angles to the main runners. To obtain border tiles of equal size, begin at the center of the room using the same procedure as in Step 2.

Step 4 If using a 2' × 2' grid pattern, locate the 2' cross tees by bisecting each 2' × 4' module.

Step 5 Estimate the materials for the grid system, using the information shown on the ceiling sketch in the same manner as described previously.

3.1.2 Establishing Room Center Lines

In order for a grid system to be installed square within a room, it is necessary to lay out two center lines (north-south and east-west) for the room. When correctly laid out, these center lines will intersect each other at right angles at the exact center of the room. If the ceiling of the room located above the proposed suspended ceiling is solid and flat, such as with a plaster or drywall ceiling, then the center lines can be laid out on the ceiling. If the ceiling is not flat, such as an open ceiling with I-beams or joists, the center lines can be laid out on the floor. In either case, the center lines that are laid out on the ceiling or floor can be transferred down from the ceiling, or up from the floor, to the level of the suspended ceiling by use of a plumb bob and **dry lines**. The procedure given here describes one common method for establishing the center lines in a rectangular room:

Step 1 Measure and mark the exact center of one of the short walls in the room. Repeat the procedure at the other short wall.

Step 2 Snap a chalkline on the ceiling (or floor) between these two marks (see *Figure 17*, chalkline A-B). This is the first center line.

Step 3 Measure the length of the room along the chalkline. Find its center, and then place a mark on the chalkline at this point (point C).

Step 4 From point C, measure a minimum of 3' in both directions along the chalkline, then place a mark on the chalkline at these points (points D and E).

Step 5 Drive a nail at point D and attach a string to the nail. Extend the string to the side wall so that it is perpendicular to the chalkline, then attach a pencil to the line.

Step 6 Making sure to keep the string taut, draw an arc on the ceiling (floor) from the wall, across the chalkline, to the opposite wall (arc 1).

Step 7 Repeat Steps 5 and 6, starting at point E on the chalkline and draw another arc (arc 2).

Step 8 Mark the intersecting points of arcs 1 and 2 on both sides of the chalkline (points F and G).

Step 9 Snap a chalkline from wall to wall on the ceiling (or floor) that passes through points F and G. If done correctly, you should now have two center lines perpendicular to each other that cross in the exact center of the ceiling (floor).

3.2.0 Suspended Ceiling Tools and Equipment

Depending on the type of suspended ceiling being installed, a variety of tools and leveling equipment is needed. The same safety guidelines and considerations should be followed as were discussed in the *Core Curriculum*.

3.2.1 General Tools

The following tools are required when installing an exposed grid ceiling system:

- Aviation snips (tin snips)
- Clamping pliers or vise grips with plastic or rubber corners
- Chalkline
- Dry line
- Dust mask
- 50' or 100' tape
- Eye protection
- Hammer
- Hard hat

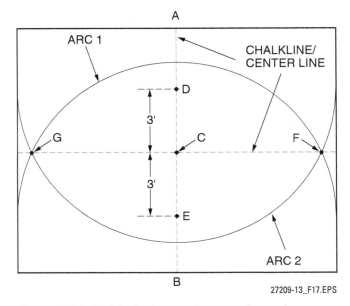

Figure 17 Method for laying out the center lines of a room.

- Awl
- Keyhole saw
- Lath nippers
- Level
- Magnetic punch
- Scribe or compass
- Plumb bob
- 6' folding rule
- Straightedge for cutting
- Board
- Tile knife
- Ladders
- Laser level
- Pop-rivet gun
- Powder-actuated tool
- Scaffold
- Special dies for cutting suspension members
- Water level
- Whitney punch

3.2.2 Ceiling Leveling Equipment

In the installation of suspended ceilings, various types of leveling devices are used to find the level plane of a ceiling. These devices include the carpenter's level, water level, and laser.

In the application of acoustical ceilings, a level, as shown in *Figure 18*, is a tool that comes in handy to check the ceiling-grid main runner and cross runner installation. When using a level to install a suspended ceiling, place it at right angles to the runners as you install them. If the tool is perfectly level, the bubble will appear centered between the crosshatches. The leveling should be checked every 6'.

By using the water level shown in *Figure 19*, a level ceiling can be installed even if the floor is not level. Benchmarks or elevation points can be established to be used as a starting point for the ceiling.

A water level is a long, clear hose, filled with water and stopped or clamped at both ends. Before using the water level, unplug both ends and tie one end to the rung of the ladder or scaffold; hold the other end at the same height. Pick up the hose approximately 3' from the U or bottom, and allow the air bubbles to escape at either end. Once this is complete, it is a true water level. Always do this prior to using the water level. If this is not done, a true and accurate reading will not be obtained.

The following suggestions will assist you in the use of the water level:

Step 1 Remove the stoppers (clamps) from each end.

Step 2 Place a benchmark on a wall near a corner at eye level (5' to 6').

Step 3 Hold one end of the water level against this mark.

Step 4 Have someone take the other end of the water level to the far end of the same wall and hold it at approximately the same height.

Step 5 Watch the benchmark and the water level as it is being raised or lowered at the other end.

Step 6 When the water level at the far end of the wall is at the same height as the benchmark, a level point has been found. Proceed to locate the other level points on the other walls.

Many types of laser levels are available that can be used to aid in the installation of suspended ceilings. To use a laser level when installing a ceil-

Figure 19 Water level.

ing, follow the manufacturer's instructions for the laser level being used. Generally, the procedure involves mounting the laser either on a wall/ceiling mount or on a tall tripod. The laser beam is rotated either at the finished ceiling height or at a reference point. A special target is snapped to a grid, then the grid is moved up or down until the laser beam crosses the target's offset mark. The grid is then secured in place.

A procedure for leveling a ceiling using a laser level attached to a wall/ceiling mount is given in this section. A laser mounted on a tall tripod can also be used.

Figure 18 Level.

> **WARNING!**
>
> OSHA (Occupational Safety and Health Administration) *CFR 1926.54* covers safety regulations for the use of lasers. Some guidelines are as follows:
> - Avoid direct eye exposure to the laser beam.
> - Only qualified and trained personnel are permitted to operate laser equipment.
> - Place a standard laser warning sign conspicuously at major approaches to the instrument use area.
> - Always turn off the laser when transmission of the beam is not required.

Step 1 Attach the laser to the wall/ceiling mount (*Figure 20*). Align the laser so that the rotating head is at the same end as the securing screws. Secure the laser firmly on the wall/ceiling mount using the securing wheel.

Step 2 Determine the height of the ceiling to be installed. Attach a 15" or longer scrap piece of wall angle about 3" above the determined ceiling height.

Step 3 Slip the slot in the wall/ceiling mount over the wall-angle flange and secure the mount by tightening the securing screws (*Figure 21*).

Step 4 Turn on the laser level and make any needed adjustments.

Step 5 Attach a laser target to the wall angle (*Figure 21*).

Step 6 Loosen the securing knob for height adjustment and center the laser beam on the white center line on the target.

Step 7 Using the center line on the laser target as your height reference, lay out the walls for the wall angle. As the laser revolves, the height of the laser target line can be penciled onto the other walls and a chalkline used to establish the wall-angle heights on all four walls. Always put the chalkline at the position of the top of the molding.

Step 8 Once the border lines (wall-angle locations) have been established, the laser level can be removed and used at another location on the job site. However, the laser level can also remain in position and be used instead of the string (dry line) while installing the ceiling grid system. The laser level is more accurate than the string method once you become familiar with the instrument.

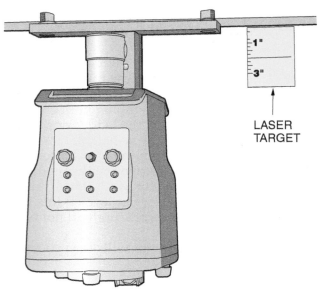

Figure 21 Wall/ceiling mount attached to wall-angle flange.

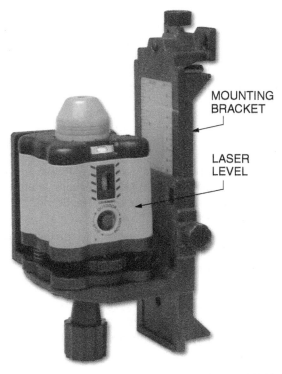

Figure 20 Laser-level wall/ceiling mount.

3.3.0 Installing Suspended Ceiling Systems

Observing the following general guidelines will help to achieve the desired level of professionalism when installing ceilings:

- All workmanship should meet the highest standards in accordance with the Ceiling and Interior Systems Construction Association policy of upgrading quality.
- Ceiling tiles should be so arranged that units less than one-half width do not occur unless otherwise directed by the reflected ceiling plans or job conditions.
- All tiles, tile joints, and exposed suspension systems must be straight and aligned.
- All acoustical ceiling systems must be level to ⅛" in 12'.

- Tile must be neatly scribed against butting surfaces and to all penetrations or protrusions where moldings are not required.
- Tile surrounding recessed lights and similar openings must be installed with a positive method to prevent movement or displacement of the tiles.
- Tiles must be installed in a uniform manner with neat hairline-fitted joints between adjoining tiles.
- Wall moldings must be firmly secured, the corners neatly mitered, or corner caps used, if preferred.
- The completed ceiling must be clean and in undamaged condition.

3.3.1 Exposed Grid Systems

The following sections describe the general procedure for installing an exposed grid ceiling system, also called a direct-hung system.

> **NOTE**
> The procedures given in this module are general in nature and are provided as examples only. Because of the differences in components made by different manufacturers, it is important to always follow the manufacturer's installation instructions for the specific system being used.

Install an exposed-grid suspended ceiling system according to the following guidelines:

Step 1 Check the room number and location.
- Ensure the correct ceiling is going into the correct room.
- Refer to the construction drawings, reflected ceiling plan, or shop drawing to determine the correct height for the ceiling.
- Check the electrical drawings to ensure the lights will be placed as indicated on the reflected ceiling plan or shop drawing.
- Check the specification sheets or shop orders for special instructions to ensure that the proper hangers and fasteners are provided.

> **WARNING!**
> Scaffolds must be used and assembled in accordance with all local, state, and federal/OSHA regulations. OSHA requires the building, moving, or dismantling of all scaffolding to be supervised by a competent person who has the training, knowledge, and experience to identify hazards on the job site and the authority to eliminate them. Mobile scaffolds, such as baker's scaffolds, should only be used on level, smooth surfaces that are free of obstructions and openings. OSHA regulations also require that mobile scaffold casters have positive locking devices to hold the scaffold in place. When moving a mobile scaffold, apply the moving force as close to the scaffold base as possible to avoid tipping it over. Never move a scaffold when someone is on it.

Step 2 If needed, install the scaffold.
- Set the scaffold at the correct height to permit the driving of inserts or other fasteners into the overhead ceiling and to permit the connection of hangers.
- Set the lower portion of the scaffold to permit installation of the grid members and ceiling tiles.

> **NOTE**
> Omit Steps 3 and 4 if a laser level is being used to establish the correct and level ceiling height, as described in the section *Ceiling Leveling Equipment*.

Step 3 Establish benchmarks. In some situations, the floor may not be level in all areas of the room you are working in. Make sure to establish benchmarks that are exactly the same height throughout the room. These benchmarks can be located with the use of a water level or laser level. The benchmarks will provide an accurate starting point for locating the desired height of the finished ceiling.
- Using the level of choice, locate benchmarks at each end of the room near the corners. If the room is extremely large and long, locate the benchmarks at 15' to 20' intervals.

- It is best to locate the benchmarks at eye level (about 5' above the floor).
- Always put benchmarks on each wall and also on protruding walls. It is better to have too many than not enough.

Step 4 Establish the height for the top of the wall angle.
- Once all benchmarks are located and marked, establish a measurement to the bottom of the wall angle.
- To that measurement, add the height of the wall angle.
- Measure from each benchmark and establish a mark at the top of the wall angle. Place a mark at various intervals on the wall.
- Snap a chalkline on those marks. This will establish the top of the wall angle. Be sure to snap the chalkline at the top of the wall angle so that the chalkline will not be visible when the wall angle is installed.

When installing the wall angle:
- Be sure that the top of the wall angle is even with the chalkline at all points.
- Nail or screw the wall angle to the wall. The wall angle should fit securely to prevent sound leaks.
- Miter the wall angle to fit at the corners, or use corner caps to cover the joints.

> **NOTE**
> The following procedure assumes that the center lines have been laid out on the floor. The procedure would be done in a similar manner if the center lines were laid out on the ceiling. The only difference is that the center lines would be transferred down from the ceiling, instead of up from the floor, to the level of the suspended ceiling grid.

Install the wall angle according to the following procedure:

Step 1 Square the room by first establishing length and width center lines on the floor or ceiling, as presented in the section *Establishing Room Center Lines*.

Step 2 Transfer the positions of the center lines from the floor/ceiling to the level required for the suspended ceiling grid as follows:
- At either one of the long walls, use a plumb bob to locate the center of the wall just above the wall angle. Do this by moving the plumb bob until it is directly over the center line marked on the floor (*Figure 22*). Then, mark the wall just above the wall-angle flange and insert a nail behind the wall angle at this point. Repeat the procedure at the opposite long wall. Run and secure a taut dry line between the nails in the two long walls.
- Repeat the procedures above for the two short walls and run a second dry line perpendicular to the first.
- Make a final check using the plumb bob at all four points.

When completed, you will have two dry lines intersecting at right angles at the center of the ceiling. Use the 3-4-5 method to ensure that the intersection of the dry lines is square. This is important because these intersecting dry lines will be used as the basis for all subsequent measurements used to lay out the ceiling grid system.

In some cases, reflected ceiling plans or shop drawings may indicate a center line and specify border width. In these instances, follow the suggested layout. If you have neither to follow, it will be necessary for you to establish the center line and plan the layout of the ceiling so that the border tiles adjacent to the facing wall are the same width and are not less than one-half the width of a full tile.

One method for determining the width of the border tiles is to convert the wall measurement from feet to inches, then divide this amount by

"Seeing" through the Ceiling

It is a good idea to keep a package of colored thumbtacks in your toolbox when installing a suspended ceiling. When other trades may have to get back in to complete their work, insert different colored thumbtacks in ceiling tiles to mark the locations of electrical and mechanical services. The marking will allow the other trades to locate their components without having to raise a lot of tiles, and will therefore help minimize finger smudges and damage to the tiles.

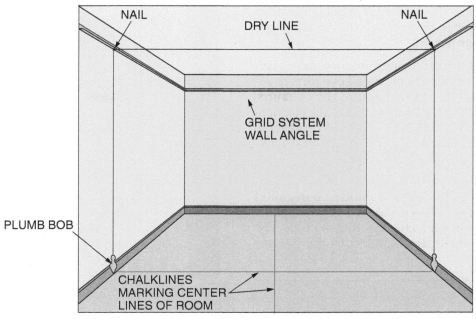

Figure 22 Transferring room center lines to the height of the suspended ceiling grid.

the width of the ceiling tiles. For example, assume the room measurements are 42'-6" × 30'-6". What will be the width of the border tiles running parallel to the two 42'-6" walls if 2' × 2' tiles are being used? To find the answer, proceed as follows:

Step 1 Take the measurement of the short wall (30'-6") and convert feet to inches:

30' × 12" = 360" + 6" = 366"

Step 2 Divide that amount by 24" (the width of a tile):

366" ÷ 24" =
15 tiles with a remainder of 6"

Step 3 If the division does not result in a whole number, add the width (in inches) of a full board to the remainder (in this case, it is 6"):

6" + 24" = 30"

Step 4 Divide this by 2; the result is 15. This would be the width of a border tile on each side. When adding the width of a tile to the remainder in Step 3, you must delete one full tile from the total. This means that there would be 14 full tiles plus a 15" border on each end (*Figure 23*):

14 × 2' + 30" = 30'-6"

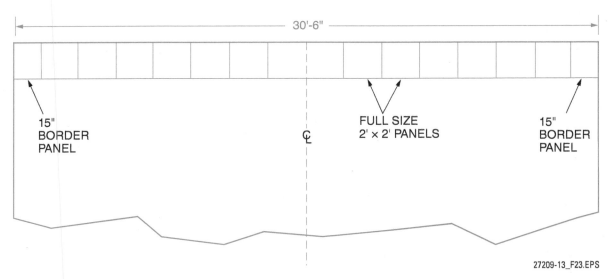

Figure 23 Fitting the tiles to the room.

Step 5 Determine the width of the border tiles for the other two remaining walls in the same manner.

Once all border unit measurements have been established, it is necessary to install eye pins to secure the hangers. Prior to installing the eye pins, you must locate the position of the main runners and cross runners. Line A in *Figure 24* represents the main runner. Line B represents the first cross runner line.

Step 1 Locate the first dry line (A) as shown in *Figure 24*. The dry line should be installed approximately 1⅛" above the flange of the wall angle. The dry line will also be used to indicate the bend in the hanger wire.

> **NOTE**
> Most main runners measure 1⅛" from the bottom of the runner to the hole for the hanger wire.

Step 2 Fasten a second dry line (B) so that it intersects line A at right angles with the proper measurement for the border tiles.

Step 3 The intersection point will be the location of the first hanger wire. Set the first eye pin directly over the intersection of dry lines A and B. Install the other eye pins every 4' on center. Be sure to follow dry line A as a guide for the locations.

> **WARNING!**
> Powder-actuated tools (PATs) are to be used only by trained operators in accordance with the operator's manual. Operators must take precautions to protect both themselves and others when using powder-actuated tools:
> - Operate the tool as directed by the manufacturer's instructions and use it only for the fastening jobs for which it was designed.
> - To prevent injury or death, make sure that the drive pin cannot penetrate completely through the material into which it is being driven.
> - To prevent a ricochet hazard, make sure the recommended shield is in place on the nose of the tool.

As you install each eye pin, suspend the hanger wire from the eye pin. Allow enough wire to ensure proper tying to the eye pin and the main runner. These two ties will require from 10" to 24" of wire. More hanger wire may be required if the ceiling is hung in part from structural members along with the eye pins. All hanger wire is pre-cut and can be obtained in various lengths and gauges. Refer to the specifications for the gauge of hanger wire to use. There are cases in which there may be cast-in-place inserts for hanger wires. In such cases, the insertion of eye pins and installation of wire will not be necessary.

The eye pins and hanger wires should be located on 4' centers in both directions. In most cases, it is best to hang both at the same time, as it requires less movement of the scaffold. To mark and bend the wires, proceed as follows:

Step 1 Mark the hangers where they touch the dry line.

Step 2 Twist the wire using side cutters and bend the other end to the mark (*Figure 25*).

> **NOTE**
> At least three turns should be made when twisting the wire. The wire should also be tight to the runner.

Install the main runners as follows:

Step 1 Measure and cut the main runner so that the cross runner will be at the proper distance for the border tile.

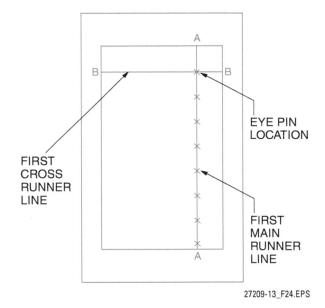

Figure 24 Locating the positions of the first main runner and cross runner.

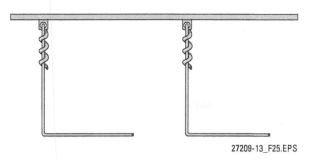

Figure 25 Bending hanger wire.

Step 2 Suspend the first length of main runner from the first row of hangers. It is important that the first few hanger wires be perpendicular to the main runner. If they are not, use a Whitney punch and make new holes.

> **NOTE**
>
> As indicated previously, in most exposed grid systems, the main runners are available with prepunched holes. The upper edges of the holes generally measure 1⅛" from the bottom of the runner; the dry line will have to be adjusted prior to bending the hanger wire.

Step 3 Continue to install the balance of the first row of main runners; splices will be needed. Note that various types of splices are used with various grid systems and are supplied by the grid system manufacturer.

Step 4 After installing the last full length of main runner, measure, cut, and install an end piece from a full length of main runner to complete the first run.

Step 5 After each end of the main runner is resting on the wall angle, insert all hanger wires and twist the wires to secure them in place (*Figure 26*).

Use the following procedure to install the cross runners:

Step 1 Measure the distance from the main runner to the wall angle. This width was determined earlier when completing the room layout.

Step 2 Using the snips, cut the cross runner to the correct border width. Be sure to cut and save the correct end or it will not slide into the main runner.

Hold-Down Clips

Some ceiling tiles require hold-down clips to secure the ceiling tiles to the grid. For example, clips are used with lightweight tiles to prevent them from reacting to drafts. One manufacturer specifies that clips be used if the tiles weigh less than 1 pound. Hold-down clips are not necessarily required for ceilings used in fire-rated applications. Check the manufacturer's instructions.

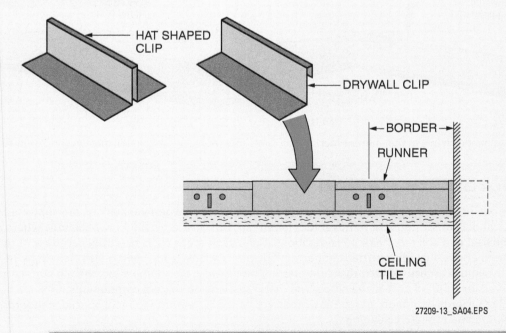

Module 27209 — Suspended Ceilings

Step 3 Insert the factory end into the main runner and let the other end rest on the wall angle (*Figure 27*).

Step 4 When the main runner and the cross runner are perpendicular to each other, lock the cross runner in place. Install the remaining cross runners between the main runners.

Step 5 In order to stabilize the grid system, install pop rivets at every other cross runner.

> **NOTE**
> Some grid systems use automatic locking devices to secure the cross runners into the main runners. Other grid systems use clips. Once all cross runners have been installed into the main runners, the grid system is ready to receive the ceiling panel or tile.

Prior to installing the ceiling tiles, it is necessary to wash your hands in order to keep the tiles clean. It is a good practice to wear white gloves when handling ceiling tiles. Start by cutting the border tiles and inserting them into position. After the border tiles are installed, proceed to install the full tiles. As the full tiles are being placed, install the hold-down clips, if required. General installation guidelines include the following:

- Install the ceiling tiles in place according to the job progression. Do not jump or scatter tiles in the grid system.
- Exercise care in installing tiles so as not to damage or mar the surface.
- Always handle tiles at the edges and keep your thumbs from touching the finished side of the tiles. If gloves are not worn, use cornstarch, powder, or white chalk on your hands.
- If the ceiling tile has a deeply textured pattern, insert it into the grid system so that the directional pattern will flow in the same direction. Such matching of ceiling tiles adds beauty to the finished suspended ceiling.

Upon completion of the ceiling, clean up the work area as follows:

Step 1 Dismantle the scaffold.

Step 2 Pick up all tools and equipment.

Step 3 Secure all equipment in a safe place.

Step 4 Pick up all trash.

Step 5 Sweep up any dust or debris.

3.3.2 Metal Pan Systems

Metal pan systems use 1½" U-shaped, cold-rolled steel furring channel members for pan support. They are normally installed 4' on center. However, it may be necessary to install them at lesser distances depending on the location of the light fixtures. The furring channels are installed by looping the hanger wires around them. Twist and secure using saddle wire. It is important that the furring channels are properly leveled to ensure a level ceiling.

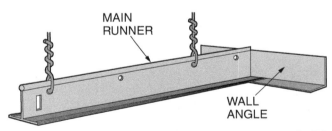

Figure 26 Inserting hanger wires in the main runner.

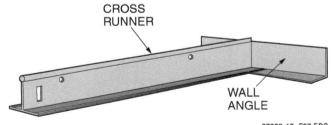

Figure 27 End of cross runner resting on wall angle.

High-Durability Ceilings

Highly durable ceiling tiles provide for long life and easy maintenance wherever ceilings are subjected to improper use, vandalism, or frequent removal for plenum access. They are used in applications where resistance to impact, scratches, and soil are major considerations. Ceilings in areas such as school corridors or gymnasiums need to withstand abuse, including surface impact. In any areas where lay-in ceiling tiles frequently need to be removed for plenum access, scratch-resistant tiles are highly desirable. Otherwise, the tile surface can be scratched, scuffed, or chipped as it is slid across the metal suspension system components. Ceilings installed in laboratories, clean rooms, and food preparation areas are normally required to meet special standards to ensure that they can withstand repeated cleaning.

Once the furring channels have been installed, the main runners (tee bars) must be placed. These bars are manufactured, cold-rolled or zinc-coated steel or aluminum. They are fastened to the furring channels using special clips. The tee bars run at right angles to the furring channels (*Figure 28*).

The tee bar has a spring-locking feature, which grips the metal pan. This feature allows the pan to be removed for access to the area above the pan. If the room plan dimensions are longer than the length of the tee bar, use a tee bar splice to couple the tee bars together, extending them to the required length (*Figure 29*).

Follow the procedure below for installing the tee bars:

Step 1 Install the first tee bar at right angles to the 1½" furring channel. Be sure it is correctly aligned.

Step 2 When in position, place tee bar clips over the furring channel and insert the ends under the flange of the tee bar. Hang additional clips in the same manner along the length of the bar.

Step 3 Install the second tee bar parallel to the first one. The spacing should be from 1' to 4', depending on the size of the metal pans.

After installing all the tee bars in one section of the room (depending on the working area of your scaffold), begin installing the metal pans.

Take care in handling the pans. Use white gloves or rub your hands with cornstarch to prevent any perspiration or grease marks from marring the surface of the pans. If care is not taken, fingerprints will be plainly visible when the units are installed. Use the following procedure to remove the pans from their container:

Step 1 Place the pan's finished surface face down on some type of raised platform, such as a table, which has been covered with a pad to protect the surface of the pan.

Step 2 Insert the wire grid into the back side of the metal pan. The wire grid is installed between the metal pan and backing pad to provide an air cushion between the two surfaces.

Step 3 Over the grid, place the paper- or vinyl-wrapped mineral wool or fiberglass batt or pad.

> **NOTE**
> In some cases, the metal pans come with the wire grid and pad already assembled.

When installing the metal pans, begin at the perimeter of the ceiling, next to the walls, since the units must fit into the channel wall angle. Follow the procedure below:

Step 1 From the room layout or reflected ceiling plan, obtain the width of the border units. Measure the pan to this width.

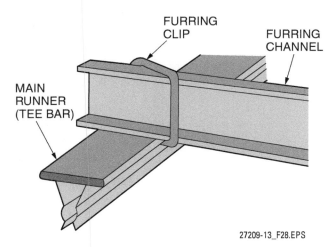

Figure 28 Main runner (tee bar) clipped to a furring channel.

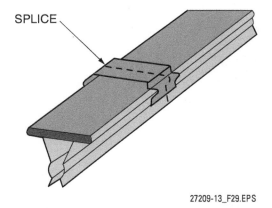

Figure 29 Tee bar splice.

Indoor Swimming Pools

All-aluminum grid systems are not recommended for use above indoor swimming pools because chlorine gases cause aluminum to corrode.

Step 2 Using a band saw, cut the pan ⅛" short of the desired width along the edge to be inserted into the wall angle.

Step 3 Slide the pan into the wall angle. Do not force the pan all the way into the molding. Leave ⅛" for expansion.

Step 4 Insert two crimped edges of the metal pan into the spring-locking tee bars.

Step 5 When the pan is in position, insert the spacer clip into the channel-molding cavity and over the cut edge of the pan. This will prevent the pan from buckling along this cut edge.

Step 6 Cut the wire grid and backing pad to fit the border unit. Place the unit into position.

Step 7 At the corners of the room, install the pans in the order shown in *Figure 30*.

Once the perimeter pans are installed in each row, the full-size pans can be put into place as follows:

Step 1 Grasping the pan at its edges, force its crimped edges into the tee-bar slots. Use the palms of your hands to seat the pan.

Step 2 After installing several of the pans as noted above, slide them along the tee bars into their final position. Use the side of your closed fist to bump the pan into level position if it does not seat readily.

Step 3 If metal pan hoods are required, slip them into position over the pans as they are installed. The purpose of the hood is to reduce the travel of sound through the ceiling into the room.

If a metal pan must be removed, a pan-removal tool is available (*Figure 31*). To pull out a pan, insert the free ends of the device into two of the perforations at one corner of the pan and pull down sharply. Repeat this at each corner of the pan. By following this removal procedure, there is no danger of bending the pan out of shape. *Figure 32* shows a finished metal-pan ceiling.

3.3.3 Direct-Hung Concealed Grid Systems

The installation of the concealed grid system begins in the same way as the conventional exposed grid system previously discussed.

Step 1 Once chalklines have been established on the walls at the required height above the floor, the next step is to install the wall angle. This molding, as for the other ceiling systems, provides support for the grid and panel or tile at the wall. The molding

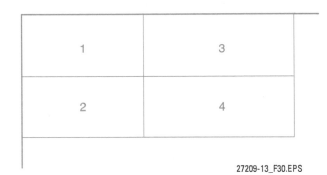

Figure 30 Tile layout.

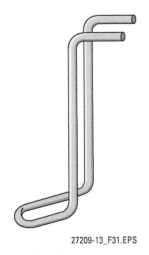

Figure 31 Pan-removal tool.

Placing Main Runners

One way to locate the position of the first main runner is to convert the width of the room to inches and divide by 48 (assuming you are installing 48" tiles). Then add 48" to any remainder and divide that result by 2 to obtain the distance from the wall to the first main runner. The remainder of the main runners are then placed at 4' intervals. Try an example assuming the width of the room is 22'.

Converting 22' into inches yields 264". Dividing that result by 48 gives 5' with a remainder of 6". Adding 48" yields 54", which is then divided by 2. The first main runner is placed 27" from the wall.

should be fastened with nails, screws, or masonry anchors. At corners, miter the inside corner and use a cap to cover the outside corner.

Step 2 Lay out the grid and install the hanger inserts and hangers according to the reflected ceiling plan or lighting layout. Once this has been completed, install the concealed main runners as done for the exposed grid system (see the section *Exposed Grid Systems*). The main runners are the primary support members. The method of coupling them to attain a specific length will vary with the manufacturer of the system, but they can all be spliced to the desired length.

Step 3 Install the cross stabilizer bars and concealed cross tees at right angles to the main runners. They rest on the flange of the main tee runners.

Step 4 Place the ceiling panel into position. Cut in the first row of panel at a line perpendicular to the main runners, as previously established. Attach the panel to the wall using spring clips.

Flat splines (*Figure 33*) are metal or fiber units that are inserted by the installer into the unfilled panel kerfs between the concealed main runners. Splines are used to prevent dust from seeping through the kerfs in adjacent panels.

If access is needed to the area above the ceiling, special systems are available, which can be incorporated into the ceiling (*Figure 34*).

3.3.4 Installing Standard Luminous Systems

The procedure for installing a standard luminous system is the same as for the exposed-grid suspended system with the exception of the border cuts. Luminous ceilings are placed into the grid members in full modules. Any remaining modules are filled in with acoustical material cut to size.

With a 2' × 2' or 2' × 4' standard exposed grid system, use luminous tiles, which provide the light-diffusing element in the system. These tiles are laid in between the runners. There are many sizes and shapes of tiles available.

3.3.5 Installing Nonstandard Luminous Systems

The procedure for installing a nonstandard luminous ceiling is the same as for the standard system with reference to room layout, hanger insert,

Figure 32 Metal-pan ceiling.

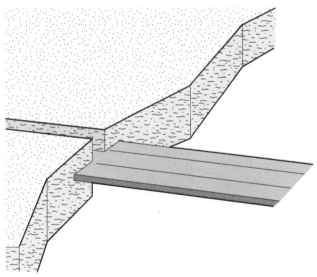

Figure 33 Ceiling-panel spline.

Insulation

It is usually recommended that insulation not be placed over mineral ceiling tiles because the additional weight could cause the tiles to sag. If the use of insulation cannot be avoided because of occupancy codes, limit the insulation to R-19 (0.26 pounds per square foot). Use only roll insulation and lay it perpendicular to the cross tees, so that the grid supports the weight of the insulation. If batts are required by code, 24" × 24" ceiling tiles should be used. Check codes carefully before applying insulation in a fire-rated system.

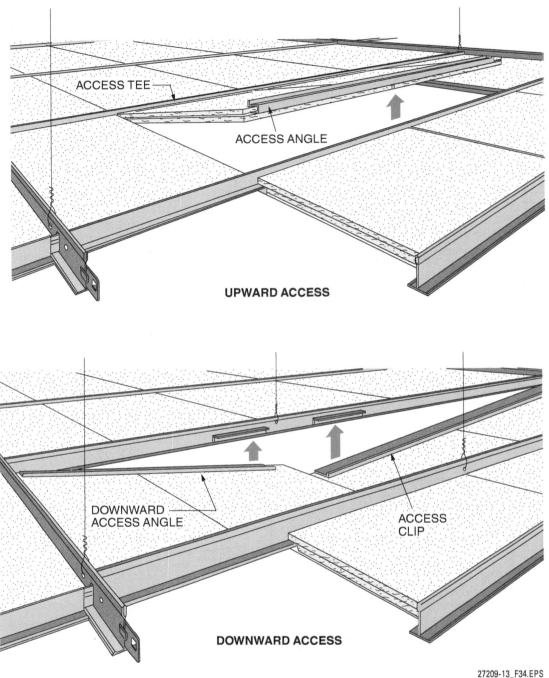

Figure 34 Access for concealed grid ceilings.

Facts about Ceiling Tiles

- Ceiling tiles are often called pads.
- Certain lighting fixtures are manufactured in the same sizes as tiles to drop directly into suspended ceilings.
- Some tiles are treated to inhibit the growth of mold, mildew, fungi, and certain bacteria.
- Special tiles are designed for use below sprinkler systems. These tiles will shrink and fall out of the grid as the room temperature rises, allowing the water to reach the fire.
- Most ceiling tiles can be recycled.

and installation of hanger wire, main supports, secondary supports, and attachment of some type of wall angle.

Since the nonstandard ceiling can deviate so much in terms of size, intricacies of the system, and the exactness of the installation, shop drawings are required. More information on nonstandard ceilings can be obtained from ceiling manufacturers.

3.3.6 Installing Suspended Drywall Ceilings

The installation of a suspended drywall ceiling begins with installing the carrying channels. A general procedure for installing the carrying channels is as follows:

Step 1 Snap chalklines on all walls.

Step 2 Install the wall angle, if used.

Step 3 Hang dry lines. From the ends of two facing walls, measure 4' along the chalklines. Secure the dry lines to the walls at these two points.

Step 4 Fasten the hanger inserts into the ceiling structure above. Using the dry line as a guide, install a row of hanger inserts 4' on center, or as specified, into the ceiling over the dry line.

Step 5 Install the hangers. Use No. 8 gauge wire or as specified. Hang the wires from the hanger inserts, then twist and secure. Allow enough wire to tie the hanger to the insert and the carrying channel. Generally, 24" to 28" of wire is adequate.

Step 6 Mark where the dry line touches the hanger wires. Mark each hanger wire where it is to be bent. Bend the wire to a 90-degree angle.

Step 7 When all the hanger wires in one row have been marked and bent, fasten the carrying channel to the hanger wires (*Figure 35*). Loop the wires around the carrying channel at the point where the wires have been bent. Twist and fasten the wires using a saddle tie. It is important that the carrying channels are installed level. If it is necessary to extend the length of a carrying channel because of the room size, face the open U of the second channel toward the U of the one already installed. Overlap for a distance of 8" to 12" and tie together with wire (*Figure 36*).

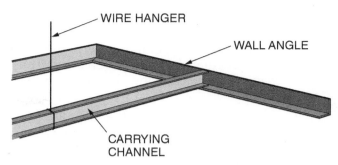

Figure 35 Carrying channel supported by hanger wire.

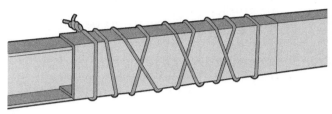

Figure 36 Splicing carrying channels.

Step 8 Move the dry line 4', in line with the next row of hangers. Continue to install the hanger inserts, wires, and carrying channels on 4' centers over the entire ceiling.

After the carrying-channel support members are installed, proceed with the installation of the cross furring members. There are different kinds of furring members that can be used at right angles to the carrying channels at 16" on center (maximum). For example, a hat furring channel (*Figure 37*) is used when it is desirable to screw the drywall board to the furring channels with self-tapping screws. Install the furring channels as follows:

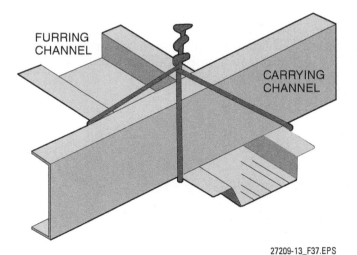

Figure 37 Hat furring channel.

Step 1 Install the furring channels perpendicular to the carrying channels. Place the first furring channel in line with the first perpendicular dry line. Attach the furring channel to the carrying channel using tie wire.

Step 2 Continue to install the furring channels 16" on center. If it is necessary to extend the furring channels beyond their normal length, lap one end of the channel over the next piece about 9" and wrap the splice well with tie wire (*Figure 38*). Avoid installing furring channels over positions where any fixtures are to penetrate the ceiling.

Step 3 Screw the drywall to the furring channels (*Figure 39*). Be sure to use screws of sufficient length. The screw length should be long enough to go through the board and extend through the furring channel about ¼" to ⅜".

3.3.7 Installing Acoustical Panel

If acoustical panel is to be used as the finish over the drywall, proceed as follows:

Step 1 Lay out the room. Measure the length of one of the two shorter walls. Divide this in half. Mark this halfway point on the drywall next to this point on the wall. Next, measure the length of the facing wall at the opposite end of the room. Divide this wall length in half and mark this point on the drywall.

Step 2 Stretch a chalkline across the room and line it up with the two marks on the drywall. Snap the line.

Step 3 Divide the chalkline in half and set a second half-line perpendicular to the first one at this midpoint.

Step 4 Install the ceiling panel per the manufacturer's instructions beginning at the junction of the two center lines.

3.3.8 Installing Furring Channels Directly to Structural Members

Sometimes furring channels are installed directly to structural members such as open-web steel joists or wood I-joists, instead of suspended carrying channels. In these instances, install the furring channels as follows:

Step 1 Mark the walls to indicate the height to be used to level the ceiling. Set the ceiling-height control marks to indicate the position of the bottom edge of the furring channels. Check the level of the structural joists. If they are not level, use the low point as the common level for installing the furring channels.

Step 2 Measure up from the benchmarks to the height of the lowest point on the joists. Using the distance of the low joist point minus the height of the furring channel, mark it on all walls to indicate a uniform level for the lower edge of all furring channels. Be sure to measure up from the benchmarks to establish this common level on all walls.

Step 3 Establish the chalklines using the ceiling control marks.

Step 4 Install the wall angles (if used). New chalklines may have to be snapped to guide the installation process.

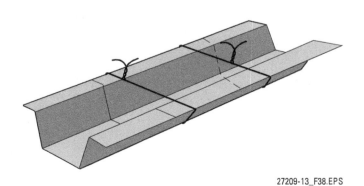

Figure 38 Splicing (lapping) furring channels.

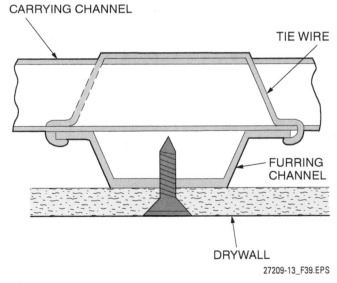

Figure 39 Drywall screwed to a furring channel.

Step 5 Hang the dry lines by securing them to facing walls at the chalkline level. Run the first line at this level right under the first joist. Run the second line at right angles to the first one at the same height. Position it to indicate where to install the first furring channel.

Step 6 Install the furring channels in the same manner as previously described, with the following exceptions:
- When attaching furring channels to steel joists, tie them together with the appropriate-gauge tie wire. Wrap the wire around the two so that it bridges the joist and supports the furring channel on either side of the joists. Tie the wire ends together at the side of the union.
- Shimming may be needed between the joists and the furring channel. Be sure to check the level of the furring channel along its entire length. Shim where needed to correct any deviation from level.

3.3.9 Drywall Suspension Systems

Drywall suspension systems are used to install conventional 2 × 2 and 2 × 4 flat tiles, as well as curved and domed ceiling systems. These systems support the use of a variety of tile types, including drywall, which provides the fire resistance needed in some applications. A variety of suspension hangers are available; examples are shown in *Figure 40*.

3.3.10 Ceiling Cleaning

Immediately after installation and after extended periods of use, suspended ceilings may require cleaning to maintain their performance characteristics and attractiveness. Dust and loose dirt can easily be removed by brushing or using a vacuum cleaner. Vacuum cleaner attachments such as those designed for cleaning upholstery or walls do the best job. Make sure to clean in one direction only as this will prevent rubbing of the dust or dirt into the surface of the ceiling.

After loose dirt has been removed, pencil marks, smudges, or clinging dirt may often be removed using an ordinary art gum eraser. Most mineral-fiber ceilings can be cleaned with a moist cloth or sponge. The sponge should contain as little water as possible. After washing, the soapy film should be wiped off with a cloth or sponge slightly dampened in clean water. Vinyl-faced fiberglass ceilings and Mylar™-faced ceilings can be cleaned with mild detergents or germicidal cleaners.

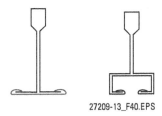

Figure 40 Examples of drywall suspension system hangers.

Special Furring Systems

Some manufacturers make furring-system cross tees in 14", 26", and 50" lengths. These sizes can reduce the time it takes to install an F-type lighting fixture from as long as 30 minutes to less than 1 minute.

Cleaning Ceiling Tiles

The materials and methods used for cleaning ceiling tiles vary widely, depending on the finish and texture of the tiles. One manufacturer prescribes eight different cleaning methods for their line of ceiling tiles. It is very important to check the manufacturer's instructions before proceeding. When cleaning grids, the ceiling tiles should be removed to prevent cleaning solution or dirt from getting on the tiles.

3.0.0 Section Review

1. If a room does not have a flat ceiling, the center lines can be laid out on the floor and then transferred upward using _____.

    a. a laser level
    b. a plumb bob and dry lines
    c. a story pole
    d. triangulation (the 3-4-5 rule)

2. When installing the grid for a suspended ceiling, you should use a carpenter's level to check the installation at intervals of _____.

    a. 3'
    b. 4'
    c. 6'
    d. 8'

3. Hanger wires used to install ceiling grids should be tight to the runner and twisted at least _____.

    a. two turns
    b. three turns
    c. four turns
    d. six turns

SUMMARY

Although other trades may be responsible for ceiling installations on some jobs, the current trend is for carpenters to do the work. This module covered the major types of acoustical suspended ceiling systems. It is important that you not only understand how the various systems differ from each other, but also how they are alike. In addition, the basic procedures for installing suspended ceiling systems should be understood.

Review Questions

1. From their source, sound waves travel _____.
   a. on a line-of-sight
   b. in all directions
   c. vertically
   d. at right angles

2. The intensity of a sound wave refers to _____.
   a. the number of wave cycles it completes in a second
   b. its loudness
   c. its frequency
   d. the degree of loudness and softness

3. A sound level of 100 dBA is considered _____.
   a. very loud
   b. moderately loud
   c. loud
   d. quiet

4. The term used by manufacturers to compare the noise absorbency of their materials is the _____.
   a. ceiling attenuation class (CAC)
   b. articulation class (AC)
   c. noise reduction coefficient (NRC)
   d. sound transmission classification (STC)

5. An integrated ceiling system commonly uses a tile (module) of size _____.
   a. 12" × 12"
   b. 12" × 24"
   c. 24" × 24"
   d. 30" × 60"

6. All ceilings, walls, pipe, and ductwork in the space above a luminous ceiling will normally be painted with a _____.
   a. light-refracting finish
   b. 75 to 90 percent reflective matte white finish
   c. 90 to 100 percent reflective matte finish
   d. luminous matte finish

7. In an exposed grid ceiling system, hangers are used to support the _____.
   a. 4' cross tees
   b. 2' cross tees
   c. wall angle
   d. main runners

8. Acoustical panels and tiles used in suspended ceilings stop the transmission of unwanted sounds by the process of _____.
   a. reflection
   b. reverberation
   c. absorption
   d. refraction

9. On some large construction jobs, the set of construction drawings often includes a(n) _____.
   a. projected ceiling plan
   b. reflected ceiling plan
   c. reversed ceiling plan
   d. orthographic ceiling plan

10. The points at which electrical and plumbing risers penetrate a ceiling are shown on _____.
    a. MEP drawings
    b. floor plans
    c. section views
    d. elevation drawings

11. For a suspended ceiling that is 10' × 16', using ceiling tiles that are 2' × 4', with the main runners parallel with the ceiling's long dimension and the 4' length of the tiles parallel with the ceiling's short dimension, the first main runner should be installed at a distance from the wall of _____.
    a. 18"
    b. 24"
    c. 36"
    d. 48"

12. For the ceiling described in Question 11, the dimensions (width and length) of the border ceiling tiles to be installed along the long dimension of the ceiling will be _____.
    a. 18" × 24"
    b. 24" × 30"
    c. 24" × 36"
    d. 36" × 48"

13. When laying out eye pins for hanger inserts, two dry lines are used to determine the _____.
    a. exact center of the room
    b. positions of all cross ties
    c. center line of the first ceiling tile
    d. location of the first pin

14. When installing the metal pans in a metal-pan ceiling system, begin by installing them _____.
    a. in the middle of the ceiling
    b. at the perimeter of the ceiling next to the walls
    c. at the corners of the room
    d. at the most convenient location

15. Splines installed into the unfilled panel kerfs in a direct-hung concealed grid ceiling system _____.
    a. prevent buckling of the ceiling
    b. are used in place of stabilizer bars
    c. prevent shifting of the panels
    d. prevent dust from seeping through the kerfs in adjacent panels

Trade Terms Quiz

Fill in the blank with the correct term that you learned from your study of this module.

1. The measuring unit that is used to express differences in sound power is the _____.

2. The space between a suspended ceiling and the roof or floor above is often used in commercial buildings as the conduit for HVAC return air. In such situations, the space is referred to as a _____.

3. _____ is the science concerned with the characteristics of sound.

4. When installing a suspended ceiling, _____ (strings stretched between two points) serve as guides for the installer.

5. _____ are sound-absorbing boards laid into an exposed grid mounting system.

6. The term used to describe a ceiling-tile surface pattern of fine parallel grooves is _____.

7. _____ are plaster, ceiling tiles, or other materials with a high capacity for absorbing sound waves.

8. A material with a surface design that appears to have cracks and splits is said to be _____.

9. Equal to one full cycle per second, the _____ is the measuring unit for frequency.

10. _____ are acoustical materials in board form that are fastened to the ceiling or suspended in a grid to provide an attractive appearance and absorb sound.

11. A sound measurement that is based upon the decibel but biased to emphasize the frequency response of the human ear is the _____.

12. _____ is a term, usually expressed in hertz (Hz), that describes the number of cycles occurring in a specified period of time.

13. A device that scatters light or breaks up the flow of air is referred to as a _____.

Trade Terms

Acoustical materials
Acoustics
A-weighted decibel (dBA)
Ceiling panels
Ceiling tiles
Decibel (dB)
Diffuser
Dry lines
Fissured
Frequency
Hertz (Hz)
Plenum
Striated

Trade Terms Introduced in This Module

Acoustical materials: Types of ceiling panel, plaster, and other materials that have high absorption characteristics for sound waves.

Acoustics: A science involving the production, transmission, reception, and effects of sound. In a room or other location, it refers to those characteristics that control reflections of sound waves and thus the sound reception in the area.

A-weighted decibel (dBA): A single number measurement based on the decibel but weighted to approximate the response of the human ear with respect to frequencies.

Ceiling panels: Acoustical ceiling boards that are suspended by a concealed grid mounting system. The edges are often kerfed and cut back.

Ceiling tiles: Any lay-in acoustical board that is designed for use with an exposed grid mounting system. Ceiling tiles normally do not have finished edges or precise dimensional tolerances because the exposed grid mounting system provides the trim-out.

Decibel (dB): An expression of the relative loudness of sounds in air as perceived by the human ear.

Diffuser: An attachment for duct openings in air distribution systems that distributes the air in wide flow patterns. In lighting systems, it is an attachment used to redirect or scatter the light from a light source.

Dry lines: A string line suspended from two points and used as a guideline when installing a suspended ceiling.

Fissured: A ceiling-panel or ceiling-tile surface design that has the appearance of splits or cracks.

Frequency: Cycles per unit of time, usually expressed in hertz (Hz).

Hertz (Hz): A unit of frequency equal to one cycle per second.

Plenum: A chamber or container for moving air under a slight pressure. In commercial construction, the area between the suspended ceiling and the floor or roof above is often used as the HVAC return air plenum.

Striated: A ceiling panel or ceiling tile surface design that has the appearance of fine parallel grooves.

Additional Resources

This module presents thorough resources for task training. The following resource material is suggested for further study.

CISCA Ceiling Systems Handbook. 1999. St. Charles, IL: The Ceiling and Interior Systems Contractors' Association.
The Gypsum Construction Handbook. 2009. Chicago, IL: R.S. Means.

Figure Credits

Courtesy of **CeilingsPlus.com**, CO01, Figure 2, Figure 4, Figure 5, Figure 8, Figure 12, SA01, SA02

Chicago Metallic Corporation, Figure 6

Courtesy of **CeilingsPlus.com**, © Keith Peterson, Figure 7

Courtesy of **CeilingsPlus.com**, © Benny Chan, Figure 9

ZIRCON CORPORATION, Figure 19

CST/berger, Figure 20

Armstrong World Industries, Inc., Figure 32

Section Review Answer Key

Answer	Section Reference	Objective
Section One		
1.c	1.1.4	1a
2.a	1.2.0	1b
3.b	1.3.1	1c
Section Two		
1.a	2.1.0	2a
2.d	2.2.0	2b
Section Three		
1.b	3.1.2	3a
2.c	3.2.2	3b
3.b	3.3.1	3c

NCCER CURRICULA — USER UPDATE

NCCER makes every effort to keep its textbooks up-to-date and free of technical errors. We appreciate your help in this process. If you find an error, a typographical mistake, or an inaccuracy in NCCER's curricula, please fill out this form (or a photocopy), or complete the online form at **www.nccer.org/olf**. Be sure to include the exact module ID number, page number, a detailed description, and your recommended correction. Your input will be brought to the attention of the Authoring Team. Thank you for your assistance.

Instructors – If you have an idea for improving this textbook, or have found that additional materials were necessary to teach this module effectively, please let us know so that we may present your suggestions to the Authoring Team.

NCCER Product Development and Revision
13614 Progress Blvd., Alachua, FL 32615

Email: curriculum@nccer.org
Online: www.nccer.org/olf

❏ Trainee Guide ❏ Lesson Plans ❏ Exam ❏ PowerPoints Other _____

Craft / Level: _____ Copyright Date: _____

Module ID Number / Title: _____

Section Number(s): _____

Description: _____

_____

_____

_____

_____

Recommended Correction: _____

_____

_____

_____

_____

Your Name: _____

Address: _____

_____

Email: _____ Phone: _____

This page is intentionally left blank.

Window, Door, Floor, and Ceiling Trim

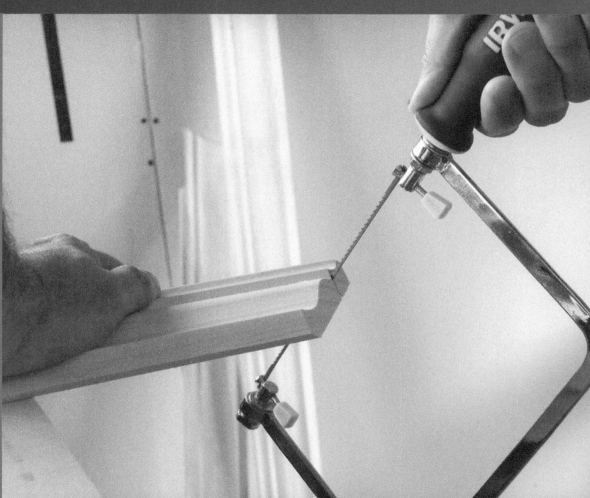

Overview

Skilled trim carpenters are always in demand to install floor and ceiling moldings, as well as the finish trim around doors and windows. Quality trim work requires the ability to measure accurately, calculate angles, and make precise, clean cuts. Even a tiny error will be visible in the finished product, so an accomplished trim carpenter is someone who has developed a reputation for careful, accurate work.

Module 27211

Trainees with successful module completions may be eligible for credentialing through NCCER's National Registry. To learn more, go to **www.nccer.org** or contact us at **1.888.622.3720**. Our website has information on the latest product releases and training, as well as online versions of our *Cornerstone* magazine and Pearson's product catalog.

Your feedback is welcome. You may email your comments to **curriculum@nccer.org**, send general comments and inquiries to **info@nccer.org**, or fill in the User Update form at the back of this module.

This information is general in nature and intended for training purposes only. Actual performance of activities described in this manual requires compliance with all applicable operating, service, maintenance, and safety procedures under the direction of qualified personnel. References in this manual to patented or proprietary devices do not constitute a recommendation of their use.

Copyright © 2013 by NCCER, Alachua, FL 32615, and published by Pearson, New York, NY 10013. All rights reserved. Printed in the United States of America. This publication is protected by Copyright, and permission should be obtained from NCCER prior to any prohibited reproduction, storage in a retrieval system, or transmission in any form or by any means, electronic, mechanical, photocopying, recording, or likewise. To obtain permission(s) to use material from this work, please submit a written request to NCCER Product Development, 13614 Progress Blvd., Alachua, FL 32615.

27210 V5

From *Carpentry, Trainee Guide*. NCCER.
Copyright © 2013 by NCCER. Published by Pearson. All rights reserved.

27210
Window, Door, Floor, and Ceiling Trim

Objectives

When you have completed this module, you will be able to do the following:

1. Describe the safety hazards related to working with window, door, floor, and ceiling trim.
    a. Identify the proper personal protection equipment required when working with window, door, floor, and ceiling trim.
    b. Identify tool and equipment safety guidelines when working with window, door, floor, and ceiling trim tools.
2. Identify the different types of standard moldings and materials.
    a. Identify the different types of base moldings.
    b. Identify the different types of wall moldings.
    c. Identify the different types of ceiling moldings.
    d. Identify the different types of window and door trim.
3. Explain how to install different types of molding.
    a. Explain how to properly cut trim.
    b. Explain how to properly fasten trim.
    c. Explain how to properly install base molding.
    d. Explain how to properly install ceiling molding.
    e. Explain how to properly install door trim.
    f. Explain how to properly install window trim.
4. Explain how to estimate window, door, floor, and ceiling trim.

Performance Tasks

Under the supervision of your instructor, you should be able to do the following:

1. Make square and miter cuts to selected moldings using a hand miter box.
2. Make square and miter cuts to selected moldings using a power miter/compound miter saw.
3. Make a coped joint using a coping saw.
4. Install interior trim using a finish nailer and hand nailing methods.
    - Door trim
    - Window trim
    - Base trim
    - Ceiling trim
5. Estimate the quantities of different trim materials required for selected rooms.

Trade Terms

Apron	Finger-jointed stock	Scarf joints	Trim
Compound cut	Moldings	Square cuts	Wainscoting
Coped joint	Reveal	Stool	

Code Note

Codes vary among jurisdictions. Because of the variations in code, consult the applicable code whenever regulations are in question. Referring to an incorrect set of codes can cause as much trouble as failing to reference codes altogether. Obtain, review, and familiarize yourself with your local adopted code.

Industry-Recognized Credentials

If you're training through an NCCER-accredited sponsor, you may be eligible for credentials from NCCER's Registry. The ID number for this module is 27210. Note that this module may have been used in other NCCER curricula and may apply to other level completions. Contact NCCER's Registry at 888.622.3720 or go to **www.nccer.org** for more information.

Contents

Topics to be presented in this module include:

1.0.0 Safety when Working with Trim 1
  1.1.0 Personal Protective Equipment 1
  1.2.0 Tool and Equipment Safety 1
    1.2.1 Finish Nailer Safety 2
    1.2.2 Router Safety 3
    1.2.3 Miter Saw Safety 3
2.0.0 Types of Molding and Trim Materials 4
  2.1.0 Base Moldings 5
  2.2.0 Wall Moldings 6
  2.3.0 Ceiling Moldings 7
  2.4.0 Window and Door Trim 7
    2.4.1 Stools 8
3.0.0 Installing Molding 10
  3.1.0 Cutting Trim 10
    3.1.1 Making Square and Miter Cuts 11
    3.1.2 Making a Coped Joint 12
  3.2.0 Fastening Trim 13
  3.3.0 Installing Base Molding 14
  3.4.0 Installing Ceiling Molding 16
  3.5.0 Installing Door Trim 17
  3.6.0 Installing Window Trim 18
4.0.0 Estimating Trim Quantities 28

Figures and Tables

Figure 1 Finger-jointed molding 4
Figure 2 Moldings used to trim the interior of a room 5
Figure 3 Typical base moldings 6
Figure 4 Typical chair-rail, inside-corner, quarter-round,
 outside-corner, and wainscot-cap moldings 6
Figure 5 Cove, crown, and bed moldings 7
Figure 6 Typical casing and casing-stop moldings 8
Figure 7 Rabbeted stool 8
Figure 8 Special trim saws 10
Figure 9 Compound miter saw 11
Figure 10 Typical butt and miter joints 12
Figure 11 Positioning molding in a miter box for cutting 13
Figure 12 Position of crown molding 13
Figure 13 Making a coped joint 14
Figure 14 Typical pneumatic finish nailers 14
Figure 15 Simplified baseboard molding installation 15
Figure 16 Procedure for fitting and cutting an outside miter joint 16
Figure 17 Scarf joint, cut and nailed directly over a wall stud 17
Figure 18 Window trimming methods 19
Figure 19 Installation of jamb extensions 20
Figure 20 Method of determining the length of the stool 20
Figure 21 Window stool showing cutout for one side
 of the window opening 21
Figure 22 How to mark and cut a return 21
Figure 23 Marking the head casing for miter cuts 23
Figure 24 Mark apron return profile with a scrap piece of molding 23
Figure 25 Mullion trim used with adjacent double-hung windows 27

Table 1 Compound Miter Saw Miter and Bevel Settings for
 US Standard Crown Molding 13

SECTION ONE

1.0.0 Safety when Working with Trim

Objective

Describe the safety hazards related to working with window, door, floor, and ceiling trim.
a. Identify the proper personal protection equipment required when working with window, door, floor, and ceiling trim.
b. Identify tool and equipment safety guidelines when working with window, door, floor, and ceiling trim tools.

Trade Terms

Moldings: Decorative strips of wood or other material used for finishing purposes.

Trim: Finish materials such as molding placed around doors and windows and at the top and bottom of a wall.

This module covers the materials, equipment, and procedures used to install trim. Trim is installed to create an attractive finished look and to hide joints at the intersection of walls, floors, ceilings, and different wall coverings. Trim also protects vulnerable edges, holds window sashes in place, and stops the swing of a door in its frame. The specific trim used should blend with the overall design of the building or room. For modern residential and commercial buildings, the interior trim typically involves the use of moldings of simple contemporary designs. On the other hand, buildings of traditional design tend to use moldings with more complex shapes and in much greater quantities.

When considering the construction sequence of a building, trim installation is one of the finishing touches. Everybody is going to see the trim; therefore, it has to be installed carefully with tight-fitting joints in order to present a suitable appearance.

1.1.0 Personal Protective Equipment

Proper personal protective equipment (PPE) is required when installing trim. While general safety guidelines are important to follow, respiratory, hearing, and eye protection are especially important when installing trim.

Respiratory protection is required for sanding and finishing operations. Respiratory protection should be used whenever there is an inhalation hazard. For sanding small amounts of trim in a fairly open area, a dust mask might be the only respiratory protection required. For applying finish to trim in an enclosed area, other means of respiratory protection will likely be required. OSHA (Occupational Safety and Health Administration) specifies the type of respiratory protection that is required for different types of hazards.

Bodily damage usually results in pain. However, repeated exposure to loud noises, like those created by finish nailers and compressors, can cause hearing loss without any resulting pain. Contractors are required to follow OSHA regulations to determine when personal protective equipment such as earplugs or earmuffs is required. When in doubt, consult your supervisor.

Eye protection must be worn whenever working on the job site. Eye and face protection must meet requirements specified in American National Standards Institute (ANSI) *Standard Z87.1*. Standard prescription eyewear is not an acceptable substitution for eye protection. While standard safety glasses are acceptable for many operations, face shields should be employed when performing operations that generate chips or shavings, such as when using a router to create a hinge recess.

1.2.0 Tool and Equipment Safety

A good carpenter is always aware of the safety rules in every area of the construction field. The interior of a building nearing completion sometimes gives the worker a false sense of security. However, the element of danger is always present.

- *Keep work area clean* – Cluttered areas and benches invite injuries.
- *Avoid dangerous equipment* – Do not expose power tools to rain. Do not use power tools in damp or wet locations. Keep the area well lit. Avoid chemical or corrosive environments. Do not use tools in the presence of flammable liquids or gases.
- *Guard against electric shock* – Avoid body contact with grounded surfaces such as pipes, radiators, ranges, or refrigerator enclosures.
- *Keep visitors away* – Do not let visitors come in contact with tools or extension cords.
- *Store idle tools* – When not in use, tools should be stored.
- *Secure the work* – Use clamps or a vise to hold the work. It is safer than using your hand, and it frees both hands for operating the tool.

- *Use the right tools* – Do not force a smaller tool or attachment to do the job of a heavier tool. Do not use a tool for any purpose for which it was not intended.
- *Dress properly* – Do not wear loose clothing or jewelry. Loose clothing, drawstrings, and jewelry can be caught in moving parts. Rubber gloves and nonskid footwear are recommended. Wear a protective covering to protect long hair.
- *Use proper eye and respiratory protection* – Wear safety glasses or goggles while operating power tools. Wear a face or dust mask if the operation creates dust. All persons in the area where power tools are being operated should also wear safety glasses and face or dust masks.
- *Do not abuse extension cords or power tool cords* – Never carry any tool by the cord or yank it to disconnect it from the receptacle. Keep the cord away from heat, oil, and sharp edges. Have damaged or worn power cords and strain relievers replaced immediately.
- *Do not overreach* – Keep proper footing and balance at all times.
- *Maintain tools with care* – Keep tools sharp and clean for safer performance. Follow instructions for lubricating and changing accessories. Inspect tool cords periodically, and if they are damaged, have them repaired by an authorized service facility. Have all worn, broken, or lost parts replaced immediately. Keep handles dry, clean, and free from oil and grease.
- *Disconnect (unplug) tools* – Always disconnect tools when not in use, before servicing, and when changing accessories such as blades, bits, and cutters.
- *Remove keys and adjusting wrenches* – Check to make certain that keys and adjusting wrenches are removed from a tool before turning it on.
- *Avoid unintentional starting* – Do not carry a plugged-in tool with a finger on the switch. Be sure the switch is off when plugging in a tool.
- *Outdoor extension cords* – When a tool is used outdoors, use only extension cords marked as suitable for use with outdoor appliances and store them indoors when not in use.
- *Check damaged parts* – Before using any tool, inspect it to be sure that it will operate properly and perform its intended function. Check for alignment of moving parts, binding of moving parts, breakage of parts or mountings, and any other conditions that may affect its operation. A guard or other part that is damaged should be properly repaired or replaced by an authorized service center unless otherwise indicated in the instruction manual. Inspect tool cords periodically, and if they are damaged, have them repaired by an authorized service facility. Have all worn, broken, or lost parts replaced immediately. Have defective switches replaced by an authorized service center.
- *Stay alert* – Watch what you are doing and use common sense. Do not operate a tool when you are tired or while under the influence of medication, alcohol, or drugs.
- *Inspect extension cords* – Inspect extension cords periodically and replace them if damaged.

> **WARNING!** Remember that an extension cord with cracked or cut outer insulation or a loose or missing ground pin can kill you. Be careful. Always unplug power tools when changing accessories or making adjustments. Plan ahead.

1.2.1 Finish Nailer Safety

Finish nailers, or nail guns as they are often called, are commonly used for installing trim. Finish nailers greatly speed up the installation time for trim and result in less damage to the trim pieces.

Finish nailers are designed to fire when the trigger is pressed and the tool is pressed against the building material being fastened. An important safety feature of all finish nailers is that they will not fire unless pressed against the material. Follow these guidelines to ensure safety for yourself and co-workers:

- Always wear appropriate personal protective equipment, including proper eye and hearing protection.
- Never aim a nail gun toward your body or anyone else.
- Review the operator's manual prior to using a finish nailer.
- Do not load a finish nailer with the hose attached.
- Never leave the nailer connected to the hose when not in use.
- If the nailer is misfiring, disconnect the hose before attempting to repair it.
- Keep all body parts and co-workers away from the nail path to avoid serious injury. Nails can go through trim and strike someone on the other side if the nail does not hit the stud.
- Check for pipes, electrical wiring, vents, and other materials behind trim before nailing.
- Do not operate a finish nailer in the presence of flammable liquids such as adhesives, gasoline, or paint thinner.
- Do not use oxygen, combustible gases, or any other bottled gas to operate finish nailers.

- Do not exceed manufacturer's recommended pressure for the nailer being used.

1.2.2 Router Safety

Routers may be used when installing trim to create recesses for hardware or other items. The sharp bit—whether moving or stationary—presents a safety hazard. Follow these guidelines to ensure safety when using a router:

- Always disconnect a router from its power source when inserting and tightening bits, as well as when making depth adjustments.
- Always wear appropriate PPE, including proper eye, face, and hearing protection. A face shield should be worn.
- Keep both hands on the router when using it and until the motor comes to a complete stop.
- Ensure electrical cords are not in the cutting path.
- Maintain proper footing and balance when using a router.

1.2.3 Miter Saw Safety

Miter saws and compound miter saws are used to a large extent when cutting trim before installation. A miter box is commonly attached to a stand or a couple of sawhorses to prevent back injuries due to excessive bending. Safety guidelines to follow when using a miter box or compound miter box include the following:

- Always wear appropriate PPE, including eye and hearing protection.
- Keep hands out of the blade path.
- Ensure that all wrenches are removed before operation.
- Allow the motor to come to full speed before cutting trim.
- Ensure the switch is in the off position prior to plugging in the saw.

1.0.0 Section Review

1. Hearing protection, such as earplugs, should be worn when working with _____.
   a. hand saws
   b. coping saws
   c. finish nailers
   d. block planes

2. A router bit is considered a safety hazard only when the tool is being operated.
   a. True
   b. False

Section Two

2.0.0 Types of Molding and Trim Materials

Objective

Identify the different types of standard moldings and materials.
a. Identify the different types of base moldings.
b. Identify the different types of wall moldings.
c. Identify the different types of ceiling moldings.
d. Identify the different types of window and door trim.

Trade Terms

Apron: A piece of window trim that is used under the stool of a finished window frame. It is sometimes referred to as undersill trim.

Finger-jointed stock: Paint-grade moldings made in a mill from shorter lengths of wood joined together.

Stool: The bottom horizontal trim piece of a window.

Wainscoting: A wall finish, usually made of wood, stone, or ceramic tile, that is applied partway up the wall from the floor.

Wood moldings are sometimes referred to as woodwork, trim, finish, or millwork. To manufacture a given wood molding profile, lumber of a preselected thickness is sawn or ripped lengthwise into strips of various widths for maximum use of the board. The rips are then sorted by width. Appropriate widths are run through a band saw or resawn where they are converted into shapes called blanks that are ideally suited for specific molding patterns with a minimum of waste. These blanks are sent through a molder, which has four or more cutter heads to shape the profile. The cutter head holds a series of identical knives, each shaped to conform to the exact molding pattern profile. As the blanks are fed into the molder, the cutter heads rotate at high speed, carving or planing the blanks into finished molded shapes. At the outfeed end of the molder, moldings are inspected to ensure conformance with industry quality standards. The finished moldings are then bundled by length, ready for shipment.

Most wood moldings are made of pine because pine can easily be milled into the desired profiles. Premium pine is clear and knot free and milled for staining. Paint-grade pine moldings, including finger-jointed stock, represent the next step down. Finger-jointed moldings (*Figure 1*) are assembled from shorter lengths. In finger jointing, two additional steps take place between the ripping and resawing operations. First, unacceptable characteristics or defects are trimmed out of lumber rips. Then, after sorting, the remaining strips of clear wood are passed through a finger jointer, which cuts small fingers in the ends of each piece. These short, fingered pieces are then glued and joined into long lengths of clear finger-jointed lumber, ready to be resawn and molded. Finger-jointed moldings are only used for paint or opaque finishes because the joints show through a stained or natural finish.

Because moldings made of pine can be nicked, dented, or otherwise damaged relatively easily, moldings are also manufactured out of harder woods for use in high-wear areas. Poplar is a clear, slightly more expensive wood that holds an edge better than pine and is a little more resistant to abuse. Moldings are also made out of hardwoods such as oak, cherry, and maple. These moldings are much more expensive than those made of pine. They are also harder to work with than pine because their installation usually requires that all the nail holes be predrilled. Normally, moldings made of hardwoods are used when it is desired to have a natural or stained finish that shows the grain and other characteristics of the hardwood.

Prefinished moldings are often made of medium-density fiberboard (MDF), a stable product that is made of wood fiber. MDF is less likely to split, warp, and cup. For this reason, it is sometimes used in areas where there are big fluctuations in humidity that would cause solid wood moldings to gap.

Standard moldings come in a variety of widths and are normally available in even lengths of 8', 10',

27210-13_F01.EPS

Figure 1 Finger-jointed molding.

12', 14', and 16'. Some are available in odd lengths, such as moldings used for door casing, which are made in 7' lengths to reduce waste. Many types of manufactured molding profiles are available.

Moldings that are to be joined must be carefully selected. Each mill makes moldings of different shapes (profiles), and slight variances occur even in the same profile from the same mill. Make sure that each piece is consistent from end to end and that the profiles of all like pieces to be used in the same room or project are matched. *Figure 2* shows the various places within a room where moldings are typically used for interior trim. The types of moldings widely used for this purpose include the following:

- Base, base cap, and base shoe moldings
- Casing and casing stop moldings
- Crown, bed, and cove moldings
- Quarter-round, corner-guard, chair-rail, and wainscot-cap moldings
- Window stools and aprons

2.1.0 Base Moldings

Base moldings (*Figure 3*) are used at the floor level to conceal the joint between a wall and floor. They protect the lower wall surfaces from damage by vacuum cleaners, brooms, furniture, and feet. Base moldings are always installed after all door casing moldings are nailed in place. Base moldings come in a variety of sizes and a multitude of contemporary and traditional profiles.

In general, base moldings have an unmolded lower edge, allowing a base shoe molding to be added to conceal any unevenness between the base and floor or to hide edges of carpeting and other flooring. The top profile of the base molding is often decorative, making the use of an additional base cap molding on top unnecessary. A square-edged molding (S4S [surfaced on four sides] molding) can be used alone as a simple base molding, or it can be combined with a decorative base cap molding and base shoe molding to create a more traditional look.

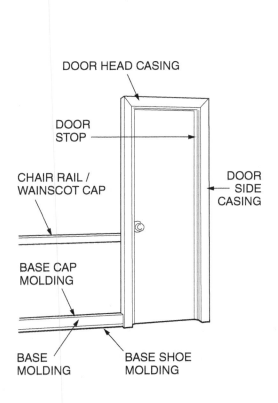

Figure 2 Moldings used to trim the interior of a room.

Custom Moldings

When only a few custom moldings are needed, they are often produced on a shaper. A shaper is a machine with a single head that can profile only one surface at a time. However, its ease of operation makes it desirable for small jobs. Shapers are ideal for producing curved moldings.

Figure 3 Typical base moldings.

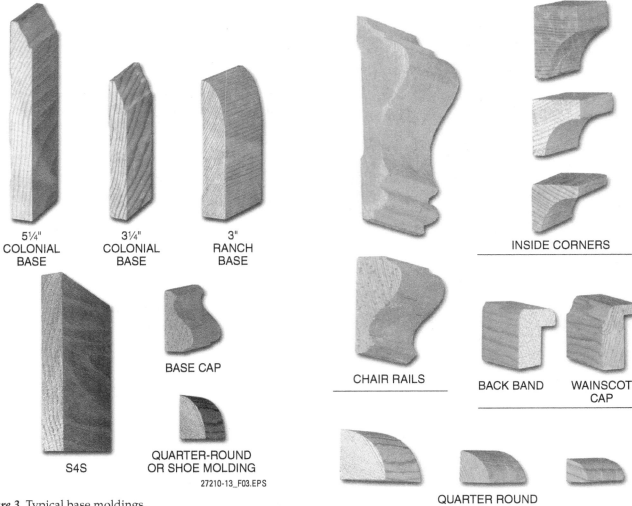

Figure 4 Typical chair-rail, inside-corner, quarter-round, outside-corner, and wainscot-cap moldings.

> **NOTE:** Wider base moldings tend to make a room look smaller, so it is best to avoid their use unless the room is spacious. However, a short base is out of proportion in a room with high ceilings, so a wider base should be used in those instances.

2.2.0 Wall Moldings

Quarter-round moldings (*Figure 4*) are sometimes used in place of base shoe moldings to hide the unevenness between the base molding and the floor. They are also used as inside-corner trim to hide the seams in corners or under wall cabinets.

Quarter-round moldings have a pie-shaped quarter-circle profile. They vary only in size, from ¼" to 1⅛". The most common sizes are ⅜", ½", and ¾".

Outside-corner moldings are used to protect outside corners, especially in high-traffic areas. They are usually plain wood corner pieces, but some highly decorative styles are made.

Wainscot caps cover exposed end grain in **wainscoting**. Wainscoting is a wall finish consisting of panels applied partway up the wall from the floor. Wainscot cap may include a lipped profile that covers the top of the wainscoting. Back-band moldings are similar to wainscot caps.

Chair-rail molding is used at chair-back height, 30" to 35" above the floor, to protect walls from chair backs. It may also be used as a cap for wainscoting or as a horizontal dividing line between two surfaces such as wallcovering and paint. Chair-rail molding comes in various styles that typically have a bulb-like feature in the top third of the molding. The back side is always flat for installation against walls.

2.3.0 Ceiling Moldings

Crown, bed, and cove moldings are usually installed at the joints formed by walls and ceilings. Cove moldings, with their concave profile (*Figure 5*), are the simplest and most common molding used at the wall/ceiling. A base cap molding is sometimes used beneath a cove molding at the ceiling to create a more intricate built-up profile.

Crown and bed moldings have decorative profiles that provide a more traditional appearance. Crown moldings come in several profiles with the top-third portion having a concave profile. The lower side of a crown molding fits against the wall. It has a wider surface than the top side, which rests against the ceiling. Because of their greater widths, these moldings can conceal large gaps at the wall and ceiling joint.

Bed moldings have basically the same profiles as crown moldings, except they are usually smaller with a top profile that is rounded.

2.4.0 Window and Door Trim

Casing moldings (*Figure 6*) come in several widths and are used to trim around windows and doors to cover gaps between the frames (jambs) and the wall. Common profiles include colonial, clam shell, and wedge. Casing profiles often match base moldings, except casing is rounded on both front edges, while the lower edge of a base mold-

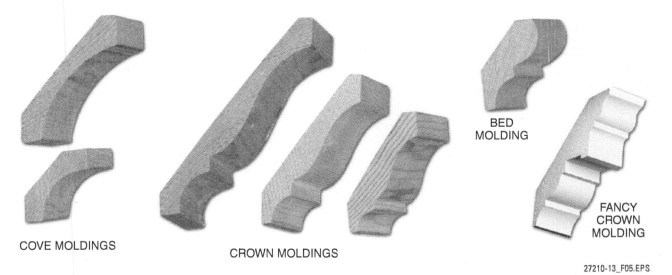

Figure 5 Cove, crown, and bed moldings.

Typical Mill Trim versus Custom Trim

Typical lumberyard or warehouse wood trim must be carefully selected. Much of it is mass produced by specialty mills that churn out lineal miles of relatively mediocre-grade material. Some inexpensive trim is sanded with shaped abrasive wheels that produce a rounded-edge profile that lacks crispness, as shown in the photo. Most lumberyard or warehouse trim is softwood trim used only for paint finish and is called paint-grade trim. For natural or stained wood finish, custom millwork shops or yards that carry custom windows and doors should carry the better-quality hardwood trim known as stain-grade trim.

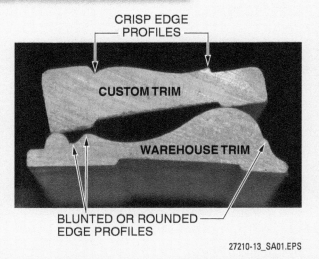

ing is square to keep dust out of the joint between the molding and the floor.

Casing stop, commonly called a door stop, is added to door frames to stop the door from overswinging and tearing out of its hinges. In window frames, casing stops hold a sliding or hung sash in place. Clam shell, colonial, and quarter round are common casing-stop profiles with one side square.

2.4.1 Stools

Stools are a type of window trim. The bottom side of a standard stool (*Figure 7*) is rabbeted at an angle, typically 10 or 14 degrees, to fit on the angled sill of the window frame so that its top side will be level.

Figure 7 Rabbeted stool.

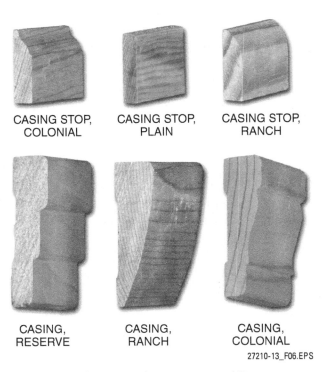

Figure 6 Typical casing and casing-stop moldings.

Polymer Trim

Ornate prefinished polymer moldings made of polystyrene or other synthetic materials are also available. These cut and nail as easily as pine. Many are glued in place. They will not rot, swell, shrink, or splinter and do not require sanding, priming, or painting.

When cutting polymer trim, place tape over the cut area and cut from the finished side to avoid splintering.

Trim Relief

The back side of most baseboard, casing, and chair-rail trim molding is relieved to allow the trim to carry over any minor drywall/plaster irregularities. This allows the edges of the trim to lie flat against most surfaces. If major irregularities exist, they must be sanded down to allow the trim to lie as flush as possible against the surfaces. For paint-grade trim, any minor gaps can be filled with a flexible painting caulk.

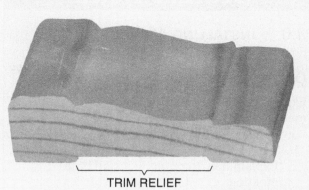

TRIM RELIEF

2.0.0 Section Review

1. Base moldings generally have _____.
   a. square edges on three sides
   b. pre-drilled nail holes
   c. a permanent protective finish
   d. an unmolded lower edge

2. Wainscot caps are a type of _____.
   a. wall molding
   b. window trim
   c. custom molding
   d. chair rail

3. The type of ceiling molding with a concave profile is _____.
   a. bed molding
   b. base cap molding
   c. cove molding
   d. crown molding

4. Wedge, clam shell and colonial are types of _____.
   a. custom trim
   b. casing moldings
   c. stools
   d. casing stops

Section Three

3.0.0 Installing Molding

Objective

Explain how to install different types of molding.
 a. Explain how to properly cut trim.
 b. Explain how to properly fasten trim.
 c. Explain how to properly install base molding.
 d. Explain how to properly install ceiling molding.
 e. Explain how to properly install door trim.
 f. Explain how to properly install window trim.

Performance Tasks 1 through 4

Make square and miter cuts to selected moldings using a hand miter box.

Make square and miter cuts to selected moldings using a power miter/compound miter saw.

Make a coped joint using a coping saw.

Install interior trim using a finish nailer and hand nailing methods.

- Door trim
- Window trim
- Base trim
- Ceiling trim

Trade Terms

Compound cut: A combined bevel and miter cut.

Coped joint: A joint made by cutting the end of a piece of molding to the shape it will fit against.

Reveal: The distance that the edge of a casing is set back from the edge of a jamb.

Scarf joints: End joints made by overlapping two pieces of molding with 22.5- or 45-degree angle cuts.

Square cuts: Cuts made at a right (90-degree) angle.

Since trim is visible to people using the house or building, it must be treated carefully during preparation and installation. Accurate cuts must be made at the intersection or termination of trim. Miter joints must be made precisely so the adjoining pieces perfectly match. Trim must be kept clean during installation. Some carpenters may wear a separate pair of gloves when handling and installing trim. This section provides some general installation guidelines for cutting and installing trim moldings.

3.1.0 Cutting Trim

Trim carpentry involves making different types of cuts to moldings using special saws and a miter saw or a compound miter saw. While most cuts are made using power saws, handsaws may also be used to cut trim. The most common saws used for cutting moldings are the backsaw, dovetail saw, and coping saw (*Figure 8*). These saws have very fine-toothed blades to produce a fine and clean cut. A backsaw is used with a miter box to make very fine right-angle cuts called square cuts and angular cuts called miter cuts. A dovetail saw is similar to a backsaw, but is smaller with a straight handle. The blade is typically 10" long with 16 to 20 teeth per inch, making it ideal for cutting very fine joints.

Coping saws are used for cutting the curves and irregular lines necessary when cutting along the profile of a trim molding in order to make a

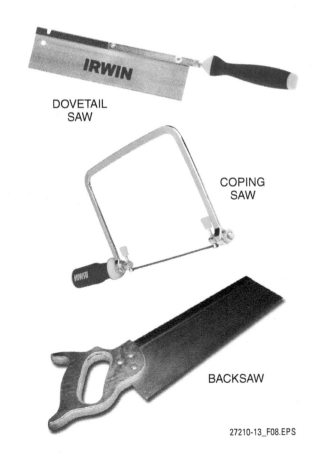

Figure 8 Special trim saws.

coped joint. A coped joint is made by cutting the end of one piece of molding to the shape of the mating piece of molding. A coping saw has a steel frame and a narrow, flexible 6"-long blade that can be rotated at any angle to cut small curves. The blade, which has 12 to 18 teeth per inch, is pulled taut by turning the handle of the frame.

Power miter saws and compound miter saws (*Figure 9*) are used for cutting straight and angled cuts on trim. The saw blade pivots horizontally from the rear of the table and locks into position to cut angles from 0 to 45 degrees right and left. Stops are set for common angles. The difference between the power miter saw and a compound miter saw is that the blade on the compound miter saw can be tilted vertically, allowing the saw to be used to make a compound cut (combined bevel and miter cut).

3.1.1 Making Square and Miter Cuts

Square cuts are made to square uneven ends of moldings and to make butt joints with walls, door casings, and similar places (*Figure 10*).

Miter cuts are made whenever an angular cut is needed. In basic trim work, 45-degree miter cuts are commonly made at the tops of doors and windows at the joints where the side and head casing meet, at outside corners such as those required when installing baseboards, and when scarf joints are needed. A scarf joint is a joint made by overlapping two pieces of molding, with one piece having an open miter cut and the other piece having a closed miter cut of 45 degrees. A scarf joint is used whenever it is necessary to join two sections of molding on a wall or ceiling in order to achieve the required length.

Cutting molding using a miter saw begins with the proper placement of the molding in the miter box. All moldings should be cut with their face side or edges up or toward you so that the saw splinters out the back side, not the face side.

Flat miters, such as those made in casing for doors and windows, are cut by holding the molding with its face side up and its thicker edge against the fence (back side) of the miter saw (*Figure 11*). Some moldings, such as the base, base cap and shoe, or chair rail, are held right side up. The bottom edge should be against the table (bottom) of the miter saw and the back against the fence. Once the molding is properly positioned, the required square cut or miter cut can be made with the saw.

When working with crown, bed, and cove moldings, some carpenters position the molding upside down in the miter saw and at the same 45-degree angle at which it will be installed (*Figure 12*). The lower side of the molding fits against the wall. It has a wider surface than the top side, which rests against the ceiling. When placing the

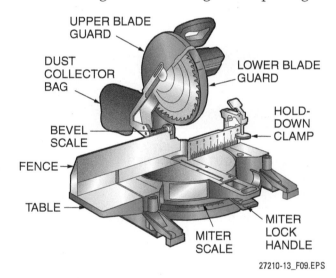

Figure 9 Compound miter saw.

Saber Saws

A saber saw, also known as a bayonet saw or portable jigsaw, can be substituted for a coping saw. Special, narrow scrolling blades that are taper ground and have fine, straight teeth that allow tight turns and produce smooth finishes are required. This saw is equipped with the Collins coping foot, which replaces the normal flat baseplate with a curved baseplate. With this coping foot, the saw can then cut through the material at any angle.

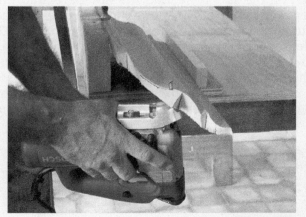

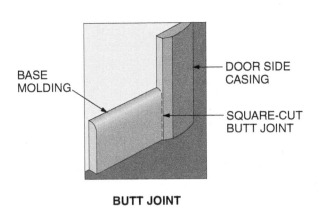
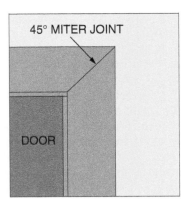
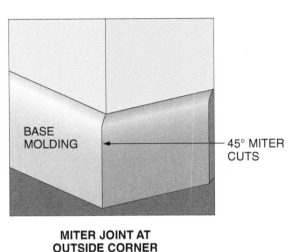
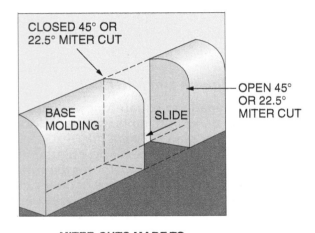

Figure 10 Typical butt and miter joints.

molding in the miter saw, position it as if the fence were the wall and as if the table were the ceiling. To aid in keeping the molding positioned correctly in the miter saw while cutting, carpenters use a jig to support the back side of the molding.

If using a compound miter saw to cut crown molding, the molding's broad back surface is laid flat on the saw table, and the saw-bevel and miter-angle controls are set to achieve the desired cut. All standard US crown moldings have a 52-degree top rear angle that fits next to the ceiling and a 38-degree bottom rear angle that fits against the wall. Most miter saw manufacturers provide a chart in the operator's manual that shows the bevel and miter settings to use with a compound miter saw for cutting different inside and outside corner cuts on standard US moldings.

A typical chart of this type is shown in *Table 1*. Before making compound miter cuts to crown molding, always make test cuts on scrap material to make sure that the bevel and miter settings being used will produce the desired results.

3.1.2 Making a Coped Joint

When installing baseboard and ceiling moldings, the use of a coped joint on inside corners is recommended rather than a miter joint because mitered joints tend to open at the inside corners when being nailed. Also, eventual shrinkage of the wood may cause the joint to open. Another reason for using a coped joint is that it makes it easier to adjust the angle if the corner is not square.

A coped joint is made by joining a piece of molding with a square cut that is butted against a wall, as shown in *Figure 13A*, with a second piece of molding cut to match the profile of the first piece, as shown in *Figure 13B*. To cut the sec-

> **NOTE**
> A stop block can keep the molding from slipping when the workpiece is at an angle on the saw blade.

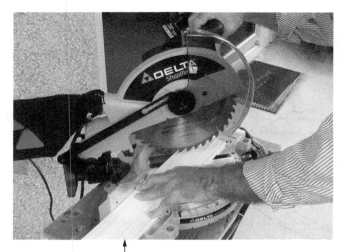

BACK SIDE DOWN

BACK SIDE AGAINST FENCE

CROWN MOLDING   STOP BLOCK

Figure 12 Position of crown molding.

Figure 11 Positioning molding in a miter box for cutting.

ond piece to the profile of the first piece, first cut a 45-degree open miter at the mating end of the piece. Then cut the piece at 90 degrees or less with a coping saw, following the edge profile of the first cut, as shown in *Figure 13C*. If the corner is greater than 90 degrees, back-cut the cope. After the coped cut is made, use a rat-tail file to finish contouring the edge.

3.2.0 Fastening Trim

Traditionally, interior trim was typically fastened in place using a lightweight (10 oz.) claw hammer and a nail set to drive finishing nails. Today, trim is commonly installed using pneumatic finish nailers (*Figure 14*). Finish nailers and pneumatically driven finish nails are available in three diameters, 15 gauge, 16 gauge, and 18 gauge, and the nail lengths range from ¾" to 2¾". The 15-gauge nails are larger in diameter than 18-gauge nails. Typically, 15-gauge nails are used for fastening thicker trim materials, and 16-gauge or 18-gauge nails are used for thinner materials. The thinner the nail, the less likely it is to split the wood. Therefore, a 16-gauge or 18-gauge nailer is commonly used for installing door and window casing, window aprons, and similar trim. Stools, hardwood trim, and similar heavier materials would be installed using a 15-gauge nailer.

Table 1 Compound Miter Saw Miter and Bevel Settings for US Standard Crown Molding

Type of Cut	Bevel Setting	Miter Setting	Remarks
Left side – inside corner	33.8°	Right, 31.6°	Position top of molding against miter box fence. Left side of molding is finished piece.
Right side – inside corner	33.8°	Left, 31.6°	Position bottom of molding against miter box fence. Left side of molding is finished piece.
Left side – outside corner	33.8°	Left, 31.6°	Position bottom of molding against miter box fence. Right side of molding is finished piece.
Right side – outside corner	33.8°	Right, 31.6°	Position top of molding against miter box fence. Right side of molding is finished piece.

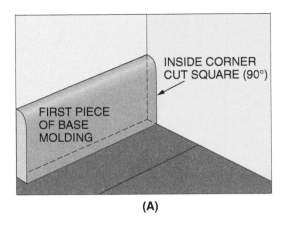

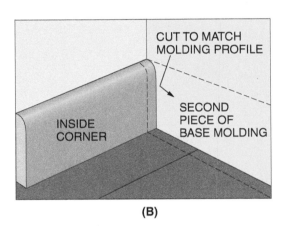

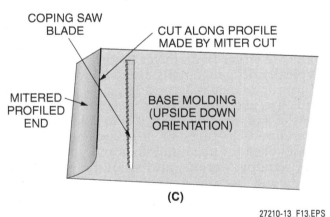

Figure 13 Making a coped joint.

Figure 14 Typical pneumatic finish nailers.

The depth to which the nail is driven can be set using a flush stop on some nailers or by adjusting the air pressure. If the nails are set flush by the nailer, then the final set can be accomplished by hand with a nail set.

3.3.0 Installing Base Molding

Base molding runs continuously around the room. It is one of the last items of interior trim to be installed since it must be installed around door casings, supply and return grilles, cabinetwork, and so forth. Baseboard joints made at inside corners should be coped instead of mitered, because mitered inside corners tend to open when being nailed. Joints at outside corners are mitered.

Before installing moldings, locate the studs within the room, and mark their locations on the wall just above the baseboard height or on the subfloor if the walls are finished. Then select and place lengths of base molding around the sides of the room. Sort the pieces so that there will be the least amount of cutting and waste. Try to select lengths that will allow you to make complete runs without joints; if you cannot do so, add 2 feet to any length that will be joined so that you can cut the joint over a stud.

Most carpentry procedures can be accomplished in more than one way. One typical procedure for installing base and related shoe molding is described here. If shoe molding is being installed, it is installed in the same basic manner as the base trim, except that it is nailed into the floor instead of the baseboard. This prevents the joint under the shoe molding from opening should shrinkage occur in the base molding. Also, since shoe molding is a small molding and has a solid backing, both inside and outside corners are usually mitered. Where it meets a door casing with nothing to butt against, its end is normally back-mitered and sanded smooth.

Typically, the installation of the base trim starts on a long wall of the room and/or on the side of the room opposite the door. Starting on a long wall makes it easier to get a good fit with a long piece of base trim. Starting on a wall opposite the door helps minimize any possibility that people will see a poorly fitted joint (should one happen for whatever reason) when entering the room.

Figure 15 shows an example of a simplified baseboard installation, with the numbers 1 through 7

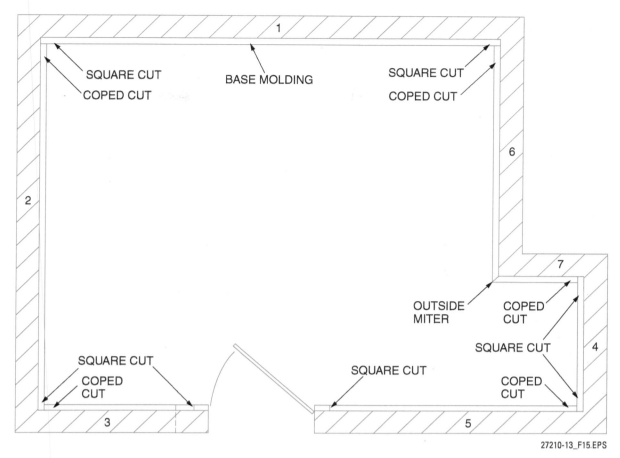

Figure 15 Simplified baseboard molding installation.

representing a typical sequence of molding installation. As shown, installation of the baseboard is started on the unbroken rear wall facing the door. This is because it can be fitted with a single length of baseboard. Begin by measuring the wall, making sure to take your measurements at floor level. Following this, pick a straight length of baseboard molding and make a square cut on one end. Measure the required length, then make a square cut on the other end. Make the length slightly longer than required so that the cut will allow a spring fit. Test the fit on the wall and trim, if necessary, to allow a spring fit. Nail the piece in place at the stud locations along the wall. If carpeting is to be used for the finished floor, fasten the base trim above the subfloor to allow for the pad and carpet, unless the molding is to be installed after the carpet.

Next, install the second piece of baseboard on the adjoining (left-hand) wall. This piece of baseboard is joined to the first with a coped cut at the inside corner. If necessary, file the contour to get a fit. Once the coped cut is made, measure from the bottom of the first piece of base to the next corner, then square-cut the second piece at the other end so that it butts tightly into the intersect-

Guidelines for Finishing Trim

- If the trim is to be stained, make sure all traces of glue are removed from exposed surfaces, using a damp cloth. If allowed to dry, glue will seal the surface and not allow the stain to penetrate, resulting in a blotchy finish.
- All pieces of interior trim should be sanded smooth after they are cut and fitted, and before they are nailed in place. Always sand with the grain, never against the grain. Sanding of interior trim provides a smooth base for the application of stains, paints, and clear coatings.
- All sharp, exposed corners of trim should be rounded slightly. Use a block plane to make a slight chamfer, then round it over with sandpaper.

ing wall surface. Nail the second piece of trim in place. Keep working your way around the room in this manner until all the baseboard molding is cut and nailed in place.

As shown in *Figure 15*, all the inside corners have coped joints and the outside corner has a mitered joint. Where the baseboard molding meets the door casing, square-cut butt joints are used. Be careful when fitting baseboard to the door casing. Do not make the joint so tight that you push the casing out of position when you spring the baseboard into place.

When making the outside miter joint for our example, make the coped cut for the inside corner on one piece first, then hold the baseboard in position and mark the back edge for the miter cut, as shown in *Figure 16A*. Make a 45-degree miter cut and fasten the piece to the wall, as shown in *Figure 16B*. Repeat the procedure for the second piece, as shown in *Figure 16C* and *Figure 16D*.

In the event that the job requires two lengths of baseboard be spliced to span a long wall, make the splice using a scarf joint. Set the first piece of molding in place and mark the center point of the stud nearest to the end of the piece. Subtract half the thickness of the molding, then cut a closed 45-degree miter at its end. Install the first piece, but do not nail into the last stud where the piece is mitered. Cut a 45-degree open miter at the end of the second piece. Measure from the face of the miter on the first piece to the corner of the wall, then cut the second piece to this length. Set it in place with the closed miter on the first piece overlapping the open miter on the second piece (*Figure 17*). Apply glue to the joint and nail through both pieces into the stud, then continue nailing the second piece to the corner. It should be pointed out that many car-

> **NOTE**
> Few outside corners are exactly 90 degrees, so it is best to use a sliding T-bevel to find the exact angle made by the outside corner, then divide this angle by two and cut a closed miter at each piece of joining base to that angle.

penters use a biscuit joint cut with a biscuit (plate) joiner to splice long or wide moldings.

3.4.0 Installing Ceiling Molding

The general approach, sequence, and procedures for installing ceiling trim are basically the same as that for installing baseboard trim. However, when using crown/bed moldings, the moldings are not applied flat against the wall. Instead, they cover the wall/ceiling joint at a 45-degree angle.

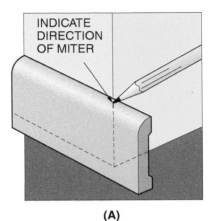
(A)

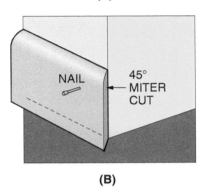
(B)

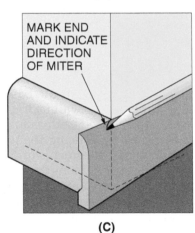
(C)

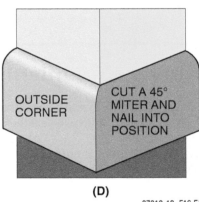
(D)

27210-13_F16.EPS

Figure 16 Procedure for fitting and cutting an outside miter joint.

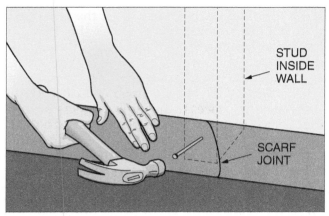

Figure 17 Scarf joint, cut and nailed directly over a wall stud.

If the ceiling has only inside corners, install each piece of trim with a square cut at one end and a coped cut of matching profile at the other end, if possible. As with base trim in rooms where a long wall is opposite the door, run square-cut crown molding from corner to corner and cope the molding of the adjoining walls. This will help make the coped joints inconspicuous if they open up later due to wood shrinkage. Unfortunately, this may result in a side that requires close fitting and coping of both ends of the molding. Like base trim, always cut crown molding slightly longer than required and spring it into place when installing and nailing. If outside corners are required, compound miter cuts are involved. General procedures for placement and cutting of crown, bed, and cove moldings using a miter saw were covered in the section *Making Square and Miter Cuts*.

When installing crown or cove molding, you must find the distance from the top edge (ceiling) to the bottom edge (wall) where the molding should be installed. One way to do this is to hold a scrap piece of molding against the ceiling and wall to determine the distance from the ceiling to its bottom edge. Once this distance is known, all walls can be marked at both ends, and lines snapped between the marks that represent the installation point for the bottom edge of the molding around the room. Another way to find this distance is to measure the molding using a framing square. With the tongue and blade of the framing square representing the wall and ceiling, respectively, align a piece of crown molding in the corner of the square and measure the tongue distance.

3.5.0 Installing Door Trim

Interior door trim should always be applied after

> The procedures in this section assume that the doors are installed and are plumb and square with the wall.

the finished floor is laid, but before the base trim is installed. Also, the casing material used to trim a door should be the same as that used to trim the windows in the same room.

Step 1 On the inside face of the side and top door jambs, use a combination square or measure and mark the distance from the jamb edge to the desired reveal setback. This is typically ¼".

Step 2 Hold a section of case molding in place at the top of the door, then mark the length of the head casing. This is the distance between the reveal marks on the door side jambs.

Step 3 Using the points marked on the head casing as a guide for where the heels of the miter cuts begin, cut left-hand and right-hand miters on the ends of the head casing.

Guidelines for Fastening Trim

- Make sure any pencil marks left along the edge of a cutline are removed before fastening the trim. Pencil marks in interior trim corners are difficult to remove after the pieces are fastened in position. Pencil marks show through a stained or clear finish and make the joint appear open. When marking interior trim, make light, fine pencil marks.
- Make sure all joints are tight fitting. Measure, mark, and cut carefully. Do not leave a poor fit. Do it over, if necessary.
- Be careful not to leave finish nailer or hammer marks when fastening the trim material.
- Set finish nails below the surface of the trim, and fill the nail holes with putty that matches the paint or stain finish. Finish nails should be set to a depth that is at least the width of the head. In the case of prefinished molding, nails of matching color are used, and they are not set below the surface.

Step 4 Set the head casing back the distance of the reveal from the edge of the door jamb and nail it in place using 4d finish nails at the jamb edge and 8d finish nails at the outer edge of the casing. Unless a finish nailer is used, drill holes near miters to prevent splitting the casing.

Step 5 Cut two pieces of case molding slightly longer than needed for the door's two side-case moldings. Cut a left-hand miter on one side-case molding and a right-hand miter on the other side-case molding.

Step 6 Turn one of the side-case moldings upside down so that its mitered end rests on the floor, then mark the other end at the top edge (miter toe) of the head casing. Cut it off square, allowing the pencil mark to remain. If the finished floor is to be carpeted, locate the casing above the subfloor to allow for the carpet and pad.

Step 7 Apply wood glue on the miters, set the side casing back the distance of the reveal from the edge of the jamb, and nail it in place.

Step 8 Repeat Steps 6 and 7 for the remaining side casing.

Step 9 If it has not already been done, set all finish nails. If necessary, sandpaper the high edge at mitered corners, making sure not to sand across the grain.

3.6.0 Installing Window Trim

When trimming out a room, start with the window and door casing. There are two basic methods used to trim windows: conventional and picture frame (*Figure 18*). As shown, the conventional method uses casing at the top and both sides. The bottom of the side casing rests on the

> **NOTE:** The procedures in this section assume that the windows are installed and are plumb and square with the wall.

stool. Installed below the stool on the face of the wall is a horizontal member called the apron. In the picture-frame method, the top, bottom, and sides of the window are all trimmed with casing.

The first thing to check before trimming a window is that the edge of the jamb is flush with the inside wall surface. If the jamb projects beyond the surface, you will have to plane or sand it flush. If the jamb is recessed, wood strips called jamb extensions must be nailed over the jamb edge to bring it flush with the inside wall surface (*Figure 19*). If the wall projects slightly beyond the jamb, the inside wall may require sanding if it does not allow the casing to lie flush against the jamb and wall surface.

Jamb extensions may be supplied by the manufacturer with the window unit, they can be ordered from the mill, or they can be made on site. Jamb extensions are not normally installed when the window is installed, but later when the window is trimmed. If supplied with the window, the extensions should be stored in a safe place until it is time to install them. Manufactured jamb extensions are usually precut to length and need only to be cut to the correct width. This is done by ripping the extension to the required width with a slight back-bevel on the inside edge.

The casing material used to trim a window should have the same profile as the door trim. As with most carpentry procedures, installing window and other types of interior trim can be accomplished in more than one way. The conventional method of trimming a window follows. Trimming a window using the picture-frame method is done in basically the same way as the conventional method, except instead of using a stool and apron, a piece of casing is installed across the bottom of the window.

To cut and install the stool, proceed as follows:

Step 1 Determine the length of the stool. Typically, the length of the stool is equal to the distance between the inside edges of the window jambs plus the amount of reveal on both sides (typically ¼" on each side), plus twice the casing width, plus twice the casing thickness (*Figure 20*). The reveal is the distance that the edge of a casing is set back from the edge of a jamb. For exam-

---

Adjusting a Mitered Angle

Once a correct miter angle is established on a miter saw or compound miter saw for a job, avoid changing the angle setting, if possible. If the angle of one piece of stock must be adjusted due to an out-of-square casing or wall, it is easier to incrementally block the stock away from the backstop or surface of the miter box or saw, using thin scraps of wood on a trial-and-error basis, rather than change the angle setting.

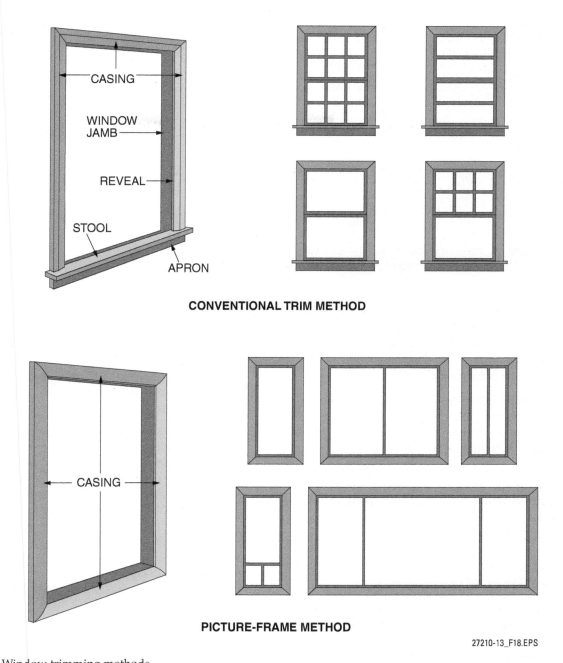

Figure 18 Window trimming methods.

Sagging or Uneven Ceilings

Most ceilings are not completely even. When crown moldings or cornices are installed in a room, gaps may show up along their length. In the case of sagging ceilings, the gaps occur at both ends of the ceiling treatment. If the gaps are ⅛" or less and the ceiling treatment is paint grade, the gaps may be caulked after installation of the treatment. If the sag or a long gap is less than ¼" and the ceiling trim is a molding, the molding may simply be bent around the sag or force-nailed tight to the gap. Another solution is to scribe the gap or sag onto the top edge of the molding with a pair of dividers and trim the molding to fit the ceiling. The simplest and best solution for fitting molding or cornices is to install the ceiling treatment straight and level and then float the gaps out using drywall compound.

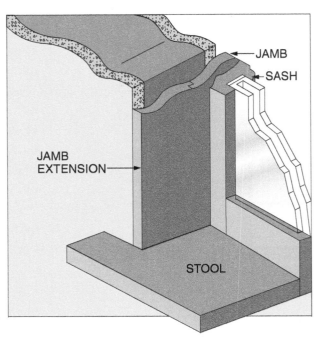

Figure 19 Installation of jamb extensions.

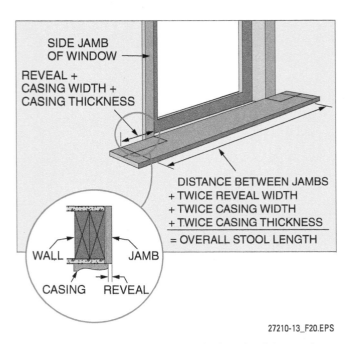

Figure 20 Method of determining the length of the stool.

ple, if the jamb-to-jamb measurement of the window is 36" and you are using 2¼" × ¾" casing, the length of the stool would be 42½" (36" + ½" + 4½" + 1½").

Step 2 If it is desired that the wood grain at the ends of the stool be covered with returns to provide a finished appearance, lay out and mark the length of the stool, but do not cut the stool. If return ends are not being used, the stool can be cut to length at this time.

Step 3 Center the stool on the window, hold it level with the sill, and mark the location of the inside edges of the side jambs on the top of the stool, using a square and a pencil. Also, mark a line on the face where the stool will fit against the wall surface.

Irregular or Wavy Baseboard Mounting Surfaces

If baseboard trim will be installed against a final finished floor without shoe molding, it is important that any noticeable gap between the bottom of the base trim and the floor is eliminated. Place the final cut length of baseboard for each side of the room on the floor against the wall, and check that it is flush with the floor over its length. If the floor is warped or sags and noticeable gaps exist along the bottom of any portion of a baseboard for each wall, use a pair of dividers set to the widest gap width for any baseboard, and scribe a mark along the bottom edge over the length of each wall baseboard. Trim the bottom of the base molding to the line with a jigsaw or band saw, or sand to the line with a belt sander. As pointed out earlier, in the section *Window and Door Trim*, if the walls behind baseboards are uneven and the base molding is paint grade, high spots on the wall can be sanded down, and gaps may be caulked after baseboard installation if they are not more than ⅛". If the gaps are larger or the baseboard is stain grade, use a two-piece baseboard with a small cap and force-nail the cap into the gaps; otherwise float the wall using drywall compound to level the gaps, and then nail the cap and/or the base molding.

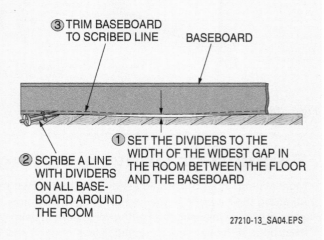

For a double-hung window, this line usually falls directly above the square edge of the stool rabbet.

Step 4 At both ends of the stool, carefully cut out that portion of the stool that will butt flush against the wall (*Figure 21*). Then open the lower sash and slide the stool in position.

Step 5 Lower the sash carefully on top of the stool, and draw or scribe the cutoff line so that the sash will clear the stool. Cut off the excess material. When finished, the stool should fit flush against the wall when the back edge of the stool has 1/16" clearance between it and the bottom rail of the window sash. This clearance is necessary to prevent binding of the sash when raising or lowering it.

Step 6 If using returns on the ends of the stool to provide a finished appearance, mark and cut a return at each end of the stool. The returns are formed by making 45-degree miter cuts at each end of the stool, as shown in *Figure 22*. If returns are being used, nail and glue the return pieces to the ends of the stool.

Step 7 Sand all sawed edges, then nail the stool in place, making sure that it is level and perpendicular. Note that some specifications require that the underside of the stool be embedded in caulking compound before being nailed in place.

To cut and install the side and top (head) casing, proceed as follows:

Step 1 On the inside face of the side and top jambs, use a marking gauge or measure and mark the distance from the jamb edge to the desired reveal setback. For our example, a ¼" setback for the reveal was used.

Step 2 Hold a section of case molding in place at the top of the window, then mark the length of the head casing. This is the distance between the reveal marks on the window side jambs (*Figure 23*).

Step 3 Using the points marked on the head casing as a guide for where the heel of your miter cuts begin, cut left-hand and right-hand miters on the ends of the head casing.

Step 4 Set the head casing back the distance of the reveal from the edge of the jamb and nail it in place. Near miters, drill holes to prevent splitting the casing unless a nail gun is used.

Step 5 Cut two pieces of case molding slightly longer than needed for the two side-case moldings. Cut a left-hand miter on one side-case molding and a right-hand miter on the other side-case molding.

Step 6 Turn one of the side-case moldings upside down so that its mitered end rests on top of the stool, then mark the other end at the top edge (miter cut toe) of the head casing. Cut it off square, allowing the pencil mark to remain.

Step 7 Apply wood glue on the miters, set the side casing back the distance of the reveal from the edge of the jamb, and nail it in place.

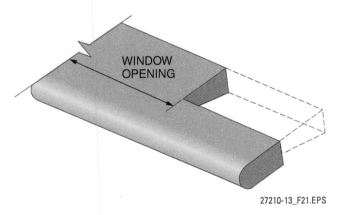

Figure 21 Window stool showing cutout for one side of the window opening.

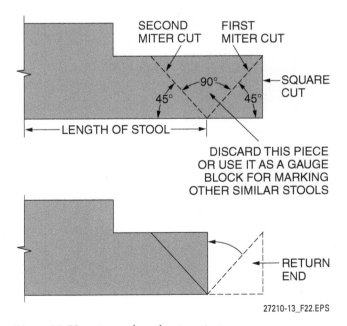

Figure 22 How to mark and cut a return.

Backing Boards

If wide wood crown molding (4" or more) is to be installed, wide backing boards (nailers) should have been installed above the edges of ceilings that are parallel with the ceiling joists (A). These boards must be fastened to the top of the top plate in order to supply a nailing support for the upper crown molding edge. If wide backing boards haven't been installed, triangular wood-support blocks can be installed behind the crown molding and secured to the wall studs (B). In this case, the crown molding will have to be nailed at the bottom edge and about two-thirds of the way up from the bottom. If a narrow nailer exists, a solution is to fasten two pieces of baseboard trim at right angles and secure the assembly to the wall studs and narrow nailer (C). The crown molding is then installed and nailed to the base molding pieces. Other solutions involve formal cornice construction using both crown and bed molding (D).

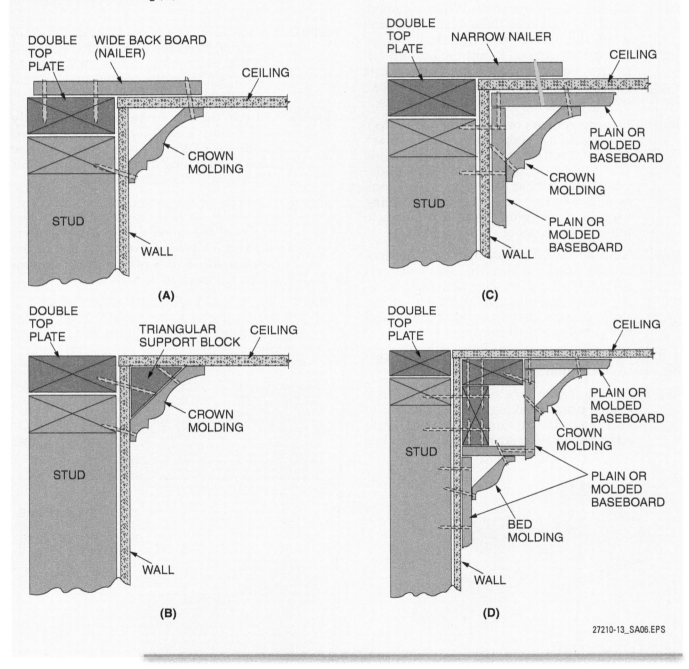

Step 8 Repeat Steps 6 and 7 for the remaining side casing.

To cut and install the apron under the stool and perform other finishing tasks, proceed as follows:

Step 1 Cut the apron to a length that brings its ends into vertical alignment with the outside edges of the side casing. When the apron is molded and returns at the ends are desired, use a scrap piece of molding as a template (*Figure 24*). Draw its profile flush with the ends of the apron, then cut out the profile with a coping saw and sand the ends. Glue and/or nail the returns in place at both ends. An alternative method is to cut the returns in the same way as for making the returns for a stool, as described in the previous procedure.

Step 2 Nail the apron in place under the stool with 8d finish nails. Make sure not to force the stool upward.

Step 3 If it has not already been done, set all finish nails in the apron and all casing. If necessary, sandpaper the high edge at mitered corners, making sure not to sand across the grain.

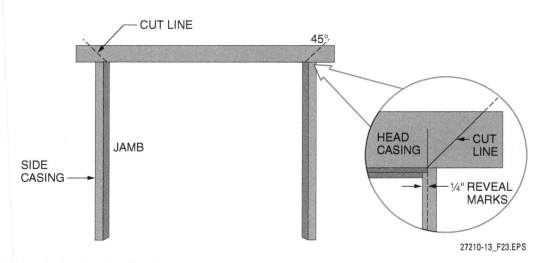

Figure 23 Marking the head casing for miter cuts.

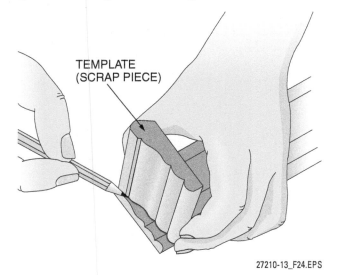

Figure 24 Mark apron return profile with a scrap piece of molding.

Nail Sizes

Nails should be sized so that two-thirds of the length of the nail penetrates the supporting material. For example, if you are attaching a ½" molding to ½" drywall that is attached to the framing, the nails need to be at least 3" long.

Matching a Base Molding and Door Casing

The door casing must be thicker than the base molding so that the base molding can butt into the casing without having the end grain of the base molding visible. The casing is usually 1/16" to 1/8" thicker than the base molding. In those instances where the casing is not thicker than the base molding, a plinth block can be used as shown in the figure.

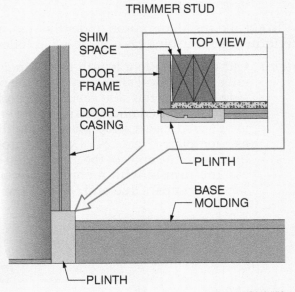

Marking Gauge

A commercial marking gauge can be set and used to scribe a mark for the setback on the edges of window or door jambs for the reveal. If a commercial gauge is not available, a job-site gauge can be made by rabbeting the four edges of a square block of wood for the desired amount(s) of reveal. The block is placed inside the jamb with the correct reveal width overlapping the edge of the jamb. Use a pencil to mark the reveal while sliding the block around the side jambs and head jamb.

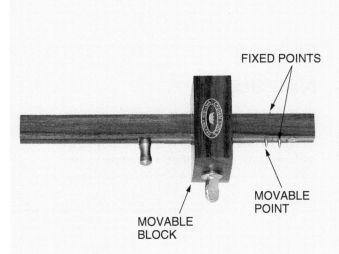
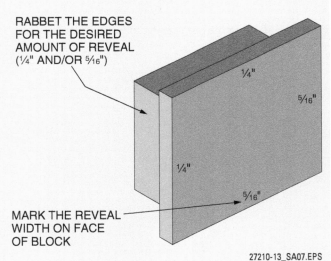

3.0.0 Section Review

1. On the inside corners of baseboard and ceiling moldings, it is best to use _____.
   a. coped joints
   b. scarf joints
   c. butt joints
   d. miter joints

2. Pneumatically-driven finish nails come in three diameters, with 15-gauge being the largest in diameter.
   a. True
   b. False

3. When installing base molding, miter joints are made _____.
   a. at the inside corners
   b. at the outside corners
   c. where it meets the door casing
   d. when splicing two lengths of molding

4. Crown molding should be cut slightly longer than needed to allow a _____.
   a. spring fit
   b. snugged fit
   c. custom fit
   d. trimmed fit

5. Interior door trim is applied _____.
   a. after the base trim is installed
   b. to make the door functional
   c. to match the window trim
   d. before the finish floor is laid

6. If a window jamb is recessed, it is necessary to _____.
   a. trim using the picture-frame method
   b. use jamb extensions
   c. adjust the stool
   d. omit the apron

SECTION FOUR

4.0.0 ESTIMATING TRIM QUANTITIES

Objective

Explain how to estimate window, door, floor, and ceiling trim.

Performance Task 5

Estimate the quantities of different trim materials required for selected rooms.

To do a material takeoff of the trim materials required for a job, begin by checking the construction drawings and specifications for the specific types of trim material to be used for each application, including doors, windows, baseboards, and ceilings. To find the total trim material needed for a typical job, it is necessary to determine how many linear feet of each type of trim will be required.

Trim is generally divided into two categories:

- *Running trim* – This type of trim is installed in random lengths for such applications as baseboards, crown molding, and chair rails.
- *Standing trim* – This trim is specified in particular lengths for purposes such as door and window casings or other special applications where splicing is not acceptable.

Estimating running trim is relatively easy:

- *Base trim* – This trim is the sum of the perimeter measurements of all rooms, minus the width of the doorways, plus 10 percent for waste. Order the shortest standard mill lengths that will minimize splicing on all but the longest walls. If a very long wall is encountered, order the maximum mill lengths for the wall to minimize splicing, but be aware that they will cost more per foot than standard lengths.
- *Chair rail* – This trim is the sum of the perimeter measurements of all rooms using the chair rail, minus the width of the doorways and any windows that will interrupt the rail, plus 10 percent for waste. Order the shortest standard mill lengths that will minimize splicing on all but the longest walls.
- *Crown molding and/or other cornice materials* – This trim is the sum of perimeter measurements of all rooms, plus 10 percent waste. Order the shortest standard mill lengths that will minimize splicing on all but the longest walls. If a very long wall is encountered, order the maximum mill lengths for the wall to minimize splicing, but be aware that they will cost more per foot than standard lengths.

Estimating standing trim is more difficult:

- *Door casing* – A typical 6'-8" door that is 2'-10" wide requires four 7' pieces of side casing and two 42" head casings, all of which include an extra 4" for each of the ends that require mitering. Therefore, five 7' pieces of casing are required for each door. Since standard lengths are usually 6', 8', 10', 12', 14', and 16', the length to use that yields the most with a minimum of scrap is the 14' length. Three 14' lengths cut in half will case one door with a 7' length left over for the next door. For every two doors then, only five 14' lengths are required. If all the doors are the same, divide the number of doors by two and multiply the result by five to obtain the total number of 14' lengths to order. For wider or narrower doors, a similar analysis must be performed to obtain the optimum ordering lengths. For waste purposes, add one extra of the widest door to the count. Store any leftover casing in the attic for future client use.
- *Window casing* – Windows are analyzed the same as doors except that only one side receives interior trim casing. If the bulk of the windows are the same size, measure the side jambs and the header jamb, and add 4" for each end that is mitered. If the windows are picture framed, add another casing length equivalent to the header jamb plus 8". Determine which standard lengths yield the least amount of scrap when cut to fit one or two of the windows. Then, order the appropriate number of standard lengths to case the windows. Odd-size windows will require a separate analysis.
- *Stool and apron* – The length of material required for the stool and apron is the width of each window plus the width of two side casings, plus 4" for ears and returns. Adjacent double-hung windows (*Figure 25*) are treated as one window for total stool and apron lengths. Determine which standard lengths yield the least amount of scrap for two or three windows and order the required quantity for all the remaining windows.

- *Mullion* – For adjacent double-hung windows, the length of mullion for the center of the windows must be measured and multiplied by the number of sets of adjacent windows. Determine which standard lengths yield the least amount of scrap and order the required quantity.

When ordering the calculated amount of trim, make sure that you specify whether paint-grade or stain-grade trim is required. Remember, paint-grade trim may be finger jointed from smaller pieces and will be unsuitable for a stained or natural finish.

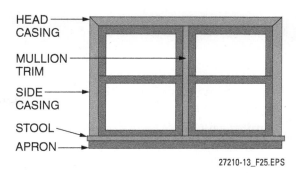

Figure 25 Mullion trim used with adjacent double-hung windows.

Mill-Calculated and Prepackaged Trim

If standard sizes of doors are used, single-sided sets of precut trim are available, and it is only necessary to order the desired number of sets. If the construction drawings are provided, some millwork shops will perform a material takeoff and supply precut trim kits for each door and window, regardless of size.

4.0.0 Section Review

1. You should allow 10 percent for waste when estimating material needs for the following, *except* _____.
   a. base trim
   b. cove molding
   c. door casing
   d. chair rail

SUMMARY

Interior trim involves the installation of molding around windows, doors, walls, floors, ceilings, and other inside surfaces. Moldings are strips of material shaped in numerous patterns for use in specific locations. Wood is used for most moldings, but some molding is made of polystyrene or other synthetic material.

When considering the construction sequence of a building, the trim installation is one of the finishing touches. Everybody is going to see the trim; therefore, it has to be installed carefully with tight-fitting joints in order to present a suitable appearance.

Review Questions

1. When cutting large quantities of trim, most carpenters use a _____.
   a. miter saw
   b. backsaw
   c. coping saw
   d. veneer saw

2. Moldings can be nicked, dented, or otherwise damaged relatively easily, especially those made from _____.
   a. poplar
   b. maple
   c. pine
   d. oak

3. Moldings made of finger-jointed wood are normally _____.
   a. stained
   b. painted
   c. clear finished
   d. stained or clear finished

4. A type of trim used at the bottom edge of a wall where the wall meets the floor is _____.
   a. base casing
   b. bed base
   c. cap shoe
   d. base shoe

5. A typical height above the floor for installation of chair-rail moldings is _____.
   a. 20" to 25"
   b. 26" to 29"
   c. 30" to 35"
   d. 36" to 40"

6. Wainscoting is sometimes capped with a type of molding called _____.
   a. quarter round
   b. chair rail
   c. cove
   d. corner guard

7. When cove molding is used at the ceiling/wall joint, a more intricate built-up profile can be achieved by adding a(n) _____.
   a. base shoe
   b. colonial base
   c. base cap
   d. S4S molding

8. The top third of a crown molding has a profile that is _____.
   a. convex
   b. corrugated
   c. complicated
   d. concave

9. Molding used to trim around windows and doors is referred to as _____.
   a. casing
   b. bed molding
   c. quarter round
   d. base molding

10. To fit on the angled sill of the window frame so that its top side will be level, the bottom of a standard stool is rabbeted to an angle of 10 degrees or _____.
    a. 12 degrees
    b. 13 degrees
    c. 14 degrees
    d. 15 degrees

11. To shape the end piece of molding so that it matches the profile of the mating molding member, you would use a _____.
    a. dovetail saw
    b. backsaw
    c. coping saw
    d. miter saw

12. A dovetail saw can be used to make very fine cuts, since it has a blade with _____.
    a. 11-14 teeth per inch
    b. 13-18 teeth per inch
    c. 15-24 teeth per inch
    d. 16-20 teeth per inch

13. To keep crown molding properly positioned when cutting it with a power miter saw, support the back of the molding with a(n) _____.
    a. armature
    b. jig
    c. clamp
    d. stop block

14. Flat miter joints are made when cutting _____.
    a. chair rail
    b. base cap
    c. crown molding
    d. door and window casing

15. The joint between two pieces of baseboard molding in inside corners is usually _____.
    a. mitered
    b. butted
    c. coped
    d. bisected

16. Before finish nailers came into wide use, trim was fastened in place using a _____.
    a. tack hammer
    b. lightweight claw hammer
    c. ball peen hammer
    d. framing hammer

17. If using both 15-gauge and 16-gauge pneumatic finish nailers to fasten trim, the 16-gauge nailer typically is used to fasten _____.
    a. thinner trim materials
    b. thicker trim materials
    c. hardwood trim
    d. window stools

18. The type of joint used where the baseboard molding meets the door casing is a _____.
    a. coped
    b. compound
    c. square-cut butt
    d. mitered

19. The space between a door or window casing and the edge of the door or window jamb is called the _____.
    a. setback
    b. backset
    c. interval
    d. reveal

20. If the jamb edges of a window are lower than the adjoining wall surface, it must be brought flush with the wall by installing _____.
    a. jamb adjusters
    b. window spacers
    c. jamb extensions
    d. casing blocks

21. The linear feet of door case molding (not including waste) needed to trim both sides of two 2'-10" × 6'-8" interior doors is approximately _____.
    a. 30'
    b. 40'
    c. 60'
    d. 70'

Trade Terms Quiz

Fill in the blank with the correct term that you learned from your study of this module.

1. A(n) _____ is made to join two pieces of molding. Pieces are overlapped and cut at a 45-degree miter.

2. A combined bevel and miter cut is known as a _____.

3. Sometimes referred to as undersill trim, a(n) _____ is a trim piece fastened beneath the stool of the window frame.

4. The general term used to describe molding placed around wall openings and at the points where walls meet floors or ceilings is _____.

5. _____ is made up of shorter lengths of molding joined at the factory. It is usually painted to hide the joints.

6. A finish method that extends partway up the wall, _____ is most commonly made from wood, but may be ceramic tile or stone.

7. A(n) _____ used on trim is cut with flexible, thin-bladed saw to match the profile of the piece it must fit against.

8. The horizontal trim piece attached at the bottom of a window is the _____.

9. _____ are shaped decorative strips, usually made from wood, that are used for finish work.

10. Saw cuts made at a 90-degree (right) angle to the length of the trim piece are called _____.

11. The _____ is the distance casing is set back from a window or door jamb's face side.

Trade Terms

Apron
Compound cut
Coped joint
Finger-jointed stock
Moldings
Reveal

Scarf joints
Square cuts
Stool
Trim
Wainscoting

Trade Terms Introduced in This Module

Apron: A piece of window trim that is used under the stool of a finished window frame. It is sometimes referred to as undersill trim.

Compound cut: A combined bevel and miter cut.

Coped joint: A joint made by cutting the end of a piece of molding to the shape it will fit against.

Finger-jointed stock: Paint-grade moldings made in a mill from shorter lengths of wood joined together.

Moldings: Decorative strips of wood or other material used for finishing purposes.

Reveal: The amount of setback of the casing from the face side of window and door jambs or similar pieces.

Scarf joints: End joints made by overlapping two pieces of molding with 22.5- or 45-degree angle cuts.

Square cuts: Cuts made at a right (90-degree) angle.

Stool: The bottom horizontal trim piece of a window.

Trim: Finish materials such as molding placed around doors and windows and at the top and bottom of a wall.

Wainscoting: A wall finish, usually made of wood, stone, or ceramic tile, that is applied partway up the wall from the floor.

Additional Resources

This module presents thorough resources for task training. The following resource material is suggested for further study.

Finish Carpentry. 1997. Newtown, CT: Taunton Press.

Figure Credits

Courtesy of Irwin Tools, CO01
CANAMOULD.COM TRIMROC INTERIOR MOULDINGS BY CANAMOULD EXTRUSIONS INC. 2006, Figure 2 (photo)

Section Review Answer Key

Answer	Section Reference	Objective
Section One		
1.c	1.1.0	1a
2.b	1.2.2	1b
Section Two		
1.d	2.1.0	2a
2.a	2.2.0	2b
3.c	2.3.0	2c
4.b	2.4.0	2d
Section Three		
1.a	3.1.1	3a
2.a	3.2.0	3b
3.b	3.3.0	3c
4.a	3.4.0	3d
5.c	3.5.0	3e
6.b	3.6.0	3f
Section Four		
1.c	4.0.0	4

This page is intentionally left blank.

NCCER CURRICULA — USER UPDATE

NCCER makes every effort to keep its textbooks up-to-date and free of technical errors. We appreciate your help in this process. If you find an error, a typographical mistake, or an inaccuracy in NCCER's curricula, please fill out this form (or a photocopy), or complete the online form at **www.nccer.org/olf**. Be sure to include the exact module ID number, page number, a detailed description, and your recommended correction. Your input will be brought to the attention of the Authoring Team. Thank you for your assistance.

Instructors – If you have an idea for improving this textbook, or have found that additional materials were necessary to teach this module effectively, please let us know so that we may present your suggestions to the Authoring Team.

NCCER Product Development and Revision
13614 Progress Blvd., Alachua, FL 32615

Email: curriculum@nccer.org
Online: www.nccer.org/olf

❏ Trainee Guide   ❏ Lesson Plans   ❏ Exam   ❏ PowerPoints   Other _____

Craft / Level: _____   Copyright Date: _____

Module ID Number / Title: _____

Section Number(s): _____

Description: _____

_____

_____

_____

_____

Recommended Correction: _____

_____

_____

_____

_____

Your Name: _____

Address: _____

_____

Email: _____   Phone: _____

This page is intentionally left blank.

Cabinet Installation

OVERVIEW

In residential work, the majority of cabinet installation work involves kitchen cabinets and bathroom vanities. Vanities are simply small base cabinets. Wall cabinets present the most difficult challenge. They must be perfectly level and they must be carefully and securely attached to the wall framing. Cabinet installation requires knowledge of hardware such as door hinges and catches, as well as the ability to select fasteners appropriate to the job. A wider variety of cabinetry is used in commercial work as display cases, reception desks, and storage cabinets.

Module 27211

Trainees with successful module completions may be eligible for credentialing through NCCER's National Registry. To learn more, go to **www.nccer.org** or contact us at **1.888.622.3720**. Our website has information on the latest product releases and training, as well as online versions of our *Cornerstone* magazine and Pearson's product catalog.

Your feedback is welcome. You may email your comments to **curriculum@nccer.org**, send general comments and inquiries to **info@nccer.org**, or fill in the User Update form at the back of this module.

This information is general in nature and intended for training purposes only. Actual performance of activities described in this manual requires compliance with all applicable operating, service, maintenance, and safety procedures under the direction of qualified personnel. References in this manual to patented or proprietary devices do not constitute a recommendation of their use.

Copyright © 2013 by NCCER, Alachua, FL 32615, and published by Pearson, New York, NY 10013. All rights reserved. Printed in the United States of America. This publication is protected by Copyright, and permission should be obtained from NCCER prior to any prohibited reproduction, storage in a retrieval system, or transmission in any form or by any means, electronic, mechanical, photocopying, recording, or likewise. To obtain permission(s) to use material from this work, please submit a written request to NCCER Product Development, 13614 Progress Blvd., Alachua, FL 32615.

From *Carpentry, Trainee Guide*. NCCER.
Copyright © 2013 by NCCER. Published by Pearson. All rights reserved.

27211
Cabinet Installation

Objectives

When you have completed this module, you will be able to do the following:

1. Describe the safety hazards when installing cabinets.
    a. Identify tool and material hazards that may be present when installing cabinets.
    b. Explain how to prevent back injuries through proper ergonomics.
2. Identify the different types of cabinets.
    a. Identify wall cabinets.
    b. Identify base cabinets.
    c. Describe the purpose of a countertop.
3. Identify cabinet components and hardware and describe their purpose.
    a. Identify cabinet components.
    b. Describe various types of hardware used on cabinets.
4. Explain how to lay out and install a basic set of cabinets.
    a. Describe the surface preparation needed before cabinet installation.
    b. Explain how to install wall cabinets.
    c. Explain how to install base cabinets and countertops.

Performance Task

Under the supervision of your instructor, you should be able to do the following:

1. Lay out and identify various types of base and wall units following a specified layout scheme.

Trade Terms

Acclimate	High-pressure laminate (HPL)	Soffit
Backsplash	Lipped	Stock cabinet
Base cabinet	Matte	T-brace
Custom cabinet	Overlay	Vanity
Ergonomics	Panel-and-frame door	Veneer
Escutcheon plate	Plastic laminate	Wall cabinet
Flush	Semicustom cabinet	

Code Note

Codes vary among jurisdictions. Because of the variations in code, consult the applicable code whenever regulations are in question. Referring to an incorrect set of codes can cause as much trouble as failing to reference codes altogether. Obtain, review, and familiarize yourself with your local adopted code.

Industry-Recognized Credentials

If you're training through an NCCER-accredited sponsor, you may be eligible for credentials from NCCER's Registry. The ID number for this module is 27211. Note that this module may have been used in other NCCER curricula and may apply to other level completions. Contact NCCER's Registry at 888.622.3720 or go to **www.nccer.org** for more information.

Contents

Topics to be presented in this module include:

1.0.0 Cabinet Installation Safety 1
  1.1.0 Tool and Equipment Safety 1
  1.2.0 Ergonomics 2
2.0.0 Types of Cabinets 4
  2.1.0 Base Cabinets 4
    2.1.1 Cabinet Islands 5
  2.2.0 Wall Cabinets 5
  2.3.0 Countertops 6
    2.3.1 Plastic-Laminate-Covered Countertops 6
    2.3.2 Solid-Surface Countertops 7
3.0.0 Cabinet Components and Hardware 11
  3.1.0 Cabinet Components 11
    3.1.1 Woods and Materials Used in Cabinet Construction 12
    3.1.2 Cabinet Doors 13
    3.1.3 Cabinet Drawers 13
    3.1.4 Cabinet Shelves 14
  3.2.0 Fasteners and Hardware 15
    3.2.1 Hinges 15
    3.2.2 Door Catches 16
    3.2.3 Knobs and Pulls 17
    3.2.4 Cabinet Shelf Hardware 18
4.0.0 Installing Cabinets 19
  4.1.0 Surface Preparation 19
  4.2.0 Installing Wall Cabinets 19
  4.3.0 Installing Base Cabinets and Countertops 21
    4.3.1 Scribing Adjoining Pieces 23
    4.3.2 Tight-Joint Fasteners 23

Figures

Figure 1	Damaged extension cords	2
Figure 2	Components of typical kitchen cabinets	5
Figure 3	Freestanding cabinet installation	6
Figure 4	Island layout	6
Figure 5	Typical construction of a plastic-laminate-covered countertop	7
Figure 6	Solid-surface countertop being machined	9
Figure 7	Cabinet construction methods	11
Figure 8	Typical vanity cabinet	12
Figure 9	Typical cabinet door designs	13
Figure 10	Types of cabinet door installations	13
Figure 11	Joints used to join drawer pieces	14
Figure 12	Drawer construction	14
Figure 13	Typical drawer-guide mechanism	15
Figure 14	Common cabinet door hinges	15
Figure 15	Euro-hinge	16
Figure 16	Self-closing hinge	17
Figure 17	Cabinet door catches	17
Figure 18	Examples of cabinet door and drawer pulls and knobs	17
Figure 19	Rails, clips, and pins used with adjustable cabinet shelves	18
Figure 20	Simplified cabinet design sketch	19
Figure 21	Marking stud locations on a wall	20
Figure 22	Marking cabinet layout lines.	20
Figure 23	T-brace used to support wall cabinets.	21
Figure 24	Shimming wall cabinets	21
Figure 25	Joining base-cabinet stiles	22
Figure 26	Shimming cabinets so that they are plumb and level	22
Figure 27	Filler strip	23
Figure 28	Tight-joint fasteners	24

This page is intentionally left blank.

SECTION ONE

1.0.0 CABINET INSTALLATION SAFETY

Objective

Describe the safety hazards when installing cabinets.
  a. Identify tool and material hazards that may be present when installing cabinets.
  b. Explain how to prevent back injuries through proper ergonomics.

Trade Terms

Ergonomics: The science of fitting workplace conditions and job demands to the capabilities of the craftworkers; also called human engineering.

Plastic laminate: Product, typically made of several layers of plastic material bonded together under intense heat and pressure, commonly used for countertops and shelves.

As with other specialty areas, some unique safety hazards are inherent to cabinet installation. Special attention should be given to back injuries that result from improper lifting techniques, lifting too much weight, or lifting in an awkward position. General safety and housekeeping guidelines, as well as general tool and equipment safety guidelines should be followed.

1.1.0 Tool and Equipment Safety

A good carpenter is always aware of the safety rules in every area of the construction field. The interior of a building nearing completion sometimes gives the worker a false sense of security. However, the element of danger is always present. Consider the following tool and equipment safety rules when installing cabinets and countertops.

- *Keep work area clean* – Cluttered areas and benches invite injuries.
- *Avoid dangerous equipment* – Do not expose power tools to rain. Do not use power tools in damp or wet locations. Keep the area well lit. Avoid chemical or corrosive environments. Do not use tools in the presence of flammable liquids or gases.
- *Guard against electric shock* – Avoid body contact with grounded surfaces such as pipes, radiators, ranges, or refrigerator enclosures.
- *Keep visitors away* – Do not let visitors come in contact with tools or extension cords.
- *Store idle tools* – When not in use, tools should be stored.
- *Secure the work* – Use clamps or a vise to hold the work. It is safer than using your hand, and it frees both hands for operating the tool.
- *Use the right tools* – Do not force a smaller tool or attachment to do the job of a heavier tool. Do not use a tool for any purpose for which it was not intended.
- *Dress properly* – Do not wear loose clothing or jewelry. Loose clothing, drawstrings, and jewelry can be caught in moving parts. Rubber gloves and nonskid footwear are recommended. Wear a protective covering to protect long hair.
- *Use appropriate personal protective equipment (PPE)* – Wear safety glasses or goggles while operating power tools. Wear a face or dust mask if the operation creates dust. All persons in the area where power tools are being operated should also wear safety glasses and face or dust masks.
- *Do not abuse the cord* – Never carry any tool by the cord or yank it to disconnect it from the receptacle. Keep the cord away from heat, oil, and sharp edges. Have damaged or worn power cords and strain relievers replaced immediately.
- *Use a ground fault circuit interrupter (GFCI)* – Per OSHA (Occupational Safety and Health Administration) regulations, any time a power tool is used on a job site, it must be plugged into a GFCI.
- *Do not overreach* – Keep proper footing and balance at all times.
- *Maintain tools with care* – Keep tools sharp and clean for safer performance. Follow instructions for lubricating and changing accessories. Inspect tool cords periodically, and if they are damaged, have them repaired by an authorized service facility. Have all worn, broken, or lost parts replaced. Keep handles dry, clean, and free from oil and grease.
- *Disconnect (unplug) tools* – Always disconnect tools when not in use, before servicing, and when changing accessories such as blades, bits, and cutters.
- *Remove keys and adjusting wrenches* – Check to make certain that keys and adjusting wrenches are removed from a tool before turning it on.
- *Avoid unintentional starting* – Do not carry a plugged-in tool with a finger on the switch. Be sure the switch is off when plugging in a tool.
- *Check damaged parts* – Before using any tool, inspect it to be sure that it will operate properly

and perform its intended function. Check for alignment of moving parts, binding of moving parts, breakage of parts or mountings, and any other conditions that may affect its operation. A guard or other part that is damaged should be properly repaired or replaced by an authorized service center unless otherwise indicated in the instruction manual. Have defective switches replaced by an authorized service center.
- *Stay alert* – Watch what you are doing and use common sense. Do not operate a tool when tired or while under the influence of medication, alcohol, or drugs.
- *Inspect extension cords* – Inspect extension cords periodically and replace them if damaged (*Figure 1*).

> **WARNING!** Remember that an extension cord with cracked or cut outer insulation or a loose or missing ground pin can kill you. Be careful. Always unplug power tools when changing accessories or making adjustments. Plan ahead.

Many cabinets are installed with their doors attached and/or their drawers installed. Unless the doors or drawers are secured in place, the potential for pinching hazards may be present when moving or installing the cabinets. Ensure that doors and drawers are properly secured, to prevent pinching fingers or hands. Another potential hazard when moving or installing cabinets is when doors or drawers open unexpectedly, resulting in the door or drawer striking carpenters and injuring them.

Another potential safety hazard may be present when repairing plastic laminate countertops. Plastic laminate is typically adhered to the base material using contact cement. The use of contact cement in an enclosed area is dangerous and must be avoided. Always provide adequate ventilation in the area where the contact cement is being used. If this is not possible, ensure that proper respiratory protection is used to prevent the inhalation of vapors from the contact cement.

1.2.0 Ergonomics

Back injuries may occur if cabinets and countertops are not handled properly during installation. Your back is at most risk from an injury when bending, lifting, reaching, and twisting. Learn to recognize situations where you may be performing tasks that involve these motions. Ergonomics, also called human engineering, is the process of designing and arranging items, such as cabinets and countertops, to prevent injuries when working on them.

If tasks involve heavy lifting or awkward positions, ensure that proper muscle-stretching exercises are performed prior to the task. Stretch leg, arm, and back muscles slowly and deliberately to avoid strains and sprains. Consider the following guidelines when installing cabinets and countertops:

- Reduce the amount of weight lifted. If possible, remove the doors and drawers before installing cabinets and countertops.
- If the cabinet is too heavy (even with the doors/drawers removed) or the cabinet or countertop is too awkward to handle individually, ask for help.
- When performing a lot of lifting, take it slowly if possible. Allow yourself more recovery time as needed between lifts.
- Rest your back frequently and restretch if needed before the next lift.
- If possible, prior to installing the cabinets, place cabinets and countertops up off the floor on a table or set of sawhorses to prevent excess bending when removing or installing hardware.

Figure 1 Damaged extension cords.

1.0.0 Section Review

1. If the ground pin of a tool's electrical plug is loose, it should be removed so you can continue to use the tool.
    a. True
    b. False

2. All of the following are good safety practices when installing cabinets, *except* _____.
    a. Do not wear loose clothing or jewelry when using power tools.
    b. Provide adequate ventilation when using contact cement.
    c. Avoid stretching back muscles before lifting heavy items.
    d. Do not lift or carry a power tool by its electrical cord.

Section Two

2.0.0 Types of Cabinets

Objective

Identify the different types of cabinets.
a. Identify wall cabinets.
b. Identify base cabinets.
c. Describe the purpose of a countertop.

Trade Terms

Backsplash: A trim piece that is attached along the back edge of a countertop and covers the wall-to-countertop abutment to conceal any irregularities.

Base cabinet: Cabinet that is made to sit on the floor with its back against a wall.

Custom cabinet: Tailor-made cabinet produced in a cabinet shop, millwork plant, or on the job site.

High-pressure laminate (HPL): Countertop material produced by pressing layers of plastic material under intense heat and pressure to form a laminate.

Matte: A surface finish that has no shine or luster.

Semicustom cabinet: Cabinet produced in a mill or cabinet shop, which is available in many materials, sizes, and finishes.

Soffit: The enclosed section (which is usually decorated by paint or wallpaper) between the ceiling and the top of the wall cabinets; also called a furr down.

Stock cabinet: Cabinet mass produced in specialized plants, with somewhat limited design options; also called modular cabinets.

Vanity: A small cabinet, commonly installed in bathrooms to store personal hygiene products.

Wall cabinet: Cabinet that is made to be attached to the upper part of a wall.

Kitchen cabinets and other cabinetry are classified as three major grades: stock, semicustom, and custom units.

Stock cabinets, also called modular cabinets, are the least expensive kind of cabinet. They are mass produced in specialized plants and are typically sold off-the-shelf at home centers and builder supply centers. They provide somewhat limited design options and can sometimes be of marginal quality. Stock cabinets are made in three forms: disassembled, assembled but not finished, and assembled and finished. Disassembled cabinets are known as RTA, or ready to assemble.

Semicustom cabinets are the next grade up from stock cabinets. Semicustom cabinets may be produced in a mill or a cabinet shop. Semicustom cabinets are usually available in several styles and in a variety of standard sizes. They are offered with many choices in materials, size, and finish. Semicustom cabinets may be assembled in either unfinished or finished form. They are usually sold by home centers, independent distributors, and midrange kitchen and bath dealerships.

Custom cabinets are tailor-made units built in a cabinet shop, millwork plant, or on the job site to satisfy a specific application. With these units, a carpenter or cabinetmaker goes to a job site, measures the cabinet area, consults with the home or building owner, and then builds each cabinet unit to the desired specifications.

The next part of this module introduces types of cabinets and countertops, as well as their components, such as hardware. This background information will allow you to knowledgeably discuss cabinetry with customers, cabinet dealers, manufacturers, and others in the trade. The remainder of this module focuses on the basic procedures for installing stock cabinets at the job site. It also covers the materials and methods involved in the construction and installation of countertops used with kitchen and bathroom cabinets.

2.1.0 Base Cabinets

Figure 2 shows the components of a basic kitchen cabinet arrangement. The lower unit, called the base cabinet, is formed by fastening a series of individual base cabinet units together to achieve the desired configuration. Most base cabinets are manufactured 34½" high and 24" deep. As shown, a countertop is installed on top of the base cabinet. By adding the usual countertop thickness of 1½", the work surface of the base cabinet is at a standard height of 36" from the floor. The countertop is typically 25½" deep if a preformed unit or 25" deep if job built.

Standard base cabinets come in several widths that vary in 3" increments. Single-door cabinets typically range from 9" to 24" in width. Most consist of one door, one drawer, and an adjustable shelf. Some cabinets have stationary shelves. However, some base units have no drawers, while others contain all drawers. Double-door base units range from 24" to 48" in width. They provide access to the entire cabinet from both sides. Corner units are also available with round rotating

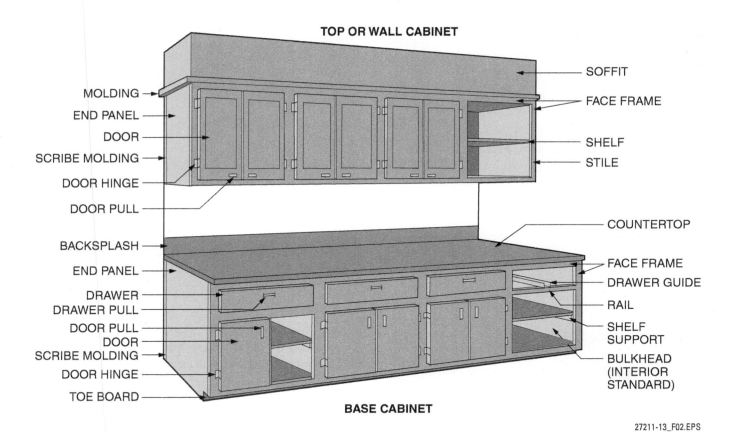

Figure 2 Components of typical kitchen cabinets.

shelves (tiered lazy Susans) that make corner storage more accessible. A variety of floor-mounted tall cabinets are also made for use as oven, utility, and pantry units.

2.1.1 Cabinet Islands

Although most cabinet systems are attached to walls, it is not uncommon to see freestanding cabinets, especially in residential kitchens (*Figure 3*). Placement of an island is critical. There should be at least 36" between the outside edge of the island and any other cabinet or appliance; 42" is preferable. Be sure the island does not interfere with door swing or access to large appliances.

In some installations, a frame is built to provide an attaching point for cabinets, as shown in *Figure 3A*. The cabinets are then attached to the frame, and the outside and ends of the frame are finished with drywall (*Figure 3B*).

In other applications, the center island is a prefabricated drop-in unit. Layout carpenters will mark the location on the slab or floor (*Figure 4*). Piping and electrical wiring may also be provided. Depending on the installation requirements and the manufacturer's instructions, cabinet installers may attach a 2 × 4 frame to the floor. They will then place the island over the frame and attach it to the frame with screws. A frame is not always used, however. If ceramic tile or hardwood flooring is to be installed against the island, the flooring material, combined with the weight of the island, is usually enough to secure the island in place.

2.2.0 Wall Cabinets

Similar to the base cabinet, the wall cabinet (*Figure 2*) is formed by fastening a series of individual wall cabinet units together to achieve the desired configuration. Wall cabinets are usually 12" deep and are available in single-door and double-door units having the same widths as similar base cabinets. They are made in several standard heights, with a 30" height being typical. Shorter cabinets that are 24", 18", 15", and 12" in height are made for use above sinks, refrigerators, and stoves. Wall cabinets are available with adjustable or stationary shelves. In kitchens without a soffit, 36"- and 42"-high cabinets are frequently used when more storage space is desired.

The vertical distance between the countertop on the base cabinet and the bottom of the wall cabinets typically ranges from 15" to 18". However, building codes normally require that there be at least 24" to 30" between the cooking unit and

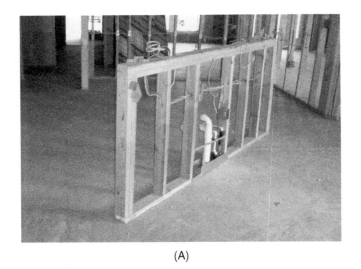
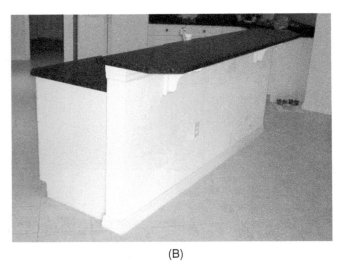

(A)   (B)

Figure 3 Freestanding cabinet installation.

Figure 4 Island layout.

any overhead wall cabinet and a minimum of 22" between the sink and an overhead wall cabinet.

2.3.0 Countertops

Numerous types of countertops can be used with cabinets. These include countertops covered with plastic laminate or those made from or covered with natural or synthetic solid-surface materials.

2.3.1 Plastic-Laminate-Covered Countertops

Plastic laminate is one of the most widely used materials to cover countertops. Made of layers of resin-impregnated kraft paper with a top layer of colored melamine, plastic laminates provide a durable, affordable, and easy-to-clean surface. Plastic laminate material is available in a wide range of colors and patterns. Sheets of plastic laminate are available in widths of 30", 36", 48", and 60", and lengths of 6', 8', 10', and 12'. Plastic laminate is referred to as HPL, or high-pressure laminate. HPL for cabinetry is available in three grades:

- *Horizontal (0.048" thick)* – Used for flat countertops.
- *Vertical (0.028" thick)* – Used for walls and doors. It is a lighter weight and not as durable.
- *Postform (0.039" thick)* – Used on factory-built countertops. It is not completely thermoset and can be heated and bent around radii. These preformed countertops are made by bonding flexible postform HPL to a core.

A plastic laminate countertop is usually 1½" thick and overhangs the front and sides of the cabinet(s). Counter overhangs vary, but 1" is common for standard kitchen cabinets that are 24" wide. Tops for peninsulas or islands can be much wider. Bathroom vanity tops range from 19" to 22" wide.

Some plastic-laminate countertops are factory built. These preformed countertops are made of plastic laminate bonded to a core material, generally medium-density fiberboard (MDF), plywood, or particleboard. The front edge of the countertop is rounded, and the rear edge often curves up to form a backsplash. The countertops are available in standard widths and are cut to length by the supplier or by the carpenter at the job site. Note that some L-shaped countertops are manufactured in two sections with a mitered corner. These sections are joined with hardware supplied with the countertop. Also supplied are preglued strips of laminate called end caps. These are applied to the ends of the countertop after it has been cut to the proper length.

A typical base for a job-built plastic-laminate-covered countertop is built up in two layers, with a total thickness of 1½". The top layer (core) is a sheet of solid ¾" MDF, particleboard, or plywood (*Figure 5*). The second layer is made up of ¾" × 3" strips of plywood or particleboard attached to the front, back, and side edges. These strips reinforce and thicken the top. Cross strips installed at 2' intervals and at sink and appliance locations provide the top with additional strength. Any seams in the core material are reinforced underneath with large squares of plywood or particleboard.

After the countertop base is built, oversized strips of plastic laminate are cut and adhered to the exposed edges of the base using contact cement. The strips are trimmed flush with a laminate trimmer or router, allowing an oversized top piece of laminate to be cemented in place. After trimming the top piece flush with the laminate trimmer, the sink cutout(s) can be made.

A backsplash that is attached to the back of the countertop is made as a separate piece. Plastic laminate is glued onto the ¾" × 4" backsplash core in exactly the same way as for the countertop core. It is fastened to the back edge of the countertop with screws. When the countertop is installed, the backsplash is fastened to the countertop and adhered to the wall with construction adhesive. The joint between the countertop and backsplash is sealed with matching caulk or a waterproof sealer. For a proper fit, the countertop may need to be scribed to the wall.

Custom countertops are usually built in specialty shops and then delivered to the installation contractor. They can be identified by mitered rather than formed (bent) edges.

2.3.2 Solid-Surface Countertops

Countertops made of solid-surface materials are popular because they offer numerous benefits, including an attractive appearance and durability. Recent studies show that about 40 percent of professionally designed kitchens and baths use solid-surface countertops.

Solid-surface countertops are available in many colors, have the solid look and feel of stone, resist stains, are easy to renew, and are available in matte, satin, semigloss, and high-gloss finishes. A matte or satin finish is best for kitchen countertops because it hides scratches better than shiny surfaces. Light colors and speckled patterns also tend to hide scratches better than dark colors.

The fabrication of countertops using solid-surface materials is labor intensive. Every countertop is usually custom made. All edges, inlays, feature strips, and backsplashes must be created individ-

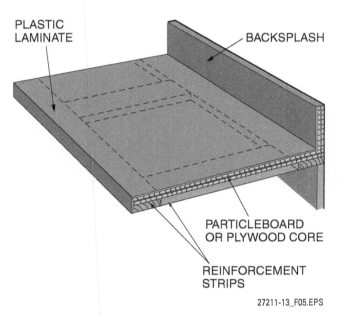

Figure 5 Typical construction of a plastic-laminate-covered countertop.

Professionalism

A professional cabinet installation like the one shown here allows a carpenter to take pride in the work.

ually (*Figure 6*). Often, the same person who fabricates a solid-surface countertop will also install it.

For installation of solid-surface countertops, some manufacturers require that dabs of silicone adhesive be applied to the top of the base cabinet frames, then the countertop placed on top of it. Others require that 1" wood strips be attached to the cabinets with the countertop glued to the strips. Some guidelines for a quality installation of a solid-surface countertop include:

- The counter should be leveled with shims rather than simply hiding any gaps with silicone.
- Any seams made at the job site should be inconspicuous and directly supported underneath.
- Inside corners should be cut on a radius and sanded to form a perfect curve, rather than a right angle. This is because corners are stress points, and a radius withstands stress better than sharp angles.

Commercial Cabinets

Commercial cabinets are likely to be exposed to more abuse and use than residential cabinets. Doors are frequently large, to allow the fullest possible access to storage. Laminates are very popular because of ease of cleaning and relative economy. Relatively high-end materials, such as Corian® and marble, are fairly common in countertops, both for appeal to the customers and for ease of cleaning.

Cabinet framing material is frequently stronger in commercial work than in residential work; that is, 2 × 4 top frames instead of 2 × 2. Keep in mind that if the cabinets are for a store or business, someone may use them to stand or sit on.

Obviously, the customers decide what their businesses need. A carpenter needs to find a sensible approach to the construction. The use of modular units is one form of economy and may be necessary because of shop space considerations. In the case of retail food and restaurant work, wiring and pipe chases need to be built in, allowing reconfiguration as the customer's needs change. Wiring is an issue that the carpenter needs to consider in the planning stage. Remember to plan with the electricians and plumbers, so that everybody knows what to do to help each other get the job right.

Most commercial contracts, especially the larger ones, are entirely specification driven and are obtained by a detailed bid. Materials will be specified and everything down to the fasteners will be listed by type or brand name. Those jobs are awarded on competitive pricing and very few changes will be permitted. They are frequently jobs worked in cooperation between a large cabinet shop and an installation crew, or by the customer's employees.

Some are modules fastened to each other at the ends, some are freestanding display units, and some are on casters. Wall displays can be mounted to the structure, and usually are, because of the possible legal consequences if a customer pulls the display over and is hurt. Remember that construction in retail stores and public access buildings is exposed to any person who comes in; don't assume that everyone who interacts with your work will do so intelligently.

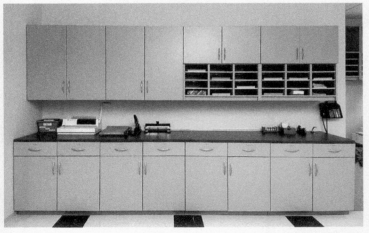

27211-13_SA02.EPS

Figure 6 Solid-surface countertop being machined.

- The finish should be consistent over the entire surface. Check for consistency by examining the surface from several angles.

Many manufacturers of solid-surface countertops require that the installer be trained in the installation of their product. If an uncertified person does the installation, the warranty is voided.

> **NOTE**
> One thing to look for when installing solid-surface countertops is proper color matching. When two sections of countertop material are placed side by side, it will be immediately obvious if they are mismatched in color or pattern.

Solid-surface countertops should be maintained according to the manufacturer's instructions. Most stains and spills wipe up with soap and water. Stubborn stains can be removed using a mild cleanser and a nylon scrubbing pad, which will not harm the surface. Minor scratches can usually be removed with very fine sandpaper, followed by buffing with a nylon scrubbing pad to revive the finish.

Going Green

Recycled Glass for Countertops

Recycled-glass countertop surfaces provide a beautiful and lasting appearance. Architectural glass, art glass, beer bottles, and jars are a few of the glass products that are recycled to make the countertop surface. Liquid binders are added to the glass fragments and allowed to set. The surfaces and edges are then ground to a high gloss.

Plastic Laminates

Some common names of plastic laminates are Formica®, Wilsonart®, and Nevamar®. Each laminate manufacturer produces laminates in a multitude of colors and patterns. To help customers compare and select among their different laminate products, laminate manufacturers readily provide samples for each of their products, like the ones shown here.

27211-13_SA03.EPS

Solid-Surface Materials

While all solid-surface manufacturers have their own formulations, most solid-surface products are made of resins and fillers, along with pigments, ultraviolet (UV) inhibitors, and other additives. The most common filler is alumina trihydrate (ATH), which can account for up to 70 percent of the product. It helps countertops resist impact as well as water, stains, and chemicals.

2.0.0 Section Review

1. Standard base-cabinet widths vary in increments of _____.
   a. 1½"
   b. 3"
   c. 4½"
   d. 6"

2. Typical height for a wall cabinet is _____.
   a. 18"
   b. 24"
   c. 30"
   d. 36"

3. A type of finish preferred for kitchen countertops is _____.
   a. extra-high-gloss
   b. high gloss
   c. semigloss
   d. matte

SECTION THREE

3.0.0 CABINET COMPONENTS AND HARDWARE

Objective

Identify cabinet components and hardware and describe their purpose.
  a. Identify cabinet components.
  b. Describe various types of hardware used on cabinets.

Trade Terms

Escutcheon plate: A decorative metal plate that is placed against a door or drawer face behind the pull.

Flush: Style of cabinet door that fits into and is flush with the face of the cabinet opening.

Lipped: Style of cabinet door in which the all edges are rabbeted, so when it is installed on a cabinet, the edges overlap and conceal the opening.

Overlay: Style of cabinet door mounted on the outside frame that overlays the cabinet opening so that the opening is concealed; also called surface door.

Panel-and-frame door: Cabinet door made of plywood or solid wood panels fitted between side stiles and horizontal rails; sometimes referred to as panel doors.

Veneer: A thin layer or sheet of wood intended to be overlaid on a surface to provide strength, stability, and/or an attractive finish. Thicknesses range between $1/16"$ and $1/8"$ for core plies and between $1/128"$ and $1/32"$ for decorative faces.

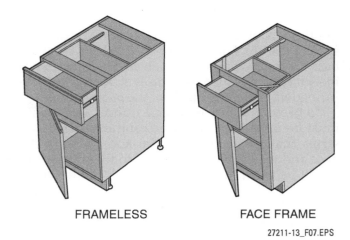

Figure 7 Cabinet construction methods.

In addition to the cabinets themselves, a carpenter should be able to install cabinet doors, drawers, and hardware. This section presents the various cabinet components and hardware used to finish cabinets.

3.1.0 Cabinet Components

All cabinets, whether designed for commercial or residential use, consist of a case fitted with shelves, doors, and/or drawers. Two types of cabinet construction are frameless and face frame (*Figure 7*).

Frameless cabinets, also called European or Eurostyle cabinets, have no support framework or face frame. Support for the shelves and work surface is provided by the cabinet panels, which are heavy enough to carry the weight of the assembly and materials that will be stored in the cabinet. This style is widely used in applications where a contemporary look is desired. One advantage of frameless cabinets is that they tend to provide more storage capacity because they are built without a space-consuming framework. For example, a frameless four-drawer base cabinet contains about 1 cubic foot more usable space than a similar size face-frame model.

Face-frame cabinets do not have framework. They are similar in construction to frameless, but are usually made of lighter stock because the face frame supports the doors. The edges of the cabinet front are covered with a solid lumber face that has openings for doors and drawers. Face-frame cabinets are widely used in applications where a traditional look is desired.

Cabinet manufacturers commonly assign catalog code numbers to their product line of manufactured cabinets to identify each type of cabinet and describe its dimensions. For example, a cabinet manufacturer might assign a cabinet the code W361824. The letter(s) preceding the cabinet nomenclature refer to the type of cabinet, such as W for wall, B for base, SB for sink base, CW for corner wall, DB for drawer base, and PB for peninsula base. The first number or pair of numbers after the letter(s) refers to the cabinet width. In the example, the number is 36. This means that the cabinet is 36" wide. If there is a second pair of numbers in the cabinet nomenclature, it refers to the cabinet height. In our example, the cabinet height is 18". If a third pair of numbers is included, it indicates the depth of the cabinet. In our example, the cabinet is 24" deep. If the third pair of numbers is omitted, the cabinets are standard depth; in the case of wall cabinets, 12" deep.

If the cabinet nomenclature has letters following the sets of numbers, they generally indicate a special feature. For example, a B24SS is a 24"-wide base cabinet with sliding shelves. The cabinet nomenclature may also have an L or R after it, indicating left or right hinging of the cabinet doors.

To provide a finished appearance, accessories must be used with the basic cabinet units. Filler strips that match the material and finish of the base and wall cabinets are used to fill any gaps in width between the end base or wall cabinet units and the room walls, or between adjacent cabinet units when no combination of standard sizes can fill an existing space. These filler pieces are usually supplied in 3" or 6" widths and then ripped to the required width on the job. Other accessories include finished cabinet end panels and face panels for exposed ends and openings such as at dishwashers, refrigerators, and at the end of cabinet runs. Attractive matching moldings are often used to trim stock or semicustom cabinets in order to make them more distinctive.

With the exception of the size, vanity cabinets (*Figure 8*) used in bathrooms are constructed in the same way as kitchen base-cabinet units.

Most vanity cabinets are 31½" high and 21" deep. Usual widths range from 18" to 36" in increments of 3", then 42", 48", and 60". They are available in a wide variety of door, drawer, and shelf configurations.

3.1.1 Woods and Materials Used in Cabinet Construction

A wide range of materials is used in manufactured cabinets. Low-priced cabinets are usually made from panels of particleboard with a vinyl film or melamine applied to exposed surfaces. The surface is printed with either a solid color or with a wood-grain pattern to give the appearance of real wood.

High-quality cabinets are made from **veneers** and solid hardwoods, typically oak, cherry, birch, ash, or hickory. Hardboard, particleboard, plywood, or medium-density fiberboard (MDF) is commonly used for interior panels, drawer bottoms, and as the base for plastic-laminate countertops. Some points to look for in higher-grade cabinets include:

Figure 8 Typical vanity cabinet.

- Face-frame cabinets should have a ½" to ¾" hardwood face frame, ⅜" to ½" plywood or particleboard side frames, and a ¼" back panel.
- Frameless cabinets should use ⅝" to ¾" particleboard or plywood for the entire cabinet. Wood veneer or high-pressure laminates are preferable to melamine for the exterior finish.
- Drawers should slide smoothly with little play. They should close quietly and solidly. It is typical for metal ball-bearing slides to be used. They can be regular or full-extension drawers, based on the specification. Look for a ½" to ¾" solid wood or 9-ply solid birch plywood drawer box with integral wood dovetail, shoulder, or dado joints. The box is typically attached to the drawer front with screws.
- Door hinges and catch mechanisms should be well made and work without binding.
- Rabbeted joints should be used where the top, bottom, back, and side pieces join.

A wide variety of softwoods and hardwoods are used in the construction of visible exterior cabinet surfaces such as frames, doors, and draw-

Repairing Manufactured Cabinets

Manufactured cabinets that are marred or otherwise damaged may be repaired with a variety of putties or other substances. Take care to match the cabinet's color as closely as possible. Whenever possible, purchase a repair kit from the cabinet manufacturer.

ers. Among the woods commonly used in cabinet construction are ash, beech, birch, cherry, oak, maple, and pine.

Plywood is an excellent building material and one of the most important modern cabinetmaking materials. It has several advantages over lumber, including availability in very large sheets, exceptional strength, and high resistance to warping. It is lightweight, yet very strong in comparison to solid wood. It is also great for cabinets. Plywood does not split as easily as solid wood and is more stable across the grain. Plywood is the best material for most job-built units because of its working ease and availability.

The face veneers on both softwood and hardwood plywood can be damaged easily because the veneer is so thin. Sanding a light scratch or marred surface in the veneer can be tricky. If done incorrectly, an expensive piece of plywood can be ruined. Before attempting to sand a scratch or mar in the plywood veneer, first moisten the wood and allow it to dry. This will raise the veneer grain, allowing it to be sanded with less chance of damage.

3.1.2 Cabinet Doors

Cabinet doors may be merely functional, or they can provide the major decoration of the cabinet unit. Both swinging and sliding doors can be used with cabinets. However, swinging doors are used extensively for kitchen and vanity units. Doors can be made of solid plywood or of panel-and-frame construction. Panel-and-frame doors consist of a solid wood frame mounted around a panel of plywood or solid wood. The frame is joined by mortise-and-tenon, mitered spline, or stile joints. Some cabinet doors have a frame covered on each side with thin plywood, similar to a hollow-core interior door. Many different trim designs are used with cabinet swinging doors. *Figure 9* shows some widely used styles.

Three main types of hinged swinging doors are used with cabinets: lipped, overlay, and flush (*Figure 10*). These terms refer to the way the door is mounted on the cabinet. As shown, a lipped door is rabbeted along all edges so that when it is mounted on the cabinet, the edges overlap and conceal the cabinet opening. Lipped doors are relatively easy to install because they do not require exact fitting in the opening and the rabbeted edges stop against the face frame of the cabinet.

Overlay doors, also called surface doors, are mounted on the outside frame and overlay the cabinet opening so that the opening is concealed. The overlay door is also easy to install because it does not require exact fitting in the opening and

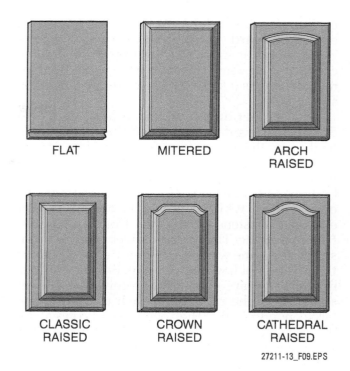

Figure 9 Typical cabinet door designs.

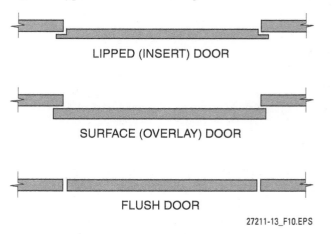

Figure 10 Types of cabinet door installations.

the face frame of the cabinet acts as a stop for the door. For frameless (Eurostyle) cabinets, this type of door completely overlays the front edges of the cabinet.

Flush doors fit into and flush with the face of the cabinet opening. Flush doors are the most difficult to install because they require fitting in the cabinet opening. A clearance of about 1/16" must be made between the opening and the door edges. Stops must be provided in the cabinet against which the door will close.

3.1.3 Cabinet Drawers

Drawer fronts are generally made of the same material as the cabinet doors. The sides and backs are typically made of ½"-thick solid lumber, plywood, or particleboard. The bottom is usually made of ¼" plywood or hardboard. The front and sides can be joined using dovetail, lock-shouldered, or rabbeted joints (*Figure 11*). The dovetail joint is the strongest.

Like cabinet doors, cabinet drawers are also classified as overlay, lipped, and flush, depending on the way the drawer front fits the cabinet drawer opening. The overlay drawer (*Figure 12*) is formed by a self-contained drawer unit consisting of the bottom, two sides, a back, and a false front. An overlay front, which conceals the cabinet drawer opening, is fastened to the false front with screws from the inside. The lipped drawer is made in basically the same way as an overlay drawer, except that it does not have a false front. Both ends of the drawer front are rabbeted to receive the drawer sides. Also, the rabbets on each end are made large enough so that they overlap and conceal the drawer opening in the cabinet. The flush door is constructed similarly to the lipped drawer, except the front is cut to fit the overall height and width of the drawer opening. For this reason, a drawer stop is provided at the back of the drawer so that the drawer is flush with the cabinet frame when closed.

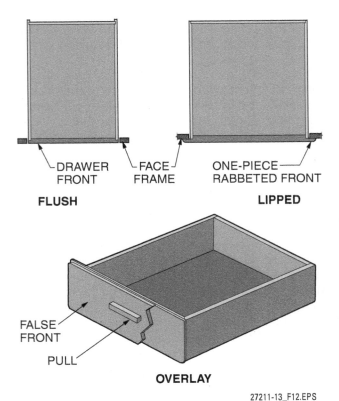

Figure 12 Drawer construction.

Several types of metal drawer guides can be used to support the drawer and limit lateral movement. They also serve to ensure that the drawer does not tilt downward when it is opened. *Figure 13* shows a typical drawer-guide mechanism. Drawer guides are designed to handle loads of various weights. Generally, side-mounted drawer guides with ball-bearing rollers carry more weight than bottom-mounted, single-rail drawer guides. Wood drawer guides are seldom used on manufactured kitchen and vanity cabinets.

3.1.4 Cabinet Shelves

Most cabinets have some kind of shelves or dividers installed. Cabinet shelves can be made from a number of materials, including solid wood, glass, plywood, and particleboard covered with plastic laminate. Most wood shelves should be made of ¾" stock and be no longer than 3' without some sort of bracing.

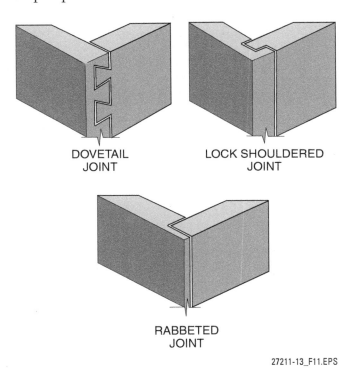

Figure 11 Joints used to join drawer pieces.

Door Sizes

When specifying the door-opening size or door size, it is common practice in the trade to list the width first and the height second.

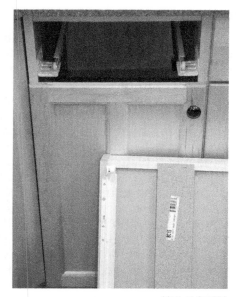

Figure 13 Typical drawer-guide mechanism.

3.2.0 Fasteners and Hardware

There is a variety of types, shapes, sizes, and finishes of door and drawer hardware available. Using the proper hardware for a cabinet is important. The style of a unit can be changed by merely switching the door and drawer hardware. For example, there is hardware designed specifically for French Provincial, Early American, Traditional, Mediterranean, and Dutch cabinets, just to name a few. In addition to using the correct style, the proper-size hardware is also important. If the hardware is too large for a cabinet, it will make the cabinet appear smaller. If small hardware is used on a large piece, it will look out of scale.

3.2.1 Hinges

Numerous types of decorative hinges are available for use with cabinet doors. *Figure 14* shows a small sample of some common hinges. Cabinet door hinges can be divided into several categories, including:

- *Surface hinges* – Fastened to the exterior surface of the door and frame. The back side of hinge leaves can be straight for use with flush doors or offset for use with lipped doors.
- *Offset hinges* – Used with lipped doors. A semi-concealed offset hinge has one leaf bent to a 3/8" offset that is screwed to the back of the door. A concealed-offset type is one in which only the pin is exposed when the door is closed.
- *Overlay hinges* – Available in semiconcealed and concealed configurations. For semiconcealed types, the amount of overlay is variable. Concealed types are generally made with a specific amount of overlay, such as ¼", ⁵⁄₁₆", ⅜", and ½".

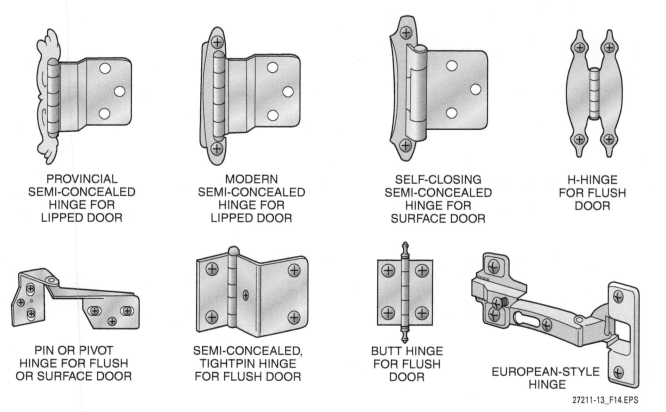

Figure 14 Common cabinet door hinges.

- *Pivot hinges* – Normally used on overlay doors. They are fastened to the top and bottom of the door and to the inside of the case.
- *Butt hinges* – Used on flush doors when it is desired to conceal most of the hardware. The leaves of the hinge are set into the edges of the frame and the door.
- *Eurostyle hinges* – The Euro-hinge is commonly used for concealed-hinge applications, which represent the majority of installations. One of the main reasons for the popularity of the Euro-hinge is its adjustability. As shown in *Figure 15*, the depth, height, and side-to-side positioning can be adjusted to obtain a level, flush match from door to door, and to compensate for any irregularities in the cabinetry.

Self-closing hinges (*Figure 16*) are used in some applications. These hinges contain a spring mechanism that will automatically pull a door shut.

3.2.2 Door Catches

Swinging doors without self-closing hinges generally require the use of door catches to hold the door closed. Catches are normally installed where they are not in the way. Magnetic, roller, flex, and bullet catches (*Figure 17*) are common. Magnetic catches are made with single or double magnets of varying holding power. The adjustable magnet plate is fastened to the inside of the case, and the metal plate is fastened to the door at the desired location. Roller and flex (friction) catches are installed in a similar way as magnetic catches. The adjustable section is fastened to the case and the other section to the door. Ball-point (bullet) catches are spring-loaded catches that fit into the edge of the door. When the door is closed, the catch fits into a recessed plate mounted on the frame.

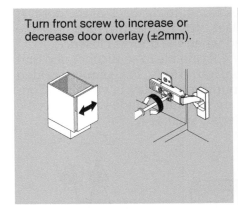
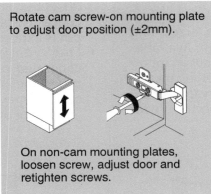
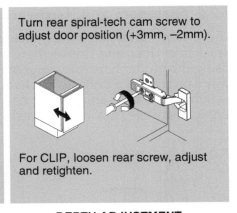

SIDE ADJUSTMENT — Turn front screw to increase or decrease door overlay (±2mm).

HEIGHT ADJUSTMENT — Rotate cam screw-on mounting plate to adjust door position (±2mm). On non-cam mounting plates, loosen screw, adjust door and retighten screws.

DEPTH ADJUSTMENT — Turn rear spiral-tech cam screw to adjust door position (+3mm, –2mm). For CLIP, loosen rear screw, adjust and retighten.

Figure 15 Euro-hinge.

Full-Extension Drawer Slides

Full-extension drawer slides or guides are a little more expensive than ordinary drawer guides, but they are much more convenient. A full-extension drawer slide permits access to the entire drawer.

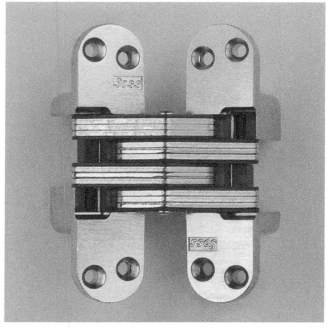

Figure 16 Self-closing hinge.

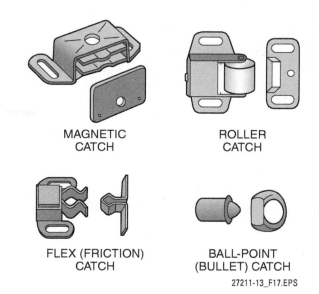

Figure 17 Cabinet door catches.

3.2.3 Knobs and Pulls

Many cabinet doors and drawers require that pulls or knobs (*Figure 18*) be installed in order to open and close them. They are available in many styles and designs, and are made of decorative metal, plastic, wood, porcelain, or other material. Knobs and/or pulls used on the cabinet doors should be the same design as those used for the cabinet drawers.

Knobs and pulls are usually installed by boring a hole through the door or drawer front and fastening them using a threaded bolt or screw that is normally supplied with the hardware. If the bolt or screw supplied is not long enough, the hole must be countersunk, or longer bolts provided.

Some smaller knobs and pulls have wood-screw threads fastened to their ends and are simply screwed in place after predrilling a slightly smaller hole. Knobs may also have **escutcheon plates** that can be placed behind them to add a decorative look to the hardware. This is especially true on the larger hardware.

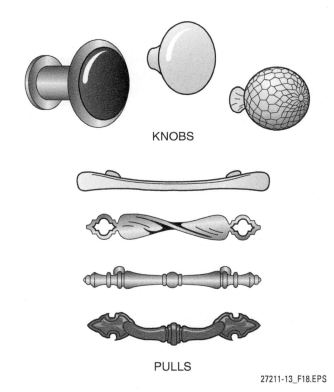

Figure 18 Examples of cabinet door and drawer pulls and knobs.

Touch Catches

Door catches are made that incorporate both a catch to hold the door closed and a spring-loaded pushing device to open the door when a slight pushing pressure is applied to the door. These door catches, called touch catches, eliminate the need to use door pulls.

3.2.4 Cabinet Shelf Hardware

Many shelves used in cabinets are adjustable. A common method of installing adjustable shelves is the use of slotted metal shelf rails (standards) and clips, which are available in a number of finishes (*Figure 19*). Four standards are used, two installed on each side of the cabinet interior. The standards are cut to the required length and can be either surface mounted on the sides of the cabinet or recessed in dadoes. If mounted on the sides of the cabinet, the shelves are cut short enough that they can be slipped down between the rails.

Another method of adjustable shelf support commonly used in kitchen cabinets involves two rows of stopped, drilled holes at the front and rear of each side of the cabinet interior. The number of such holes drilled in each row depends on whether the cabinet uses multiple shelves or a single shelf. Typically, a series of holes spaced about 1½" to 2" apart is drilled to cover an area of about 6" in the vicinity of each shelf's location. Two L-shaped plastic or metal pegs or spade pins are inserted into these holes at identical heights to support each end of the related shelf.

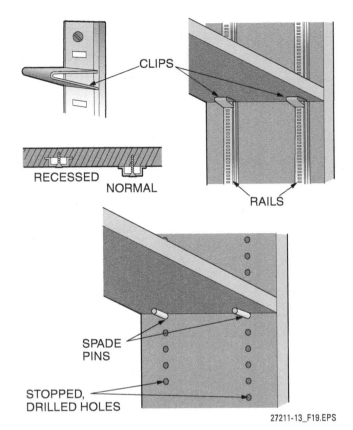

Figure 19 Rails, clips, and pins used with adjustable cabinet shelves.

3.0.0 Section Review

1. Frameless cabinets are also known as _____.

   a. contemporary cabinets
   b. Eurostyle cabinets
   c. modern-design cabinets
   d. Scandinavian cabinets

2. Cabinet door catches that eliminate the need for door pulls are called _____.

   a. touch catches
   b. magnetic catches
   c. spring-loaded catches
   d. flex catches

SECTION FOUR

4.0.0 INSTALLING CABINETS

Objective

Explain how to lay out and install a basic set of cabinets.

a. Describe the surface preparation needed before cabinet installation.
b. Explain how to install wall cabinets.
c. Explain how to install base cabinets and countertops.

Performance Task 1

Lay out and identify various types of base and wall units following a specified layout scheme.

Trade Terms

Acclimate: Adjust to a new temperature, humidity level, or environment.

T-brace: Type of brace used to support wall cabinets during installation.

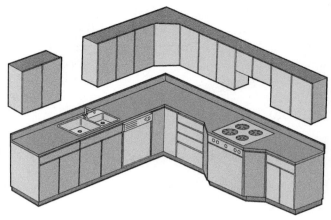

Figure 20 Simplified cabinet design sketch.

This section provides general procedures for installing stock kitchen cabinets and countertops. Since kitchens usually require more cabinets than all other rooms combined, cabinet installation in general is best explained from the viewpoint of installing kitchen cabinets. The installation of other types of cabinets will use the same basic methods and procedures.

Cabinets are installed in accordance with the construction drawings for the building and/or the design sketch or shop drawings that show their exact location and configuration within the room (*Figure 20*).

As with most carpentry procedures, the installation of kitchen cabinets can be accomplished in more than one way. Some carpenters prefer to start by installing the wall-mounted cabinets first; others choose to install the base units first.

4.1.0 Surface Preparation

To allow base and wall cabinets to be installed square and plumb, the underlying surface must be properly prepared. Proper surface preparation actually starts prior to the drywall or other surface finish material being applied. Backing boards (for wood frame construction) or 6-inch-wide, 20-gauge sheet-metal straps (for steel frame construction) should be installed in the proper location to support wall cabinets. Backing locations are typically shown on the construction drawings.

Check the wall and floor surfaces with a straightedge for unevenness. Unevenness can cause cabinets to be misaligned, resulting in twisting of doors and drawer fronts. Wall high spots should be removed by shaving or sanding off excess plaster or drywall. Low spots can be shimmed later during cabinet installation.

Thoroughly clean the area in which the cabinets are to be placed. It is much easier to work in a room free of sawdust and scattered tools. If required, remove any existing baseboard, window, or other trim from the walls in the work area.

Ensure the room in which the cabinets are to be installed is heated and the temperature and humidity level is similar to what would be expected when the building is completed. Move the cabinets into the room and allow them to acclimate to the environment before installing them.

4.2.0 Installing Wall Cabinets

To install wall cabinets, proceed as follows:

Step 1 Locate and mark the position of all wall studs in the area where cabinets are to be installed (*Figure 21*). At each stud location, draw plumb lines on the wall. Locate the marks where they can easily be seen after the cabinets are in position.

Step 2 Using a long straightedge and a 4' level, check the floor area for high spots where the base cabinets are to be installed. Find the highest point on the floor, then snap a level chalkline on the wall at this high

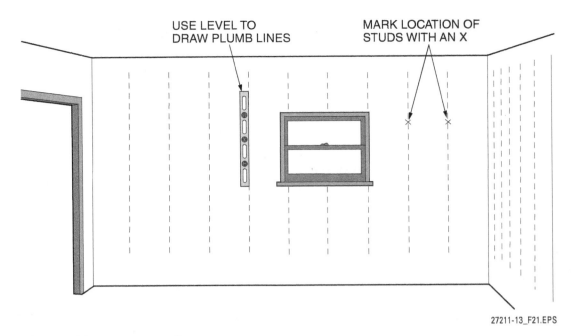

Figure 21 Marking stud locations on a wall.

point around the wall as far as the cabinets will extend (*Figure 22*). Using this line as a reference, measure up and mark the required distances for the tops of the base cabinets (normally 34½"), bottoms of the wall cabinets (normally 54"), and tops of the wall cabinets (normally 84"). Snap level chalklines around the walls at these heights. Many carpenters use a water level or laser level to lay out the horizontal high-point line.

Step 3 Mark the outline for all cabinets on the wall using a story pole (layout stick). A story pole is a narrow strip of wood that is cut to the exact length or height of a wall and marked with the different widths or heights of the cabinets to be installed on that wall. It is used to lay out and mark the locations of the cabinets on the wall by transferring the location marks from the story pole to the wall.

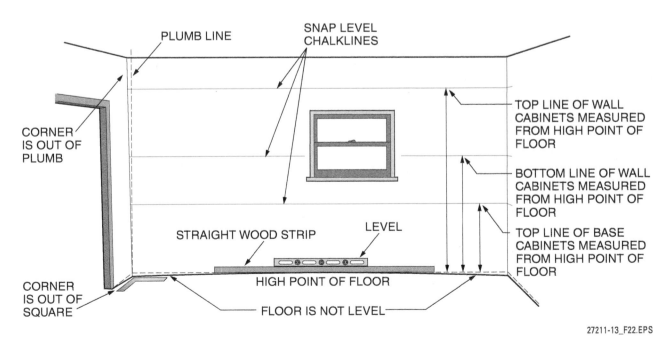

Figure 22 Marking cabinet layout lines.

Step 4 Remove the shipping skids, braces, and other packaging from the cabinets. Mark, then remove all the doors, drawers, and adjustable shelves and store them in a remote place where they cannot be damaged.

Step 5 Start the wall cabinet installation with a corner unit. First, lay out and mark the back of the cabinet for the stud locations, then predrill through the top and bottom cabinet mounting rails at these points. Countersink the drilled holes on the mounting rails.

Step 6 Position the corner cabinet in place with its bottom on the chalkline and temporarily hold it in place using a T-brace, as shown in *Figure 23*, or support it with a temporary cleat. Ensure this cabinet is installed plumb and level, using shims as necessary. Fasten the cabinet in place with screws of sufficient length to hold the cabinet securely against the wall, but do not fully tighten the screws.

Step 7 Position, brace, and fasten the adjoining wall cabinet so that leveling may be done without removing it. Align the adjoining stiles so that their faces are flush with each other, then clamp them together. Drill through the stiles and screw them together. Continue this procedure for the remaining wall cabinets.

Step 8 Use a level to check the horizontal and vertical cabinet surfaces. Shim between the cabinets and the wall until the cabinets are plumb and level (*Figure 24*). This is necessary if the doors are to fit, swing, and close properly. When the cabinets are level and plumb, tighten all mounting screws. If required, cut and install any filler strips.

4.3.0 Installing Base Cabinets and Countertops

Base cabinets are commonly installed after the wall cabinets to provide better access for the carpenter during wall cabinet installation and to prevent back injuries resulting from overreaching. The following general procedure for installing base cabinets and countertops assumes that the countertop used with the base cabinets is a manufactured unit that is ready to be installed.

Step 1 Start the installation with a corner base unit. Slide it into place, then continue to slide the remaining base cabinets into position. Make sure that the types of cabinets and their placement match those shown on the construction drawings or cabinet layout sketch.

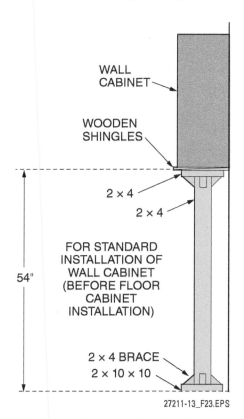

Figure 23 T-brace used to support wall cabinets.

Figure 24 Shimming wall cabinets.

NOTE: Most frame-style cabinets have a ³⁄₁₆" or ¼" overlay on the stiles at the sides for applying finished sides. When joining two cabinets together, a gap of about ½" between the sides is created. A good practice is to shim and screw this gap to hold the cabinet square.

Step 2 When all of the base cabinets are in their correct positions, shim the corner cabinet so it is level and plumb and its top is aligned with the cabinet-top level line marked on the wall. Fasten the cabinet to the wall by driving 2½" to 3" screws through the cabinet mounting rail into the studs or blocking in the wall.

Step 3 Use suitable clamps to clamp the stiles (frame members) of the corner and the adjoining cabinet together (*Figure 25*). Shim as necessary to make sure that the horizontal frame members form a level and straight line and that the frame faces are flush. After drilling countersunk pilot holes, fasten the cabinets together using screws. One screw should be installed close to the top of the cabinet end stile and one close to the bottom. If using screws, lubricate the screws with wax or soap to prevent the screws from binding or snapping in hardwood.

Step 4 Continue to install the remaining base cabinets in the same manner as described above. Check the cabinet tops from front to back and across the front edges with a level (*Figure 26*). Shim as necessary between the wall and cabinet backs and the floor and cabinet bottoms until the cabinets are plumb and level and the tops are aligned with the mark on the wall.

Step 5 Fasten the cabinets to the wall by driving 2½" to 3" screws through the cabinet mounting rails into the studs in the wall. Use a chisel or utility knife to cut off any shims flush with the edges of the cabinet.

Step 6 If required, scribe and cut a filler strip to fit the remaining space between the cabinet and end wall. (Scribing is discussed in the following section.) Clamp, then fasten the filler strip to the cabinet stile with countersunk screws. Use glue with thin filler strips. These strips can be attached at the end of a run, or at a corner, to provide drawer and door-handle clearance, or to eliminate gaps between the cabinet and wall (*Figure 27*). While filler strips are usually not needed for custom cabinets, they are commonly needed for stock cabinets. The filler strip will need to be ripped to the required width and may need to be scribed to compensate for irregularities in the wall surface or to fit around baseboards.

Step 7 Place the countertop on top of the base cabinets and firmly against the wall. If the countertop backsplash does not fit tightly

Figure 25 Joining base-cabinet stiles.

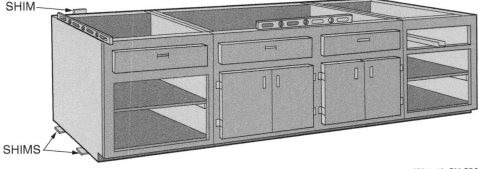

Figure 26 Shimming cabinets so that they are plumb and level.

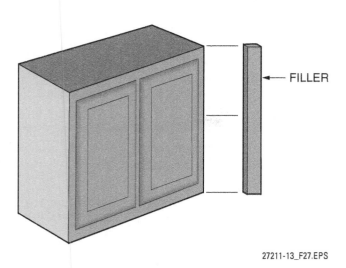

Figure 27 Filler strip.

against the wall, scribe the backsplash to match the irregular wall surface. Following this, remove the countertop, place it on sawhorses, then plane or belt-sand the backsplash to the scribed line. It is important to make sure the overhang is uniform before scribing the backsplash.

Step 8 Reposition the countertop on top of the base cabinets and tightly against the wall and check the fit. Fasten the countertop from underneath by drilling and screwing through the corner blocks and/or diagonal bracing in the top of the base cabinets. Be careful not to drill through the countertop surface.

4.3.1 Scribing Adjoining Pieces

Scribing is fitting a piece against a surface that is curved or not perfectly straight and smooth. Use the following procedure to scribe one piece to another:

Step 1 Cut a piece of the appropriate material to length for a snug fit. Use a piece that is slightly wider than the space needs.

Step 2 Place the outside edge of the piece against the surface, so that the edge to be contoured is away from the surface.

Step 3 Use a piece of material that is the width of the finished piece as a marking block. Slide the block along the surface, flat against the piece, with a pencil held against the block to mark the shape of the surface along the edge of the piece.

Be sure to hold the point of the pencil firmly against the corner between the filler and the marking block, and keep the block firmly against the surface.

Step 4 Once the piece has been marked with the surface contour, use a belt sander; hand plane; portable power plane; or if the contour is not a straight line, a band saw or jigsaw to duplicate the marked contour.

4.3.2 Tight-Joint Fasteners

Tight-joint fasteners (*Figure 28*) are fasteners used at the joint of two mating pieces, such as between long countertop sections, the junction of 45-degree countertop miter joints, or other pieces that cannot be installed in one section. Tight-joint fasteners are installed so that they are concealed from the finished side and placed into a drilled hole in the two pieces to be joined. The fastener has a built-in bolt that is tightened with a wrench to pull the two pieces together to form a tight joint.

To complete the installation, replace the doors, drawers, and shelves in their respective cabinets. Ensure that the doors open and close properly, and that they are installed level and properly aligned with the doors on the other cabinets.

Figure 28 Tight-joint fasteners.

Fastening Cabinets

When preparing to fasten cabinets to a wall, predrill screw holes in the cabinet mounting rails that are slightly larger than the screw threads. Otherwise, the cabinet will pull away from the wall when the screws are installed and tightened.

What If There Are No Studs?

In some homes and other buildings, wall studs are few and far between. Of course, if a wall is brick, there won't be any studs at all. For some installations, it may be necessary to build a new stud wall in front of the existing wall to accommodate the cabinets. For masonry that is in good condition, concrete anchors may be used.

4.0.0 Section Review

1. Unevenness of the wall surface can result in installed cabinets being _____.
   a. unsafe for use
   b. unusable
   c. misaligned
   d. damaged

2. Carpenters commonly mark the outlines of each of the cabinets on the wall, using a _____.
   a. story pole
   b. builder's transit
   c. laser level
   d. folding rule

3. Most frame-style cabinets have a small overlay on the stiles to _____.
   a. allow for the use of shims to square the cabinet
   b. allow for installation of a filler strip
   c. create a flush fit to a wall
   d. allow for application of finished sides on end cabinets

SUMMARY

Cabinets and countertops are almost always fabricated in a shop and installed somewhere else. The minor variations in angles and lines common to the work of all trades become critically visible in cabinet installation and a test of the installers' skills. The work is difficult and customers will appreciate good results. Exact measurements, precise layout and cutting, and careful assembly are required to produce a professional installation.

Review Questions

1. Tool parts that are worn, broken, or lost should be _____.
   a. reported to your supervisor
   b. replaced
   c. pointed out to fellow workers
   d. used without hesitation

2. The standard height of a base cabinet, with a countertop installed, is _____.
   a. 32¼"
   b. 34"
   c. 36"
   d. 37½"

3. A freestanding kitchen cabinet is referred to as a(n) _____.
   a. peninsula
   b. floater
   c. divider
   d. island

4. Most wall cabinets have a depth of _____.
   a. 12"
   b. 14"
   c. 16"
   d. 24"

5. When referring to kitchen countertops, the letters HPL stand for _____.
   a. high-priced laminate
   b. heated plastic laminate
   c. high-pressure laminate
   d. hydrostatic polymer laminate

6. One of the advantages of solid-surface countertops over laminate countertops is that solid-surface countertops are _____.
   a. less expensive
   b. more durable
   c. easier to fabricate
   d. available as prefabricated units

7. A separate piece attached to the rear edge of a countertop is called a _____.
   a. return
   b. back trim
   c. closing block
   d. backsplash

8. Two common types of cabinet construction are known as face frame and _____.
   a. unitized
   b. frameless
   c. integral frame
   d. contemporary

9. Low-priced cabinets are typically constructed of particleboard covered with a finish layer of _____.
   a. vinyl film or melamine
   b. high-pressure laminate
   c. epoxy enamel
   d. wood veneer

10. The only one of the following natural woods not normally used in high-quality cabinets is _____.
    a. oak
    b. birch
    c. sweet gum
    d. cherry

11. An important advantage of lipped cabinet doors is _____.
    a. low cost
    b. efficient use of material
    c. availability of hardware
    d. ease of installation

12. A nonrabbeted cabinet door that mounts on the outside frame of the cabinet and conceals the cabinet opening is called a(n) _____.
    a. overlay door
    b. lipped door
    c. flush door
    d. oversized door

13. A pivot hinge is designed for use with _____.
    a. panel-and-frame doors
    b. flush doors
    c. lipped doors
    d. overlay doors

14. Most wooden cabinet shelves should be made from _____.
    a. ¼" stock
    b. ⅜" stock
    c. ½" stock
    d. ¾" stock

15. A spring-loaded catch mounted in the edge of a cabinet door is referred to as a _____.
    a. concealed catch
    b. bullet catch
    c. snap catch
    d. bayonet catch

16. For shelf adjustability, some cabinet designs use a combination of drilled holes and _____.
    a. spring-loaded pins
    b. captive fasteners
    c. metal or plastic pegs
    d. extension rods

17. For a period of time before they are installed, cabinets should be moved into the room so they will _____.
    a. be easier to unpack
    b. acclimate
    c. not be damaged
    d. dehydrate

18. Before cabinets are installed, use a level and straightedge to find the kitchen floor's _____.
    a. low point
    b. average level
    c. midpoint
    d. high point

19. When measuring up from the high point of the floor, a line representing the tops of base cabinets would normally be drawn at _____.
    a. 25 ½"
    b. 32"
    c. 34 ½"
    d. 36"

20. For a finished appearance, space between the last cabinet and an intersecting wall should be closed with a(n) _____.
    a. bridge piece
    b. filler strip
    c. expansion joint
    d. trim board

Trade Terms Quiz

Fill in the blank with the correct term that you learned from your study of this module.

1. A cabinet installation component that is perpendicular to the back edge of the countertop is the _____.
2. A(n) _____ is a decorative metal plate used behind a drawer or cabinet pull.
3. Also called a furr down, the _____ encloses the space between the top of a wall cabinet and the ceiling.
4. A cabinet unit designed for installation on the wall above the base cabinets is called a _____.
5. Used to support wall cabinets while they are being installed, a _____ is often fabricated on the job site.
6. _____ is a term used to describe a style of cabinet door that has been rabbeted on all four sides.
7. A cabinet that has been made to measure for the installation, whether in a cabinet shop or on the job site, is referred to as a _____.
8. To _____ is to adjust to an environment's conditions, such as temperature or humidity level.
9. A _____ door is fitted into a cabinet opening and does not protrude beyond the cabinet face.
10. A cabinet door that is designed with solid wood or plywood panels enclosed by side stiles and horizontal rails is the _____.
11. Sometimes called a modular cabinet, a _____ is a mass-produced unit with limited design options.
12. Mounted on the outside frame of a cabinet, the _____ door conceals the cabinet opening. It is also called a surface door.
13. Often used to provide an attractive surface for a cabinet, a _____ is a thin sheet of wood applied as an overlay.
14. A cabinet unit that is installed on the room's floor with its back against a wall is called a _____.
15. A nonreflective laminate finish that is sometimes preferred to a shiny finish is referred to as _____.
16. Another term for human engineering, _____ is the science of matching craftworkers' capabilities to the job demands and conditions of the work site.
17. _____ is a countertop surfacing material made by using heat and pressure to bond together layers of plastic material into thin sheets.
18. A small cabinet that is installed in a bathroom to store personal hygiene products is called a _____.
19. _____ are the next step up in price and available features from stock cabinets.
20. Often referred to as HPL, _____ is widely used as a surfacing material for both countertops and cabinets.

Trade Terms

Acclimate
Backsplash
Base cabinet
Custom cabinet
Ergonomics
Escutcheon plate
Flush
High-pressure laminate
Lipped
Matte
Overlay
Panel-and-frame door
Plastic laminate
Semicustom cabinet
Soffit
Stock cabinet
T-brace
Vanity
Veneer
Wall cabinet

Trade Terms Introduced in This Module

Acclimate: Adjust to a new temperature, humidity level, or environment.

Backsplash: A trim piece that is attached along the back edge of a countertop and covers the wall-to-countertop abutment to conceal any irregularities.

Base cabinet: Cabinet that is made to sit on the floor with its back against a wall.

Custom cabinet: Tailor-made cabinet produced in a cabinet shop, millwork plant, or on the job site.

Ergonomics: The science of fitting workplace conditions and job demands to the capabilities of the craftworkers; also called human engineering.

Escutcheon plate: A decorative metal plate that is placed against a door or drawer face behind the pull.

Flush: Style of cabinet door that fits into and is flush with the face of the cabinet opening.

High-pressure laminate (HPL): Countertop material produced by pressing layers of plastic material under intense heat and pressure to form a laminate.

Lipped: Style of cabinet door in which the all edges are rabbeted, so when it is installed on a cabinet, the edges overlap and conceal the opening.

Matte: A surface finish that has no shine or luster.

Overlay: Style of cabinet door mounted on the outside frame that overlays the cabinet opening so that the opening is concealed; also called surface door.

Panel-and-frame door: Cabinet door made of plywood or solid wood panels fitted between side stiles and horizontal rails; sometimes referred to as panel doors.

Plastic laminate: Product, typically made of several layers of plastic material bonded together under intense heat and pressure, commonly used for countertops and shelves.

Semicustom cabinet: Cabinet produced in a mill or cabinet shop, which is available in many materials, sizes, and finishes.

Soffit: The enclosed section (which is usually decorated by paint or wallpaper) between the ceiling and the top of the wall cabinets; also called a furr down.

Stock cabinet: Cabinet mass produced in specialized plants, with somewhat limited design options; also called modular cabinets.

T-brace: Type of brace used to support wall cabinets during installation.

Vanity: A small cabinet, commonly installed in bathrooms to store personal hygiene products.

Veneer: A thin layer or sheet of wood intended to be overlaid on a surface to provide strength, stability, and/or an attractive finish. Thicknesses range between $1/16"$ and $1/8"$ for core plies and between $1/128"$ and $1/32"$ for decorative faces.

Wall cabinet: Cabinet that is made to be attached to the upper part of a wall.

Additional Resources

This module presents thorough resources for task training. The following resource material is suggested for further study.

Cabinet Makers Association website. **www.cabinetmakers.org**
Kitchen Cabinet Manufacturers Association website. **www.kcma.org**
Mill's Pride Cabinetry website. **www.millspridekitchens.com**

Figure Credits

Courtesy of Artistic Stone Design, CO01, Figure 6
Strasser Woodenworks, Inc., Figure 8
Blum, Inc., Figure 15

Photo courtesy of SOSS Invisible Hinges, Figure 16
Formica Corporation, SA01
Polycor Vetrazzo, SA04

Section Review Answer Key

Answer	Section Reference	Objective
Section One		
1.b	1.0.0	1a
2.c	1.2.0	1b
Section Two		
1.b	2.1.0	2a
2.c	2.2.0	2b
3.d	2.3.2	2c
Section Three		
1.b	3.1.0	3a
2.a	3.2.2	3b
Section Four		
1.c	4.1.0	4a
2.a	4.2.0	4b
3.d	4.3.0	4c

NCCER CURRICULA — USER UPDATE

NCCER makes every effort to keep its textbooks up-to-date and free of technical errors. We appreciate your help in this process. If you find an error, a typographical mistake, or an inaccuracy in NCCER's curricula, please fill out this form (or a photocopy), or complete the online form at **www.nccer.org/olf**. Be sure to include the exact module ID number, page number, a detailed description, and your recommended correction. Your input will be brought to the attention of the Authoring Team. Thank you for your assistance.

Instructors – If you have an idea for improving this textbook, or have found that additional materials were necessary to teach this module effectively, please let us know so that we may present your suggestions to the Authoring Team.

NCCER Product Development and Revision
13614 Progress Blvd., Alachua, FL 32615

Email: curriculum@nccer.org
Online: www.nccer.org/olf

❏ Trainee Guide ❏ Lesson Plans ❏ Exam ❏ PowerPoints Other _____

Craft / Level: _____ Copyright Date: _____

Module ID Number / Title: _____

Section Number(s): _____

Description: _____

_____

_____

_____

_____

Recommended Correction: _____

_____

_____

_____

_____

Your Name: _____

Address: _____

_____

Email: _____ Phone: _____

This page is intentionally left blank.

Glossary

A-weighted decibel (dBA): A single number measurement based on the decibel but weighted to approximate the response of the human ear with respect to frequencies.

Access: A passageway or a corridor between rooms.

Acclimate: Adjust to a new temperature, humidity level, or environment.

Acoustical materials: Types of ceiling panel, plaster, and other materials that have high absorption characteristics for sound waves.

Acoustics: A science involving the production, transmission, reception, and effects of sound. In a room or other location, it refers to those characteristics that control reflections of sound waves and thus the sound reception in the area.

AISC: American Institute of Steel Construction.

AISI: American Iron and Steel Institute.

All-purpose compound: Combines the features of taping and topping compounds. It does not bond as well as taping compound but finishes better.

Apron: A piece of window trim that is used under the stool of a finished window frame. It is sometimes referred to as undersill trim.

Asphalt roofing cement: An adhesive that is used to seal down the free tabs of strip shingles. This plastic asphalt cement is mainly used in open valley construction and other flashing areas where necessary for protection against the weather.

Astragal: A piece of molding attached to the edge of an inactive door on a pair of double doors. It serves as a stop for the active door.

Backsplash: Trim piece that is attached along the back edge of a countertop and covers the wall-to-countertop abutment to conceal any irregularities.

Base cabinet: Cabinet that is made to sit on the floor with its back against a wall.

Base flashing: The protective sealing material placed next to areas vulnerable to leaks, such as chimneys.

Base steel thickness: The thickness of bare steel exclusive of all coatings.

Beams: Loadbearing horizontal framing members supported by walls or columns and girders.

Benchmark: A known elevation on the site used as a reference point during construction.

Board-and-batten siding: A type of vertical siding consisting of wide boards with the joint covered by narrow strips known as battens.

Bracing: Structural elements that are installed to provide restraint or support (or both) to other framing members so that the complete assembly forms a stable structure.

Brown coat: A coat of plaster with a rough face on which a finish coat will be placed.

Building paper: A heavy paper used for construction work. It assists in weatherproofing the walls and prevents wind infiltration. Building paper is made of various materials and is not a vapor barrier.

Bull nose: A metal corner bead with rounded edges.

Bundle: A package containing a specified number of shingles or shakes. The number is related to square-foot coverage and varies with the product.

Butt: Any kind of hinge, except a strap or T-hinge.

Callouts: Markings or identifying tags describing parts of a drawing; callouts may refer to detail drawings, schedules, or other drawings.

Cap flashing: The protective sealing material that overlaps the base and is embedded in the mortar joints of vulnerable areas of a roof, such as a chimney.

Casing: The exposed finish material around the edge of a door or window opening.

Catches: Spring bolts used to secure a door when shut.

Ceiling joist: A horizontal structural framing member that supports ceiling components and may be subject to attic loads.

Ceiling panels: Acoustical ceiling boards that are suspended by a concealed grid mounting system. The edges are often kerfed and cut back.

Ceiling tiles: Any lay-in acoustical board that is designed for use with an exposed grid mounting system. Ceiling tiles normally do not have finished edges or precise dimensional tolerances because the exposed grid mounting system provides the trim-out.

Civil drawings: A drawing that shows the overall shape of the building within the confines of the site. Also referred to as site plans.

Cold-formed sheet steel: Sheet steel or strip steel that is manufactured by press braking of blanks sheared from sheets or cut lengths of coils or plates, or by continuous roll forming of cold- or hot-rolled coils of sheet steel. Both forming operations are performed at ambient room temperature, that is, without any addition of heat such as would be required for hot forming.

Cold-formed steel: See *cold-formed sheet steel*.

Columns: Vertical structural members that support the load of other members.

Component assembly: A fabricated assemblage of cold-formed steel structural members that is manufactured by the component manufacturer, which may also include structural steel framing, sheathing, insulation, or other products.

Component design drawing: The written, graphic, and pictorial definition of an individual component assembly, which includes engineering design data.

Component designer: The individual or organization responsible for the engineering design of component assemblies.

Component manufacturer: The individual or organization responsible for the manufacturing of component assemblies for the project.

Component placement diagram: The illustration supplied by the component manufacturer identifying the location assumed for each of the component assemblies, which references each individually designated component design drawing.

Compound cut: A combined bevel and miter cut.

Condensation: The process by which a vapor is converted to a liquid, such as the conversion of the moisture in air to water.

Contour lines: Imaginary lines on a civil drawing that connect points of the same elevation. Contour lines normally never cross one another.

Convection: The movement of heat that either occurs naturally due to temperature differences or is forced by a fan or pump.

Coordinator: A device for use with exit features to hold the active door open until the inactive door is closed.

Coped joint: A joint made by cutting the end of a piece of molding to the shape it will fit against.

Corner bead: A metal or plastic angle used to protect outside corners where drywall panels meet.

Cornice: The construction under the eaves where the roof and side walls meet.

Course: One row of brick, block, or siding as it is placed in the wall.

Cripple stud: A stud that is placed between a header and a window or door head track, a header and a wall top track, or a window sill and a bottom track to provide a backing to attach finishing and sheathing material.

Custom cabinet: Tailor-made cabinet produced in a cabinet shop, millwork plant, or on the job site.

Cylindrical lockset: A lock in which the keyhole and tumbler mechanism are in a cylinder that is separate from the lock case and can be removed to change the keying of the lock.

Deadbolt: A square-head bolt in a door lock that requires a key to move it in either direction.

Decibel (dB): An expression of the relative loudness of sounds in air as perceived by the human ear.

Design thickness: The steel thickness used in design that is equal to the minimum base steel thickness divided by 0.95.

Detail drawings: Drawings shown at a larger scale in order to show specific features or connections.

Dew point: The temperature at which air becomes oversaturated with moisture and the moisture condenses.

Diffuser: An attachment for duct openings in air distribution systems that distributes the air in wide flow patterns. In lighting systems, it is an attachment used to redirect or scatter the light from a light source.

Diffusion: The movement, often contrary to gravity, of molecules of gas in all directions, causing them to intermingle.

Door closer: A mechanical device used to check a door and prevent it from slamming when it is being closed. It also ensures the closing of the door.

Door frame: The surrounding case of a door into which a door closes. It consists of two upright pieces called jambs and a horizontal top piece called the head.

Door stop: The strip against which a door closes on the inside face of the frame or jamb. It can also be a hardware device used to hold the door open to any desired position or a hardware device placed against the baseboard to keep the door from marring the wall.

Dry lines: A string line suspended from two points and used as a guideline when installing a suspended ceiling.

Drywall: A generic term for paper-covered gypsum-core panels; also known as gypsum drywall.

Dust-proof strike: A metal piece used to receive the latch bolt when a door is closed. The piece has a spring-action cover to keep particles out of the strike when the door is open.

Duty rating: Load capacity of a ladder.

Eave: The lower part of a roof, which projects over the side wall.

Edge stiffener: That part of a C-shape framing member that extends perpendicular from the flange as a stiffening element.

Elevation drawings: Drawings showing a view from the front, rear, or side of a structure.

Erection drawings: See *installation drawings*.

Erector: *See installer*.

Ergonomics: The science of fitting workplace conditions and job demands to the capabilities of the craftworkers; also called human engineering.

Escutcheon plate: A decorative metal plate that is placed against a door or drawer face behind the pull.

Exposure: The distance (in inches) between the exposed edges of overlapping shingles.

Fascia: The exterior finish member of a cornice on which the rain gutter is usually hung.

Feathering: Tapering joint compound at the edges of a drywall joint to provide a uniform finish.

Field notes: A permanent record of field measurement data and related information.

Finger-jointed stock: Paint-grade moldings made in a mill from shorter lengths of wood joined together.

Finish coat: The final coat of plaster or paint.

Finish hardware: The exposed hardware in a building such as doorknobs, door hinges, door locks, door closers, window hardware, shelf and clothing storage hangers, and bathroom hardware.

Finishing sawhorse: A pair of trestles used to support lumber, molding, doors, stair parts, and other materials while they are being fitted and shaped for installation.

Fissured: A ceiling-panel or ceiling-tile surface design that has the appearance of splits or cracks.

Flange: That portion of the C-shape framing member or track that is perpendicular to the web.

Floating interior angle construction: A drywall installation technique in which no fasteners are used at the edge of the panel in order to allow for structural stresses.

Floor joist: A horizontal structural framing member that supports floor loads and superimposed vertical loads.

Flush: Style of cabinet door that fits into and is flush with the face of the cabinet opening.

Flush bolt: A sliding bolt mechanism that is mortised into a door at the top and bottom edge. It is used to hold an inactive door in a fixed position on a pair of double doors.

Flush door: A door of any size that has a totally flat surface.

Framing contractor: See *installer*.

Framing material: Steel products, including but not limited to structural members and prefabricated structural assemblies, ordered expressly for the requirements of the project.

Frequency: Cycles per unit of time, usually expressed in hertz (Hz).

Frieze board: A horizontal finish member connecting the top of the sidewall, usually abutting the soffit. Its bottom edge usually serves as a termination point for various types of siding materials.

General contractor: See *installer*.

Girders: Large steel or wood beams supporting a building, usually around a perimeter.

Hanging stile: The door stile to which the hinges are fastened.

Hardware: Components, such as hinges, locksets, and closers, used to attach a door to its frame or operate the door.

Harsh environments: Coastal areas where additional corrosion protection may be necessary.

Head: The horizontal member at the top of a door or window opening.

Hertz (Hz): A unit of frequency equal to one cycle per second.

High-pressure laminate (HPL): Countertop material produced by pressing layers of plastic material under intense heat and pressure to form a laminate.

Hinge: The hardware fastened to the edge of a door that allows the frame to pivot around a steel pin, permitting the operation of the door.

In-line framing: Framing method where all vertical and horizontal load-carrying members are aligned.

Installation drawings: Field installation drawings that show the location and installation of the cold-formed steel structural framing.

Installer: Party responsible for the installation of cold-formed steel products.

Invert: The lowest point of a pipe through which liquid flows.

Isometric drawings: Three-dimensional drawings in which the object is tilted so that all three faces are equally inclined to the picture plane.

Jack stud: A stud that does not span the full height of the wall and provides bearing for headers. Also called a trimmer stud.

Jamb: One of the vertical members on either side of a door or window opening.

Joint: A place where two pieces of material meet.

Joint compound: Patching compound used to finish drywall joints, conceal fasteners, and repair irregularities in the drywall. It dries hard and has a strong bond. Sometimes called mud or taping compound.

Joists: Horizontal wood or steel members supported by beams; joists support floor sheathing.

Kerf: A slot or cut made with a saw.

King stud: A stud, adjacent to a jack stud, that spans the full height of the wall and supports vertical and lateral loads.

Knob lockset: A lock for a door, with the locking cylinder located in the center of the knob.

Landscape drawings: A drawing that shows proposed plantings and other landscape features.

Latch bolt: A spring-loaded bolt in a lock, with a beveled head that is retracted when hitting the strike.

Ledger: A board to which the lookouts are attached and which is placed against the outside wall of the structure. It is also used as a nailing edge for the soffit material.

Liability: An obligation or responsibility, typically financial.

Lightweight compound: An all-purpose compound having less weight than standard compounds.

Lip: See *edge stiffener*.

Lipped: Style of cabinet door in which the all edges are rabbeted, so when it is installed on a cabinet, the edges overlap and conceal the opening.

Lockset: The entire lock unit, including locks, strike plate, and trim pieces.

Lookout: A member used to support the overhanging portion of a roof.

Louver: A slatted opening used for ventilation, usually in a gable end or a soffit.

Material supplier: An individual or entity responsible for furnishing framing materials for the project.

Matte: A surface finish that has no shine or luster.

Molding: Long strips of material used for finishing and decorative trim.

Mortise: A measured portion of wood removed to receive a piece of hardware such as a lock or butt.

Mortise lockset: A rectangular metal box that houses a lock. It usually has a latch and deadbolt as part of the unit. Options consist of a locking cylinder and/or thumb turn by which it can be secured. It is used on residential entry doors and commercial doors.

Mud: See *joint compound*.

Nail pop: The protrusion of a nail above the wallboard surface that is usually caused by shrinkage of the framing or by incorrect installation. Also applies to screws.

Nonstructural member: A member in a steel framed assembly that is limited to a transverse load of not more than 10 lb/ft^2 (480 Pa); a superimposed axial load, exclusive of sheathing materials, of not more than 100 lb/ft (1,460 N/m); or a superimposed axial load of not more than 200 lb (890 N).

Overhang: The part that extends beyond the building line. The amount of overhang is always given as a projection from the building line on a horizontal plane.

Overlay: Style of cabinet door mounted on the outside frame that overlays the cabinet opening so that the opening is concealed; also called surface door.

Panel-and-frame door: Cabinet door made of plywood or solid wood panels fitted between side stiles and horizontal rails; sometimes referred to as panel doors.

Panic hardware: Hardware that provides an emergency escape exit.

Perm: The measure of water vapor permeability. It equals the number of grains squared of water vapor passing through a 1-square-foot piece of material per hour, per inch of mercury difference in vapor pressure.

Permeability: The measure of a material's capacity to allow the passage of liquids or gases.

Permeable: Porous; having small openings that permit liquids or gases to seep through.

Permeance: The ratio of water vapor flow to the vapor pressure difference between two surfaces.

Pitch: The ratio of the rise to the span, indicated as a fraction. For example, a roof with a 6' rise and a 24' span will have a ¼ pitch.

Plan view: A drawing that represents a view looking down on an object.

Plancier: Similar to a soffit, but the member is usually fastened to the underside of a rafter rather than the lookout.

Plastic laminate: Product, typically made of several layers of plastic material bonded together under intense heat and pressure, commonly used for countertops and shelves.

Plenum: A chamber or container for moving air under a slight pressure. In commercial construction, the area between the suspended ceiling and the floor or roof above is often used as the HVAC return air plenum.

Plumb: A true vertical position.

Prefinished: Material such as molding, doors, cabinets, and paneling that has been stained, varnished, or painted at the factory.

Prehung door: A door that is delivered to the job site from the mill already hung in the frame or jamb. In some instances, the trim may be applied on one side.

Punchout: A hole made during the manufacturing process in the web of a steel framing member.

R-value: The resistance to conductive heat flow through a material or gas.

Rabbet: A groove cut in the edge of a board so as to receive another board; a rectangular groove cut in the edge of a board.

Rail: A horizontal member of a door or window sash.

Rake: The slope or pitch of the cornice that parallels the roof rafters on the gable end.

Reveal: The amount of setback of the casing from the face side of window and door jambs or similar pieces.

Ridge: The horizontal line formed by the two rafters of a sloping roof that have been nailed together. The ridge is the highest point at the top of the roof where the roof slopes meet.

Ridges: Slight protrusions in the center of a finished drywall joint that are usually caused by insufficient drying time. Also known as beads.

Riser diagram: Type of drawing that depicts the layout, components, and connections of a piping system.

Roof sheathing: Usually 4 × 8 sheets of plywood, but can also be 1 × 8 or 1 × 12 roof boards, or other new products approved by local building codes. Also referred to as decking.

Rough opening: Any unfinished door or window opening in a building.

Saddle: An auxiliary roof deck that is built above the chimney to divert water to either side. It is a structure with a ridge sloping in two directions that is placed between the back side of a chimney and the roof sloping toward it. Also referred to as a cricket.

Safety data sheet (SDS): A document that must accompany any hazardous substance. The SDS identifies the substance and gives the exposure limits, the physical and chemical characteristics, the kind of hazard it poses, precautions for safe handling and use, and specific control measures.

Sanitary stop: Door stop with a 45-degree angle cut at the bottom on the vertical jambs.

Scarf joints: End joints made by overlapping two pieces of molding with 22.5- or 45-degree angle cuts.

Schedule: Tables that describe and specify the various types and sizes of construction materials used in a building. Door schedules, window schedules, and finish schedules are the most common types.

Scratch coat: The first coat of cement plaster consisting of a fine aggregate that is applied through a diamond mesh reinforcement or on a masonry surface.

Scrim: A loosely knit fabric.

Selvage: The section of a composition roofing roll or shingle that is not covered with an aggregate.

Semicustom cabinet: Cabinet produced in a mill or cabinet shop, which is available in many materials, sizes, and finishes.

Shim: A thin, tapered piece of wood such as a wooden shingle used to fill in gaps and to level or plumb structural components.

Shop drawings: Drawings for the production of individual component assemblies for the project.

Side lap: The distance between adjacent shingles that overlap, measured in inches.

Sill: The lowest member at the bottom of a door or window opening. It also refers to the lowest member of wood framing supporting the framing of a building.

Skim coat: A thin coat of joint or topping compound that is applied over the entire drywall surface. Sometimes required under a high-gloss finish.

Slope: The ratio of rise to run. The rise in inches is indicated for every foot of run.

Slurry: A thin mixture of water or other liquid with any of several substances such as cement, plaster, or clay.

Smoke gasket: A rubber strip that goes all the way around the door to keep smoke from penetrating an area in case of fire.

Soffit: The enclosed section (which is usually decorated by paint or wallpaper) between the ceiling and the top of the wall cabinets; also called a furr down; The underside of a roof overhang.

Sound attenuation: The reduction of sound as it passes through a material.

Sound transmission class (STC): The rating by which the sound attenuation of a door is determined. The higher or greater the number, the better the sound reduction.

Span: The clear horizontal distance between bearing supports.

Square: The amount of shingles needed to cover 100 square feet of roof surface. For example, square means 10' square, or 10' × 10'.

Square cuts: Cuts made at a right (90-degree) angle.

Standard cold-formed steel structural shapes: Cold-formed steel structural members that meet the requirements of the SSMA Product Technical Guide.

Stile: The vertical edge of a door.

Stock cabinet: Cabinet mass-produced in specialized plants, with somewhat limited design options; also called modular cabinets.

Stool: The bottom horizontal trim piece of a window.

Strap: Flat or coil sheet steel material typically used for bracing and blocking that transfers loads by tension and/or shear.

Striated: A ceiling panel or ceiling tile surface design that has the appearance of fine parallel grooves.

Strike plate: A metal plate screwed to the jamb of a door so that when the door is closed, the bolt of the lock strikes against it. The bolt is then retracted and slides along the metal plate. When the door is fully closed, the bolt inserts itself into a hole in the plate to hold the door securely in place.

Structural engineer-of-record: The design professional who is responsible for sealing the contract documents, which indicates that the structural engineer-of-record has performed or supervised the analysis, design, and document preparation for the structure and has knowledge of the requirements for the load-carrying structural system.

Structural member: A floor joist, rim track, structural stud, wall track in a structural wall, ceiling joist, roof rafter, header, or other member that is designed or intended to carry loads.

Structural stud: A stud in an exterior wall or an interior stud that supports superimposed vertical loads and may transfer lateral loads, including full-height wall studs, king studs, jack studs, and cripple studs.

Stud: A vertical framing member in a wall system or assembly.

Substrate: The underlying material to which a finish is applied.

Sweep: A type of weather stripping. A sweep is a felt or rubber flap mounted in a metal channel to seal door bottoms to prevent air infiltration.

T-brace: Type of brace used to support wall cabinets during installation.

Tape: A strong paper or fiberglass tape used to cover the joint between two sheets of drywall.

Tapered joint: A joint where tapered edges of drywall meet.

Taping compound: See *joint compound*.

Template: A thin piece of material such as plastic or heavy paper with a shape cut out of it, or the whole of it cut to a shape on its perimeter. A template is used to transfer that shape to another object by tracing it with a pencil or scribe.

Threshold: A piece of wood, metal, or stone that is set between the door jamb and the bottom of a door opening.

Top lap: The distance, measured in inches, between the lower edge of an overlapping shingle and the upper edge of the lapping shingle.

Topping compound: A joint compound used for second and third coats. It dries soft and smooth and is easier to sand than taping compound.

Transom: A panel above a door that lets light and/or air into a room or is used to fill the space above a door when the ceiling heights on both sides of the door opening allow.

Trim: Finish materials such as molding placed around doors and windows and at the top and bottom of a wall.

Trimmer: See *jack stud*.

Truss: A coplanar system of structural members joined together at their ends, usually to construct a series of triangles that form a stable beam-like framework.

Underlayment: Asphalt-saturated felt protection for sheathing; 15-lb roofer's felt is commonly used. The roll size is 3' × 144', or a little over four squares.

Valley: The internal part of the angle formed by the meeting of two roofs.

Valley flashing: Watertight protection at a roof intersection. Various metals and asphalt products are used; however, materials vary based on local building codes.

Vanity: A small cabinet, commonly installed in bathrooms to store personal hygiene products.

Vapor barrier: A material used to retard the flow of vapor and moisture into walls and prevent condensation within them. The vapor barrier must be located on the warm side of the wall.

Veneer: A thin layer or sheet of wood intended to be overlaid on a surface to provide strength, stability, and/or an attractive finish. Thicknesses range between $1/16"$ and $1/8"$ for core plies and between $1/128"$ and $1/32"$ for decorative faces; a brick face applied to the surface of a frame structure.

Vent: A small opening to allow the passage of air.

Vent-stack flashing: Flanges that are used to tightly seal pipe projections through the roof. They are usually prefabricated.

Wainscoting: A wall finish, usually made of wood, stone, or ceramic tile, that is applied partway up the wall from the floor.

Wall cabinet: Cabinet that is made to be attached to the upper part of a wall.

Wall flashing: A form of metal shingle that can be shaped into a protective seal interlacing where the roof line joins an exterior wall. Also referred to as step flashing.

Water stop: Thin sheets of rubber, plastic, or other material inserted in a construction joint to obstruct the seepage of water through the joint.

Water vapor: Water in a vapor (gas) form, especially when below the boiling point and diffused in the atmosphere.

Weather stripping: Strips of metal or plastic used to keep out air or moisture that would otherwise enter through the spaces between the outer edges of doors and windows and the finish frames.

Yield strength: A characteristic of the basic strength of the steel material defined as the highest unit stress that the material can endure before permanent deformation occurs, as measured by a tensile test in accordance with ASTM A370, *Standard Test Methods and Definitions for Mechanical Testing of Steel Products*.

This page is intentionally left blank.

Index

A

Absorption, (27209):2
Abuse-resistant drywall, (27206):5, 29
AC. *See* Articulation class (AC)
Access, (27208):39, 45, 77
Acclimate, (27211):19, 28
Acoustical materials, (27209):1, 40
Acoustical panel installations, (27209):34
Acoustical tile, (27209):6
Acoustics, (27209):1, 2, 40
Activation device locksets, (27208):41–42
Adhesives
    asphalt roofing cement, (27202):12, 16, 74
    drywall
        applications, (27206):5, 9
        applying, (27206):9, 10–12
        for architectural boards, (27206):11
        bridging ability, (27206):10
        for ceiling installations, (27206):10, 12
        for concrete, (27206):11–12
        contact adhesives, (27206):9–10
        disadvantages, (27206):10
        drying time, (27206):10
        dry-powder, (27206):9
        function, (27206):8–9
        for furring, (27206):12
        laminating adhesives, (27206):9
        for masonry, (27206):11
        modified contact adhesives, (27206):10
        multi-ply systems, (27206):38
        for rigid/semirigid insulation, (27206):10
        sheet lamination, (27206):38
        sound-isolation construction, (27206):38
        spot lamination, (27206):38
        for steel framing, (27206):10, 11
        strip lamination, (27206):38
        stud adhesives, (27206):9, 10
        supplemental fasteners, (27206):9–10, 38–39
    roofing, (27202):26, 42
Aerial lifts, (27204):2–3
Aerial lift safety, (27204):2–3
Air infiltration and heat loss, (27203):4
Air infiltration control, (27203):28–31
AISC. *See* American Institute of Steel Construction (AISC)
AISI. *See* American Iron and Steel Institute (AISI)
Algae-resistant shingles, (27202):18–19
All-purpose compound, (27207):3, 6, 7, 8, 52
Aluminum foil, (27203):25, 28
Aluminum ladders, (27202):7
American Institute of Steel Construction (AISC), (27205):37
American Iron and Steel Institute (AISI), (27205):37
American National Standards Institute (ANSI)
    direct-hold electromagnetic lock grades, (27208):46
    Standards
        ANSI/UL 263 (fire-resistant tile standard ratings), (27209):8
        Z87.1 (eye and face protection), (27210):1
American Society for Testing and Materials (ASTM) International Standards
    A370, Standard Test Methods and Definitions for mechanical Testing of Steel Products, (27205):38
    C557, Standard Specification for Adhesives for Fastening Gypsum Wallboard to Wood Framing, (27206):9, 10
    C630 (drywall finishing), (27207):1
    cold-formed steel materials, (27205):2
    E84 (drywall testing), (27206):2
    E84 (flame-spread rating), (27209):8
    E119 (fire-resistant tile standard ratings), (27209):8
    E814, Fire Test, (27206):36–37
    E1110, E111, AC ratings, (27209):2, 8
    E1264, CAC minimums, (27209):2, 8
    E1477, light reflectance (LR) values, (27209):9
    slate quality, (27202):23
Anchorage points, (27202):5
Anchoring devices, (27202):4
Anchors, steel framing, (27205):12
Angles, (27205):18
Annular drywall nails, (27206):5
ANSI. *See* American National Standards Institute (ANSI)
APP modified bitumen products. *See* Atactic polypropylene (APP) modified bitumen products
Apron, (27210):4, 5, 26, 31
Architectural drawings
    callouts, (27201):13
    detail drawings, (27201):10, 12, 14
    detailing instructions, (27201):9–10
    elevation drawings and sections, (27201):12–13, 14
    finish schedule, (27201):9, 12, 13
    floor plans, (27201):9–10
    grid lines, (27201):10
    legends, (27201):9
    notes, (27201):10
    roof plans, (27201):10, 11
    scale, (27201):9
    schedules, (27201):10, 12
Architectural drywall, (27206):2, 3, 4, 11
Architectural Graphic Standards (AIA), (27201):4
Architectural shingles, (27202):18, 19, 37
Area separation walls, (27205):20
Arresting force, (27202):2
Arrows, (27201):4–5
Articulation class (AC), (27209):2, 8
Asbestos, (27207):7
Asbestos-fiber shingles, (27202):18
As built drawings, (27201):4
Asphalt-coated, fiberglass mat shingles, (27202):18
Asphalt roofing cement, (27202):12, 16, 74
Association of the Wall & Ceiling Industries-International, *A Recommended Specification for Levels of Gypsum Board Finish*, (27207):1
ASTM International. *See* American Society for Testing and Materials (ASTM) International
Astragal, (27208):39, 50, 77

Atactic polypropylene (APP) modified bitumen products, (27202):25–26
Attics
　insulation for ice-dam control, (27202):33–34
　ventilation and moisture control in, (27202):29–30, (27203):23
Automatic flush bolts, (27208):48
Aviation snips, (27204):38, 40, (27205):8
a-weighted decibel (dBA), (27209):1, 3, 40

B

Backing, steel framed structural walls, (27205):24
Backing boards, (27210):22
Backsaw, (27210):10
Backsplash, (27211):4, 6, 7, 28
Balloon framing, (27205):19, 23
Ball-point catches, (27211):16, 17
Banjo, (27207):16
Bar clamps, (27205):7
Base cabinets, (27211):4–5, 21–23, 28
Base connections, (27205):19–20
Base flashing, (27202):32, 57, 74
Basements, moisture control in, (27203):22–23
Base moldings, (27210):5–6, 11, 14–16, 20, 24
Base steel thickness, (27205):37
Base trim, (27210):26
Basic oxygen furnace (BOF) process, (27205):4
Bathroom drywall installation, (27206):26, 28–29
BAZOOKA®, (27207):17
Beams, (27201):1, 13, 37
Bed moldings, (27210):7, 11
Benchmark, (27201):1, 8, 37
Beveled wood siding, (27204):10, 11, 25–28
Bevel settings, crown molding, (27210):13
BHMA finish numbering system. See Builders Hardware Manufacturer's Association (BHMA) finish numbering system
Bifold doors, (27208):8, 10
BIM. See Building information modeling (BIM)
Blocking, (27205):14, 16, 36
Board-and-batten siding
　basics, (27204):10, 11
　defined, (27204):6, 59
　installation, (27204):28
　material takeoffs, (27204):54
Board blisters, drywall, (27207):43
Board-on-board siding, (27204):11–12, 55
Body belts, (27202):2
BOF process. See Basic oxygen furnace (BOF) process
Bolt locks, electric, (27208):45, 46
Bolts, (27205):12, (27209):7
Bowing, drywall, (27207):43
Box cornice, (27204):44
Box cutters, (27206):17
Box vents, (27202):29, 30, 65
Bracing
　defined, (27205):37
　shear walls, (27205):24, 25
　steel construction, (27205):16
Breakout key, (27208):42
Brick veneer, (27204):19
Bridging, (27205):16
Broad knife, (27207):13
Brown coat, (27204):6, 17, 59
Brushes, texturing, (27207):16
Bugle-head screw, (27205):10
Builders Hardware Manufacturer's Association (BHMA) finish numbering system, (27208):70

Building information modeling (BIM), (27201):27, 28
Building materials, R-values, (27203):7–8
Building paper, (27204):6, 17, 59
Building wraps, (27203):28–31, (27204):6, 7
Built-up roofing (BUR), (27202):12
Built-up roofing (BUR) systems
　built-up roofing (BUR) membrane, (27202):24–25
　modified bitumen membrane roofing systems, (27202):25–26
Built-up shapes, steel construction, (27205):16
Bulb rivet guns, (27202):45
Bullet catches, (27211):16, 17
Bull nose, (27207):3, 52
Bundle, (27202):1, 9, 74
Bungalow beveled siding, (27204):11, 27
BUR. See Built-up roofing (BUR)
Butt, (27208):3, 77
Butt hinges, (27211):16
Butt-hinge template, (27208):59
Butt joints, (27210):11, 12
Butt-to-butt swing, (27208):6
Bypass doors, (27208):7, 9
Bypass framing, (27205):19

C

Cabinet codes, (27211):11–12
Cabinet construction
　doors, (27211):13, 14
　drawers, (27211):12, 13–14
　face frame, (27211):11, 12
　frameless, (27211):11, 12
　materials, (27211):12–14
　rabbeted joints, (27211):12, 14
　shelves, (27211):14
　vanity cabinets, (27211):12
　woods commonly used, (27211):13
Cabinet doors
　construction, (27211):13, 14
　hardware, (27211):12, 13, 15–16, 17
Cabinet drawers
　construction, (27211):12, 13–14
　hardware, (27211):12, 14, 15, 16
Cabinet hardware
　catch mechanisms, (27211):12, 16, 17
　door hinges, (27211):12
　drawer guides/slides, (27211):12, 14, 15, 16
　hinges, (27211):12, 15–16
　knobs, door and drawer, (27211):17
　pulls, door and drawer, (27211):17
　selecting, (27211):15
　for shelving, (27211):18
Cabinet installation
　accessories, (27211):12
　base cabinets, (27211):21–23
　fastening to walls, (27211):24
　filler pieces, (27211):12
　moldings, (27211):12
　no studs available, (27211):24
　safety rules, (27211):1–2
　scribing adjoining pieces, (27211):23
　surface preparation, (27211):19
　tight-joint fasteners, (27211):23
　wall cabinets, (27211):19–21
Cabinets
　classifications, (27211):4
　components, (27211):5, 11
　repairing, (27211):12, 13
　shelves, (27211):14

storage space in, (27211):11
types of
    base cabinets, (27211):4–5, 28
    commercial cabinets, (27211):8
    islands (freestanding), (27211):5, 6
    wall cabinets, (27211):4, 5–6, 19–21, 28
CAC. *See* Ceiling attenuation class (CAC)
CAD. *See* Computer-aided design (CAD)
Calcined gypsum, (27206):1
Calcium sulfate dihydrate (gypsum), (27206):1
Callouts, (27201):1, 9, 13, 37
Callout sequence markings, (27201):15
Cap flashing, (27202):32, 56, 74
Cap row shingles, (27202):62–63, 65, 68
Carbide cutter, (27206):16
Carbide-tipped power shears, (27204):3–4
Casing, (27208):3, 77
Casing moldings, door and window, (27210):5, 7–8, 11–13, 14, 16–24, 26
Casing stop, (27210):7–8, 8
Catches, (27208):39, 40, 77
Catch mechanisms, (27211):12, 16, 17
Cathedral ceilings, (27203):17
Caulking
    drywall, (27206):29
    for sound-isolation, (27206):20–22
    vinyl/metal siding, (27204):36, 43–44
Ceiling attenuation class (CAC), (27209):2, 8
Ceiling joist, (27205):37
Ceiling panels, (27209):1, 7, 8–9, 40
Ceilings
    cold-formed steel framing, (27205):32
    drywall installation, (27206):10, 12, 26, (27207):5
    moldings, (27210):7, 16–17
    sagging or uneven, (27210):19
Ceilings & Interior Systems Construction Association, *A Recommended Specification for Levels of Gypsum Board Finish*, (27207):1
Ceiling tiles
    acoustical, (27209):8
    ceiling panels vs., (27209):7
    characteristics, (27209):8
    defined, (27209):1, 7, 40
    exposed grid systems, (27209):6
    facts about, (27209):32
    fire-resistant, (27209):8
    grid interfaces, (27209):8
    high-durability, (27209):28
    high-humidity-resistant, (27209):8–9
    light reflectance (LR) values, (27209):9
    materials, (27209):8
    sagging, (27209):9
Celotex Thermax™, (27203):4
Celotex Tuff-R™, (27203):4
Cement board, (27206):26
Cement board drywall, (27206):4
Certi-Guard™, (27202):21
Chair rail, (27210):26
Chair-rail molding, (27210):6–7, 11
Chalkline, (27204):25
Chase walls, (27205):20
Chimneys, (27202):57, 59–62, 63
Chop saw, (27205):8
Circle cutter, (27206):17
Circular patterned sponge, (27207):17
Circular saw, (27205):8
Civil drawings
    defined, (27201):1, 37

drainage plans, (27201):9
labeling, (27201):4
site improvement plans, (27201):9
site plan, (27201):6–9
utility plans, (27201):9
Climate zones, US, (27203):10
Clinchers, (27207):13–14
Clinching, (27205):11–12, (27207):13–14
Clip angle, (27205):14, 18, 36
Clip angle connections, (27205):19–20
Clips
    drywall, (27206):5
    suspended ceiling systems, (27209):7
Closed cornice, (27204):44, 48
Closed-cut valley shingle pattern, (27202):43–44, 45
Closed valleys, (27202):29
Closed valley shingles, (27202):42–43
Closed-woven valley shingle pattern, (27202):42–43
Clothing, appropriate
    cabinet installation, (27211):1
    door installation, (27208):1
    trim, working with, (27210):1
Coarse-fiber vents, (27202):30
Cold-cement fully-adhered roll roofs, (27202):12
Cold-formed sheet steel, defined, (27205):37
Cold-formed steel
    defined, (27205):1, 36, 37
    safety guidelines when working with, (27205):2
Cold-formed steel curtain walls, (27205):18–20
Cold-formed steel framing
    advantages, (27205):1, 24
    ceiling systems, (27205):32
    fasteners
        anchors, (27205):12
        bolts, (27205):12
        clinching, (27205):11–12
        pins, (27205):10, 11
        powder-actuated, (27205):10–11
        screws, (27205):9–10
    floor assemblies, (27205):31
    loadbearing, (27205):18–19
    material takeoffs
        commercial projects, (27205):12
        construction projects, (27205):12
        joists and joist headers, (27205):12
        studs, (27205):13
        tracks, bottom and top, (27205):12
    nonstructural walls
        area separation walls, (27205):20
        chase walls, (27205):20
        components, (27205):17–18
        curtain walls, (27205):18–20
        fire-rated assemblies, (27205):20
        head-of-wall conditions, (27205):20–21
        members, (27205):18
        radius (curved) walls, (27205):20
        shaft wall systems, (27205):21
    roof assemblies, (27205):31
    structural walls
        accessories, (27205):21
        assembly, (27205):22
        backing, (27205):24
        bracing, (27205):24, 26
        function, (27205):21
        header assembly, (27205):27–29
        installation, (27205):22–24
        jambs, (27205):29
        layout, (27205):21–22

Cold-formed steel framing
  structural walls (*continued*)
      members, (27205):21
      shear walls, (27205):24–25
      sills, (27205):29
      tracks, (27205):21
  thermal considerations, (27205):16
  tools
      aviation snips, (27205):8
      bar clamps, (27205):7
      chop saw, (27205):8
      circular saw, (27205):8
      for cutting, (27205):7–8
      dry-cut metal-cutting saw, (27205):8
      hammer drills, (27205):6–7
      for hole-cutting, (27205):8–9
      hole punch, (27205):9
      hole saw, (27205):9
      locking C-clamps, (27205):7
      powder-actuated (PATs), (27205):6–7
      screwguns, (27205):6
      steel stud punch, (27205):9, 15
      stud drivers, (27205):6–7
      swivel-head shears, (27205):8
Cold-formed steel framing materials
  advantages, (27205):2
  coating methods, (27205):2
  customization, (27205):2–3
  furring, (27205):4–5
  identification markings, (27205):3
  manufacturing tolerances, standards, (27205):2
  minimum base thickness, (27205):4, 18
  nonstructural members, (27205):17–18
  recycled content, (27205):5
  slip connectors, (27205):5, 6
  thermal considerations, (27205):16
Cold-storage vapor barriers, (27203):28
Columns, (27201):1, 10, 37
Commercial cabinets, (27211):8
Commercial construction, contractor qualifications to bid, (27201):32
Commercial construction drawings. *See also* Architectural drawings; Civil drawings; Plumbing drawings; Structural drawings
  computer generated, (27201):27
  contents, (27201):3–4
  coordination drawings, (27201):27
  duplicate, (27201):4
  electrical drawings, (27201):22–24
  elevation drawings, (27201):14
  legends, (27201):19
  lines commonly used, (27201):4–5
  mechanical drawings, (27201):20–22
  reading, (27201):5–6
  residential drawings vs., (27201):2–3
  revisions, (27201):6
  section identification formats, (27201):4
  section views, (27201):4
  symbols, (27201):4
  three-dimensional models, (27201):27
Commercial doors. *See also* Doors
  delivery, (27208):55
  exterior, (27208):15–16
  hardware
      exterior door stops, holders, closers, (27208):43–44
      external door stops, holders, closers, (27208):45
      hinges, (27208):41, 57–59, 61, 62
      interior locksets, (27208):41–42

      keying systems, (27208):42
      security hardware, (27208):44–47
  installation tools
      butt-hinge template, (27208):59
      fore plane, (27208):57
      hand plane, (27208):57, 58
      power plane, (27208):56–57, 58
      power saw, (27208):57
      routers, (27208):60
      router selection and preparation, (27208):60
      template saw guide, (27208):57
  installing
      bottom and edge sealing, (27208):58
      clearance, (27208):57
      door closers, (27208):67–68
      fitting in prepared openings, (27208):56–57
      handling, (27208):55, 56
      hanging the door, (27208):61
      hinges, (27208):57–59, 61, 62
      instructions, (27208):55
      locksets, (27208):61–66
      mortise lockset, (27208):66–67
  job finishing, (27208):55
Component assembly, (27205):37
Component design drawing, (27205):37
Component designer, (27205):37
Component manufacturer, (27205):37
Component placement diagram, (27205):37
Composition shingle knife, (27202):12, 13
Composition shingles
  applications, (27202):18–20
  bundle weight, (27202):36
  characteristics, (27202):37
  courses, (27202):36
  hip row/ridge row/cap row, (27202):62–63, 65, 68
  installation patterns
      4" pattern, (27202):41
      6" pattern, (27202):39, 44
      closed-cut valley, (27202):43–44, 45
      closed-woven valley, (27202):42–43
  installing
      at chimneys, (27202):57, 60, 62
      in closed valleys, (27202):42–43
      in dormer valley, (27202):57
      on gable roofs, long- and short-runs, (27202):37–40
      in high-wind areas, (27202):44
      on hip roofs, (27202):40–41, 42
      nailing, (27202):40
      in open valleys, (27202):41–42
      ribbon courses, (27202):41
      on ridges, (27202):42
      starter strips, (27202):41
      in valleys, (27202):41–44
      at vent stacks, (27202):55, 56
      at vertical walls, (27202):55, 56
  standard, (27202):37
  types of, (27202):18–20
Compound cut, (27210):10, 11, 31
Compound miter saws, (27210):11–12
Compound miter saw settings, (27210):13
Computer-aided design (CAD), (27201):27
Concealed nail roll roofing, (27202):50–53
Concealed tee tiles, (27209):8
Concrete adhesives, (27206):11–12
Condensation, (27203):22, 38
Construction drawings. *See also* Commercial construction drawings
  residential vs. commercial, (27201):2–3, 6, 9

suspended ceiling systems
   mechanical, electrical, plumbing (MEP) drawings, (27209):13–15
   reading, (27209):12
Construction key, (27208):42
Construction Specifications Canada (CSC), (27201):30
Construction Specifications Institute (CSI), (27201):30
Contact adhesives, (27206):9–10
Continuous angle, (27205):18
Continuous angle connections, (27205):19–20
Continuous flashing, (27202):56–57
Contour lines, (27201):1, 6, 7, 8, 37
Contractor qualifications, (27201):32
Control points, (27201):7–8
Convection, (27203):22, 24, 38
Convection vents, (27202):29
Coordinator, (27208):39, 41, 50, 52, 77
Coped joint, (27210):10–11, 12–13, 14, 31
Coping saw, (27210):10
Corner applicators and finishers, (27207):20–21
Corner beads, (27206):1, 45, (27207):3–4, 5, 13
Corner caps
   vinyl/metal siding installation, (27204):38–40
   wood siding installation, (27204):30
Corner posts, (27204):34, 35
Corners, wood siding installation, (27204):27–30
Corner strips, exterior finishing, (27204):10
Corner trowels, (27207):13–14
Cornice, (27204):6, 59
Cornice installation, (27204):44, 48–53
Corrugated light panels, (27202):51
Corrugated metal roofing, (27202):44–45, 46
Countertops
   custom, (27211):7
   installing, (27211):21–23
   plastic-laminate-covered, (27211):6–7
   recycled glass, (27211):9
   solid-surface, (27211):7–9, 10
Course, (27204):6, 13, 59
Cove moldings, (27210):7, 11, 17
Crack, drywall, (27207):38
Cracks, drywall, (27207):43–44
Crawl spaces
   vapor barrier installation, (27203):25
   ventilation, (27203):25
Cricket flashing, (27202):57, 59–61
Cripple stud, (27205):37
Cross runners, exposed grid systems, (27209):6
Crown moldings, (27210):7, 11–12, 13, 17, 22, 26
CSC. *See* Construction Specifications Canada (CSC)
C-shape, (27205):1, 8, 20, 36
CSI. *See* Construction Specifications Institute (CSI)
Cupped-head drywall nails, (27206):5
Curtain walls, steel-framed, (27205):14, 18–19, 36
Custom cabinets, (27211):4, 28
Cutting plane lines, (27201):4–5
Cutting table, (27204):4
Cylinder knob, (27208):41
Cylindrical lockset, (27208):65–66, 77
Cylindrical lockset jig, (27208):65

D
dB. *See* Decibel (dB)
dBA. *See* a-weighted decibel (dBA)
Deadbolt, (27208):39, 41, 63, 77
Deceleration devices, (27202):2, 3–4
Deceleration distance, (27202):2
Decibel (dB), (27209):1, 40

Deep foundations, (27201):18
Delayed-exit alert locks, (27208):47, 48
Denshield™, (27206):4
Depressions, drywall, (27207):37
Design thickness, (27205):37
Detail drawings
   defined, (27201):1, 37
   electrical plans, (27201):22
   example, (27201):14
   function, (27201):10, 12
   structural plans, (27201):20
Detailing instructions, (27201):9–10
Dew point, (27203):22, 27, 38
Diagonal brace, (27205):18, 19
Diaphragm, (27205):14, 24, 36
Diffuser, (27209):1, 3, 40
Diffusion, (27203):22, 24, 38
Dimension lines, (27201):4–5
Direct-hold electromagnetic locks, (27208):46–47
Direct-hung ceiling systems
   components, (27209):9, 10
   concealed grid, (27209):30–31, 32
   function, (27209):3
Discoloration, drywall, (27207):37
Display key, (27208):42
Distribution plans, plumbing drawings, (27201):24, 26
Dolly, (27206):25
Dolly Varden siding, (27204):10, 11
Door catches, (27211):12, 16, 17
Door closer, (27208):17, 25, 43–44, 45, 77
Door closer installation, (27208):67–68
Door coordinator, (27208):41, 48, 50, 52, 77
Door frame, (27208):3, 77
Door hardware
   cabinet doors, (27211):12, 15–16, 17
   commercial doors
      coordinator, (27208):50, 52
      dust-proof strike, (27208):50
      flush bolt, (27208):48
      smoke gasket, (27208):50
      threshold, (27208):47, 49
      touch-bar/crossbar hardware, (27208):47–48, 49, 50, 51
      weather stripping, (27208):47
   residential doors
      hinges, (27208):39–40
      knobs and knob functions, (27208):40–41
      locksets, (27208):40
   swing and, (27208):6
Door holders, (27208):43
Door jack, (27208):53
Door lights, (27208):6, 9
Doors. *See also* Commercial doors; Residential doors
   cabinet, (27211):13, 14
   commercial exterior, (27208):13
   direction of swing, (27208):4, 6, 7
   hardware finish, (27208):69
   installation
      interior doors, (27208):3
      safety, (27208):1–2
      schedules, (27208):4, 6
   types of
      fire doors, (27208):15–16
      flush doors, (27208):3–4
      hollow-core doors, (27208):4, 5
      prehung door, (27208):6, 7
      solid-core doors, (27208):4, 5
      sound transmission class (STC) doors, (27208):4
      stile-and-rail doors, (27208):4

Doors. *See also* Commercial doors; Residential doors
   types of (*continued*)
      wood panel doors, (27208):4
   with veneer faces, (27208):57
   vinyl/metal siding installation around, (27204):40–41
   wood siding installation around, (27204):27, 29–30
Door schedules, (27201):10, 12, (27208):4, 6, 69–71
Door stop, (27208):3, 43, 77, (27210):8
Door swing, (27208):4, 6, 7
Door trim
   casing moldings, (27210):5, 7–8, 11–13, 14, 16–24, 26
   vinyl/metal siding installation, (27204):35–37
Dormer flashing, (27202):58
Dormer valley flashing, (27202):57, 58, 59
Dormer valley shingles, (27202):57
Double-coverage roll roofing, (27202):20, 53–54
Double doors, (27208):50
Dovetail saw, (27210):10
Downspouts, (27202):75–79
Dow Styrofoam™, (27203):4
Drainage plans, (27201):9
Drawer guides/slides, (27211):12, 14, 15, 16
Drawers, cabinet, (27211):12, 13–14, 17
Dri-Deck™, (27202):27
Drip edges, (27202):28
Dry-cut metal-cutting saw, (27205):8
Drying time
   drywall adhesives, (27206):10
   drywall finishing, (27207):27
   exterior finishing materials, (27204):19
Dry lines, (27209):17, 19, 40
Dry-mix safety, (27207):9–10
Dry-powder adhesives, (27206):9
Drywall
   accessories
      casings, (27206):12–13
      control (expansion) joints, (27206):13–14
      corner beads, (27206):12–13
      moldings, (27206):12–13
   advantages, (27206):2, 16–17
   applications, (27206):2
   characteristics, (27206):1, 2
   defined, (27206):1, 45
   fungus-resistance, (27206):5
   history, (27206):2
   materials comprising, (27206):1
   patterns, (27206):5
   plaster applications, (27206):5
   prebowing, (27206):11
   radius applications, (27206):5
Drywall fasteners
   adhesives
      applications, (27206):5, 9
      applying, (27206):9, 10–12
      for architectural boards, (27206):11
      bridging ability, (27206):10
      for ceiling installations, (27206):10, 12
      for concrete, (27206):11–12
      contact adhesives, (27206):9–10
      disadvantages, (27206):10
      drying time, (27206):10
      dry-powder, (27206):9
      function, (27206):8–9
      for furring, (27206):12
      laminating adhesives, (27206):9
      for masonry, (27206):11
      modified contact adhesives, (27206):10
      multi-ply systems, (27206):38
      for rigid/semirigid insulation, (27206):10
      sheet lamination, (27206):38
      sound-isolation construction, (27206):38
      spot lamination, (27206):38
      for steel framing, (27206):10, 11
      strip lamination, (27206):38
      stud adhesives, (27206):9, 10
      supplemental fasteners, (27206):9–10, 38–39
   attachments, (27206):5
   clips, (27206):5
   guards, (27206):5
   material takeoffs, (27206):40
   multi-ply systems, (27206):37
   nails
      annular type, (27206):5
      application procedure, (27206):6–7
      casing, (27206):5
      common, (27206):5
      cupped-head, (27206):5
      hammers used to drive, (27206):5
      head shape and size, (27206):5
      nailing method, (27206):6
      penetration, (27206):5
      resurfacing existing construction, (27206):29
      spacing, (27206):6
   placement, (27206):5
   screws
      for furring, (27206):7, 8
      head shape, (27206):7
      penetration, (27206):8
      resurfacing existing construction, (27206):29
      selecting, (27206):7
      spacing, (27206):8
      for steel framing, (27206):8
      Type G, (27206):8
      Type S, (27206):8
      Type S-12, (27206):8
      Type W, (27206):7–8
   shields, (27206):5
   staples, (27206):5
Drywall finishing
   drying time, (27207):27
   job-site conditions, (27207):23
   levels of finish and applications, (27207):1–2
   personal protective equipment (PPE), (27207):12
   procedure, (27207):2
   responsibility for, (27206):1
   taping, (27207):28–32, 33
   temperature and, (27207):23, 25
   texturing, (27207):9, 10, 16–17, 18
   trim
      corner beads, (27207):3–4, 5, 13
      expansion joints, (27207):4
      reveal trim, (27207):4
Drywall finishing compounds
   all-purpose compound, (27207):6, 7, 8
   dry-mix safety, (27207):9–10
   joint compound, (27207):5–6
   lightweight compound, (27207):6, 7
   powder compounds, (27207):5–7, 8, 10
   premix compounds, (27207):7–8
   quick-setting compounds, (27207):8, 10
   shrinkage, (27207):8
   specialty compounds, (27207):8
   storage, (27207):6–7, 8
   taping compound, (27207):6, 7, 8
   topping compound, (27207):6, 7, 8
Drywall finishing materials
   sanding, (27207):10, 11, 25

storage, (27207):23
tape, (27207):4–5, 15–16
texture, (27207):9, 10
Drywall finishing procedures
  corners, outside and inside, (27207):27, 32–33, 35–36
  fastener heads, spotting, (27207):26–27, 30
  hand finishing, (27207):24–26
  housekeeping, (27207):27–28
  inspection prior to finishing, (27207):23–24
  safety, (27207):27–28
  sanding, (27207):26
  taping and finishing process, (27207):24
Drywall finishing tools
  automatic
    corner applicator, (27207):35–36
    corner applicators and finishers, (27207):20–21
    corner-plow, (27207):33
    corner roller, (27207):32
    flat finishers, (27207):19–20, 34–35
    loading pumps, (27207):21
    nail spotters, (27207):19, 30
    pumping, (27207):29–30
    pump loading, (27207):29
    spray-texturing machines, (27207):11
    taping tool, (27207):28–32, 33
    taping tools, (27207):17–19
    vacuum sanders, (27207):21–22
  hand tools
    4" straightedge, (27207):12
    clinchers, (27207):13–14
    corner trowels, (27207):13–14
    finishing knives, (27207):12–13
    finishing trowels, (27207):13–14
    glitter gun, (27207):12
    hawk, (27207):14–15, 26
    mud masher, (27207):15
    mud pan, (27207):14, 26
    sanders, (27207):10, 11, 15
    saws, (27207):12, 13
    taping knife, (27207):13
    T-square, (27207):12
  maintenance, (27207):12, 27
  mixing tools, (27207):15–16
  power tools
    mud mixer, (27207):15–16
    spray-texturing machines, (27207):11
    vacuum sanders, (27207):21–22
  safety, (27207):12
  stilts, (27207):28, 30
  storage, (27207):28
  tape dispensers, (27207):15–16
  texture tools
    circular patterned sponge, (27207):17
    flat blade knife, (27207):17
    glitter gun, (27207):16
    mud paddle, (27207):16
    paddle, (27207):17
    roller pan, (27207):17
    rollers, (27207):17, 18
    sea sponge, (27207):17
    spray-texturing machines, (27207):11
    stucco brush, (27207):16
    texture brush, (27207):16
    trowel, (27207):17
    whisk broom, (27207):17
    wipe-down blade, (27207):17
Drywall furring ceiling system, (27209):4
Drywall hammer, (27206):17, 18
Drywall installation
  application defects, (27206):5, 6, 8
  base requirements, (27206):23
  bathrooms, (27206):26, 28–29
  caulking, (27206):29
  ceilings, (27206):26, (27207):5, (27209):33–34
  commercial interiors, (27206):34
  cutting and fitting procedures, (27206):25–26
  electrical boxes, locating, (27206):20
  fastening schedules, (27206):16
  fire-rated walls, (27206):32–37
  floating interior angle construction, (27206):26, 27
  furring for, (27204):31, (27206):24
  materials handling and storage, (27206):24–25, 26
  material takeoffs, (27206):40, (27207):47
  moisture-resistant construction, (27206):26, 28–29
  multi-ply systems
    adhesives, (27206):38
    base-ply attachment, (27206):37–38
    corners, (27206):37
    face-ply attachment, (27206):37–38
    fasteners, (27206):37, 38–39
    floating-angle construction, (27206):37
    floating-corner treatment, (27206):37
    sound-isolation construction, (27206):31
  parallel vs. perpendicular, (27206):31
  prioritizing walls, (27206):39
  radius walls, (27207):5
  residential interiors, (27206):24, 31, 33–34
  responsibility for, (27206):1
  resurfacing existing construction, (27206):29
  single-ply applications, (27206):31, 32
  site preparation, (27206):24–25
  sound-isolation construction
    adhesive application, (27206):38
    air leaks, closing, (27206):18–19
    caulking, (27206):20–22
    double-ply applications, (27206):23–24
    flanking paths, closing, (27206):18–19
    multi-ply systems, (27206):31
    residential interiors, (27206):33
    resilient mountings, (27206):22–23
    separated partitions, (27206):20–21
    single-ply applications, (27206):23–24
    sound-isolating materials, (27206):23–24
    steel stud walls, (27206):32–33
    wallboard applications, (27206):23
  special applications
    abuse-resistant drywall, (27206):29
    exterior sheathing panels, (27206):29–30
    lead-lined drywall, (27206):29–30
    paperless drywall, (27206):29
  steel studs, (27206):32–33
  stilts, (27207):28, 30
  temperature requirements, (27206):24
  tools
    4' T-square, (27206):17
    adhesive guns, (27206):9
    box spreader, (27206):38
    carbide cutter, (27206):16
    circle cutter, (27206):17
    dolly, (27206):25
    drywall hammer, (27206):17, 18
    drywall lift, (27206):18, 19
    drywall lifter, (27206):17, 18
    drywall saw, (27206):17, 18
    drywall stripper, (27206):19
    hammers, (27206):5

Drywall installation
  tools (*continued*)
    hook-bill knife, (27206):17
    jab saw, (27206):17, 18
    lead pencil, (27206):17
    light box cutter, (27206):17, 18
    notched spreader, (27206):38
    rasp, (27206):17
    rubber mallet, (27206):10
    saws, (27206):25
    screw gun, (27206):8, 17, 19
    shims, (27206):26
    T-brace, (27206):18, 20
    utility knife, (27206):17, 25
Drywall lift, (27206):18, 19
Drywall lifter, (27206):17, 18
Drywall panels
  attaching, (27206):6
  cutting, (27206):25, (27207):12
  edges, (27206):3, 4, 25
  patching, (27207):45–46
  production, (27206):1–2
  sizes, (27206):3
  types and applications
    abuse-resistant, (27206):5
    architectural panels, (27206):2, 3, 4
    cement board, (27206):4
    drywall substrate, (27206):4–5
    flexible ¼," (27206):3, 5
    foil-backed, (27206):2
    gypsum backing board, (27206):4
    gypsum base for veneer plaster, (27206):5
    gypsum coreboard, (27206):3, 4
    gypsum form board, (27206):4
    gypsum lath, (27206):3, 5
    gypsum sheathing, (27206):4
    high-strength, (27206):3
    moisture resistant, (27206):3
    predecorated drywall, (27206):4
    regular, paper faced, (27206):3
    regular with foil back, (27206):3
    special-use, (27206):3
    by thickness, (27206):2–3
    tile backer, (27206):4
    Type X, fire retardant, (27206):2, 3
    water-resistant (green board), (27206):4
    weather-resistant, (27206):3
Drywall problems
  board blisters, (27207):43
  bowing, (27207):43
  chipping, (27207):38–39
  cracking, (27207):38
  cracks and fractures, (27207):43–44
  debonding, (27207):38–39
  depressions, (27207):37, 42
  discoloration, (27207):37
  edge damage, (27207):43
  finishing compounds, (27207):38–39
  flaking, (27207):38–39
  high joints, (27207):37
  holes, (27207):45–46
  loose panels, (27207):44
  manufacturing defects, (27207):43
  mold, (27207):39
  nail pops, (27207):40–42
  photographing, (27207):37
  pitting, (27207):39
  ridging, (27207):36
  sagging, (27207):39
  shrinkage, (27207):39–40
  tape blisters, (27207):37–38
  water damage, (27207):43
Drywall saw, (27206):17, 18, (27207):12, 13
Drywall stripper, (27206):19
Drywall substrate, (27206):4–5
Drywall suspension systems, (27209):35
Duckbill snips, (27204):38, 40
Dummy trim, (27208):40–41
DuPont® flashing system, (27203):32
Durock™, (27206):4
Dust-proof strike, (27208):39, 50, 77
Duty rating, (27204):1, 2, 59

E

e. *See* Exposure (e) overlapping shingles
EAF process. *See* Electric arc furnace (EAF) process
Eave, (27204):23, 24, 43, 59
Eave vent, (27203):24
Edge pull, (27208):10
Edges, drywall, (27206):3, 4, 25, (27207):43
Edge stiffener, (27205):37
EIFSs. *See* Exterior insulation finish systems (EIFSs)
Electrical cords, (27210):2, (27211):1, 2
Electrical plans
  detail drawings, (27201):22
  fixture schedules, (27201):24
  labels, (27201):22
  lighting plan, (27201):24
  panel schedules, (27201):23
  plan view, (27201):22
  power plan, (27201):22–23
  sample, (27201):25–26
  symbols used on, (27201):22, 42
Electric arc furnace (EAF) process, (27205):4
Electric bolt locks, (27208):45, 46
Electric locksets/latches, (27208):46
Electric plunger strike, (27208):48
Electric shock, (27210):1, (27211):1
Electric strikes, (27208):44–45, 48
Electromagnetic locks, (27208):46–47
Elevated surfaces safety
  aerial lifts, (27204):2–3
  fall protection, (27204):1
  ladders, (27202):4, 7–9, (27204):1–2
  scaffolds, (27202):36, (27204):2, 3
Elevation drawings, (27201):1, 12–13, 14, 37
Embed, (27205):18
Emergency-exit touch bars, (27208):46–47, 51
Emergency release knob, (27208):41
Emergency/shutout key, (27208):42
Empire State Building, (27201):3, 23, 24
EMR. *See* Experience modification rate (EMR)
Endothermic fire-stopping materials, (27206):36
Entrance lockset, (27208):40–41, 61, 62
EPD polymer membrane. *See* Ethylene-propylene-diene
  monomer (EPD) polymer membrane
Equipment connectors, (27202):4
Erection drawings, (27205):37
Erector, (27205):37
Ergonomics, (27211):1, 2, 28
Escutcheon plate, (27211):11, 17, 28
Ethylene-propylene-diene monomer (EPD) polymer
  membrane, (27202):26
European (Eurostyle) cabinets, (27211):11
Eurostyle hinges, (27211):16
Exit doors, (27208):45, 47

Expansion joints, (27207):4
Experience modification rate (EMR), (27201):32
Exposed grid systems, (27209):2, 6–7, 23–28
Exposed nail roll roofing, (27202):49–50
Exposure, (27202):74
Exposure (e) overlapping shingles, (27202):38
Extension cords, (27208):1, 2, (27210):2, (27211):2
Extension lines, (27201):4–5
Exterior finishing installation. *See also Specific types of finishes*
   cornices, (27204):44, 48–53
   equipment
      cutting table, (27204):4
      portable brake, (27204):4
   fascia, aluminum or vinyl, (27204):49–52
   fiber-cement siding, (27204):44, 45–47
   function, (27204):6
   furring and insulation techniques
      aluminum foil underlayment, (27204):24
      furring, (27204):23–24
      undereave furring, (27204):24–25
      undersill furring, (27204):24
      window and door buildout, (27204):24
   pre-installation checks, (27204):6, 7
   process, (27204):6
   reference line, (27204):25, 26
   reference marks, (27204):27
   soffits, aluminum or vinyl, (27204):49–52
   surface preparation, (27204):23
   tools
      aviation snips, (27204):38, 40
      chalkline, (27204):25
      duckbill snips, (27204):38, 40
      hacksaw, (27204):38
      pneumatic carbide-tipped power shears, (27204):42
      pneumatic nailers, (27204):7
      portable brake, (27204):36
      power saw, (27204):37, 40, 42
      power tools, (27204):7
      score-and-snap knife, (27204):42
      siding gauge (preacher), (27204):27, 28
      snap-lock punch, (27204):41
      tin snips, (27204):37, 41, 42
      utility knife, (27204):38, 40–41
   ventilation, (27204):42, 44
Exterior finishing materials. *See also Specific types of finishes*
   common, (27204):6
   corner strips, (27204):10
   drying time, (27204):19
   fiber-cement siding, (27204):16, 17
   flashing, (27204):20–21
   material takeoffs, (27204):54
   nails, (27204):6, 10, 49, 55
   specialty systems
      direct-applied exterior finish systems (DEFS), (27204):19–20
      exterior insulation finish systems (EIFSs), (27204):19–20
Exterior finishing safety
   clothing, (27204):1
   elevated surfaces
      aerial lifts, (27204):2–3
      fall protection, (27204):1
      ladders, (27204):1–2
      scaffolds, (27204):2, 3
   equipment, (27204):4–5
   materials, (27204):5
   tools, hand and power, (27204):3–4
   weather hazards, (27204):1

Exterior insulation finish systems (EIFSs), (27203):28–29
Exterior sheathing panels, (27206):29–30
Exterior walls, insulating, (27203):4–5
Eye protection, (27210):1, 2

F

Face frame cabinet construction, (27211):11, 12
Face protection, (27210):1, 2
Fail-safe/fail-secure strikes, (27208):45–46
Fall protection, (27204):1
Fall protection plans, (27202):4
Fall rescue and retrieval plans, (27202):5–6
Fans, (27202):29
Fascia
   aluminum or vinyl installation, (27204):49–52
   defined, (27204):1, 59
   metal, (27202):23–24, 25
Fast-joint hinges, (27208):40
Feathering, (27207):52
Felt underlayment, (27202):27–28
Fiber-cement siding, (27204):3, 16, 17
Fiber-cement siding installation, (27204):44, 45–47
Fiberglass ingredients, (27203):13
Fiberglass insulation, (27203):8–9, 11, 13
Fiberglass joint tape, (27207):5
Fiberglass ladders, (27202):7
Fiberglass mesh tape, (27207):5, 6
Fiberglass shingles, (27202):18
Field notes, (27201):1, 7, 37
Field set, (27201):4
Finger-jointed stock, (27210):4, 31
Finish coat, (27204):6, 17, 59
Finish hardware, (27208):77
Finishing knives, (27207):12–13
Finishing sawhorse, (27208):77
Finishing trowels, (27207):13–14
Finish nailers, (27210):2–3, 13–14
Finish numbering system, (27208):70
Finish schedule, (27201):9, 12, 13
Finish symbols, (27208):69
Fire alarms, (27208):43, 44, 47
Fire doors, (27208):15–16, 45
Fire-rated assemblies, (27205):20
Fire-rated construction, (27205):29, (27206):34–35
Fire ratings, (27205):29, (27208):15–16
Fire-resistance
   ceiling tiles, (27209):8
   drywall, (27206):2, 3
   slate roofs, (27202):23
   synthetic tiles, (27202):24
   tile roofing, (27202):22
   wood shakes and shingles, (27202):21
Fire-stopping, (27206):34–37
Fire-stopping ratings, (27206):36–37
Fissured, (27209):1, 6, 40
5-gallon pail, (27207):8
5/8" drywall, (27206):3
5/16" drywall, (27206):2
Fixed ladders, (27202):7
Fixed-pin hinges, (27208):40
Fixture schedules, (27201):24
Flame heating equipment, (27202):26, 31
Flame-spread rating, (27209):8
Flange, (27205):37
Flashing
   bending, (27202):12, 57
   function, (27204):20

Flashing (continued)
   roll roofing for, (27202):20
   slate roofing, (27202):23
   types of
      cap flashing, (27202):60–62, 63
      continuous flashing, (27202):56–57
      counter flashing, (27202):60–61
      step-flashing shingles, (27202):55–56
   vinyl/metal siding installation, (27204):37
   wood siding installation, (27204):31
Flashing installation
   chimneys, (27202):59–62, 63
   dormers, (27202):58
   dormer valley, (27202):57, 58, 59
   gable dormer valley, (27202):58
   horizontal abutments, (27202):56–57
   ice edge, (27202):34, 36
   nailing, (27202):56, 57
   open valleys, (27202):42
   tools, (27202):12
   typical, (27204):21
   valleys, (27202):28–29, 32
   vent-stack, (27202):54–55
   vertical wall, (27202):55–56
Flat blade knife, (27207):17
Flat finishers, (27207):19–20
Flat-head screw, (27205):10
Flat miters, (27210):11
Flat-roll valley flashing, (27202):29
Flex catches, (27211):16, 17
Flexible ¼" drywall, (27206):3, 5
Flexible-hem vinyl siding, (27204):39
Flexible insulation, (27203):8–9, 11, 16–17
Flexible tape, (27207):4
Floating-angle construction, (27206):37
Floating-corner treatment, (27206):37
Floating interior angle construction, (27206):16, 26, 27, 45
Floor assemblies, cold-formed steel framing, (27205):31
Floor joist, (27205):37
Floor plans, (27201):9–10, 12, 24
Flush, (27211):11, 13, 28
Flush bolt, (27208):3, 12, 48, 77
Flush doors, (27208):3–4, 77, (27211):13
Flush drawers, (27211):14
Foamed-in-place insulation, (27203):15, 16
Foil-backed drywall, (27206):2
Folding doors, (27208):10–12
Fore plane, (27208):57
45- to 60-minute mineral core door, (27208):5
Foundation plans, (27201):18–19
Foundation vents, (27203):25
Foundation walls, (27203):27, (27206):12
4" composition shingle pattern, (27202):41
4" straightedge, (27207):12
4' T-square, (27206):17
Fractures, drywall, (27207):43
Frameless cabinet construction, (27211):11, 12
Framing contractor, (27205):37
Framing material, (27205):37
Framing plans, (27201):16, 19–20
Free-fall distance, (27202):2
Freestanding cabinets (islands), (27211):5, 6
Frequency, (27209):1, 40
Frieze board, (27204):6, 17, 59
Full-body harness, (27202):2–5
Full-weave design, (27202):42–43
Fungus-resistant shingles, (27202):18–19

Furring
   drywall installation and, (27204):31, (27206):12, 24
   drywall screws for, (27206):7, 8
   pressure-treated, (27204):41
   under siding, (27204):23–24
   for sound isolation, (27205):4–5, (27206):23
   substitutes for, (27204):41
   undereave, (27204):24–25
   undersill, (27204):24
   vinyl/metal siding installation, (27204):41, 43
   window and door buildout, (27204):24
Furring channels, installing, (27209):34–35

G

Gable dormer valley, (27202):58
Gable ends, (27204):41–42
Gable end trim, (27204):37
Gable louver vent, (27203):24
Gable roof shingles, (27202):37–40
Gable vents, (27202):29
Galvanized metal roofs, (27202):44–45
Gauge, (27205):1, 4, 36
General contractor, (27205):37
GFCI. *See* Ground fault circuit interrupter (GFCI)
Girders, (27201):1, 15, 37
Girts, (27205):18
Glass-fiber insulating batts and blankets, (27206):23
Glass-fiber tiles, (27209):6
Glazing, (27208):6, 9
Glitter gun, (27207):16
Grand master key, (27208):42
Green board, (27206):4, 12, 26
Grid lines, (27201):10, 15, 18
Ground fault circuit interrupter (GFCI), (27211):1
Guardrails, (27202):2
Guards, drywall, (27206):5
Guest's key, (27208):42
Guides, drawer, (27211):12, 14, 15, 16
Guide strips, (27208):10
Gutters, (27202):75–79
Gypsum, (27206):1, 24
Gypsum Association, *A Recommended Specification for Levels of Gypsum Board Finish*, (27207):1
Gypsum backing board, (27206):4, 8
Gypsum base for veneer plaster, (27206):5
Gypsum board screw, (27205):10
Gypsum coreboard, (27206):3, 4
Gypsum-core sound-insulating board, (27206):23
Gypsum drywall. *See* Drywall
Gypsum form board, (27206):4
Gypsum lath, (27206):3, 5
Gypsum sheathing, (27206):4

H

Hacksaw, (27204):38
Half-laced valley design, (27202):43–44
Half-weave design, (27202):43–44
Hammer drills, (27205):6–7
Hammers, drywall, (27206):5
Hand grinders, (27202):12
Hand of the door, (27208):4
Hand plane, (27208):57, 58
Hangers, exposed grid systems, (27209):7
Hanging stile, (27208):3, 6, 77
Hardibacker™, (27206):4
Hardware, (27208):3, 77
Hardware finish, (27208):69
Harsh environments, (27205):37

Hawk, (27207):14–15
Head, (27208):77
Header, (27205):1, 7, 36
Header assembly, loadbearing, (27205):27–29
Head lap (HL) overlap, (27202):38
Head-of-wall conditions, (27205):20–21
Head to jamb connections, (27205):19–20
Health hazards
    back injuries, (27211):2
    hearing loss, (27209):2, (27210):1
    laminate installation, contact cement vapors, (27211):2
    sanding dust, (27207):28
    siding materials dust, (27204):5
    western red cedar dust, (27204):7
Hearing loss, (27209):2, (27210):1
Heat loss and gain, (27203):4, 11–12, (27206):2
Heat welding, (27202):26, 31
Heavy roller, (27202):12, 13, 32
Hertz (Hz), (27209):1, 40
Hex washer-head screw, (27205):10
High-durability ceiling tiles, (27209):28
High-humidity-resistant ceiling tiles, (27209):8–9
High joints, drywall, (27207):37
High-pressure laminate (HPL), (27211):4, 6, 28
High-strength drywall, (27206):3
High-wind-load vinyl siding, (27204):39
Hinges
    cabinet hardware, (27211):12, 15–16
    cabinets, (27211):12, 15–16
    commercial doors, (27208):41, 57–59, 61, 62
    defined, (27208):3, 77
    door swing and, (27208):6, 8
    installing, (27208):57–59, 61, 62
    reversible, (27208):59
    wood folding doors, (27208):11
Hinge screw lubrication, (27208):61
Hip roof shingles, (27202):40–41, 42
Hip row shingles, (27202):62–63, 65, 68
HL overlap. *See* Head lap (HL) overlap
Hold-down clips, (27209):7, 27
Hole covers, (27202):2
Hole punch, (27205):9
Holes, drywall, (27207):45–46
Hole saw, (27205):9
Hollow-core doors, (27208):4, 5
Hook-bill knife, (27206):17
Hot-air welding, (27202):31
Hot mud, (27207):8
Housekeeper's master key, (27208):42
House wrap, (27203):28–31, (27204):6, 7
HPL. *See* High-pressure laminate (HPL)
HVAC drawings, (27201):20–22, 23, 41
Hz. *See* Hertz (Hz)

I

ICC. *See* International Code Council (ICC)
Ice buildup behind chimneys, (27202):59
Ice dams, (27202):33–35
Ice edge, (27202):34, 36
Impact wrenches, (27202):45
Infill method, curtain wall construction, (27205):19
In-line framing, (27205):14–15, 37
Installation drawings, (27205):37
Installer, (27205):37
Insulation
    building codes, (27203):7–8
    condensed moisture in, (27203):22
    design standards, (27203):7–8
    for drywall below grade, (27206):12
    estimating requirements for, (27203):33–34
    excessive, (27203):8
    fire-resistance, (27206):2
    installing
        flexible insulation, (27203):16–17
        loose-fill insulation, (27203):17–20
        requirements, (27203):8
        rigid/semirigid insulation, (27203):20
    materials
        by classification, (27203):5
        flexible insulation, (27203):8
        loose-fill insulation, (27203):9
        rigid/semirigid insulation, (27203):9–10, 13
        weight, (27203):14
    R-values, (27203):4–8, 10, 39
    safety guidelines when working with, (27203):1–2, 13
    types of
        flexible, (27203):8–9, 11, 16–17
        foamed-in-place, (27203):15, 16
        lightweight aggregate, (27203):15
        loose-fill, (27203):9, 12, 17–20
        reflective, (27203):13, 14
        rigid/semirigid, (27203):9–12, 12, 20
        sprayed-in-place, (27203):15, 16
    wood shakes and shingles, (27202):21
Insulation sheathing, (27203):4
Integrated ceiling systems, (27209):3, 4
Interior door components, (27208):3
Interior locksets, (27208):41–42
Interlocking threshold, (27208):47, 49
Intermediate foundations, (27201):18
Intermediate stud bracing, (27205):24
International Code Council (ICC), R-value
    recommendations, (27203):8, 10, 39
Intumescent fire-stopping materials, (27206):35–36
Invert, (27201):1, 9, 37
Islands (freestanding cabinets), (27211):5, 6
Isometric drawings, (27201):1, 24, 26, 37

J

Jab saw, (27206):17, 18
Jack stud, (27205):37
Jambs, (27205):29, (27208):3, 77
J-beads, (27206):12, (27207):3–4
J-channel, (27204):36–37
Joint compound, (27207):1, 5–6, 52
Joints
    defined, (27206):1, 45
    drywall, (27206):6, 24
    waterproofing, (27203):28
Joist headers, (27205):12
Joists
    defined, (27201):1, 37
    on drawings, (27201):20
    material takeoffs, (27205):12

Kerf, (27208):17, 77
Keying systems, (27208):42
Kicker, (27205):18, 19
King stud, (27205):22, 37
Knee-wall base connection, (27205):19–20
Knives
    broad knife, (27207):13
    composition shingle knife, (27202):12, 13
    drywall, (27207):12–13, 17
    finishing knives, (27207):12–13
    flat blade knife, (27207):17

Knives (*continued*)
   hook-bill knife, (27206):17
   light box cutter, (27206):17
   putty knife, (27207):12–13
   score-and-snap knife, (27204):42
   taping knife, (27207):13
   utility knife, (27204):38, 40–41, (27206):17
Knob lockset, (27208):39–41, 77
Knobs
   door and drawer, (27211):17
   knurled, (27208):66
Knurled, (27205):1, 6, 36
Knurled knobs, (27208):65–66
Kraft paper, (27203):4, 8, 24–25, 28

L

Labels, electrical plans, (27201):22
Laced valley design, (27202):42–43
Ladder conveyors, (27202):11
Ladder duty ratings, (27204):2
Ladder jacks, (27202):7
Ladders, (27204):1–2
Ladder safety, (27202):4, 7–9, (27204):1–2
Laminating adhesives, (27206):9
Landscape drawings, (27201):1, 4, 37
Lanyards, (27202):3
Laser levels, (27209):21–22
Latch bolt, (27208):39, 40, 77
Latches, electric, (27208):46
Lateral, (27205):14, 36
Lath and plaster process, (27206):2
Lay-in tiles, (27209):8
L-beads, (27206):12, (27207):3–4
Leaders, (27201):4–5
Lead for sound isolation, (27206):23
Lead-lined door, (27208):5
Lead-lined drywall, (27206):29–30
Lead pencil, (27206):17
Ledger, (27204):23, 49, 59
Legends, (27201):8, 9, 19
Leveling equipment, (27209):20–22
Lever locksets, (27208):62
Liability, (27201):1, 2, 37
Lifelines, (27202):4
Lifting, proper methods of, (27211):2
Light box cutter, (27206):17, 18
Lighting plan, (27201):24
Light reflectance (LR) values, (27209):9
Lightweight aggregate insulation, (27203):15
Lightweight compound, (27207):3, 6, 7, 52
Lip, (27205):37
Lipped, (27211):11, 13, 28
Lipped doors, (27211):13
Lipped drawers, (27211):14
Liquid joint compound, (27207):5, 7
Loadbearing sills, (27205):29
Loadbearing studs, (27205):21–22
Loadbearing walls, (27201):15, (27205):18–19. *See also* Structural steel-framed walls
Loading pumps, (27207):21
Locking C-clamps, (27205):7
Locksets
   defined, (27208):3, 77
   door swing and, (27208):6
   installing, (27208):61–67
   types of
      activation device locksets, (27208):41–42
      cylindrical lockset, (27208):65–66, 77
      electric, (27208):46
      entrance lockset, (27208):61, 62
      interior locksets, (27208):41–42
      knob-type, (27208):39–41, 77
      lever locksets, (27208):62
      mortise lockset, (27208):3, 12, 42, 66–67, 77
      privacy lockset, (27208):61, 62
Lookout, (27204):23, 49, 59
Loose-fill insulation, (27203):9, 12, 17–20
Loose panels, drywall, (27207):44
Loose-pin hinges, (27208):40
Louver, (27204):23, 42, 59
LR values. *See* Light reflectance (LR) values
Luminous ceiling systems, (27209):4, 31, 33

M

Magnetic catches, (27208):40, (27211):16, 17
Magnetic door holder, (27208):43
Magnetic pass card locks, (27208):42
Maid's master key, (27208):42
Main runners, exposed grid systems, (27209):6
Manufacturing defects, drywall, (27207):43
Marking gauges, (27210):24
Masonite™, (27206):5
Masonry adhesives, (27206):11
MasterFormat™, (27201):30, 31
Master key systems, (27208):42
Material movement devices, (27202):9
Material supplier, (27205):37
Matte, (27211):4, 7, 28
MDF. *See* Medium-density fiberboard (MDF)
Mechanical, electrical, plumbing (MEP) drawings, (27209):13–15
Mechanical fire-stops, (27206):35
Medium-density fiberboard (MDF), (27210):4
Membrane roofing, (27202):12, 19
Membrane roofing systems
   modified bitumen membrane roofing systems, (27202):25–26
   single-ply systems, (27202):26–27, 31
MEP drawings. *See* Mechanical, electrical, plumbing (MEP) drawings
Metal door frames
   grouting, (27208):25
   plumb and level, (27208):25
   sound attenuation, (27208):25
   support, (27208):27
   typical, (27208):23, 26–27
   unassembled, installing in drywall construction, (27208):30–31, 33–34
   welded, installing
      in existing masonry construction, (27208):34–36
      in new masonry construction, (27208):36–38
      in steel-framed construction, (27208):27, 31–34
      in wood-framed construction, (27208):24–25, 29, 30
Metal doors, (27208):12–13, 14
Metal edge tape, (27207):5
Metal insulation materials, (27203):5
Metallic ceiling system, (27209):5
Metal pan ceiling systems, (27209):2–3, 9, 28–30
Metal ridge vents, (27202):67
Metal roofs
   baked-enamel steel, (27202):24
   corrugated metal, (27202):44–45, 46
   engineered/preformed architectural, (27202):23–24, 25
   shingle and shake, (27202):22
   simulated standing-seam metal, (27202):45–47, 53
   snug-rib systems, (27202):48

standing-seam metal, (27202):45–47, 53
vinyl-coated aluminum, (27202):24
Metal siding. *See* Vinyl/metal siding
Metal-stud curtain walls, (27205):18–20
Mil, (27205):1, 4, 36
Mineral fiberboard for sound isolation, (27206):23
Mineral-fiber tiles, (27209):6
Mineral insulation materials, (27203):5
Mineral-wool insulation, (27206):23
Miter cuts, (27210):11–13
Miter saws, (27210):3, 11
Modified bitumen membrane roofing systems, (27202):25–26
Modified contact adhesives, (27206):10
Modular cabinets. *See* Stock cabinets
Moisture control. *See also* Vapor barriers
   in attics, (27202):29–30, (27203):23
   in basements, (27203):22–23
   flashing for, (27203):32
   function, (27203):22
   mold and, (27203):23
   slabs, (27203):23
   underlayment for, (27202):27–28
   ventilation for, (27202):29–30, (27203):23–24, 25
   waterproofing for, (27203):27–31
   waterproof membranes, (27202):27–28, 30
Moisture-resistant construction, (27206):26, 28–29
Moisture-resistant drywall, (27206):3
Mold, (27203):23, (27209):9
Moldings. *See also* Trim
   cabinet installation, (27211):12
   custom, (27210):5
   defined, (27208):3, 77, (27210):1, 31
   installing
      backing boards for, (27210):22
      base moldings, (27210):14–16, 20, 24
      ceiling moldings, (27210):16–17
      ceilings, sagging or uneven, (27210):19
      crown moldings, (27210):22
      door trim, (27210):24
      irregular mounting surfaces, (27210):20
      marking gauges, (27210):24
   interior trim, (27210):5
   materials
      medium-density fiberboard (MDF), (27210):4
      wood, (27210):4
   prefinished, (27210):4
   selecting for joining, (27210):5
   sizes, (27210):4–5
   types of
      base moldings, (27210):5–6, 14–16, 20, 24
      ceiling moldings, (27210):7, 16–17
      crown moldings, (27210):22
      paint-grade, (27210):4
      wall moldings, (27210):6–7
   wood
      finger-jointed, (27210):4
      shaping, (27210):4
   wood folding doors, (27208):11
Molly bolts, (27209):7
Monsanto Fome-Cor™, (27203):4
Mortise, (27208):3, 4, 77
Mortise lockset
   cut-away view, (27208):42
   defined, (27208):3, 77
   installation, (27208):66–67
   metal doors, (27208):12

Mortise lockset jig, (27208):67
Mud. *See* Joint compound
Mud masher/mixer, (27207):16
Mud paddle, (27207):16
Mud pan, (27207):14
Mullion, (27210):27
Multi-ply drywall systems
   adhesives, (27206):38–39
   base-ply attachment, (27206):37–38
   corners, (27206):37
   face-ply attachment, (27206):37–38
   fasteners, (27206):37, 38–39
   floating-angle construction, (27206):37
   floating-corner treatment, (27206):37
   sound-isolation construction, (27206):31

N

Nail pop, (27206):1, 6, 45
Nail ripper, (27202):12, 13
Nails
   drywall
      annular type, (27206):5
      application procedure, (27206):6–7
      casing, (27206):5
      common, (27206):5
      cupped-head, (27206):5
      head shape and size, (27206):5
      nailing method, (27206):6
      penetration, (27206):5
      resurfacing existing construction, (27206):29
      spacing, (27206):6
   exposed grid systems, (27209):7
   exterior finishing materials, (27204):6, 10, 49
   roofing, (27202):16, 17, 22–23, 28
Nail spotters, (27207):19
National Electrical Code® (*NEC*®), (27205):17
National Fire Protection Association (NFPA) Standard *251*, fire-resistant tile standard ratings, (27209):8
Natural fiber insulation materials, (27203):5
Natural gas piping, (27201):26
NFPA. *See* National Fire Protection Association (NFPA)
Nibblers, (27202):12
90-minute mineral core door, (27208):5
90-minute mud, (27207):8, 9
Noise distractions, workplace, (27209):8
Noise reduction coefficient (NRC), (27209):2, 8
Noises, sound levels of common, (27209):3
Nonmechanical fire-stops, (27206):35–36
Nonremovable pin (NRP) hinges, (27208):40
Nonstructural member, (27205):38
Nonstructural rigid foam board, (27203):9–10, 12
Nonstructural steel-framed walls
   area separation walls, (27205):20
   chase walls, (27205):20
   components, (27205):17–18
   curtain walls, (27205):18–20
   fire-rated assemblies, (27205):20
   head-of-wall conditions, (27205):20–21
   members, (27205):18
   radius (curved) walls, (27205):20
   shaft wall systems, (27205):21
Notes on architectural drawings, (27201):10
NRC. *See* Noise reduction coefficient (NRC)
NRP hinges. *See* Nonremovable pin (NRP) hinges

O

Object lines, (27201):4–5

Occupational Safety and Health Administration (OSHA) incidence rate, (27201):32
Occupational Safety and Health Administration (OSHA) regulations
   distance from power lines, (27202):1
   ground fault circuit interrupter (GFCI), (27211):1
   ladders, (27204):1
   personal fall arrest systems, (27202):2, 4
   personal positioning systems, (27202):2
   personal protective equipment (PPE), (27210):1
   respiratory protection, (27210):1
   safety data sheets (SDSs), (27203):1–2
   scaffolds, (27204):2, (27209):23
Occupational Safety and Health Administration (OSHA) Standards
   CFR1926 Safety and Health Standards for the Construction Industry
      Subpart L (guardrails), (27202):2
      Subpart M (fall protection), (27202):2, (27204):1
      Subpart M, Appendices C and Appendices D, (27202):6
   CFR 1926.453 (aerial lifts), (27204):3
   Communications, (27203):1
Offset hinges, (27211):15
¼" drywall, (27206):2
½" drywall, (27206):2–3
1" drywall, (27206):3
Open valley, (27202):29
Open valley flashing, (27202):42
Open valley shingles, (27202):41–42
Organic fiber shingles, (27202):18
OSHA. *See* Occupational Safety and Health Administration (OSHA)
Outrigger clips, (27205):19–20
Outside-corner moldings, (27210):6
Oval-head screw, (27205):10
Overhang, (27202):74
Overinsulating, (27203):8
Overlay, (27211):11, 13, 28
Overlay doors, (27211):13
Overlay drawers, (27211):14
Overlay hinges, (27211):15

P

Paddles, texturing, (27207):16, 17
Painting and Decorating Contractors of America, *A Recommended Specification for Levels of Gypsum Board Finish*, (27207):1
Panel-and-frame door, (27211):11, 13, 28
Panelization
   clinching systems and, (27205):12
   curtain wall construction, (27205):19
   defined, (27205):1, 36
   shakes/shingles, (27202):21–22, (27204):14
   steel framed structural walls, (27205):21, 22
Panel schedules, (27201):23
Pan-head screw, (27205):10
Panic hardware, (27208):39, 47, 77
Paperless drywall, (27206):29
Paper tape, (27207):4–5
Particleboard core door, (27208):5
Passage latch, (27208):40–41, 61, 62
Patio lock, (27208):40–41
PATs. *See* Powder-actuated tools (PATs)
Patterns, (27204):8–9
Perforated paper tape, (27207):5
Perm, (27203):22, 24, 38
Permeability, (27203):22, 24, 38
Permeable, (27203):22, 28, 38

Permeance, (27203):22, 24, 38
Personal fall arrest systems
   anchorage points, selecting and tying off, (27202):5
   anchoring devices (tie-off points), (27202):4
   body belts, (27202):2
   deceleration devices, (27202):3–4
   equipment connectors, (27202):4
   full-body harness, (27202):2–5
   inspecting and testing, (27202):4–5, 6
   lanyards, (27202):3
   lifelines, (27202):4
   OSHA Standards, (27202):2
   rope grabs, (27202):3–4
   self-retractable lifelines, (27202):4
   terminology, (27202):2
   training requirements, (27202):4
Personal positioning systems, (27202):2
Personal protective equipment (PPE)
   for cabinet installation, (27211):1
   for door installation, (27208):1
   for drywall finishing, (27207):12
   for insulation projects, (27203):1
   for power nailers, (27202):15
   for roofing safety, (27202):6
   for sanding, (27207):28
   for steel frame projects, (27205):2
   for trim, working with, (27210):1
Photographing, drywall, (27207):37
Pinching hazards, (27211):2
Pins, steel framing, (27205):10, 11
Piping, protecting, (27205):16–17
Pitch, (27202):1, 74
Pivot hinges, (27211):16
Plain beveled siding, (27204):11, 26, 27
Plain knob, (27208):41
Planar ceiling system, (27209):6
Plancier, (27204):23, 44, 59
Plan view
   defined, (27201):1, 37
   drainage and utility plans, (27201):9
   electrical plans, (27201):22
   plumbing drawings, (27201):24
   structural drawings, (27201):15
Plaster application drywall, (27206):5
Plastic composition vents, flexible, (27202):30
Plastic insulation materials, (27203):5
Plastic-laminate-covered countertops, (27211):6–7
Plastic laminates, (27211):1, 2, 10, 28
Plastic ridge vents, (27202):67
Plenum, (27205):14, 36, (27209):1, 2, 40
Plenum ceilings, (27209):16
Plumb, (27208):3, 8, 77
Plumbing drawings
   distribution plans, (27201):24, 26
   floor plans, (27201):24
   isometric drawings, (27201):24
   natural gas piping, (27201):26
   plan view, (27201):24
   riser diagram, (27201):24
   sample, (27201):27, 28
   site plans, (27201):24
   symbols used on, (27201):24, 42
   waste disposal plans, (27201):24, 26
   water systems, (27201):24
Plywood, (27211):13
Plywood siding, (27204):14, 16, 33, 34
Plywood veneers, (27211):13
Pneumatic carbide-tipped power shears, (27204):42

Pneumatic nailers, (27204):7
Pocket doors, (27208):8–10
Polyethylene vapor barriers, (27203):25, 26
Polystyrene forms, (27203):17
Polyvinyl chloride (PVC) membrane, (27202):26–27
Portable brake, (27202):12, (27204):4, 36
Powder-actuated fasteners, (27205):10–11, (27209):7
Powder-actuated tools (PATs), (27205):6–7, (27209):26
Powder compounds, (27207):5–7, 8, 10
Powder drywall texture, (27207):9
Powder joint compound, (27207):5–6
Powder joint-compound textures, (27207):9
Powder load, (27205):1, 11, 36
Power nailers, (27202):15
Power plan, (27201):22–23
Power plane, (27208):56–57, 58
Power saw, (27204):37, 40, 42, (27208):57
PPE. *See* Personal protective equipment (PPE)
Predecorated drywall, (27206):4
Pre-engineered steel wall assembly, (27205):22
Prefinished, (27208):3, 10, 77
Prehung door, (27208):3, 4, 6, 7, 77
Premix compounds, (27207):7–8
Premixed drywall textures, (27207):9
Premixed joint-compound textures, (27207):9
Privacy lockset, (27208):40–41, 61, 62
Profiled edge tiles, (27209):8
ProWrap®, (27203):28
Pulls, door and drawer, (27211):17
Pump jacks, (27202):7, 8, 10
Punchout, (27205):15–16, 38
Purlins, (27202):45, 46
Putty knife, (27207):12–13
PVC membrane. *See* Polyvinyl chloride (PVC) membrane

Q

Quarter-round moldings, (27210):6
Quick-setting compounds, (27207):8, 10

R

Rabbet, (27204):6, 59, (27208):17, 78
Rabbeted, (27204):10, (27208):22
Rabbeted beveled siding, (27204):10, 11
Rabbeted stool, (27210):8
Racking, (27205):14, 24, 36
Radius walls, (27205):20, (27207):5
Rail, (27208):3, 4, 78
Rake, (27204):6, 59
Rasp, (27206):17
Rawls, (27209):7
Ready to assemble (RTA) cabinets, (27211):4
Rebar drawings, (27201):20, 21
A Recommended Specification for Levels of Gypsum Board Finish, (27207):1
Recycled glass countertops, (27211):9
Recycling steel, (27205):1, 5
Reference line, (27204):25, 26
Reference marks, (27204):27
Reflected ceiling plans, (27209):12
Reflection, (27209):2
Reflective ceiling, (27209):6
Reflective insulation, (27203):13, 14
Regular, paper faced drywall, (27206):3
Regular with foil back drywall, (27206):3
Request-to-exit (RTE/REX) devices, (27208):47
Rescue and retrieval plans, (27202):5–6
Residential doors. *See also* Doors
 bifold doors, (27208):8, 10
 bypass doors, (27208):7, 9
 installation tools, (27208):53
 manufactured prehung door-unit, (27208):53–55
 metal doors, (27208):12–13, 14
 pocket doors, (27208):8–10
 wood folding doors, (27208):10–12
Respiratory protection, (27207):28, (27210):1, 2
Reveal, (27210):10, 17, 31
Reveal trim, (27207):4
Reverberation, (27209):2
Reverse batten, (27204):10–11
Ribbon courses, (27202):41
Ridge row shingles, (27202):62–63, 65, 68
Ridges, (27202):42, 74, (27207):1, 52
Ridge vents, (27202):29, 30, 65–67, (27203):24
Ridging, (27207):36
Rigid expanded polystyrene, (27203):13
Rigid extruded polystyrene, (27203):13
Rigid plastic foam furring systems, (27206):23
Rigid polyisocyanurate, (27203):13
Rigid polyurethane, (27203):13
Rigid/semirigid insulation
 adhesives used with, (27206):10
 basics, (27203):9–12
 common types, (27203):12
 drywall installation over, (27206):23, 24
 installation, (27203):20
 secondary roof systems, (27202):35
Rim track, (27205):31, 36
Riser diagram, (27201):1, 24, 37
Roller catches, (27211):16, 17
Roller pan, (27207):17
Rollers, (27207):17, 18
Roll roofing
 characteristics, (27202):49
 concealed nail installation method, (27202):50–53
 double-coverage, (27202):53–54
 exposed nail installation method, (27202):49–50
 exposures recommended, (27202):49
 installation preparations, (27202):49
 preparations, (27202):49
 removing the curl, (27202):51–52
 single-coverage, installing, (27202):49
 slope requirements, (27202):19, 20, 49
 weights, typical, (27202):49
Roof assemblies, (27205):31
Roof drainage systems, (27202):75–79
Roof framing plans, (27201):17
Roofing accessories, (27202):26–27
Roofing brackets, (27202):9–10
Roofing fasteners, (27202):16
Roofing hammer, (27202):12
Roofing materials. *See also* Flashing; *Specific materials*; *Specific roof types*
 drip edges, (27202):28
 fasteners, (27202):45–46
 function, (27202):1
 material takeoffs, (27202):69
 for moisture control, (27202):34–35, (27203):26–27
 nails, (27202):16, 17, 22–23, 28
 purlins, (27202):45, 46
 roll roofing, (27202):19, 20
 sag rods, (27202):46
 single-ply systems, (27202):26–27
 underlayment, (27202):1, 6, 27–28, 34, 74
 vents, (27202):29–30
 waterproof membranes, (27202):27–28, 28, 30, 34
Roofing plans, (27201):10, 11

Roofing safety
  brackets, (27202):9–10
  ergonomic injuries, (27202):1
  fall protection, (27202):2–6
  guardrails, (27202):2
  hazard control, (27202):6
  hole covers, (27202):2
  injuries and deaths, primary causes of, (27202):2
  ladders, (27202):7–9
  material movement devices, (27202):9
  overhead power lines, (27202):1
  personal fall arrest systems, (27202):2–6
  personal positioning systems, (27202):2
  personal protective equipment (PPE), (27202):6
  safety nets, (27202):2
  scaffolding, (27202):6–7
  scaffold platforms, (27202):9
  scaffolds, (27202):36
  slate roofs, (27202):25
  staging, (27202):6–7
  staging platforms, (27202):9
  tools and materials, (27202):1, 13–15
Roofing symbols and terminology, (27202):38
Roofing systems
  built-up roofing (BUR), (27202):24–26, 49
  membrane systems, (27202):25–27, 31
  secondary, (27202):35
  standing-seam metal, (27202):45–47
Roof installation
  corrugated light panels, (27202):51
  downspouts, (27202):75–79
  drainage systems, (27202):75–79
  gutters, (27202):75–79
  ice dams, (27202):33–35
  methods
    adhesives, (27202):26, 42
    heat welding, (27202):26, 31
    hot-air welding, (27202):31
  terminology, (27202):36, 37
  tools
    bulb rivet guns, (27202):45
    composition shingle knife, (27202):12, 13
    flame heating equipment, (27202):26, 31
    hand grinders, (27202):12
    hand tools, (27202):12–13
    heavy roller, (27202):12, 13, 32
    impact wrenches, (27202):45
    nail ripper, (27202):12, 13
    nibblers, (27202):12
    portable brake, (27202):12
    power nailers, (27202):15
    power tools, (27202):45
    roofing hammer, (27202):12
    screw guns, (27202):45
    seaming machines, (27202):45
    shingle hatchet, (27202):12, 13
    slate cutter, (27202):12, 13
    slater's hammer, (27202):12, 13
    tile cutters, (27202):12
    wet saw, (27202):12
  typical, (27202):33
  wind considerations, (27202):36
Roof installation deck preparations
  asphalting nail heads, (27202):32, 33
  debris removal, (27202):34
  drip edge installation, (27202):32
  flashing, (27202):32
  protruding nails, (27202):34
  underlayment, (27202):32
  waterproof membrane, (27202):32
Roof rafter, (27205):14, 21, 36
Roof sheathing, (27202):1, 10, 74
Roof stack, (27203):24
Roof ventilation
  box vents, (27202):29–30, 65
  chimneys, (27202):57
  convection vents, (27202):29
  eave vent, (27203):24
  gable louver vent, (27203):24
  gable vents, (27202):29
  ice-dam control, (27202):33–34
  metal ridge vents, (27202):67
  plastic ridge vents, (27202):67
  residential construction, (27202):29–30
  ridge vents, (27202):29–30, 65–67, (27203):24
  roof stack, (27203):24
  soffit vents, (27202):29, (27203):24
  turbine vents, (27202):29–30
  vent stacks, (27202):55–56
Roosevelt, Franklin D., (27201):23
Rope grabs, (27202):3–4
Rough hardware, (27208):69
Rough opening, (27208):3, 8, 78
Routers, (27208):60, (27210):3
RTA cabinets. *See* Ready to assemble (RTA) cabinets
RTE/REX devices. *See* Request-to-exit (RTE/REX) devices
Running trim, (27210):26
R-values
  of common materials, (27203):7, 8
  defined, (27203):4, 38
  ICC recommendations, (27203):8, 10, 39
  loose-fill insulation, (27203):7
  requirements, determining, (27203):4–8
  rigid/semirigid insulation, (27203):13

S
S4S molding. *See* Surfaced on four sides (S4S) molding
Saber saws, (27210):11
Saddle, (27202):12, 16, 56–57, 74
Safety. *See also* Roofing safety
  cabinet installation, (27211):1–2
  door installation, (27208):1–2
  dry-mix compounds, (27207):9–10
  drywall finishing tools, (27207):12
  elevated surfaces
    aerial lifts, (27204):2–3
    fall protection, (27204):1
    ladders, (27202):4, 7–9, (27204):1–2
    scaffolds, (27202):36, (27204):2, 3
  ergonomics for, (27211):2
  fiberglass insulation, (27203):13
  finish nailers, (27210):2–3
  insulation projects, (27203):1–2, 13
  ladder safety, (27202):4, 7–9, (27204):1–2
  miter saws, (27210):3
  power nailers, (27210):13–15
  power tools, (27202):13–15
  routers, (27210):3
  steel frame projects, (27205):2
  suspended ceiling system installation, (27209):10–11
  trim, working with, (27210):1–3
Safety data sheets (SDSs), (27203):1–2, (27204):1, 5, 59
Safety nets, (27202):2
Sag rods, (27202):46
Sanders, (27207):10, 15
Sanding materials, (27207):10, 11, 21–22

Sanitary stop, (27208):17, 23, 78
Saws
   backsaw, (27210):10
   chop saw, (27205):8
   circular saw, (27205):8
   compound miter saws, (27210):11–12
   coping saw, (27210):10
   dovetail saw, (27210):10
   dry-cut metal-cutting saw, (27205):8
   drywall saw, (27206):17, 18, (27207):12, 13
   hacksaw, (27204):38
   hole saw, (27205):9
   jab saw, (27206):17, 18
   miter saws, (27210):3, 11
   power saw, (27204):37, 40, 42, (27208):57
   saber saws, (27210):11
   wet saw, (27202):12
SBS products. *See* Styrene-butadiene-styrene (SBS) products
Scaffolding safety, (27202):6–7
Scaffold platforms, (27202):9
Scaffolds, (27204):2, 3
Scaffold safety, (27202):36, (27204):2, 3, (27209):23
Scale, architectural drawings, (27201):9
Scarf joints, (27210):10, 11, 31
Schedules
   architectural drawings, (27201):10, 12
   defined, (27201):1, 37
   door schedules, (27201):10, 12, (27208):4, 6, 69–71
   electrical drawings, (27201):23, 24
   electrical plans, (27201):23, 24
   fastening schedules, drywall, (27206):16
   finish schedule, (27201):9, 12, 13
   fixture schedules, (27201):24
   legality of, (27201):2
   panel schedules, (27201):23
   window schedules, (27201):10, 12
Score-and-snap knife, (27204):42
Scratch coat, (27204):6, 17, 59
Screw eyes, (27209):7
Screw guns, (27202):45, (27206):17, 19
Screws
   drywall
      for furring, (27206):7, 8
      head shape, (27206):7
      penetration, (27206):8
      resurfacing existing construction, (27206):29
      selecting, (27206):7
      spacing, (27206):8
      for steel framing, (27206):8
      Type G, (27206):8
      Type S, (27206):8
      Type S-12, (27206):8
      Type W, (27206):7–8
   exposed grid systems, (27209):7
   steel framing, (27205):9–10
Scrim, (27202):74
SDSs. *See* Safety data sheets (SDSs)
Seaming machines, (27202):45
Sea sponge, (27207):17
Section identification formats, (27201):4
Section views
   cutting plane lines showing, (27201):4
   elevation drawings, (27201):12–13, 14
   symbols, (27201):39
Security hardware, (27208):44–47
Self-closing hinges, (27211):16, 17
Self-drilling screw, (27205):9–10
Self-piercing screws, (27205):9–10

Self-retractable lifelines, (27202):4
Selvage, (27202):74
Selvage roll roofing, (27202):20
Semi-automatic flush bolts, (27208):48
Semicustom cabinets, (27211):4, 28
Semirigid fiberglass, (27203):13
Semirigid insulation. *See* Rigid/semirigid insulation
Shaft wall systems, (27205):21
Shakes
   baked-enamel steel, (27202):24
   installation, (27202):17, (27203):4, (27204):31–33, 44
   installation tools, (27202):12
   metal, (27202):22
   sidewall finish, (27204):13
   slope requirements, (27202):49
   vinyl-coated aluminum, (27202):24
   wood
      advantages, (27202):20–21
      dimensions, (27204):13
      installation, (27204):31–33, 44
      manufacturing, (27202):21, (27204):13
      panelized, (27202):21–22, (27204):14
      wood-fiber composition hardboard, (27202):21–22
Shallow foundations, (27201):18
Shear-hold electromagnetic locks, (27208):46–47
Shear wall, (27205):14, 19, 24, 36
Sheet lamination, (27206):38
Shelf rails and clips, (27211):18
Shelves, cabinets, (27211):14, 18
Shields drywall, (27206):5
Shims, (27208):17, 18, 78
Shingle hatchet, (27202):12, 13
Shingle knife, (27202):12, 13
Shingles. *See also* Composition shingles
   algae-resistant, (27202):18–19
   baked-enamel steel, (27202):24
   fungus-resistant, (27202):18–19
   installation tools, (27202):12
   installing over insulation sheathing, (27203):4
   metal, (27202):22
   metric, (27202):21
   slope requirements, (27202):49
   vinyl-coated aluminum, (27202):24
   wood
      advantages, (27202):20–21
      installation, (27204):31–33, 44
      manufacturing, (27202):21
      panelized, (27202):21–22, (27204):14
Shingle siding, (27204):13, 14
Shiplap siding, (27204):13, 30–31
Shoe molding, (27210):6, 11
Shop drawings, (27201):20, (27205):38
Shrinkage, (27207):8
Side lap, (27202):32, 49, 74
Siding, (27203):4, 5. *See also Specific types of*
Siding gauge (preacher), (27204):27, 28
Sills, (27205):29, (27208):17, 24, 78
Sill to jamb connections, (27205):19–20
Simulated standing-seam metal roofs, (27202):45–47, 53
Single-coverage roll roofing, (27202):49
Single-ply drywall applications, (27206):23–24, 31, 32
Single-ply roofing systems, (27202):26–27, 31
Site improvement plans, (27201):9
Site plans, (27201):6–9, 24
6" composition shingle pattern, (27202):39, 44
Skim coat, (27207):1, 52
Slabs, vapor barrier installation, (27203):23, 26
Slate cutter, (27202):12, 13

Slate roofs
    basics, (27202):23
    drip edges, (27202):28
    installation tools, (27202):12
    safety, (27202):25
    synthetic, (27202):23, 24
Slater's hammer, (27202):12, 13
Slide clip, (27205):18
Slides, drawer, (27211):12, 14, 15, 16
Slip connectors, (27205):5, 6
Slip tract, (27205):18
Slope, (27202):1, 9, 74
Slurry, (27206):1, 45
Smoke gasket, (27208):39, 50, 78
Snap-lock punch, (27204):41
Snug-rib metal roof systems, (27202):48
Soffits
    aluminum or vinyl installation, (27204):49–52
    cabinet installation and, (27211):5
    defined, (27204):1, 59, (27211):4, 28
Soffit vents, (27202):29, (27203):24
Solid-core doors, (27208):4, 5
Solid-surface countertops, (27211):7–9, 10
Sound attenuation, (27208):6, 17, 25, 78
Sound intensity, (27209):1–2
Sound isolation, (27205):4–5, (27209):1
Sound-isolation construction
    adhesive application, (27206):38
    air leaks, closing, (27206):18–19
    caulking, (27206):20–22
    double-ply applications, (27206):23–24
    flanking paths, closing, (27206):18–19
    resilient mountings, (27206):22–23
    separated partitions, (27206):20–21
    single-ply applications, (27206):23–24
    sound-isolating materials, (27206):23–24
    wallboard applications, (27206):23
Sound levels for common noises, (27209):3
Sound transmission, controlling, (27206):2
Sound transmission class (STC), (27206):18–19, 33, (27208):3, 4, 78, (27209):2
Sound transmission class (STC) doors, (27208):4
Sound transmission loss, (27209):2
Sound waves, (27209):1–2
Spade pins, (27211):18
Span, (27205):38
Special-use drywall, (27206):3
Specifications
    discrepancies in, (27201):30
    format of, (27201):30, 31
    writing, (27201):30–31
Sponges, texturing, (27207):17
Spot lamination, (27206):38
Sprayed-in-place insulation, (27203):15, 16
Spray-texturing machines, (27207):11
Square, (27202):12, 16, 74
Square cuts, (27210):10, 11–12, 31
Square-edged molding (S4S [surfaced on four sides]), (27210):5–6
SSMA. *See* Steel Stud Manufacturers Association (SSMA)
Stacked wall framing, (27205):19
Staging platforms, (27202):9
Staging safety, (27202):6–7
Stairway doors, (27208):45
Standard cold-formed steel structural shapes, (27205):38
Standards and clips, (27211):18
Standing-seam metal roofs, simulated, (27202):45–47, 53
Standing-seam valley flashing, (27202):28–29

Standing trim, (27210):26
Staples, drywall, (27206):5
Star anchors, (27209):7
Starter strip, (27202):41, (27204):34–35
Staved-wood core door, (27208):5
STC. *See* Sound transmission class (STC)
Steel doors, (27208):13
Steel frame construction
    components, (27205):2–3
    drywall adhesives for, (27206):10, 11
    drywall screws for, (27206):8
    structural drawings, (27201):15, 19
Steel frame construction methods
    blocking, (27205):16
    bracing, (27205):16
    bridging, (27205):16
    built-up shapes, (27205):16
    in-line framing, (27205):14–15
    piping, protecting, (27205):16–17
    thermal considerations, (27205):16
    web holes (punchouts) and patches, (27205):15–16
    web stiffeners, (27205):15
    wiring, protecting, (27205):16–17
Steel Framing Alliance, *Thermal Design and Code Compliance for Cold-Formed Steel Walls*, (27205):16
Steel production technologies, (27205):4
Steel recycling, (27205):1, 5
Steel Stud Manufacturers Association (SSMA)
    Product Technical Guide, (27205):38
    universal designator (S-T-U-F) system, (27205):4
Steel stud punch, (27205):9, 15
Stick building, (27205):22
Stiffening lip, (27205):1, 4, 21, 36
Stile, (27208):3, 4, 78
Stile-and-rail doors, (27208):4
Stilts, (27207):30, (27209):10–11
Stock cabinets, (27211):4, 28
Stone veneer, (27204):19
Stool, (27210):4, 5, 8, 26, 31
Storage, (27207):6–7, 8
Storm-Guard™, (27202):27
Straightedge shake/shingles, (27204):15
Strap, (27205):38
Stretching to avoid injury, (27211):2
Striated, (27209):1, 6, 40
Strike plate, (27208):3, 78
Strikes
    dust-proof, (27208):50
    electric, (27208):44–45
    electric plunger, (27208):48
Strip lamination, (27206):38
Structural drawings
    callout sequence markings, (27201):15
    detail sheets, (27201):20
    foundation plans, (27201):18–19
    framing plans, (27201):16, 19–20
    function, (27201):13–15
    grid lines, (27201):15, 18
    loadbearing walls, (27201):15
    plan view, (27201):15
    roof framing plans, (27201):17
    shop drawings, (27201):20
    steel frame construction, (27201):15, 19
    timber frame construction, (27201):15
    typical, (27201):15
Structural engineer-of-record, (27205):38
Structural insulating boards, (27203):9
Structural member, (27205):38

Structural sheathing, (27205):25, 26
Structural steel-framed walls
   accessories, (27205):21
   assembly, (27205):22
   backing, (27205):24
   bracing, (27205):24, 26
   function, (27205):21
   header assembly, (27205):27–29
   installation, (27205):22–24
   jambs, (27205):29
   layout, (27205):21–22
   members, (27205):21
   shear walls, (27205):24–25
   sills, (27205):29
   tracks, (27205):21
Structural steel notations, (27201):19
Structural stud, (27205):38
Stucco, (27203):4, 28, (27204):17–19
Stucco brush, (27207):16
Stud adhesives, (27206):9, 10
Stud bracing, (27205):24, 26
Stud drivers, (27205):6–7
Studs
   defined, (27205):38
   loadbearing, (27205):21–22
   material takeoffs, (27205):13
   punch openings, (27205):15
Stud-to-stud connection, (27205):19–20
S-T-U-F identification system, (27205):4
Styrene-butadiene-styrene (SBS) products, (27202):25–26
Substrate, (27206):1, 4, 45
Surfaced on four sides (S4S) molding, (27210):5–6
Surface hinges, (27211):15
Suspended ceiling systems
   acoustical, (27209):2
   applications, (27209):1
   cleaning post-installation, (27209):35
   components
      ceiling panels/tiles, (27209):8–9
      direct-hung systems, (27209):9, 10
      exposed grid systems, (27209):6–7
      installing, (27209):4
      metal pan systems, (27209):9
   construction drawings
      mechanical, electrical, plumbing (MEP) drawings, (27209):13–15
      reading, (27209):12
   function, (27209):1
   furring, (27209):4, 34–35
   installation guidelines, (27209):22–23
   installation tools, (27209):10–11, 20–22, 23
   layout and takeoff procedures, (27209):17–20
   material handling and storage, (27209):10
   safety guidelines, (27209):10–11
   types of
      acoustical panel, (27209):34
      direct-hung concealed grid systems, (27209):30–31, 32
      direct-hung systems, (27209):3
      drywall, (27209):33–34, 35
      exposed grid systems, (27209):2, 23–28
      integrated systems, (27209):3, 4
      luminous systems, (27209):4, 31, 33
      metal pan systems, (27209):2–3, 28–30
      planar, (27209):6
      reflected, (27209):12
      reflective, (27209):6
      special metallic, (27209):5
      special wood, (27209):5
      transparent tile, (27209):6

Sweep, (27208):39, 47, 78
Swimming pools, indoor, (27209):29
Swing of the door, (27208):4
Switch bars, (27208):46–47
Swivel-head shears, (27205):8
Symbols
   electrical, (27201):22, 42
   electrical plans, (27201):22
   general plan, (27201):4, 39
   HVAC, (27201):41
   plumbing drawings, (27201):24, 42
   section views, (27201):39
   topographic, (27201):40
Synthetic slate, (27202):23

T

Tape, (27207):1, 4–5, 15–16, 52
Tape blisters, drywall, (27207):37–38
Tapered joint, (27207):52
Taping compound, (27207):6, 7, 8. *See also* Joint compound
Taping drywall, (27207):28–32, 33
Taping knife, (27207):13
T-brace, (27206):18, 20, (27211):19, 21, 28
Temperature, average US lows, (27203):39
Temperature and drywall, (27206):24, (27207):6–7, 23, 25
Template, (27208):78
Template saw guide, (27208):57
Temporary bracing, (27205):24, 26
Terminology, roofing, (27202):36, 37
Texture brush, (27207):16
Texture materials, drywall finishing, (27207):9, 10
Texture tools
   circular patterned sponge, (27207):17
   flat blade knife, (27207):17
   glitter gun, (27207):16
   mud paddle, (27207):16
   paddle, (27207):17
   roller pan, (27207):17
   rollers, (27207):17, 18
   sea sponge, (27207):17
   spray-texturing machines, (27207):11
   stucco brush, (27207):16
   texture brush, (27207):16
   trowel, (27207):17
   whisk broom, (27207):17
   wipe-down blade, (27207):17
Texturing process, (27207):16–18
Thermal Design and Code Compliance for Cold-Formed Steel Walls (Steel Framing Alliance), (27205):16
Thermally efficient doors, (27208):13, 15
Thermoplastic membranes, (27202):26
Thermoplastic polyolefin (TPO) membrane, (27202):26–27
Thermoset membranes, (27202):26
30-minute mud, (27207):8
⅜" drywall, (27206):2
Three-tab shingles, (27202):18, 19
Threshold, (27208):39, 47, 49, 78
Tie-off points, (27202):4
Tile backer drywall, (27206):4
Tile cutters, (27202):12
Tile roofs
   baked-enamel steel, (27202):24
   basics, (27202):22–23
   installation tools, (27202):12
   vinyl-coated aluminum, (27202):24
Timber frame construction, (27201):15
Tin snips, (27204):37, 41, 42
TL overlap. *See* Top lap (TL) overlap

Tongue-and-groove siding, (27204):12, 14, 30–31
Top lap, (27202):32, 49, 74
Top lap (TL) overlap, (27202):38
Topographical information, (27201):6–7, 8
Topographic symbols, (27201):40
Toppets, (27209):7
Topping compound, (27207):3, 6, 7, 8, 52
Touch-bar/crossbar hardware, (27208):47–48, 49, 50, 51
Touch bars/handles, (27208):46–47
Touch catches, (27211):17
Touch-sensitive switches, (27208):46–47
TPO membrane. *See* Thermoplastic polyolefin (TPO) membrane
Tracks
   bottom and top, cold-formed steel framing, (27205):12, 21
   defined, (27205):1, 2, 36
   radius (curved) walls, (27205):20
   wood folding doors, (27208):11
Training requirements, personal fall arrest systems, (27202):4
Transom, (27208):39, 43, 78
Transparent tile ceiling system, (27209):6
Trim. *See also* Moldings
   on ceilings, sagging or uneven, (27210):19
   coped joints, (27210):10–11, 12–13, 14
   cutting saws, (27210):10–11
   defined, (27210):1, 31
   drywall finishing
      corner beads, (27207):3–4, 5, 13
      expansion joints, (27207):4
      flexible tape, (27207):4
      reveal trim, (27207):4
   estimating quantities, (27210):26–27
   fastening, (27210):13–14, 17
   installing, (27210):15, 24
   miter cuts, (27210):11–13
   safety guidelines when working with, (27210):1–3
   square cuts, (27210):11–12
   types of
      door, (27210):8
      door trim, (27210):17–18
      mill vs. custom, (27210):7
      paint-grade, (27210):9
      polymer, (27210):8
      window, (27210):8
      window trim, (27210):18–21, 23
Trim-head screw, (27205):10
Trim kits, (27210):27
Trimmer, (27205):38
Trimmer stud. *See* Jack stud
Trim relief, (27210):9
Tripolymer®, (27203):14
Trowels
   drywall, (27207):13–14
   texturing, (27207):17
Truss, (27205):38
T-square, (27207):12
Turbine vents, (27202):29, 30
Turn buttons, (27208):41
20-minute mud, (27207):8
Type G drywall screws, (27206):8
Type S-12 drywall screws, (27206):8
Type S drywall screws, (27206):8
Type W drywall screws, (27206):7–8
Type X, fire retardant drywall, (27206):2, 3
Tyvek®, (27203):28

U

UL. *See* Underwriters Laboratory (UL)
Underlayment
   aluminum foil underlayment, (27204):24
   defined, (27202):1, 74
   for hazard control, (27202):6
   removing the curl, (27202):32
   roofing, (27202):12, 27–28
Underwriters Laboratory (UL), (27206):16
   1479 (fire-stopping ratings), (27206):36–37
Urea formaldehyde foamed-in-place insulation, (27203):14
Urethane foam, (27203):14
USA Standard for Organic Adhesives for Installation of Ceramic Tile, (27206):29
US Finish Symbols, (27208):69
Utility knife, (27204):38, 40–41, (27206):17
Utility plans, (27201):9

V

Vacuum sanders, (27207):21–22
Valley, (27202):12, 16, 74
Valley flashing, (27202):28–29, 32, 74
Valley shingles, (27202):41–44
Vanity, (27211):4, 6, 28
Vanity cabinets, (27211):12
Vapor barriers
   cold-storage, (27203):28
   defined, (27203):4, 38
   for drywall attachment, (27206):12
   function, (27203):24
   installation
      crawl spaces, (27203):25
      roofs, (27203):26–27
      slabs, (27203):23, 26
      walls, (27203):26–27
   materials, (27203):4, 24–25, 28
   veneer finishes, (27204):19
Vapor diffusion retarders (VDR), (27203):4
VDR. *See* Vapor diffusion retarders (VDR)
Veneer
   cabinet finishes, (27211):12, 13
   cutting, (27208):57
   defined, (27204):6, 59, (27211):11, 28
   plywood, (27211):13
Veneer finishes
   brick and stone, (27204):19
   doors, (27208):57
   moisture barrier, (27204):19
   stucco (cement), (27204):17–19
Veneer plaster, (27206):5
Ventilation
   for moisture control, (27202):29–30, (27203):23–24, 25, (27204):42
   roof, (27202):29–30, 65, 67, (27204):44
Vents
   box vents, (27202):65
   defined, (27204):23, 59
   highlighted in text, (27204):42
   metal ridge vents, (27202):67
   plastic ridge vents, (27202):67
   ridge vents, (27202):65–67
Vent-stack flashing, (27202):32, 54–55, 74
Vent stacks, (27202):55–56
Vertical wall flashing, (27202):55–56
Vertical wall shingles, (27202):55, 56
Vinyl/metal siding
   components, (27204):15–16, 18
   installing

around windows and doors, (27204):40–41
  caulking, (27204):36, 43–44
  cleanup, (27204):43–44
  corner cap, (27204):38–40
  corner posts, (27204):34, 35
  cutting, (27204):37–38
  eaves, (27204):43
  equipment, (27204):3–4
  expansion, (27204):35
  flashing, (27204):37
  furring, (27204):41, 43
  gable ends, (27204):41–42
  gable end trim, (27204):37
  J-channel, (27204):36–37
  pre-installation trim, (27204):33
  procedure, (27204):38
  starter strip, (27204):34–35
  wind considerations, (27204):38, 39
  window and door trim, (27204):35–37
 materials, (27204):15–16, 17

W

w. *See* Width of shingles (w)
Wafer-head screw, (27205):10
Wainscot caps, (27210):6
Wainscoting, (27210):4, 6, 31
Wall angles, (27209):6
Wall cabinets, (27211):4, 5–6, 19–21, 28
Wall flashing, (27202):32, 55, 74
Wall moldings, (27210):6–7
Walls, vapor barrier installation, (27203):26–27
Wall section drawings, (27201):12–13
Waste disposal plans, (27201):24, 26
Water damage, drywall, (27207):43
Water-Guard™, (27202):27
Water level, (27209):21
Waterproofing
  air infiltration control, (27203):28–31
  below-grade foundation walls, (27203):27
  joint treatment, (27203):28
  water stops, (27203):27–28
Waterproof membranes
  installing, (27202):28, 30
  removing the curl, (27202):32
  roofs, (27202):27–28
Water-resistant (green board) drywall, (27206):4, 12, 26
Water stops, (27203):22, 27–28, 38
Water systems, (27201):24
Water vapor, (27203):22, 38. *See also* Vapor barriers
Water vapor permeability, (27203):24
Weather hazards, (27204):1
Weather-resistant drywall, (27206):3
Weather stripping, (27208):17, 25, 47, 78
Web, (27205):1, 3, 36
Web holes and patches, (27205):15–16
Web stiffeners, (27205):15
Wet saw, (27202):12
Whisk broom, (27207):17
Width of shingles (w), (27202):38
Wind
  shingle roofs and, (27202):36, 44
  vinyl/metal siding installation and, (27204):38, 39
Wind girt connections, (27205):19–20
Wind hazards, (27204):1
Windows
  insulating glass, (27203):5–6
  vinyl/metal siding installation, (27204):40–41
  wood siding installation around, (27204):27, 29–30
Window schedules, (27201):10, 12
Window trim
  casing moldings, (27210):5, 7–8, 11–13, 14, 16–24, 26
  flashing, (27203):32
  vinyl/metal siding installation, (27204):35–37
Winged screw, (27205):10
Wipe-down blade, (27207):17
Wiring, protecting, (27205):16–17
W-metal valley flashing, (27202):28–29
Wood ceiling systems, (27209):5
Wood doors
  folding doors, (27208):10–12
  panel doors, (27208):4
  staved-wood core door, (27208):5
Wood-fiber composition hardboard, (27202):21–22
Wood jambs and frames
  commonly used, (27208):17
  door-stop strips, (27208):22–23
  double wood frames, (27208):22
  fixed- and adjustable-width jambs, (27208):18, 28
  installing, (27208):17–22
  plumb and square, (27208):20
Wood moldings, (27210):4
Wood R-values, (27203):8
Wood shakes and shingles
  advantages, (27202):20–21
  dimensions, (27204):13
  manufacturing, (27202):21, (27204):13
  panelized, (27202):21–22, (27204):14
Wood siding. *See also* Specific types of
  installing
    around windows and doors, (27204):27, 29–30
    chalkline, (27204):25
    corner caps, (27204):30
    corners, (27204):27–30
    flashing, (27204):31
  patterns, (27204):8–9
Work areas cleanliness and safety, (27210):1, (27211):1
Wrigley Building, (27208):13, 15

X

X-bracing, (27205):19, 25, 26

Y

Yield strength, (27205):38

Z

Zinc galvanizing, (27205):2
Z-purlins, (27202):46

This page is intentionally left blank.

This page is intentionally left blank.

This page is intentionally left blank.